BONE REGENERATION

Edited by **Haim Tal**

Bone Regeneration

Edited by Haim Tal

Published by InTech

Janeza Trdine 9, 51000 Rijeka, Croatia

Publishing Process Manager Ana Skalamera

Technical Editor Teodora Smiljanic

Cover Designer InTech Design Team

Additional hard copies can be obtained from orders@intechopen.com

Bone Regeneration, Edited by Haim Tal
p. cm.
ISBN 978-953-51-0487-2

For EU product safety concerns:
IN TECH d.o.o., Prolaz Marije Krucifikse Kozulić 3, 51000 Rijeka, Croatia,
info@intechopen.com or visit our website at intechopen.com.

Meet the editor

Professor Haim Tal is the Head of the Department of Periodontology and Implantology , and incumbent of the Gerald Niznick Chair of Implantology at the Maurice and Gabriella Goldschleger School of Dental Medicine of Tel Aviv University. During the years 1991-1996; 2006-2010 he served as Dean of the School. Professor Tal graduated from Hadassah School of Dental Medicine at the Hebrew University, Jerusalem, and took specialty training in Periodontics and Oral Medicine at the University of the Wits, Johannesburg, where he also received the degrees M.Dent. and PhD. in Physical Anthropology. He is the recipient of the 1984 honorary fellowship of the American Academy of Oral Medicine and Doctor Honoris Causa from the University of Bukarest. Professor Tal is a visiting Professor at NYU and at Toronto University. He published over 200 articles and abstracts and lectured world wide in the fields of Periodontal Wound Healing, Implantology and Implant Navigation. Professor Tal is a co-developer of Ossix, a leading cross-linked collagen membrane for guided bone regeneration, manufactured by Col-Bar Life Sciences, Israel.

Contents

Preface XI

Introductory Chapter 1
Haim Tal

Part 1 Tissue Engineering of Bone 9

Chapter 1 **Stem Cell Based Bone Tissue Engineering 11**
Lauren Vernon, Lee Kaplan and Chun-Yuh Charles Huang

Chapter 2 **Autologous Cell Therapies for Bone Tissue Regeneration 33**
Nevenka Kregar Velikonja, Hana Krečič Stres, Elvira Maličev, Danica Gantar, Matija Krkovič, Vladimir Senekovič, Matjaž Rode, Miomir Knežević, Gordana Vunjak Novakovic and Mirjam Fröhlich

Chapter 3 **Signals Between Cells and Matrix Mediate Bone Regeneration 59**
Ron Zohar

Chapter 4 **Tissue Engineering of Bone: Critical Evaluation of Scaffold Selection 83**
Itzhak Binderman, Avinoam Yaffe, Yuval Samuni, Hila Bahar, Joseph Choukroun and Philippe Russe

Chapter 5 **Tissuue Engineering in Maxillar Sinus Lifting: A Comparation of Differents Grafts and Confocal Laser Scanning Microscopic Evaluation 95**
Andrea Ballini, Michele Scivetti, Stefania Cantore, Biagio Rapone, Gianfranco Favia and Felice Roberto Grassi

Part 2 Regenerative Bone Therapy 109

Chapter 6 **Bioresorbable Collagen Membranes for Guided Bone Regeneration 111**
Haim Tal, Ofer Moses, Avital Kozlovsky and Carlos Nemcovsky

Chapter 7 **Augmentation and Preservation of the Alveolar Process and Alveolar Ridge of Bone 139**
Haim Tal, Zvi Artzi, Roni Kolerman, Ilan Beitlitum and Gal Goshen

Chapter 8 **Distraction Osteogenesis and Its Challenges in Bone Regeneration 185**
Reggie C. Hamdy, Juan S. Rendon and Maryam Tabrizian

Chapter 9 **Skull Expansion by Spring-Mediated Bone Regeneration 213**
Rodrigo Dornelles, Vera Cardim and Nivaldo Alonso

Chapter 10 **The Use of Cancellous Bone-Block Allograft for Reconstruction of the Atrophic Alveolar Ridge 239**
Gavriel Chaushu and Joseph Nissan

Part 3 Applied Biotechnology and Biomaterials 265

Chapter 11 **Self-Regenerative Ability of Bone and Micro Processing of Bone-Component Material in Orthopedic Surgery Healing 267**
T. Ohtani, T. Nakai, R. Mori and Y. Uchio

Chapter 12 **Microstructure and Biocompatibility of Hydroxyapatite Porous Ceramics Designed by a Partial Dissolution-Precipitation Technique with Supersonic Treatment for Bone Regeneration 283**
Toshiyuki Akazawa, Masaru Murata, Yasuhiko Tabata and Manabu Ito

Chapter 13 **Development and Evaluation of Superporous Ceramics Bone Tissue Scaffold Materials with Triple Pore Structure A) Hydroxyapatite, B) Beta-Tricalcium Phosphate 301**
Michiko Sakamoto and Toshio Matsumoto

Chapter 14 **Histocompatibility of Acellular Matrix Bone with Osteoblast and Vascular Endothelial Cells 321**
Gang Rui, Xuhong Jin and Shan Lin

Preface

Ever since the dawn of humanity, invention has inspired progress. The resourcefulness of mankind has continually led to advances in the implements we use; from time to time we have witnessed revolutionary innovations. Bone regeneration and tissue engineering therapy may well be counted among these outstanding breakthroughs.

Bone is a specialized connective tissue most prominently characterized by its mineralized organic matrix that imparts the physical properties that allow bone tissue to resist load, to support functional organs such as teeth, and to protect highly sensitive body parts such as the central nervous system. Bone loss and bone damage may occur as a result of genetic conditions, infectious diseases, tumours, and trauma. Additional causes of bone loss or bone damage specific to the oral cavity include severe periodontal disease, tooth loss, and post-extraction disuse atrophy of the jaws. While physical activity may result in formation of functionally oriented bone trabeculae , muscular disuse or muscular atrophy, hormonal changes and aging may bring about reduction in bone density and lead to osteopenia and osteoporosis. Pathological bony defects and deficiencies frequently pose therapeutic and reconstructive challenges.

Bone regeneration, which is part of the process of bone healing, involves integrative activity of native tissues and living cells, and lends itself to the incorporation of naturally derived or biocompatible synthetic scaffolds, aimed at replacing missing or damaged osseous tissues. There are several modalities of bone regeneration that take advantage of the principles of bone biology: these include tissue engineering, guided bone regeneration, distraction ontogenesis, and bone grafting.

During the past 30 years since the introduction of dental implants, lack of sufficient volume of bone at implant recipient sites has motivated the development of new regenerative therapies for promoting *de novo* formation of bone at such compromised sites. These procedures, often referred to as "implant site development", have received special attention over the past few years.

Awareness of what may be likened to bone regeneration and tissue engineering is as old as human culture. The Book of Genesis provides a detailed description of the creation of man, followed by the creation of a woman from one of the man's ribs: [21] "So the LORD GOD caused a deep sleep to fall upon the man, and while he slept took one of his ribs and closed up its place with flesh; [22] and the rib which the LORD GOD had taken from the man he made into a woman." The ancient Kabbalah refers to these divine acts as Genesis I and II, and considers the power given to Noah to select _the species to repopulate the world after the great flood, when the power of divine regeneration was granted to a human being, as Genesis III. Tissue engineering and clinical bone regeneration may one day be considered as Genesis IV, an era when a new kind of creative power is given to man, with all the profound responsibility and commitment that such creation carries with it.

Editing this book has been a work of love. It is the culmination of three decades of teaching, research and clinical activity in a rapidly changing scientific and clinical environment. I am indebted to my teachers and mentors, Prof. John Lemmer , University of the Witwatersrand , Johannesburg, and Prof. S. Sigmund Stahl, New York University, whose academic and professional dedication inspired me throughout the years. I hope that the readers will find this book informative, motivating and stimulating

Haim Tal
DMD, MDent., PhD (Rand)
Department of Periodontology and Implantology
Tel Aviv University,
Israel

Introductory Chapter

Haim Tal
Department of Periodontology and Implantology
Tel Aviv University
Israel

The fields of medicine and dentistry continue to develop in an ever-changing environment, with day-to-day innovations and discoveries. Among the many procedures that have to be implemented in the course of medical and dental therapy, guided bone repair and regeneration present vast challenges to the science, art and practice of reconstituting the shape and function of damaged skeletal structures.

Bone is a specialized connective tissue that is characterized mainly by its mineralized organic matrix. The bone matrix is composed of collagenous and non-collagenous proteins. Within this matrix, calcium and phosphate ions are laid down, ultimately to form hydroxyapatite. In most parts of the skeleton, bone formation occurs during embryogenesis, by the initial deposition of cartilaginous templates that are subsequently replaced by bone, a process referred to as *endochondral* bone formation. In the cranial vault, in the diaphysis of long bones, and in the alveolar processes of the maxilla and of the mandible, bone is primarily formed within fibrous connective tissue; this is termed *intramembranous* bone formation.

Mature bones are made up of mineralized tissue and bone marrow. The mineralized compartment comprises an outer smooth, compact portion, the cortical bone, and an inner spongy part, the trabecular bone, in the proportion, by weight, of about 80% to 20%. The cellular component of the mineralized bone tissue, including osteoblasts, osteocytes and osteoclasts, is located within and upon the cortical and the trabecular bone. This composition and structure of bone allows it to resist load, to protect vulnerable organs, such as the central nervous system, and to support functional organs, like the teeth.

As bone is a specialized connective tissue, *osteoblasts* rather than fibroblasts are the cells primarily responsible for its formation. They are located on bone surfaces, where they actively deposit organic bone matrix and control its mineralization. Osteoblasts are the direct descendants of osteoprogenitor cells (Kneser et al., 2006; Buckwalter & Hunziker, 1996; Heinegard & Oldberg, 1989), and they differentiate either into *bone lining cells* or into *osteocytes*. Among other activities, osteocytes participate in the regulation of blood-calcium homeostasis and in signalling mechanical loading to other cells within the bone. Since osteoblasts are fully differentiated stable cells that lack the ability either to migrate or to proliferate, new bone formation is entirely dependent on the presence of *osteoprogenitors,* which are undifferentiated mesenchymal cells that can migrate to target sites, proliferate and differentiate into osteoblasts.

It has been suggested that osteoprogenitor cells may be designated as *determined* or as *inducible osteogenic precursor cells* (Friedenstein 1973). The *determined* osteoprogenitor cells are located in the bone marrow, in the endosteum and in the periosteum, and they possess the capacity to proliferate and to differentiate into osteoblasts. The *inducible* osteogenic precursor cells (e.g. myoblasts or adipocytes), present in other organs and tissues, may differentiate into bone-forming cells when exposed to specific stimuli. The main source of osteoprogenitor cells is considered to be the *pericyte*, a stellate perivascular cell.

The differentiation of osteoblasts from osteoprogenitor cells is dependent upon the release of, or the presence of factors that induce or promote bone growth, among them bone growth factors and bone growth proteins, as well as insulin-like growth factor, platelet-derived growth factor, and fibroblast growth factor.

Bone formation and remodelling are consistently associated with bone resorption that is initiated and maintained by *osteoclasts,* which are multinucleated cells originating from haematopoietic precursor cells. The processes of modelling and remodelling of bone start shortly after bone formation, by resorption and apposition of new mineralized tissue resulting in changes in bone architecture and morphology. It is believed that changes brought about by bone modelling are induced by functional demands, such as muscle tension and external loads, while remodelling occurs within the bone as an ongoing maintenance process.

Experimental observations show that primary (woven) bone that appears to be more amorphous and has a low load-bearing capacity is the first to be formed in areas of bone regeneration. Woven bone is gradually replaced by lamellar bone with a structure that is more resistant to stress in general, and to functional loading in particular.

Bone is one of a few tissues that possess spontaneous regenerative capacity; but spontaneous regeneration is limited, and falls far short of the ideal ultimate goal of therapeutic reconstruction, which is complete restoration of tissues or organs that have been damaged or removed, to their original normal structure, architecture, size and function.

In the oral cavity, for example, the periodontal attachment apparatus is an organ that comprises alveolar bone, periodontal ligament and cementum. If the periodontal attachment apparatus is disrupted by inflammatory disease such as periodontitis, or by trauma, or by surgical damage, spontaneous healing is achieved mainly by repair, i.e., fibrous tissue replaces the lost tissues (Le *et al.* 2005). The capacity of alveolar bone to regenerate has been shown to improve significantly in the presence of growth factors, which are natural biological mediators that significantly increase cellular chemo-attraction, proliferation and differentiation by regulating essential cellular events (Giannobile 1996). Once growth factors bind to surface receptors of specific target cells, they induce the activation of genes that change cellular activity and phenotype (Anusaksathien & Giannobile 2002; Schilephake 2002; Ripamonti *et al.* 2005).

This book is divided into three sections: chapters 1-5 deal mainly with regenerative tissue engineering, chapters 6-10 with different techniques of enhancing and supporting bone regeneration, and chapters 11-14 concentrate on biotechnology, and on recent advances and new approaches in developing biomaterials for bone regeneration.

Regenerative tissue engineering (RTE) may be defined as a process of combining living cells with biocompatible scaffolds to generate a biological substitute capable of sustaining itself and of integrating with functional native tissue (Chapter 1). RTE addresses the discrepancy between available transplantable donor tissues and the anatomical need.

Optimized methods have improved the function and maturation of engineered cellular constructs to produce new ones with clinically useful, near-native tissue properties. Following any bone injury, a healing cascade is triggered to restore the tissue's original state. This reaction occurs in the three phases of inflammation, repair, and remodelling. Initially, the inflammatory phase follows the formation of a blood clot arising from blood flowing and cells migrating into the site of injury from the borders of the injury. These cells include fibroblasts and inflammatory cells, such as macrophages, monocytes, lymphocytes, etc., which, together with ingrowing blood vessels that are a source of pericytes, promote new formation of collagen fibers and osteoid, forming a soft callus. This process starts within hours to a few days after bone injury and continues for about 4-8 weeks, while ossification of the callus and the formation of unorganized woven bone may take an additional 2-4 months. This progression of healing events establishes a basis for formulation of the principles of guided bone regeneration and distraction osteogenesis (Chapters 6–9).

The restructuring of woven bone is responsive to muscular activity and to mechanical stresses (Kneser et al. 2006). In the dental environment, for instance, restructuring of bone is observed around teeth and dental implants which transfer functional and occlusal forces to their anchoring bone.

The range over which healing, followed by spontaneous regeneration of bone may occur is, however, of limited potential. If the zone of damage exceeds a certain critical size, a bone defect may not self-repair, in which case guided bone regeneration using tissue barriers, autologous bone grafts, allografts, xenografts or alloplasts may provide a partial solution (Chapters 11-14). The following terms are defined for the sake of clarity:

- *Autografts* - human bone is harvested from the recipient of the graft himself or herself. This is currently considered by many as being the gold standard for bone grafting.
- *Allografts* - human bone for grafting harvested from donors usually unrelated to the recipient. Such grafts do not have the recipient-compatible immunogenic properties of autografts, thus increasing the risk of rejection (Mankin et al., 2005). Allografts are generally osteoconductive, although some are osteoinductive, depending somewhat upon the source of the bone and the technique of preparation.
- *Xenografts* - deproteinized bone grafts prepared from species other than human that contain only the hydroxyapatite matrix of bone. Xenografts are osteoconductive.
- *Alloplasts* - synthetic grafts made of biocompatible and/or bioactive materials, such as ceramics, bioglasses or calcium sulphate. Some alloplastic materials have the potential to be used as carriers of growth factors, thereby improving their osteoconductive and osteoinductive properties.

The potential of regenerative tissue engineering (RTE) has attracted much interest in the field of bone research. Current studies on RTE focus mainly on stem cells. The human body has many different types of cells, each specialized for a distinct rôle. The cells are committed to specific lineages and functions, for example, cardiomyocytes in the heart, chondrocytes in cartilage, and osteoblasts in bone. Cells are assessed for their possible utility for tissue engineering, mainly by their rate of proliferation and by their potential for differentiation, both of which depend upon the speed at which the individual cells divide as well as on the cell line's capability of developing into specific lineages, in our case, an osteogenic lineage. For the engineered tissue to mirror the native tissue, it is essential for the cells to expand at a specific rate and to differentiate towards the desired lineage.

Stem cells are multipotent, not lineage-specific, with the potential to differentiate into many kinds of specialized daughter cells. Adult stem cells can be harvested from various tissues of the body and can then be cultured *in vitro,* where they can be directed to provide a potentially unlimited supply of tissue. Stem cell-based bone tissue engineering is founded upon the potential of multipotent postnatal stem cells to participate in the regenerative healing of bone defects. Postnatal stem cells can be isolated from bone marrow, from adipose tissue, from muscle, from dental pulp tissue, from oral mucosa and from umbilical cord (Zuk et al., 2001; Miura et al., 2003; Schugar et al., 2009). It is noteworthy that in spite of the fact that embryonic stem cells are considered to be the gold standard in RTE , postnatal bone marrow-derived mesenchymal stem cells are the most researched and the most frequently used.

Clinically, most efforts to increase bone volume have focused on procedures that exploit spontaneous bone regeneration. The introduction of dental implant therapy, and the well established need for adequate bone volume at the implant site, in order to foster a favourable long-term prognosis for dental implants (Lekholm 1986) have dramatically increased interest in the development of implant sites.

Four approaches to the augmentation of bone volume have been described: a. *osteoinduction,* using appropriate growth factors (Urist 1965 ; Reddi 1981); b. *osteoconduction,* using grafting materials that serve as scaffolds for new bone growth (Buch *et al.* 1986; Reddi *et al.* 1987); c. *distraction osteogenesis,* by which bone growth is induced between the fragments at a surgically created bone fracture when the fragments are pulled apart in a slow, controlled manner. (Ilizarov 1989a,b); d. *guided bone regeneration,* which allows selective growth of bone tissue into a space maintained by tissue barriers (Dahlin *et al.* 1988, 1991**a,b**; Kostopoulos & Karring 1994; Nyman & Lang 1994). The purpose of all these procedures is to deal with the problem of localized lack of bone volume resulting from congenital, post-traumatic, postsurgical or pathological defects in various parts of the skeleton (Chapters 6-10).

Guided-tissue regeneration (GTR) was introduced into dental clinical practice soon after it became understood that the alveolar bone and the periodontal ligament can be sources of progenitor cells for the regenerative repair of adjacent periodontal lesions (Melcher 1970, 1976). Karring et al. (1980) and Nyman et al. (1980) formulated the basic principles of GTR over three decades ago. These are as follows: under certain conditions, cells that originate from a tissue adjacent to a delimited space are able to grow into that space and to form new tissue identical to their tissue of origin. In order to allow exclusive migration into, and population of such a space by a specific tissue, cells of that tissue must be given preferential access to the space. This is achieved by preventing access of cells from neighbouring dissimilar tissues by means of tissue barriers, commonly referred to as membranes.

For detailed discussion of the rationale and techniques of contemporary GBR procedures and the biotechnology associated with resorbable collagen bio-barriers, the reader is referred to Chapters 6 and 7. There is a wide variety of available resorbable and non-resorbable tissue barrier materials, including polytetrafluoroethylene (PTFE), expanded PTFE (e-PTFE), polyglactin 910, polylactic acid, polyglycolic acid, polyorthoester, polyurethane, polyhydroxybutyrate, calcium sulfate, freeze-dried fascia lata and freeze-dried dura mater allografts, titanium micro-mesh, and titanium foil.

Collagen membranes are generally to be preferred to other barrier materials because they possess all the essential properties required in a bio-barrier, including biocompatibility, cell occlusiveness, integration into the host tissues, space-making capacity, and also clinical manageability and ease of application.

To secure the space necessary for bone regeneration in GBR procedures, tissue barrier membranes need to be supported, or they will collapse owing to peripheral pressure, thus possibly reducing or completely eliminating the space that had been designated for potential bone growth and regeneration. Although membrane-stiffeners, or a wide range of membrane-supporting materials have been claimed successfully to fulfill the necessary requirements for membrane support, there is an ongoing search for new materials, both osseoinductive and osseoconductive, to be placed under the membrane (Chapters 11-14). The current consensus is that the optimal osteoconductive membrane-supporting material is one that interferes least with the spontaneous bone growth replacing the submembranous clot of blood. Taking this a step further, however, it may well be that with advances in engineered bone regeneration, future materials will promote bone growth, perhaps to the extent that tissue barriers may become unnecessary.

The main limitation of current guided bone regeneration procedures is that reliance is placed upon bone's spontaneous regenerative capacity, which suffices only for defects of limited size. Furthermore, the problem of stabilization of the volume of regenerated bone is a major limiting factor in bony defects of this nature, especially when lateral and, even more so, vertical bone loss is to be treated. Other measures, such as block grafts or distraction osteogenesis, may be more suitable for those cases.

The effectiveness of bone grafts depends on their osteoinductivity and osteoconductivity as well as on their biomechanical properties (Khan et al., 2005). Synthetic block grafts have recently been added to the range of available allografts: these grafts provide sizable, stable scaffolds that encourage new bone formation originating from any bone that is in direct contact with the graft material (Chapter 10). Therefore, while autologous block grafts still remain the gold standard, allograft and alloplast blocks reduce morbidity by obviating the need for harvesting bone blocks from the patient's iliac crest, fibula, ribs, calvaria or mandible (Burchardt, 1983).

Distraction osteogenesis (DO) may offer a viable alternative when size or volume of a bone defect exceeds the capacity of grafting successfully to replace the missing tissue. DO is a surgical technique by which, through the appropriate application of traction to the bone, the intrinsic capacity of bone to regenerate is directed towards lengthening or altogether replacing segments of bone. DO allows the spontaneous *de novo* formation of native bone without bone grafts. It may be considered a type of *in vivo* bone tissue engineering and may be superior to other techniques in certain cases. The current status, future developments and applications of DO are discussed in Chapters 8 and 9.

A number of extrinsic, local or systemic factors may affect bone mass, volume, structure and density, thereby influencing skeletal function. Among these, osteopenia, osteoporosis, diabetes mellitus, smoking and periodontal disease are frequently mentioned in the context of the jaws. This subject is not discussed separately in this book since data on the influence of those conditions on bone regeneration and on osseointegration in the setting of oral and orthopaedic implants are limited, neither is there any agreement on whether or not they should be regarded as contraindications for implant placement (Shernoff et al., 1994; Farzad et al., 2002).

In view of the increasing clinical application of bone regenerative procedures to dental implant site development, the relationship between such procedures and general health conditions has attracted considerable interest. However, no conclusive data are available with respect to bone augmentation procedures in patients suffering from those systemic diseases or conditions referred to above, including smoking, or in those with poor compliance in bacterial plaque control, any of which can have the effect of impairing tissue healing.

On the other hand, in patients who have lost their teeth owing to periodontal disease, implant procedures had greater rates of failure and more complications, than implant procedures in patients in whom periodontal disease had not been the primary reason for tooth loss (Mengel et al., 2001; Hardt et al., 2002; Karoussis et al., 2003; Wennstrom et al., 2004).

References

Anusaksathien, O. & Giannobile, W.V. (2002). Growth factor delivery to re-engineer periodontal tissues. *Current Pharmaceutical Biotechnology* 3, 129–139.

Buch, F., Albrektsson, T. & Herbst, E. (1986). The bone growth chamber for quantification of electrically induced osteogenesis. *Journal of Orthopedic Research* 4, 194–203.

Buckwalter, J., & Hunziker, E. (1996, Dec). Healing of bones, cartilages, tendons, and ligaments: a new era. *Lancet* 348 Suppl 2: sII18, 21-28.

Burchardt, H. (1983). The biology of bone graft repair. Clin Orthop Relat Res, 28-42.

Dahlin, C., Linde, A., Gottlow, J. & Nyman, S. (1988). Healing of bone defects by guided tissue regeneration. *Plastic and Reconstructive Surgery* 81, 672–677.

Dahlin, C., Alberius, P. & Linde, A. (1991a). Osteopromotion for cranioplasty. An experimental study in rats using a membrane technique. *Journal of Neurosurgery* 74 (3), 487–491.

Dahlin, C., Andersson, L. & Linde, A. (1991b). Bone augmentation at fenestrated implants by an osteopromotive membrane technique. A controlled clinical study. *Clinical Oral Implants Research* 2, 159–165.

Farzad, P., Andersson, L. & Nyberg, J. (2002). Dental implant treatment in diabetic patients. *Implant Dentistry* 11,262–267.

Friedenstein, A.J. (1973). Determined and inducible osteogenic precursor cells. In: *Hand Tissue Growth Repair and Remineralisation. Ciba Foundation Symposium. New series* 11,169–181.

Giannobile, W.V. (1996). Periodontal tissue engineering by growth factors. *Bone* 19, 23S-37S.

Hardt CR, Gröndahl K, Lekholm U, Wennström JL. (2002) Outcome of implant therapy in relation to experienced loss of periodontal bone support: a retrospective 5- year study. Clin Oral Implants Res. 13,488-94.

Heinegard, D., & Oldberg, A. (1989, Jul). Structure and biology of cartilage and bone matrix noncollagenous macromolecules. *FASEB J* 3(9), 2042-2051.

Ilizarov, G.A. (1989a). The tension-stress effect on the genesis and growth of tissues: Part I. The influence of stability of fixation and soft tissue preservation. *Clinical Orthopaedics* 238, 249-281.

Ilizarov, G.A. (1989b). The tension-stress effect on the genesis and growth of tissues: Part II. The influence of the rate and frequency of distraction. *Clinical Orthopaedics* 239, 263-285.

Karoussis IK, Salvi GE, Heitz-Mayfield LJ, Brägger U, Hämmerle CH, Lang NP. (2003) Long-term implant prognosis in patients with and without a history of chronic periodontitis: a 10-year prospective cohort study of the ITI Dental Implant System. Clin Oral Implants Res. 14,329-39.

Khan, M., Sahibzada, A., Khan, M., Sultan, S., Younas, M., & Khan, A. (2005, Apr-Jun). Outcome of plating, bone grafting and shortening of non-union humeral diaphyseal fracture. *Journal of Ayub Medical College Abbottabed*, 17(2), 44-46.

Karring, T., Nyman, S. & Lindhe, J. (1980). Healing following implantation of periodontitis affected roots into bone tissue. *Journal of Clinical Periodontology* 7, 96–105.

Kneser, U., Schaefer, D., Polykandriotis, E., & Horch, R. (2006, Jan-Mar). Tissue engineering of bone: the reconstructive surgeon's point of view. *Journal of Cellular and Molecular Medicine*, 10(1), 7-19.

Kostopoulos, L. & Karring, T. (1994). Augmentation of the rat mandible using guided tissue regeneration. *Clinical Oral Implants Research* 5, 75–82.

Le, A.D., Basi, D.L. & Abubaker, A.O. (2005). Wound healing:fi ndings of the 2005 AAOMS Research Summit. *Journal of Oral and Maxillofacial Surgery* 63, 1426–1435.

Lekholm *U.* (1986) Osseointegrated implants in clinical practice. J Oral Implantol. 1986;12(3):357-64

Lugero, G.G., de Falco Caparbo, V., Guzzo, M.L., Konig, B., Jr. & Jorgetti, V. (2000). Histomorphometric evaluation of titanium implants in osteoporotic rabbits. *Implant Dentistry* 9, 303–309.

Mankin, H., Hornicek, F., & Raskin, K. (2005, Mar). Infection in massive bone allografts. *Clinical Orthopedics and Related Research* (432), 210-216.

Mengel R, Schröder T, Flores-de-Jacoby L.(2001) Osseointegrated implants in patients treated for generalized chronic periodontitis and generalized aggressive periodontitis: 3- and 5-year results of a prospective long-term study. J Periodontol. 2001 Aug;72(8):977-89

Melcher AH. (1970) Repair of wounds in the periodontium of the rat. Influence of periodontal ligament on osteogenesis. Arch Oral Biol. 1970 Dec;15(12):1183-204.

Melcher AH (1976) On the repair potential of periodontal tissues. Journal of Periodontology; 47(5):256-60.

Miura, M., Gronthos, S., Zhao, M., Lu, B., Fisher, L. W., Robey, P. G., & Shi, S. (2003). SHED: stem cells from human exfoliated deciduous teeth. *Proceeds of the National Academy of Science USA*, 100(10), 5807-5812.

Nyman, S., Karring, T., Lindhe, J. & Planten, S. (1980). Healing following implantation of periodontitis affected roots into gingival connective tissue. *Journal of Clinical Periodontology* 7, 394–401.

Nyman, S.R. & Lang, N.P. (1994). Guided tissue regeneration and dental implants. *Periodontology 2000* 4, 109–118.

Reddi, A.H. (1981). Cell biology and biochemistry of endochondral bone development. *Collagen Related Research* 1,209–226.

Reddi, A.H., Weintroub, S. & Muthukumaran, N. (1987). Biologic principles of bone induction. *Orthopedic Clinics of North America* 18, 207–212.

Ripamonti, U., Herbst, N.N. & Ramoshebi, L.N. (2005). Bone morphogenetic proteins in craniofacial and periodontal tissue engineering: experimental studies in the nonhuman primate Papio ursinus. *Cytokine and Growth Factor Review* 16, 357–368.

Schilephake, H. (2002). Bone growth factors in maxillofacial skeletal reconstruction. *International Journal of Oral & Maxillofacial Surgery* 31, 469–484.

Schugar, R., Chireleison, S., Wescoe, K., Schmidt, B., Askew, Y., Nance, J., Deasy, B. (2009). High harvest yield, high expansion, and phenotype stability of CD146 mesenchymal stromal cells from whole primitive human umbilical cord tissue. *Journal of Biomedicine & Biotechnology*, 789526.

Shernoff, A.F., Colwell, J.A. & Bingham, S.F. (1994). Implants for type II diabetic patients: interim report. VA Implants in Diabetes Study Group. *Implant Dentistry* 3, 183–185.

Urist, M.R. (1965). Bone: formation by autoinduction. *Science* 150, 893–899.

Wennström JL, Ekestubbe A, Gröndahl K, Karlsson S, Lindhe J. (2004). Oral rehabilitation with implant-supported fixed partial dentures in periodontitis-susceptible subjects. A 5-year prospective study. J Clin Periodontology 31,713-24.

Zuk, P., Zhu, M., Mizuno, H., Huang, J., Futrell, J., Katz, A., Hedrick, M. (2001). Multilineage cells from human adipose tissue: implications for cell-based therapies. *Tissue Engineering*, 7, 211-228.

Part 1

Tissue Engineering of Bone

1

Stem Cell Based Bone Tissue Engineering

Lauren Vernon, Lee Kaplan
and Chun-Yuh Charles Huang
University of Miami, Coral Gables, FL
USA

1. Introduction

Regenerative tissue engineering is defined as the combination of living cells and biocompatible scaffolds to generate a biologic substitute capable of sustaining itself and integrating with functional native tissue. By engineering and delivering tissues and/ or cells capable of replacing damaged tissue, regenerative medicine offers the potential for the treatment and possibly curing of debilitating diseases. Optimized methods for improving the function and maturation of engineered cellular constructs to produce constructs with near-native tissue properties are necessary to enable translation to clinically useful therapies.

In regenerative tissue engineering, there are many issues to consider in the creation of a functional, implantable replacement tissue. Most importantly, there must be an easily accessible, readily abundant cell source with the capacity to express the desired tissues' phenotype, and a biocompatible inert scaffold to deliver the cells to the damaged region. Currently, there are many regenerative tissue engineering studies in preclinical and clinical testing using stem cells.

What are stem cells? In the body there are many different types of cells, each specialized for a specific distinct function. These cells are committed to a specific lineage and function, for example cardiomyocytes in the heart, chondrocytes in cartilage, and osteoblasts in bone. Immature progenitor cells called stem cells are not lineage specific. Stem cells have the ability to differentiate into many different kinds of cells. Differentiation is the process by which an unspecialized parent or progenitor cell gives rise to a specialized daughter cell; this process occurs in stages with the cell becoming increasingly specialized with each step. At fertilization, a zygote is formed containing totipotent stem cells which can differentiate into any cell type. After a few days, the blastocyst develops with an outer layer and inner cell mass. Cells from the inner cell mass are pluripotent stem cells and can become cells from all of the germ layers (ectoderm, mesoderm, endoderm). As the embryo continues to develop, the cells become increasingly specialized and begin to commit to specific cell lineages (as seen in Figure 1 below). Cells committed to a specific cell type and function lose their stemness; however, a population of undifferentiated stem cells remains amongst the differentiated cells. These adult stem cells can be harvested from various tissues in the body and then cultured in-vitro. Under specific culture conditions, those undifferentiated cells can be directed towards specific lineages, providing us with a potentially limitless supply of tissue.

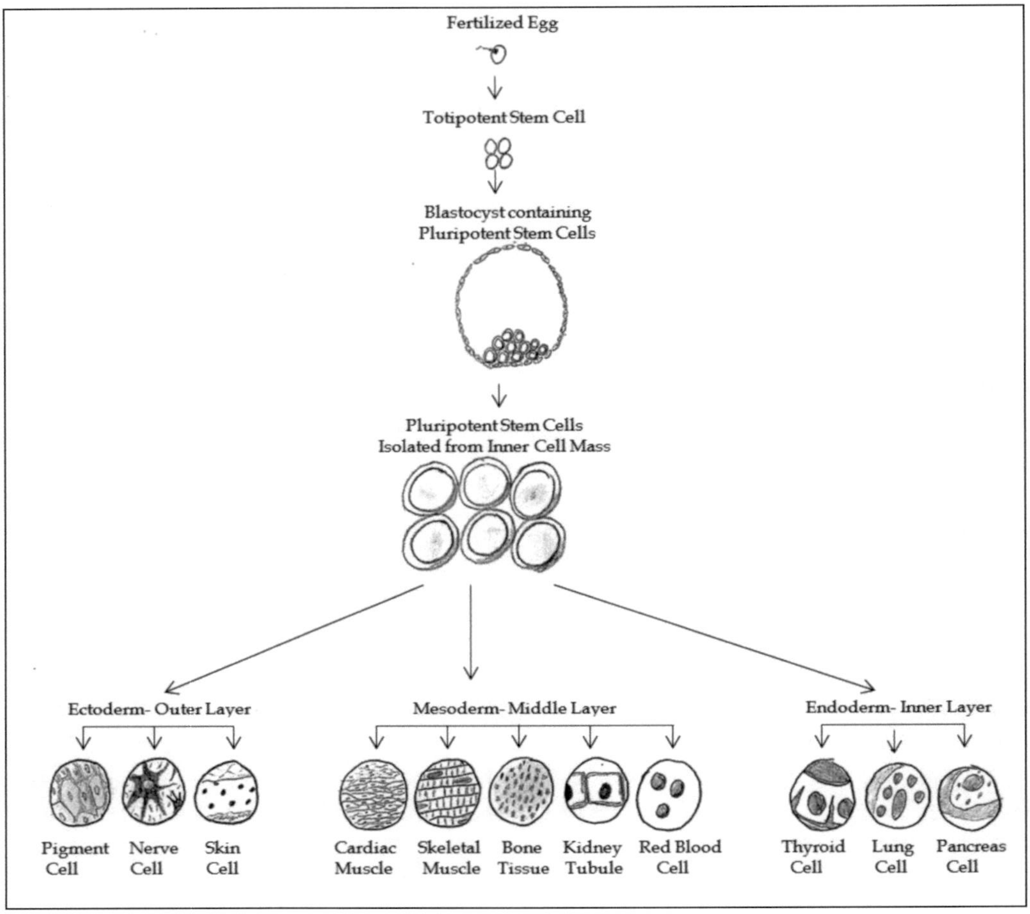

Fig. 1. Stem cell differentiation diagram

There are two parameters used to assess the tissue engineering potential of stem cells: proliferation rate and differentiation potential. The cells' proliferation rate is the speed at which the cells divide. The cell differentiation potential determines whether the cell line is capable of developing into specific lineages, in this case osteogenic. These components are critical, because if cells are not expanding at a specific rate and are not differentiating towards the lineage desired, the engineered tissue will not mirror the characteristics of native tissue and serve the desired function.

2. Bone properties

2.1 Bone composition

Bone is a multifunctional organ that plays many important roles in the body, such as providing protection and structure (shape) among others. Despite its relatively simple outer appearance, bone is not a solid homogenous tissue. Bone is made up of solid material with spaces between its hard elements. The outer smooth portion of the bone is compact or cortical bone (80% of bone), and the inner spongy part of the bone is trabecular bone (20% of bone). Furthermore, within those regions are different types of cells making up the cellular structure of the bone itself: osteoblasts, osteocytes, and osteoclasts. Osteoblasts are bone forming cells and are the direct descendants of osteoprogenitor cells (Kneser et al., 2006; Buckwalter & Hunziker, 1996; Heinegard & Oldberg, 1989). They produce a protein mixture called osteoid which mineralizes to form solid bone, and produce alkaline phosphatase (ALP), a key enzyme in bone mineralization. Osteocytes are descendants of osteoblasts which migrate into the bone matrix forming lacunae spaces. Osteocytes are responsible for bone formation, matrix maintenance, and calcium homeostasis. Osteoclasts are the bone's absorption cells. These are the cells that are responsible for bone remodeling and the overall decrease in bone mineral density (Shier et al., 2002). Bone is a connective tissue which provides structural support and protects the vital organs while allowing for movement. Bone is composed of a type I collagen matrix embedded with calcium, phosphorous, sodium, magnesium and other ions necessary for homeostasis in the body. The extracellular components of bone (organic matrix and minerals) combine to strengthen the bone, giving it the ability to withstand mechanical stresses.

2.2 Bone formation and maturation in vivo

Within the body there are two main mechanisms of bone formation and growth, intramembranous ossification for flat bones and endochondral ossification for long bones. In intramembranous ossification, connective tissue occupies the place of the future bone and is slowly replaced as ossification centers develop, calcification occurs, and the trabeculae forms. In endochondral ossification, also referred to as intracartilaginous ossification, a cartilage model template structure is used on which the primary and secondary ossification centers are established. When bone is finished forming, remnant articular cartilage often remains on the end of the bones. Although induced tissue engineering differs from either pathway, in vitro osteogenic differentiation more accurately mirrors the intramembranous ossification process.

2.3 Bone repair

Following injury, a healing cascade is triggered to restore the tissue's original state. This healing cascade occurs in three phases: inflammatory, repair, and remodeling (as seen in Figure 2). Briefly, the inflammatory phase results in hematoma formation as fibroblasts and cells from the inflammatory cascade (macrophages, monocytes, lymphocytes etc.) migrate into the injury site (hours to days). In the repair phase, collagen fibers and mineralized osteoid combine to form a soft callus around the injury site (4-6 weeks). As the callus ossifies it forms a disorganized structure known as woven bone. During the remodeling phase the disorganized woven bone is replaced by highly organized sheets of parallel collagen fibers called lamellar bone. This restructuring occurs in response to mechanical stresses signaled to the osteocytes and subsequently osteoclasts and osteoblasts (Kneser et al., 2006; Probst & Spiegel, 1997). The repair mechanism continues to change the bone properties through constant remodeling via bone deposition and resorption (Kneser et al., 2006; Kalfas, 2001).

Bone tissue can be damaged due to injury or disease. Following a traumatic event, fractured bones can be aligned by a physician, and the damaged area can heal and repair. However, bones that are damaged due to disease are not as successful at self-restoration as bones set by a physician and do not have the ability to repair and remodel properly.

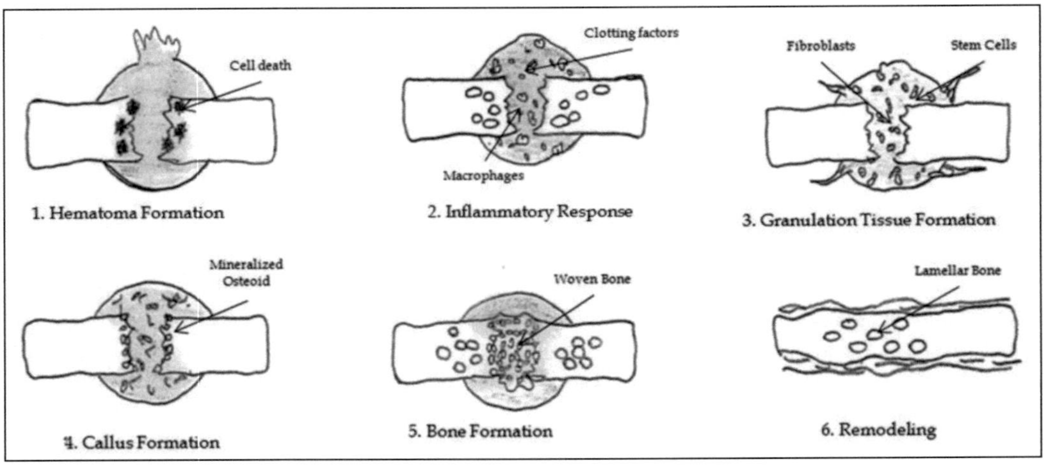

Fig. 2. Bone repair mechanism

2.4 Current treatments for damaged bone

2.4.1 Bone grafts

Bone grafts are used to enhance bone healing in specific injury zones. There are a number of factors to take into consideration when selecting a bone graft, such as osteoinductiveness, osteoconductiveness, mechanical stability, quality of transplantable bone, and preservation techniques, as well as implantation site considerations such as local disease and biomechanical properties (Khan, et al., 2005). Osteoconductive grafts function as a scaffold for native bone cells adjacent to the graft to migrate into and form new bone. Osteoinductive grafts serve to actively stimulate progenitor cells to differentiate into osteoblasts. A graft can be considered osteogenic if it already contains osteoblasts. The main types of grafts currently being utilized are autografts, allografts, xenografts, and synthetic grafts.

Conventionally, autografts are the most effective grafts since they are harvested from the patient's own body, usually from the iliac crest, fibula, ribs, or mandible. Although using the patient's own bone reduces the risk of rejection and disease transfer, it also increases pain, since the donor graft must be harvested surgically. In addition to the risk of infection from harvesting the patient's own bone, the process can be extremely painful, and as with all surgeries, there is always the risk of complications. Autografts are osteoconductive, osteoinductive, and osteogenic, which is what makes them the ideal choice (Burchardt, 1983). However, the amount of bone available for autografts is limited and the procedure creates a secondary healing site.

Allografts, on the other hand, are also natural human bone grafts, but unlike autografts which are harvested from the patient's own body, allografts are harvested from an external donor. Although the allograft eliminates the need for an additional surgery on the recipient, they also do not have the immunogenic properties of the autograft, and the risk of rejection increases (Mankin et al., 2005). Depending on how the allograft was harvested, stored, and treated, it will have osteoconductive properties, but may or may not have osteoinductive properties, and will not have any osteogenic properties (Mankin et al., 2005). For example, freeze-drying samples reduces antigenicity while maintaining biomechanical properties.

Xenografts are grafts that are harvested from another species and are stripped down to just their calcium matrix. These grafts are osteoconductive but not osteoinductive or osteogenic. Due to their limited surface properties, synthetic grafts are a preferred alternative.

Synthetic grafts are artificial bone grafts that can be made of biologically active materials such as ceramics, bioglasses, or even calcium sulphate. These grafts can be functionalized with growth factors to increase their osteoconductive and osteoinductive abilities. Along those lines, engineering a replacement bone tissue from the patients' own autologous stem cells seems to be a more practical solution. An explanation of various biomaterials that are used to make synthetic grafts will be explained later in the scaffolds section of the chapter (Section 5).

2.4.2 Bone bracing

In some cases damaged bones are incapable of properly repairing and remodelling; therefore, additional steps such as utilizing bone bracing can be taken to aid the reconstruction. These approaches include both surgical and non-surgical methods of immobilizing the injured region allowing bone repair to begin.

Non-surgical alternatives such as external casts and bracing are viable alternatives in cases with fracture displacements less than 2mm. This is the preferable method for bone healing, as it is non-invasive and minimizes risk of infection. Regenerative tissue engineering is not required in these cases and may not be applicable for these cases.

Surgical approaches, such as open reduction internal fixation in which bones are set and held in place by nails, screws, or plates to guide and facilitate the healing process. Unfortunately, as with an surgical option, there is an increased risk of infection; however, surgical approaches are necessary in fracture displacements exceeding 2mm. In cases where surgery is required, regenerative tissue engineering may provide the materials necessary for creating bone in these voids, reducing healing time and increasing bone strength. Usually these fixation devices are metals, which provide immediate mechanical support; however, metals have poor integration into the bone and can cause re-injury and implant failure over time, therefore, alternative approaches are being investigated.

2.4.3 Biomaterials and alternative approaches

In some cases, bones that are damaged are unable to repair and remodel properly and additional steps can be taken to aid the reconstruction, such as utilizing bone cements. In other cases, joint replacement surgery is needed to repair areas where the cartilage has been destroyed and the bones are exposed. When this occurs, a prosthetic joint is drilled into the bone and can either be cemented in with a polymer bone cement or left un-cemented. Traditionally, acrylate-based bone cements have been used for their high mechanical stability (Muh, et al., 2002; Lewis, 1997). When the bone is cemented in place, the recovery time is much quicker than when left un-cemented. On the other hand, when left un-cemented, the natural bone can grow into the porous surfaces of the implant, making it stronger over time. This means that the implant is less likely to fail, which would require an additional surgery. Therefore, osteoinductive alternative, such as calcium phosphate and apatite-based bone cements, have been investigated. Although osteoinductivity is increased, these cements are also biodegradable, therefore, their mechanical strength decreases over time (Cassidy, et al., 2003; Zijderveld et al., 2005).

In regenerative tissue engineering, living cells and biocompatible scaffolds are combined to generate biologic substitutes capable of sustaining themselves and mimicking functional native bone. This is accomplished by incorporating osteoinductive and osteoconductive molecules into the scaffold, as explained in the scaffolds section of the chapter (Section 5).

3. Bone tissue engineering with stem cells

Cells are the basic functional and structural unit in the body. All cells in the body have the same primary DNA sequence; however, changes in phenotypic expression result in cells with a variety of shapes and sizes which correlate to the function of each cell. This diversity comes at a cost; in general, once a cell has differentiated into a specific lineage, it cannot revert back to its primordial state (exceptions, ex: cancer and induced pluripotent stem cells). Stem cells, however, are undifferentiated cells (a characteristic referred to as stemness). These cells have not committed to any specific lineage and can give rise to a variety of specialized cell types depending on their plasticity. A cell's plasticity is its ability or inability to differentiate into any of the germ layers (ectoderm, mesoderm, endoderm) based on culture conditions.

In general, differentiated cells have a limited number of population doublings over the course of a person's life (approximately 80-90 times) (Fox, 2001). During telophase of mitosis, the telomere region of the chromosomes lose some of the DNA sequences, approximately 50-100 base pairs (Fox, 2001). Stem cells (as well as germinal cells and cancer cells) produce telomerase, an enzyme which duplicates the telomere DNA; therefore, stem cells are capable of dividing and self-renewal for long periods of time and are often referred to as immortal. Furthermore, when stem cells divide, they can undergo either symmetric division, producing two daughter stem cells identical to the parent cell, or by asymmetric division, producing one daughter stem cell and one daughter progenitor cell.

Stem cell biology has become an important topic in regenerative tissue engineering, specifically the use of multipotent mesenchymal stem cells (MSCs). Although embryonic stem cells are considered the gold standard in stem cell research, bone-marrow derived mesenchymal stem cells (BM-MSCs) are the most researched postnatal stem cells. Multipotent postnatal stem cells have been isolated from numerous tissues throughout the body such as bone marrow, adipose tissue, muscle, dental tissue, umbilical cord, etc. (Campagnoli et al., 2001; Zuk et al., 2001; Gronthos et al., 2000; Miura, et al., 2003; Seo, et al., 2004; Schugar, et al., 2009; Young, et al., 2001).These stem cells are capable of differentiating into a variety of cell lineages, including bone. Therefore, transplanting a patient's own stem cells may be a potential treatment for repairing bone defects. This chapter details progress made to date in the osteogenic differentiation potential of these cell lines and their potential use in repairing bone defects.

3.1 Embryonic stem cells

At fertilization, a zygote is formed that contains totipotent cells which are cells with the ability to form any of the 200+ cell types in the body and the cells of the placenta. After four days (about 40-150 cells), the blastocyst develops. The blastocyst is identifiable by the development of the outer trophoblastic layer and inner cell mass (ICM). The outer layer of cells becomes the placenta and other tissues necessary for fetal development and survival. The inner cell mass forms the fetus and contains pluripotent cells that go on to form all the tissues in the human body. The use of pluripotent embryonic stem cells in regenerative therapies is an attractive option with the ability to give rise to tissues from the three germ layers, including the mesodermal lineages such as bone.

Both totipotent and pluripotent (blastocyst) embryonic stem cells (ESCs) (Figure 1) are considered the gold standard in stem cells; however, controversy and debate surround ESCs. Despite the ethical debate, clinical benefits from ESCs could be numerous, but more work is needed before they can be used for clinical applications. In addition to the ethical considerations, additional concerns for using ESCs include the potential for teratoma formation when implanted in vivo.

3.1.1 Fetal stem cells

Fetal stem cells are cells obtained from an unborn fetus when the fetus has developed enough that cellular extraction does not cause fetal death. These cells are pluripotent and responsible for the development of all tissues before birth. Unlike ESCs, fetal stem cells can be obtained without completely destroying the embryo, allowing the fetus to develop into a full-term baby (Biswas & Hutchins, 2007). However, the effect of removing cells during fetal development is unknown, and fetal stem cells have many of the same ethical considerations as ESCs. These pluripotent cells can undergo osteogenic differentiation, making them a valid source for regenerative bone tissue engineering.

3.2 Bone marrow-derived mesenchymal stem cells

BM-MSCs are a heterogeneous population of multipotent cells. BM-MSCs are capable of differentiating into multiple lineages in vitro including the osteogenic lineage. BM-MSCs are a popular source of autologous adult stem cells, because they are readily available. However, the extraction procedure is extremely painful and invasive. In addition to their differentiation potential, BM-MSCs can be used directly to positively influence the repair mechanism and healing of cardiac tissue following a myocardial infarction (Amado, et al., 2005). Impressively, this is accomplished with BM-MSCs from living donor tissue, and the BM-MSCs inherent immunogenic characteristics limit the recipient's immune response to the foreign cells. This makes BM-MSCs a great source for regenerative tissue engineering applications, because they can be extracted, expanded, and banked, making them readily available when they are needed.

3.3 Umbilical cord blood stem cells

Umbilical cord blood (UCB) stem cells are cells found in the umbilical cord blood of a newborn baby, and they share the newborn's genetic material. UCB cells can be obtained with higher cell yields and without the pain and morbidity associated with BM-MSC acquisition (G. Huang et al., 2009). These cells are multiotent, hematopoietic stem cells and can differentiate into various cell lines including the osteogenic lineage (Liu, et al., 2011). In addition to their differentiation potential, UCB cells can be used directly to successfully treat leukemia, lymphoma, myelodysplasia, apalstic anemia, hemoglobinpathies, metabolic diseases, and immunodeficiencies (Brunstein et al., 2007; Ballen, et al., 2008). UCB cells have similar morphologic and immunophenotypic properties to BM-MSCs (Kern et al., 2006); however, UCBs form fewer colonies than BM-MSCs, and UCB cells form have a high proliferative capacity (G. Huang et al., 2009), perhaps due to their age and harvest location. Until recently, umbilical cords were discarded at birth, so most individuals have lost their source for autologous UCB cells. Now, parents are given the option of donating their newborn's UCB cells to public storage banks or saving them in a private bank for autologous or family member use.

3.4 Stem cells from dental tissues

Physiological similarities between dental-tissue and bone make dental-derived progenitor cells a logical source of stem cells for osteogenic differentiation. Cells from dental tissues are called ectomesenchyme cells, because they are remnant tissues derived from the cranial neural crest. Cranial neural crest cells are capable of differentiating into bone, cartilage, and ligament during embryonic development; therefore, cells derived from them possess similar abilities (Chai, et al., 2000; Huysseune & Thesleff, 2004; Le Douarin et al., 2004).

3.4.1 Stem cells from human exfoliated deciduous teeth

SHED (stem cells from human exfoliated deciduous teeth) are multipotent stem cells isolated from the remnant pulp of deciduous (baby) teeth (Miura, et al., 2003). Similar to the umbilical cord, deciduous teeth offer the opportunity for painlessly obtaining primordial cells that would otherwise have been thrown away. SHED proliferate faster than BM-MSCs, can undergo osteogenic differentiation, and express ESC markers (Kerkis, et al., 2006; Miura, et al., 2003). Furthermore, SHED appear to have osteoinductive properties, meaning they induce new bone formation by recruiting osteogenic host cells into an osteoinductive template (Miura, et al., 2003). This approach is already being used to repair critical-size calvarial defects in mice (Seo, et al., 2007).

3.4.2 Dental pulp stem cells

Similar to SHED, dental pulp stem cells (DPSCs) are heterogeneous populations of cells isolated from the human dental pulp of mature teeth. Within a colony of these multipotent cells, various cell morphologies are expressed. DPSCs proliferate faster than BM-MSCs but not as fast as SHED (Miura, et al., 2003) and are capable of differentiating into various lineages including osteogenic (Laino, et al., 2005). Interestingly, DPSCs cultured on dentin differentiate into odontoblast-like cells (Batouli, et al., 2003), suggesting that DPSCs are pre-differentiated towards soft tissue repair or even bone tissue interfaces, but not osteogenesis.

3.4.3 Periodontal ligament stem cells

Heterogeneous populations of multipotent stem cells (PDLSCs) have also been extracted from the periodontal ligament, a descendant of the cranial neural crest. PDLs express several ESC markers and have an upregulated telomerase activity, suggesting similar differentiation abilities to ESCs (C. Huang et al., 2009). Similar to other dental tissues, PDLSCs are capable of undergoing osteogenic differentiation and express osteogenic characteristics (Seo, et al., 2004; Gay et al., 2007; Isaka, et al., 2001; C. Huang et al., 2009). Furthermore, PDLSCs that are implanted in periodontal injuries regenerated a periodontal ligament-like tissue while aiding in the bone regeneration itself (Seo, et al., 2004). This suggests that they would be a viable source for regenerative bone tissue engineering.

3.5 Adipose tissue-derived stem cells

Adipose tissue-derived stem cells (ASCs) are multipotent cells located in fat that can differentiate into various cell lines including the osteogenic lineage (Levi & Longaker, 2011; Wagner, et al., 2005). ASCs can be isolated from the lipoaspirate usually discarded from liposuction treatments. ASCs have been shown to not only undergo osteogenesis, but have actually been used to heal critical-size defects in mice (Cowan, et al., 2004). Furthermore, ASCs promote angiogenesis (new blood vessel formation), which can be crucial for engineered scaffolds to properly integrate with native tissue (Schroeder & Mosheiff, 2011; Kim Y. , et al., 2007).

3.6 Induced pluripotent stem cells

As mentioned before, adult stem cells can be harvested from various tissues in the body and then, under specific culture conditions, directed towards specific lineages, thereby providing us with a potentially limitless supply of tissue. In general, ESCs are considered to be the gold standard in stem cell research due to their true pluripotency, and all other stem cells are measured in comparison. However, there are many ethical issues surrounding ESCs, most notably their source and the debate of whether or not the method used for the isolation of ESCs is murder. These ethical dilemmas and political restrictions on ESC use led researchers to investigate methods of reverting differentiated somatic cells back into their primordial pluripotent state. These reverted cells are called induced pluripotent stem (iPS) cells.

One of the earliest techniques for creating iPS cells was by retrovirus transduction of specific transcription factors to promote ESC-like characteristics and ultimately create a line of pluripotent adult stem cells (Yu, et al., 2007; Takahashi, et al., 2007; Takahashi & Yamanaka, 2006; Meissner et al., 2007; Okita et al., 2007). Once the iPS cells are generated, they are evaluated for ESC characteristics by RT-PCR, Western blot, telomerase detection, genomic sequencing, and immunohistochemistry. Most retrovirus transduction techniques use transcription factors that are naturally found up-regulated in ESCs. Once cells expressing ESC characteristics are identified (transcription factor transduction has a relatively low yield), the iPS cells' stemness must be evaluated, and the cells are tested for population clonogenicity, immortality, and pluripotency. Clonogenicity and immortality are evaluated by allowing the iPS cells to undergo significant population doublings. Next, the cells pluripotency is evaluated through in vitro culture and in vivo, and the cells are evaluated on their ability to differentiate into cells representing each of the germ layers. iPS cells represent a unique source for pluripotent adult stem cells which can serve as a source for generating patient-specific tissue for regenerative tissue engineering applications, such as repairing bone defects.

Unfortunately, many of the techniques that are used to induce the cells back into the pluripotent state have an extremely low yield, making the process both inefficient and costly. Furthermore, those same techniques use factors (such as viral vectors to transduce genes) which are not FDA approved. With those considerations in mind, research continues on traditional stem cells as well as iPS cells.

3.7 Evaluating potency

When selecting stem cells for regenerative tissue engineering applications, it is important to select cells that are capable of undergoing the necessary differentiation. To evaluate the potency of the cells, a variety of tests are performed, and specific factors are identified.

As mentioned before, telomerase activity in ESCs and some adult stem cells correlates with those cells ability to divide indefinitely (Heins, et al., 2004; Biswas & Hutchins, 2007). As cells begin to differentiate into specialized cell lines, telomerase levels decrease and ultimately disappear. Therefore, the presence of telomerase is a good benchmark for evaluating the primordial state of a cell and is often found in cells with the most potency potential (totipotent and pluripotent cells) (Heins, et al., 2004; Thomason & Marshall, 1998a; Biswas & Hutchins, 2007; Odorico et al., 2001).

Another method of evaluating potency is to allow the cells to grow in culture without passaging. When the cells become confluent, contact inhibition allows for spontaneous differentiation, and the stem cells will randomly divide into various lineages, thereby establishing their ability to form cells from specific germ layers, from which potency can be determined (Heins, et al., 2004; Itskovitz-Eldor, et al., 2000). Similarly, cells that are grown in 3-D cell suspensions form aggregates of differentiated cells called embryoid bodies. Continued culture of these embryoid bodies results in the formation of various germ layer derivatives (Itskovitz-Eldor, et al., 2000).

Furthermore, teratoma formation can be used to determine a cell's potency. Teratoma formation occurs when undifferentiated stem cells are injected into severe combined immunodeficient (SCID) mice, and the cells form tumors consisting of a variety of cell types (Heins, et al., 2004; Thomson, et al., 1998b; Park, et al., 2003; Knoepfler, 2009).

Finally, specific surface marker expression has been identified for undifferentiated and differentiated tissue. Antigen analysis can reveal a cell's primordial state, suggesting its potency (Amit & Itskovitz-Eldor, 2002; Knowles et al., 1978; Biswas & Hutchins, 2007; Lebkowski, et al., 2001).

4. Media formulations for osteogenic differentiation

All cells in our body contain the exact same DNA sequence of genes; however, the genes that are activated depend on the function of each cell. Most cell types have genes that are specific to that cell type (due to its unique functionality). These genes are responsible for coding specific proteins that are necessary to retain cell function. When cells are changed from stem cells (no specific function) into osteoblasts (cells that make our bones), genes that are specific to bone are activated. When a gene is activated, RNA is synthesized (transcription), and corresponding proteins are produced (translation).

In culture, cells can be propagated or differentiated based on the culture media. Transplanted cells that have been pre-differentiated in vitro into the osteogenic lineage form bone-like tissues better than undifferentiated transplanted cells (Cowan, et al., 2005; Conjero, et al., 2006). Currently, there are two treatments that are being used in osteogenic differentiation, dexamethasone (DEX) and retinoic acid (RA) (Ogston et al., 2002). In addition, there are osteoinductive molecules that can be added to enhance bone formation such as bone morphogenic protein (BMP) and platelet derived growth factor. The exact intra-cellular mechanism through which these treatments stimulate osteogenic differentiation is unknown.

4.1 Dexamethasone

DEX is a synthetic glucocorticoid. Osteogenic differentiation occurs when stem cells are treated with DEX (Bielby et al., 2004; Pittenger, et al., 1999; C. Huang et al., 2009; Zuk, et al., 2001; Young, et al., 2001), resulting in mineral deposits when grown in monolayer (Gronthos et al., 2000; Miura, et al., 2003; Seo, et al., 2004). However, the efficiency of the DEX treatment varies depending on the initial cell source (Chadipiralla, et al., 2010). The exact intra-cellular mechanism through which dexamethasone treatments stimulate osteogenic differentiation is unknown.

4.2 Retinoic acid

RA is an oxidized form of Vitamin A. RA was first used to up-regulate osteoblast differentiation and proliferation with great success (Song, et al., 2005; Kawaguchi et al., 2005; Skillington et al., 2002; Choong et al., 1993), then later used to induce osteogenic stem cell differentiation (Malladi et al., 2006; San Miguel, et al., 1998). The exact intra-cellular mechanism through which retinoic acid treatments stimulate osteogenic differentiation is unknown.

4.3 Bone morphogenic protein

BMP is a crucial protein in the osteoconductive process (Einhorn, 2003; Dimitriou, et al., 2005; Kain & Einhorn, 2005; Govender, et al., 2002). There are many variations of BMP, each with its own functionality; BMP-2 and BMP-7 have been shown to be crucial for bone maintenance. BMP-2 and RA have been shown to accelerate bone formation and osteoclast recruitment in vivo (Cowan, et al., 2005).

4.4 Platelet derived growth factor

Bone platelet derived growth factor is a dose-dependent bone formation enhancer (Thorwarth et al., 2006).

5. Scaffolds

For regenerative tissue engineering, cells can be expanded and differentiated in monolayer but must be delivered to the body in a 3-D scaffold. When designing a 3-D scaffold, it is important to take into account several key design elements such as the mechanical, chemical, and physiological properties of the scaffold, the environment it will be placed in, and the type of tissue it needs to support (Vacanti & Vacanti, 2000). The scaffold must have the correct surface chemistry and structure for optimizing the cell-scaffold interaction. In this case, the cells must be able to adhere to and proliferate on the scaffold. The scaffold must be biocompatible with the cells that it will support, in this case stem cells, and it must be biocompatible with the environment where it will eventually be placed in the body Additionally, the scaffold needs to have pores large enough for the cells to reside within it without being dislodged (pore size), and it needs to have well interconnected pores so nutrients can diffuse to the cells (porosity/interconnectivity). Mechanical strength is necessary for creating structurally sound replacement bone. However, the ideal scaffold would degrade as the new tissue formed. Many factors can cause polymers to break down through hydrolysis and enzymatic cleavage. Furthermore, some scaffolds are functionalized to release key agents in the healing cascade. Unfortunately, the ideal scaffold has not yet been identified, as a multitude of factors must be considered. The best scaffold for regenerative tissue engineering will vary depending on the type of injury, type of repair, and the final desired outcome.

5.1 Extracellular matrix substitutes

In-vivo, cells are suspended in a 3D scaffold called the extracellular matrix (ECM), which provides the cells with both the mechanical support and nutrients necessary for their survival. For optimal modeling of in-vivo conditions, scaffolds consisting of hydroxyapatite, naturally found in the ECM, have been investigated; unfortunately (Vial, 2008; Mastrogiacomo, et al., 2005; Salgado et al., 2004), these scaffolds have low mechanical stability.

5.2 Natural polymers

Alternatively, natural polymers such as collagen, fibrinogen, chitosan, etc. have been investigated. Natural polymers are a popular choice, because they are biodegradable, bioactive, elicit a minimal immune response, and can be chemically versatile (Vial, 2008). Furthermore, researchers have shown that these polymers can be mineralized (Salgado et al., 2004), further mimicking the natural conditions of bone. Oftentimes, natural polymers can be formed by combining two components, thereby crosslinking the polymer to form a gel. In this case, the crosslinking agent and the cells could be injected into a bone healing site arthroscopically, thereby reducing the risk of infection from traditional open surgery.

5.3 Synthetic polymers

In addition, synthetic polymers can also be utilized to form a 3-D scaffold through various techniques such as fiber bonding, emulsion freeze drying, solvent casting/ particulate leaching, high-pressure processing, gas foaming/ particle leaching, thermally induced phase separation, electrospinning, and rapid prototyping (Chung & Park, 2007; Tsang & Bhatia, 2004). Each is useful in forming a viable scaffold; however, the ideal scaffold must ultimately be selected based on the final desired outcome of the tissue repair. Ultimately, no synthetic manufactured scaffold is a perfect replacement for natural bone grafts.

5.4 Organ printing

Organ printing is a branch of regenerative medicine in which 3D living tissues are constructed from single cells printed with a thermo-reversible gel. Mironov et al (Mironov et al., 2003) define organ printing as 'a rapid prototyping computer-aided 3D printing technology, based on using layer by layer deposition of cell and/or cell aggregates into a 3D gel with sequential maturation of the printed construct into perfused and vascularized living tissue or organ.' Organ printing takes place in 3 distinct phases: preprocessing, processing, and postprocessing. In preprocessing, a computer model of the tissue is constructed based on deficiencies observed via MRI or CT-scans. The processing is the actual layer by layer printing (placement) of cells in the 3D environment. The postprocessing consists of tissue perfusion, directing tissue maturation, and, finally biomechanical conditioning. If successful, organ printing may prove to be a valuable alternative form of regenerative medicine with a significantly decreased tissue maturation period compared to traditional tissue engineering approaches.

6. In vitro studies of bone tissue engineering - bioreactor design

It has been demonstrated that a combination of biochemical agents and mechanical forces can be used to accelerate the production of a desired phenotype (Freed, et al., 2006). Mechanotransduction refers to the biochemical cascade that converts a mechanical stimulus into chemical activity; in other words, the application of mechanical stress can guide stem cells to undergo guided differentiation (Haudenschild et al., 2009; Wang & Thampatty, 2008). However, the application of mechanical stresses cannot be arbitrary, rather biomechanical models must be established to mimic in-vivo force patterns in an in-vitro bioreactor (Burdick & Vunjak-Novakovic, 2009; Butler, et al., 2008). The most important consideration in bioreactor design is mimicking in-vivo conditions, thereby mechanically conditioning tissue before it is transplanted into the body.

6.1 Mechanical strain

Bone specifically requires mechanical stimulation to maintain its strength. Astronauts who spend extended periods of time in space experience severe bone mineral loss due to the zero gravity conditions (Sikavitsas et al., 2001). Mechanical strain can be applied directly by compressing, stretching, or bending cell/scaffold constructs. Compressive bioreactors have been shown to increase both proteoglycan and matrix deposition by MSCs (Burger et al., 1992; Wartella & Wayne, 2009). Uniaxial stretching of human osteoblastics cells has been shown to increase cell proliferation and increase gene expression of ALP, osteocalcin (OCN), osteopontin (OPN), and collagen type I (Col1) (Ignatius, et al., 2004; Ignatius, et al., 2005). On the other hand, cyclic stretching results in only increased proliferation (Neidlinger-Wilke et al., 1994). With MSCs, 4-point bending bioreactors can increase ALP activity levels, increase mineralized matrix production, and increase gene expression of ALP and OPN (Mauney, et al., 2004).

6.2 Hydrodynamic shear stress

In addition to mechanical stimulation, another important consideration is nutrient supply; cells placed in a bioreactor must have access to fresh nutrients from media, as well as a means for removal of cellular waste (Rauh et al., 2011). Hydrodynamic shear stresses can be applied via spinner flask, rotation, or perfusion bioreactors: For spinner flask bioreactor systems, convective forces are applied by a stirrer which moves the media around the cell/scaffold construct. Studies using MSCs in spinner flasks have shown increased cellular proliferation, increased ALP activity levels, and increased gene expression of ALP, OPN, BSP, and Col1 (Kim H. , et al., 2007; Meinel, et al., 2004a; Mygind, et al., 2007; Stiehler, et al., 2009). Rotating bioreactor systems use laminar flow by rotating a vessel along the horizontal surface to create shear stress. Human ASCs and BM-MSCs have shown an increase in cellular proliferation and respective increases in ALP activity and OCN activity when cultured in rotating bioreactors (Diederichs, et al., 2009; Pound et al., 2007). Furthermore, additional testing in rats also shows an increase in cellular distribution and extracellular matrix formation (Goldstein et al., 2001; Qiu et al., 1999). Finally, perfusion driven bioreactors address nutrient diffusion deficiencies by providing fresh media and creating fluid shear stress. Perfusion bioreactors increase cellular proliferation and distribution and cell viability in the center of cell/ scaffold constructs (Bernhardt et al., 2008; Frohlich, et al., 2010; Grayson, et al., 2008; Jagodzinski, et al., 2008; Meinel, et al., 2004b; Rauh et al., 2011).

7. In vivo study of bone tissue engineering

Currently, there is no consensus on the best clinical model for engineered bone tissue. Comparable to how transplanted cells that have been pre-differentiated in vitro form bone-like tissues better than undifferentiated transplanted cells (Cowan, et al., 2005; Conjero, et al., 2006), engineered bone tissue that has been allowed to develop and mature in vitro promotes better bone healing after implantation than cell/ scaffold constructs that were not cultivated (Meinel, et al., 2004a; Meinel, et al., 2004b; Frohlich, et al., 2008). To assess the osteogenic potential of a cell/ scaffold construct, human cellular constructs were implanted subcutaneously in immunodeficient mice. Under these conditions, cells can develop into bone tissue in vivo (Kuznetsov, et al., 1997). Engineered bone tissue constructs have been used to repair load bearing and non-load bearing critical-size defects in various rodent models (Ohgushi et al., 1989; Puelacher et al., 1996; Bruder, et al., 1998a). Naturally critical-sized defects exhibit incomplete repair due to limited number of autologous stem cells available for use in the repair process. Research on larger animal models such as sheep and dogs has shown repair of critical-sized defects was enhanced by the use of stem cells in engineered bone tissue (Bruder et al., 1998b; He, et al., 2007; Kon, et al., 2000; Petite, et al., 2000; Shang, et al., 2001; Viateau, et al., 2007).

8. Conclusions and future directions

Regenerative tissue engineering addresses the discrepancy between the available transplantable donor tissue and the need. There are many choices for both cells and scaffolds, and the best combination will vary depending on the type of injury, type of repair, and the final desired outcome. Stem cells are an ideal cell source for bone regenerative tissue engineering applications, because they are capable of self- renewal, are undifferentiated, and can give rise to specialized tissue like bone. As our understanding of these cells improves, new engineering approaches (such as the 3D organ printer) will be developed to optimize the production of functional tissues.

9. References

Amado, L; Saliaris, A; Schuleri, K; St John, M; Xie, J; Cattaneo, S; Durand, D; Fitton, T; Kuang, J; Stewart, G; Lehrke, S; Baumgartner, W; Martin, B; Heldman, A; Hare, J. (2005). Cardiac repair with intremyocardial injection of allogenic mesenchymal stem cells after myocardial infarction. *Proc Natl Acad Sci USA, 102*(32), 11474-11479.

Amit, M., & Itskovitz-Eldor, J. (2002). Derivation and spontaneous differentiation of human embryonic stem cells. *J Anat, 200*(Pt3), 225-232.

Ballen, K., Barker, J., Stewart, S., Greene, M., Lane, T., & Transplantation, A. S. (2008). Collection and preservation of cord blood for personal use. *Biol Blood Marrow Transplant, 14*(3), 356-363.

Batouli, S; Miura, M; Brahim, J; Tsutsui, T; Fisher, L; Gronthos, S; Robey, P; Shi, S. (2003). Comparison of stem-cell-mediated osteogenesis and dentinogenesis. *J Dent Res, 82*(12), 976-981.

Bernhardt, A., Lode, A., Boxberger, S., Pompe, W., & Gelinsky, M. (2008). Mineralised collagen-an artificial, extracellular bone matrix-- improves osteogenic differentiation of bone marrow stromal cells. *J Mater Sci Mater Med, 19*(1), 269-275.

Bielby, R., Boccaccini, A., Polak, J., & Buttery, L. (2004). In vitro differentiation and in vivo mineralization of osteogenic cells derived from human embryonic stem cells. *Tissue Eng, 10*(9-10), 1518-1525.

Biswas, A., & Hutchins, R. (2007). Embryonic Stem Cells. *Stem Cells and Development, 16*, 213-221.

Bruder, S., Kraus, K., Goldberg, V., & Kadiyala, S. (1998a). The effect of implants loaded with autologous mesenchymal stem cells on the healing of canine segmental bone defects. *J Bone Joint Surg Am, 80*(7), 985-996.

Bruder, S., Kurth, A., Shea, M., Hayes, W., Jaiswal, N., & Kadiyala, S. (1998b). Bone regeneration by implantation of purified, culture-expanded human mesenchymal stem cells. *J Orthop Res, 16*(2), 155-162.

Brunstein, C., Setubal, D., & Wagner, J. (2007). Expanding the role of umbilical cord blood transplantation. *Br J Haematol, 137*(1), 20-35.

Buckwalter, J., & Hunziker, E. (1996). Healing of bones, cartilages, tendons, and ligaments: a new era. *Lancet, 348 Suppl 2: sII18*, 21-28.

Burchardt, H. (1983). The biology of bone graft repair. *Clin Orthop Relat Res*, 28-42.

Burdick, J., & Vunjak-Novakovic, G. (2009). Engineered Microenvironments for Controlled Stem Cell Differentiation. *Tissue Eng Park A, 15*(2), 205-219.

Burger, E., Klein-Nulend, J., & Veldhuijzen, J. (1992). Mechanical stress and osteogenesis in vitro. *J Bone Miner Res, 7 Suppl*, S397-S401.

Butler, D., Juncosa-Melvin, N., Boivin, G., Galloway, M., Shearn, J., Gooch, C., & Awad, H. (2008). Functional tissue engineering for tendon repair: A multidisciplinary strategy using mesenchymal stem cells, bioscaffolds, and mechanical stimulation. *J Orthop Res, 26*(1), 1-9.

Campagnoli, C., Roberts, I., & Kumar, S. (2001). Identification of Mesenchymal stem/ progenitor cells in human first-trimester blood, liver, bone and bone marros. *Blood, 98*, 2396-2402.

Cassidy, C; Jupiter, J; Cohen, M; Delli-Sant, M; Leinberry, C; Husband, J; Ladd, A; Seitz, W; Constanz, B. (2003). Norian SRS cement compared with conventional fixation in distal radial fractures. A randomized study. *J Bone Joint Surg Am, 85-A*(11), 2127-2137.

Chadipiralla, K., Yochim, J., Bahuleyan, B., Huang, C., Garcia-Godoy, F., Murray, P., & Stelnicki, E. (2010). Osteogenic differentiation of stem cells derived from human periodontal ligaments and pulp of human exfoliated deciduous teeth. *Cell Tissue Res.*

Chai, Y; Jiang, X; Ito, Y; Bringas, P Jr; Han, J; Rowitch, D; Soriano, P; MaMahon, A; Sucov, H. (2000). Fate of the mammalian cranial neural crest during tooth and mandibular morphogenesis. *Development, 127*(8), 1671-1679.

Choong, P., Martin, T., & Ng, K. (1993). Effects of ascorbic acid, calcitriol, and retinoic acid on the differentiation of preosteoblasts. *J Orthop Res, 11*, 638-647.

Chung, H. J., & Park, T. G. (2007). Surface engineered and drug releasing pre-fabricated scaffolds for tissue engineering. *Advanced Drug Delivery Reviews, 59*, 249-262.

Conjero, J., Lee, J., Parrett, B., Terry, M., Wear-Maggitti, K., Grant, R., & Breitbart, A. (2006). Repair of palatal bone defects using osteogenically differentiated fat-derived stem cells. *Plast Reconstr Surg, 117.*

Cowan, C; Aalami, O; Shi, Y; Chou, Y; Mari, C; Thomas, R; Quarto, N; Nacamuli, R; Contag, C; Wu, B; Longaker, M. (2005). Bone morphogenetic protein 2 and retinoic acid accelerate in vivo bone formation, osteoclast recruitment, and bone turnover. *Tissue Eng, 11*, 645-658.

Cowan, C; Shi, Y; Aalami, O; Chou, Y; Mari, C; Thomas, R; Quarto, N; Contaq, C; Wu, B; Longaker, M. (2004). Adipose-derived adult stromal cells heal critical-size mouse calvarial defects. *Nat Biotechnol, 22*(5), 560-567.

Diederichs, S., Roker, S., Marten, D., Peterbauer, A., Scheper, T., van Griensven, M., & Kasper, C. (2009). Dynamic cultivation of human mesenchymal stem cells in a rotating bed bioreactor system based on the Z RP platform. *Biotechnol Prog, 25*(6), 1762-1771.

Dimitriou, R., Dahabreh, Z., Katsoulis, E., Matthews, S., Branfoot, T., & Giannoudis, P. (2005). Application of recombinant BMP-7 on persistent upper and lower limb non-unions. *Injury, 35*(Suppl 4), S51-9.

Einhorn, T. (2003). Clinical applications of recombinant human BMPs: early experience and future development. *J Bone Joint Surg Am, 85-A*, Suppl 3:82-8.

Fox, 1 S. I. (2001). *Human Physiology* (7 ed.). New York, United States of America: McGraw-Hill Higher Education.

Freed, L; Guilak, X; Gray, M; Tranquillo, R; Holmes, J; Radisic, M; Sefton, M; Kaplan, D; Vunjak-Novakovic. (2006). Advanced tools for tissue engineering: scaffolds, bioreactors, and signaling. *Tissue Eng, 12*(12), 3285-3305.

Frohlich, M., Grayson, W., Marolt, D., Gimble, J., Kregar-Velikonja, N., & Vunjak-Novakovic, G. (2010). Bone grafts engineered from human adipose-derived stem cells in perfusion bioreactor culture. *Tissue Eng Part A, 16*(1), 179-189.

Frohlich, M., Grayson, W., Wan, L., Marolt, D., Drobnic, M., & Vunjak-Novakovic, G. (2008). Tissue engineered bone grafts: biological requirements, tissue culture and clinical relevance. *Curr Stem Cell Res Ther, 3*(4), 254-264.

Gay, I., Chen, S., & MacDougall, M. (2007). Isolation and characterization of multipotent human periodontal ligament stem cells. *Orthod Craniofac Res, 10*, 149-160.

Goldstein, A., Juarez, T., Helmke, C., Gustin, M., & Mikos, A. (2001). Effect of convection on osteoblastic cell growth and function in biodegradable polymer foam scaffolds. *Biomaterials, 22*(11), 1279-1288.

Govender, S; Csimma, C; Genant, H; Valentin-Opran, A; Amit, Y; Arbel, R; Aro, H; Atar, D; Bishay, M; Borner, MG; Chiron, P; Choong, P; Cinats, J; Courtenay, B; Feibel, R; Geulette, B; Gravel, C; Haas, N; Raschke, M; Hammacher, E; van her Velde, D; Hardy, P; M, Holt; Josten, C; Kettlerl, RL; Lindeque; B; Lob, G; Mathevon, H; McCoy, G; Marsh, D; Miller, R; Munting, E; Oevre, S; Nordsletten, L; Patel, A; Pohl, A; Rennie, W; Reynders, P; Rommens, PM; Rondia, J; Rossouw, WC; Daneel, PJ; Ruff, S; Ruter, A; Santavirta, S; Schildhauer, TA; Gekle, C; Schnettler, R; Segal, D; Seiler, H; Snowdowne, RB; Stapert, J; Taglag, G; Verdonk, R; Vogels, L; Weckbach, A; Wentzensen, A; Wisniewski, T; BMP-1 Evaluation in Surgery for Tibial Trauma, (BESTT) Study Group (2002). Recombinant human bone morphogenetic protein-2 for treatment of open tibial fractures:a prospective, controlled, randomized study of four hungrady and fifty patients. *J Bone Joint Surg Am, 84-A*(12), 2123-2134.

Grayson, W., Bhumiratana, S., Cannizzaro, C., Chao, P., Lennon, D., Caplan, A., & Vunjak-Novakovic, G. (2008). Effects of initial seeding density and fluid perfusion rate on formation of tissue-engineered bone. *Tissue Eng Part A, 14*(11), 1809-1820.

Gronthos, S., Mankani, M., Brahim, J., Robey, P., & S, S. (2000). Postnatal human dental pulp stem cells (DPSCs) in vitro and in vivo. *Proc Natl Acad Sci USA, 97*, 13625-13630.

Haudenschild, A., Hsieh, A., Kapila, S., & Lotz, J. (2009). Pressure and distortion regulate human mesenchymal stem cell gene expression. *Ann Biomed Eng, 37*(3), 492-502.

He, Y., Zhang, Z., Zhu, H., Qiu, W., Jiang, X., & Guo, W. (2007). Experimental study on reconstruction of segmental mandible defects using tissue engineered bone combined bone marrow stromal cells with three-dimensional tricalcium phosphate. *J Craniofac Surg, 18*(4), 800-805.

Heinegard, D., & Oldberg, A. (1989). Structure and biology of cartilage and bone matrix noncollagenous macromolecules. *FASEB J, 3*(9), 2042-2051.

Heins, N; Englund, M; Sjoblom, C; Dahl, U; Tonning, A; Bergh, C; Lindahl, A; Hanson, C; Semb, H. (2004). Derivation, characterization, and differentiation of human embryonic stem cells. *Stem Cells, 22*(3), 367-376.

Huang, C.-Y. C., Pelaez, D., Bendala, J. D., Garcia-Godoy, F., & Cheung, H. S. (2009). Plasticity of stem cells derived from adult periodontal ligament. *Regen Med, 4*(6).

Huang, G.-J., Gronthos, S., & Shi, S. (2009). Mesenchymal Stem Cells Derived from Dental Tissues vs. Those from Other Sources: Their Biology and Role in Regenerative Medicine. *Journal of Dental Research, 88*(9), 792-806.

Huysseune, A., & Thesleff, I. (2004). Continuous tooth replacement: the possible involvement of epithelial stem cells. *Bioessays, 26*(6), 665-671.

Ignatius, A., Blessing, H., Liedert, A., Kaspar, D., Kreja, L., Friemert, B., & Claes, L. (2004). Effects of mechanical strain on human osteoblastic precursor cells in type I collagen matrices. *Orthopade, 33*(12), 1386-1393.

Ignatius, A; Blessing, H; Liedert, A; Schmidt, C; Neidlinger-Wilke, C; Kaspar, D; Friemert, B; Claes, L. (2005). Tissue engineering of bone: effects of mechanical strain on osteoblastic cells in type I collagen matrices. *Biomaterials, 26*(3), 311-318.

Isaka, J; Ohazama, A; Kobayashi, M; Nagashima, C; Takiguchi, T; Kawasaki, H; Tachikawa, T; Hasegawa, K. (2001). Participation of periodontal ligament cells with regeneration of alveolar bone. *J Periodontol, 72*(3), 314-323.

Itskovitz-Eldor, J; Schuldiner, M; Karsenti, D; Eden, A; Yanuka, O; Amit, M; Soreq, H; Benvenisty, N. (2000). Differentiation of human embryonic stem cells into embryoid bodies compromising the three embryonic germ layers. *Mol Med, 6*(2), 88-95.

Jagodzinski, M; Breibart, A; Wehmeier, M; Hesse, E; Haasper, C; Krettek, C; Zeichen, J; Hankemeier, S. (2008). Influence of perfusion and cyclic compression on proliferation and differentiation of bone marrow stromal cells in 3-dimensional culture. *J Biomech, 41*(9), 1885-1891.

Kain, M., & Einhorn, T. (2005). Recombinant human bone morphogenetic proteins in the treatment of fractures. *Foot Ankle Clin, 10*(4), 639-50, viii.

Kalfas, I. (2001). Principles of bone healing. *Neurosurg Focus, 10*(4), E1.

Kawaguchi, J., Mee, P., & Smith, A. (2005). Osteogenic and chondrogenic differentiation of embryonic stem cells in response to specific growth factors. *Bone, 36*, 758-769.

Kerkis, I; Kerkis, A; Dozortsev, D; Stukart-Parsons, G; Gomes Massironi, SM Pereira, L; Caplan, A; Cerruti, H. (2006). Isolation and characterization of a population of immature dental pulp stem cells expressing OCT-4 and other embryonic stem cell markers. *Cells Tissues Organs, 184*(3-4), 105-116.

Kern, S., Eichler, H., Stoeve, J., Kluter, H., & Bieback, K. (2006). Comparative analysis of mesenchymal stem cells from bone marrow, umbilical cord blood, or adipose tissue. *Stem Cells, 24*(5), 1294-1301.

Khan, M., Sahibzada, A., Khan, M., Sultan, S., Younas, M., & Khan, A. (2005). Outcome of plating, bone grafting and shortening of non-union humeral diaphyseal fracture. *J Ayub Med Coll Abbottabed, 17*(2), 44-46.

Kim, H., Kim, U., Leisk, G., Bayan, C., Georgakoudi, I., & Kaplan, D. (2007). Bone regeneration on macroporous aqueous-derived silk 3-D scaffolds. *Macromol Biosci, 7*(5), 643-655.

Kim, Y., Kim, H., Cho, H., Bae, Y., Suh, K., & Jung, J. (2007). Direct comparison of human mesenchymal stem cells derived from adipose tissues and bone marrow in mediating neovascularization in response to vascular ischemia. *Cell Physiol Biochem, 20*(6), 867-876.

Kneser, U., Schaefer, D., Polykandriotis, E., & Horch, R. (2006). Tissue engineering of bone: the reconstructive surgeon's point of view. *J Cell Mol Med, 10*(1), 7-19.

Knoepfler, P. S. (2009). Deconstructing Stem Cell Tumorigenicity: A Roadmap to Safe Regenerative Medicine. *Stem Cells, 27*(5), 1050-1056.

Knowles, B., Aden, D., & Solter, D. (1978). Monoclonal antibody detecting a stage-specific embryonic antigen (SSEA-1) on preimplantation mouse embryos and teratocarcinoma cells. *Curr Top Microbiol Immunol, 81*, 51-53.

Kon, E; Muraglia, A; Corsi, A; Bianco, P; Marcacci, M; Martin, I; Boyde, A; Ruspantini, I; Chistolini, P; Rocca, M; Giardino, R; Cancedda, R; Quarto, R. (2000). Autologous bone marrow stromal cells loaded onto porous hydroxyapatite ceramic accelerate bone repair in critical-size defects of sheep long bones. *J Biomed Mater Res, 49*(3), 328-337.

Kuznetsov, S., Krebsback, P., Satomura, K., Kerr, J., Riminucci, M., Benayahu, D., & Robey, P. (1997). Single-colony derived strains of human marrow stromal fibroblasts form bone after transplantation in vivo. *J Bone Miner Res, 12*(9), 1335-1347.

Laino, G; d'Aquino, R; Graziano, A; Lanza, V; Carinci, F; Naro, F; Pirozzi, G; Papaccio, G. (2005) A new population of human adult dental pulp stem cells: a useful source of living autologous fibrous bone tissue (LAB). *J Bone Miner Res, 20*(8), 1394-1402.

Le Douarin, N., Creuzet, S., Couly, G., & Dupin, E. (2004). Neural crest cell plasticity and its limits. *Development, 131*(19), 4637-4650.

Lebkowski, J., Gold, J., Xu, C., Funk, W., Chiu, C., & Carpenter, M. (2001). Human embryonic stem cells: culture, differentiation, and genetic modification for regenerative medicine applications. *Cancer J, 7 Suppl 2*, S83-S93.

Levi, B., & Longaker, M. (2011). Concise review: adipose-derived stromal cells for skeletal regenerative medicine. *Stem Cells, 29*(4), 576-582.

Lewis, G. (1997). Properties of acrylic bone cement: state of the art review. *J Biomed Mater Res, 38*(2), 155-182.

Liu, G., Ye, X. Z., Li, Y., Sun, J., Cui, L., & Cao, Y. (2011). Osteogenic differentiation of GFP-labeled human umbilical cord blood derived mesenchymal stem cells after cryopreservation. *Cryobiology.*

Malladi, P., Xu, Y., Yang, G., & Longaker, M. (2006). Functions of vitamin D, retinoic acid, and dexamthasone in mouse adipose-derived mesenchymal cells. *Tissue Eng, 12*, 2031-2040.

Mankin, H., Hornicek, F., & Raskin, K. (2005). Infection in massive bone allografts. *Clin Orthop Relat Res*(432), 210-216.

Mastrogiacomo, M., Muraglia, A., Komlev, V., Peyrin, F., Rustichelli, R., Crovace, A., & Cancedda, R. (2005). Tissue engineering of bone: search for a better scaffold. *Orthod Craniofac Res, 8*(4), 277-284.

Mauney, J; Sjostorm, S; Blumberg, J; Horan, R; O'Leary, J; Vunjak-Novakovic, G; Volloch, V; Kaplan, DL. (2004). Mechanical stimulation promotes osteogenic differentiation of human bone marrow stromal cells on 3-D partially demineralized bone scaffolds in vitro. *Calcif Tissue Int, 74*(5), 458-468.

Meinel, L; Karageorgiou, V; Fajardo, R; Snyder, B; Shinde-Patil, V; Zichner, L; Kaplan, D; Langer, R; Vunjak-Novakovic, G. (2004a). Bone tissue engineering using human mesenchymal stem cells: effects of scaffold material and medium flow. *Ann Biomed Eng, 32*(1), 112-122.

Meinel, L., Karageorgiou, V., Hofmann, S., Fajardo, R., Snyder, B., Li, C., Kaplan, D. (2004b). Engineering bone-like tissue in vitro using human bone marrow stem cells and silk scaffolds. *J Biomed Mater Res A, 71*(1), 25-34.

Meissner, A., Wernig, M., & Jaenisch, R. (2007). Direct reprogramming of genetically unmodifies fibroblasts into pluripotent stem cells. *Nat Biotechnol, 25*(10), 1177-1181.

Mironov, V., Boland, T., Trusk, T., Forgacs, G., & Markwald, R. R. (2003). Organ printing: computer-aided jet-based 3D tissue engineering. *TRENDS in Biotechnology, 21*(4), 157-161.

Miura, M., Gronthos, S., Zhao, M., Lu, B., Fisher, L. W., Robey, P. G., & Shi, S. (2003). SHED: stem cells from human exfoliated deciduous teeth. *Proc Natl Acad Sci USE, 100*(10), 5807-5812.

Muh, E., Zimmermann, J., Kneser, U., Marquardt, J., Mulhaupt, R., & Stark, B. (2002). Lysineurethanedimethacrylate--a novel generation of amino acid based monomers for bone cements and tissue repair. *Biomaterials, 23*(14), 2849-2854.

Mygind, T; Stiehler, M; Baatrup, A; Li, H; Zou, X; Flybjerg, A; Kassem, M; Bunger, C. (2007). Mesenchymal stem cell ingrowth and differentiation on coralline hydroxyapatite scaffolds. *Biomaterials, 28*(6), 1036-1047.

Neidlinger-Wilke, C., Wilke, H., & Claes, L. (1994). Cyclic stretching of human osteoblasts affects proliferation and metabolism: a new experimental method and its application. *J Orthop Res, 12*(1), 70-78.

Odorico, J., Kaufman, D., & Thomson, J. (2001). Multilineage differentiation from human embryonic stem cell lines. *Stem Cells, 19*(3), 193-204.

Ogston, N., Harrison, A., Cheung, H., Ashton, B., & Hampson, G. (2002). Dexamethasone and retinoic acid differentially regulate growth and differentiation in an immortalised human clonal bone marrow stromal cell line with osteoblastic characteristics. *Steroids, 67*, 895-906.

Ohgushi, H., Goldberg, V., & Caplan, A. (1989). Repair of bone defects with marrow-cells and porous ceramic. Experiments in rats. *Acta Orthop Scand, 60*(3), 334-339.

Okita, K., Ichisaka, T., & Yamanaka, S. (2007). Generation of germline-compotent induced pluripotent stem cells. *Nature, 448*(7151), 313-317.

Park, J., Kim, S., Oh, E., Moon, S., Roh, S., Kim, C., & Yoon, H. (2003). Establishment and maintenance of human embryonic stem cells on STO, a permanently growing cell line. *Biol Reprod, 69*(9), 2007-2014.

Petite, H; Viateau, V; Bensaid, W; Meunier, A; de Pollak, C; Bourguignon, M; Oudina, K; Sedel, L; Guillemin, G. (2000). Tissue-engineered bone regeneration. *Nat Biotechnol, 18*(9), 959-961.

Pittenger, MF; Mackay, AM; Beck, SC; Jaiswal, RK; Douglas, R; Mosca, JD; Moorman, MA; Simonetti, DW; Craig, S; Marshak, DR. (1999). Multilineage potential of adult human mesenchymal stem cells. *Science, 284*, 143-147.

Pound, J., Green, D., Roach, H. M., & Oreffo, R. (2007, Jun). An ex vivo model for chondrogenesis and osteogenesis. *Biomaterials, 28*(18), 2839-2849.

Probst, A., & Spiegel, H. (1997). Cellular mechanisms of bone repair. *J Invest Surg, 10*(3), 77-86.

Puelacher, W., Vacanti, J., Ferraro, N., Schloo, B., & Vacanti, C. (1996). Femoral shaft reconstruction using tissue-engineered growth of bone. *Int J Oral Maxillofac Surg, 25*(3), 223-228.

Qiu, Q., Ducheyne, P., & Ayyaswamy, P. (1999). Fabrication, characterization and evaluation of bioceramic hollow microspheres used as microcarriers for 3-D bone tissue formation in rotating bioreactors. *Biomaterials, 20*(11), 989-1001.

Rauh, J., Milan, F., Gunther, K., & Stiehler, M. (2011). Bioreactor Systems for Bone Tissue Engineering. *Tissue Eng Part B Rev, 17*(4), 263-280.

Salgado, A., Coutinho, O., & Reis, R. (2004). Bone tissue engineering: state of the art and future trends. *Macromol Biosci, 4*(8), 743-765.

San Miguel, S., Goseki-Sone, M., Sugiyama, E., Watanabe, H., Yanagishita, M., & Ishikawa, I. (1998). The effects of retinoic acid on alkaline phosphatase activity and tissue-non-specific alkaline phosphatase gene expression in human periodontal ligament cells and gingival fibroblasts. *J Periodontal Res, 33*, 428-433.

Schroeder, J., & Mosheiff, R. (2011) Tissue engineering approaches for bone repair: Concepts and evidence. *Injry, 42*(6), 609-613.

Schugar, R; Chireleison, S; Wescoe, K; Schmidt, B; Askew, Y; Nance, J; Evron, J; Peault, B; Deasy, B. (2009). High harvest yield, high expansion, and phenotype stability of CD146 mesenchymal stromal cells from whole primitive human umbilical cord tissue. *J Biomed Biotechnol*, 789526.

Seo, B; Miura, M; Gronthos, S; Bartold, P, Batouli, S; Brahim, J; Young, M; Robey, P; Wang, C; Shi, S. (2004). Investigation of multipotent postnatal stem cells from human periodontal ligament. *Lancet, 364*, 149-155.

Seo, B; Sonoyama, W; Yamaza, T; Coppe, C; Kikuiri, T; Akiyama, K; Lee, J; Shi, S. (2007). SHED repair critical-size calvarial defects in mine. *Oral Diseases, 14*, 428-434.

Shang, Q., Wang, Z., Liu, W., Shi, Y., Cui, L., & Cao, Y. (2001). Tissue-engineered bone repair of sheep cranial defects with autologous bone marrow stromal cells. *J Craniofac Surg, 12*(6), 586-595.

Shier, D., Butler, J., & Lewis, R. (2002). *Hole's Human Anatomy & Physiology* (9 ed.). New York, NY: Mc-Graw-Hill.

Sikavitsas, V., Temenoff, J., & Mikos, A. (2001). Biomaterials and bone mechanotransduction. *Biomaterials, 22*(19), 2581-2593.

Skillington, J., Choy, L., & Derynck, R. (2002). Bone morphogenic protein and retinoic acid signaling cooperate to induce osteoblast differentiation of preadipocytes. *J Cell Biol, 159*, 135-146.

Song, H., Nacamuli, R., Xia, W., Bari, A., Shi, Y., Fang, T., & Longaker, M. (2005). High-dose retinoic acid modulates rat calvarial osteoblast biology. *J Cell Physiol, 202*, 255-262.

Stiehler, M., Bunger, C., Baatrup, A., Lind, M., Kassem, M., & Mygind, T. (2009). Effect of dynamic 3-D culture on proliferation,distribution, and osteogenic differentiation of human mesenchymal stem cells. *J Biomed Mater Res A, 89*(1), 96-107.

Takahashi, K., & Yamanaka, S. (2006). Induction of pluripotent stem cells from mouse embryonic and adult fibroblast cultures by defined factors. *Cell, 126*(4), 663-676.

Takahashi, K., Tanabe, K., Ohnuki, M., Narita, M., Ichisaka, T., Tomoda, K., & Yamanaka, S. (2007). Induction of pluripotent stem cells from adult human fibroblasts by defined factors. *Cell, 131*(5), 861-872.

Thomson, J., & Marshall, V. (1998a). Primate embyonic stem cells. *Curr Top Dev Biol, 38*, 133-165.

Thomson, J., Itskovitz-Eldor, J., Shapiro, S., Waknitz, M., Swiergiel, J., Marshall, V., & Jones, J. (1998b). Embryonic stem cell lines derived from human blastocysts. *Science, 282*(5391), 1145-1147.

Thorwarth, M., Wehrhan, F., Schultze-Mosqau, S., Wiltfang, J., & Schlegel, K. (2006). PRP modulated expression of bone matrix proteins in vivo without long-term effects on bone formation. *Bone, 38*(1), 30-40.

Tsang, V. L., & Bhatia, S. N. (2004). Three-dimensional tissue fabrication. *Advanced Drug Delivery Reviews, 56*, 1635-1647.

Vacanti, C., & Vacanti, J. (2000). The science of tissue engineering. *Orthop Clin North Am, 31*(3), 351-356.

Vial, X. (2008). Gelatin Based Scaffolds for Bone Tissue Engineering. *Open Access Theses*, Paper 163. Retrieved from http://scholarlyrepository.miami.edu/oa_theses/163

Viateau, V; Guillemin, G; Bousson, V; Oudina, K; Hannouche, D; Sedel, L; Logeart-Avramoglou, D; Petite, H. (2007). Long-bone critical-size defects treated with tissue-engineered grafts: a study on sheep. *J Orthop Res, 25*(6), 741-749.

Wagner, W; Wein, F; Seckinger, A; Frankhauser, M; Wirkner, U; Krause, U; Blake, J; Schwager, C; Echstein, V; Ansorge, W; Ho, A. (2005). Comparative characterisitcs of mesenchymal stem cells from human bone marrow, adipose tissue, and umbilical cord blood. *Exp Hematol, 33*(11), 1402-1416.

Wang, J., & Thampatty, B. (2008). Mechanobiology of adult and stem cells. *Int Rev Cell Mol Biol, 271*, 301-346.

Wartella, K., & Wayne, J. (2009). Bioreactor for biaxial mechanical stimulation to tissue engineered constructs. *J Biomech Eng, 131*(4), 044501,1-044501,5.

Young, H; Steele, T; Bray, R; Hudson, J; Floyd, J; Hawkins, K; Thomas, K; Austin, T; Edwards, C; Cuzzourt, J; Duenzl, M; Lucas, P; Black, A Jr. (2001). Human reserve pluripotent mesenchymal stem cells are present in the connective tissues of skeletal muscle and dermis derived from fetal, adults, and geriatric donors. *Anat Rec, 264*, 51-62.

Yu, J; Vodyanik, M; Smuga-Otto, K; Antosiewicz-Bourget, J; Frane, J; Tian, S; Nie, J; Jonsdottir, G; Ruotti, V; Stewart, R; Slukvin, I; Thomson, J. (2007). Induced pluripotent stem cell lines derived from human somatic cells. *Science, 318*(5858), 1917-1920.

Zijderveld, S., Zerbo, I., van den Bergh, J., Schulten, E., & ten Bruggenkate, C. (2005). Maxillary sinus floor augmentation using a beta-tricalcium phosphate (Cerasorb) alone compared to autogenous bone grafts. *Int J Oral Maxillofac Implants, 20*(3), 432-440.

Zuk, P; Zhu, M; Mizuno, H; Huang, J; Futrell, J; Katz, A; Benhaim, P; Lorenz, H; Hedrick, M. (2001). Multilineage cells from human adipose tissue: implications for cell-based therapies. *Tissue Eng, 7*, 211-228.

2

Autologous Cell Therapies for Bone Tissue Regeneration

Nevenka Kregar Velikonja[1], Hana Krečič Stres[1], Elvira Maličev[3], Danica Gantar[1], Matija Krkovič[2], Vladimir Senekovič[2], Matjaž Rode[4], Miomir Knežević[1,3], Gordana Vunjak Novakovic[5] and Mirjam Fröhlich[1,5]

[1]Educell d.o.o.,
[2]University Medical Centre Ljubljana,
[3]Blood Transfusion Centre Ljubljana,
[4]Community Health Centre, Ljubljana,
[5]Department of Biomedical Engineering, Columbia University, New York,
[1,2,3,4]Slovenia
[5]USA

1. Introduction

Bone tissue engineering using various cell sources and materials has become an intriguing field, aimed at solving the problem of treatment of numerous clinical indications requiring regeneration of damaged or deficitary bone.

During the Paleozoic period, evolution produced the skeleton. This 500 million-year-old creation has the capacity for regeneration, a term which until recently has been reserved to new tissue and organs formation as in hydra, planarians or salamanders (Braddock *et al.*, 2001). The regeneration of bone (or the stimulation of bone production) is often required to treat loss of bone tissue brought about by trauma, osteonecrosis and tumors. Among the 6 millions fractures occurring every year in the United States, 5-10% are classified unfavorable, requiring further treatment due to compromised healing (Praemer et al., 1992). The clinical and socioeconomic challenge of treatments of bone defects is staggering. For example, the number of total joint arthroplasties (TJAs) and revision surgeries in the US has increased from 700,000 in 1998 to over 1.1 million in 2005. Medical expenses relating to fracture, reattachment, and replacement of hip and knee joint was estimated to be over $20 billion (USD) in 2003, and predicted to increase to over $74 (USD) billion by the year 2015. Similar trend is observed in spinal arthrodesis (reviewed in Porter et al, 2009).

Bone tissue provides mechanical stability to the skeleton, which is needed for load bearing, locomotion and protection of internal organs. Furthermore, bone serves as a mineral reservoir and has the capacity rapidly to mobilize mineral stores if needed for homeostasis of the calcium blood level (Kneser *et al.*, 2006).

The functional integrity of bone tissue is maintained by three main different cell types: osteoblasts, osteocytes and osteoclasts, which are embedded in a highly complex matrix consisting of a mineralized (hydroxyapatite) and a non-mineralized component. The non-mineralized organic part contains mainly collagens (approx. 95%), the remaining organic component of 5% is composed of glycoproteins, proteoglycans and other numerous non-collagenous proteins (Meyer and Wiesmann, 2006).

Bone development and bone regeneration are complexly regulated processes that involve a plethora of different growth and transcription factors, which coordinate the interaction of cells and matrix in response to external or internal stimuli (Kneser *et al.*, 2006). Bone metabolism involves the resorption of existing bone by osteoclasts and the subsequent formation of a new bone matrix by osteoblasts. These activities are essential for bone remodeling, regeneration and repair (DeLong *et al.*, 2007).

Various sources of cells (periosteal cells, cortical cells, cells derived from the surrounding soft tissues, and marrow cells) and signals that set up these fields are responsible for the features of the repair tissue. The primary tissue source of cells that form repair tissue is believed to be from the periosteum. Other cells that contribute to or repair tissue formation appear to be derived from the adjacent cortical and cancellous bone. Mesenchymal stem cells, assumed to be derived from either the surrounding muscle tissue or the marrow space, are a third source of cells that participate in the formation of new bone. Cells synthesize a network of collagnenous and non-collagenous proteins. The final stage of bone repair and regeneration is the establishment of mineralized, mechanically competent tissue. Collagens, as the major constituent of the extracellular matrix network, are of major importance in the formation of a mineralized matrix. (Nakahara *et al.*, 1990; Meyer and Wiesmann, 2006)

The healing potential of bone is sufficient to restore simple fractures, which are generally treated by standard conservative or surgical therapy. However, in some cases, reparative osteogenesis does not result in structural and functional recovery of the bone (Logeart-Avramoglou *et al.*, 2005). Extended bone defects following trauma or cancer resection or non-unions of fractures may require more sophisticated treatment. In these cases, bone grafting procedures, segmental bone transport, distraction osteogenesis or biomaterials are applied for reconstruction (Meyer and Wiesmann, 2006). The repair of bone defects in reconstructive surgery is subject to significant limitations, including donor site morbidity, limited supply of autograft, risk of infection and immune rejection of allograft, and poor osteogenic effect of synthetic bone substitutes (Logeart-Avramoglou *et al.*, 2005). In addition, bridging of a large bone defect by callus distraction requires a long time and usually an external fixator, both very inconvenient for patients. Regardless of the technique used, the percentage of failure is considerable. Bone repair is therefore the subject of intensive investigation in reconstructive surgery.

2. Treatment of bone defects

The reconstruction of large bone defects is an important clinical problem and none of the approaches thus far have proved completely effective. Since there are major limitations when treating "problematic" bone tissue defects according to standard protocols, there is a great need for the development of new approaches for reparative osteogenesis. There are a number of clinical indications, such as non-unions, benign bone lesions, parodontal bone lesions, traumatic injuries, which could benefit from advances made during the past decade in bone cell therapies and tissue engineering. Cell therapies involve the use of any kind of cells to repair damaged or destroyed bone cells or tissues, and are unique in that the active component consists of living cells.

One of the biggest cell therapy areas is tissue engineering, defined by Langer and Vacanti (1993) as an interdisciplinary field that applies the principles of engineering and the life sciences to the development of biological substitutes that restore, maintain or improve tissue function. Tissue can be engineered 1) *in vivo* - by stimulating the body's own regeneration response with the appropriate biomaterial, or 2) *ex vivo* - cells can be expanded in culture, attached to a scaffold and then reimplanted into the host. Depending on the source, cells may be heterologous (different species), allogeneic (same species, different individual) or autologous (same individual). Autologous cells are preferred because they will not evoke an immunologic response and the deleterious side effects of immunosuppressive agents can thus be avoided. In addition, the potential risks of pathogen transfer are also eliminated (Hipp and Atala, 2004).

When engineering bone tissue substitutes, mechanical stability, osteoconductivity, osteoinductivity, osteogenicity and ease of handling have to be well balanced in order properly to meet clinical needs (Kneser *et al.*, 2006). According to Muschler and co-workers (2004), there are four types of cell-based tissue engineering: (1) local targeting of connective tissue progenitors where new tissue is needed, (2) transplanting autogenous connective tissue progenitors to augment the local population, (3) transplanting culture expanded or modified connective tissue progenitors and (4) transplanting fully formed tissue.

Osteogenic cells are an integral part of any bone tissue engineering strategy. These cells are either transplanted along with the appropriate scaffolds into the bone defects or attracted from the host by osteoinductive factors (Kneser *et al.*, 2006). The affectors of bone remodeling, regeneration and fracture repair in an adult organism are the cellular components. Various types of osteogenic cells, including bone marrow mesenchymal stem cells (BMSC) (Frank *et al.*, 2002; Meinel *et al.*, 2004; Meinel *et al.*, 2005), adipose-derived stem cells (ASC) (Lendeckel *et al.*, 2004; Peterson *et al.*, 2005) mesenchymal cells of the periosteum (Hutmacher and Sittinger, 2003; Schimming and Schmelzeisen, 2004, Turhani *et al.*, 2005) and alveolar bone derived osteoblasts (AO) (Xiao *et al.*, 2003; Zhu *et al.*, 2006) have been studied for bone reconstruction. Pluripotent mesenchymal stem cells are present in many adult tissues, although they are most abundant in bone marrow (Pittenger *et al.*, 1999) and adipose tissue (Zuk *et al.*, 2001). *In vitro*, BMSC are rapidly adherent, clonogenic and capable of extended proliferation (Bianco *et al.*, 2001). Isolation and expansion efficiency, stability of osteoblastic phenotype, *in vivo* bone formation capacity and long-term safety are essential requirements that must be met by any type of osteogenic cell for successful clinical application. Serum-free culture conditions or culture medium supplemented with autologous serum are preferable for cell expansion *in vitro* (Kneser *et al.*, 2006).

Nevertheless, it has still not yet been determined which type of osteogenic cell is most suitable for engineering bone tissue. At the moment, BMSC seem to be the best candidate for cell therapy to regenerate injured skeletal tissues, owing to their ease of isolation, expansion and multilineage potential. These cells can be induced to differentiate into chondrocytes or osteoblasts when subjected to specific environmental factors (Jorgensen *et al.*, 2004). Proof-in-principle for bone tissue engineering using BMSC has been demonstrated in various animal models (for review see Cancedda *et al.*, 2003); in addition, 7 human clinical studies had been conducted by 2010 (Chatterjea *et al.*, 2010). However, several studies have also shown ASC and AO to be appropriate cell sources for bone regeneration (Zuk *et al.*, 2001; Cowan *et al.*, 2004; Hattori *et al.*, 2006; Fröhlich *et al.*, 2010; Turhani *et al.*, 2005; Maličev *et al.*, 2008).

Bone tissue engineering requires not only living cells but also the use of scaffolds, which serve as a three-dimensional environment for the cells. Scaffolds for engineering bone should satisfy a number of criteria. According to Logeart-Avramoglou *et al.* (2005), such matrices should be: (i) biocompatible (non-immunogenic and non-toxic); (ii) absorbable (with rates of resorption commensurate with those of bone formation); (iii) preferably radiolucent (to allow the new bone to be distinguished radiographically from the implant); (iv) osteoconductive; (v) easy to manufacture and sterilize; and (vi) easy to handle in the operating theater, preferably without preparatory procedures (in order to limit the risk of infection). Three-dimensional scaffolds for bone tissue regeneration require an internal microarchitecture, specifically highly porous interconnected structures and a large surface-to-volume ratio, to promote cell in-growth and cell distribution throughout the matrix (Logeart-Avramoglou *et al.*, 2005). Pore sizes in the range of 200-900 μm have performed most satisfactorily in these applications because, in addition to osteoprogenitor cells, they also enable endothelial cells to migrate into the matrix and develop the vascular beds necessary to nourish the newly formed tissue (Logeart-Avramoglou *et al.*, 2005). Particle size, shape and surface roughness affect cellular adhesion, proliferation and phenotype. Specifically, cells are sensitive and responsive to the chemistry, topography and surface energy of the material substrates with which they interact. In this respect, the type, amount and conformation of specific proteins that adsorb onto material surfaces, subsequently modulate cell functions (Boyan *et al.*, 1996). Calcium based ceramics undergo dissolution and precipitation at their surfaces. These events lead to the formation of a carbonate-containing hydroxyapatite layer, which promotes the attachment of bone forming cells (i.e., osteoconductivity) (Ohgushi *et al.*, 1999). Tricalcium phosphate (TCP) and hydroxyapatite (HA) are therefore most commonly used as a scaffold in bone tissue engineering. In addition, these two materials are commercially available from various producers (DeLong *et al.*, 2007) and well accepted in clinical practice as synthetic substitutes (or bone fillers).

In order to evaluate where is a niche for autologous cell therapy in medical practice, an overview of other established treatments is necessary.

2.1 Estabished treatments of bone defects

Orthopaedic trauma surgery requires the regular use of bone grafts to help provide timely healing of muscoskeletal injuries. The "perfect" bone graft has properties categorized as: osteoconductive, osteoinductive and osteogenic (De Long *et al.*, 2007). Osteoconduction is the property of a matrix that supports the attachment of bone-forming cells for subsequent bone formation. Osteoinduction is a process that supports the mitogenesis of undifferentiated cells, leading to the formation of osteoprogenitor cells that form new bone. The terms "osteogenic" and „osteogenesis" may be reserved for the ability to generate or the generation of bone by bone-forming cells.

2.1.1 Bone grafts and substitutes

Today, autologous bone grafting is the gold standard for osteogenic replacement in osseous defects (DeLong *et al.*, 2007). Autologous bone grafts reliably fill substance deficits and induce bone tissue formation at the defect site following transplantation. These grafts exhibit some initial stability, depending on donor site, size, shape and quality (Kneser *et al.*,

2006). However, the clinical use of autologous osseous transplants is limited by a considerable donor site morbidity, which increases with the amount of harvested bone. Bleeding, hematoma, infection and chronic pain are common complications of bone graft harvest (Ebraheim *et al.*, 2001). In addition, when the bone defect is large, there may not be enough autologous bone tissue to harvest. Processed allogeneic or xenogenic bone grafts are also commonly used for repair of osseous defects when autologous transplantation is not appropriate (Gazdag *et al.*, 1995). Although the initial properties of allogeneic or xenogenic grafts resemble those of autologous bone, the lack of osteogenicity is a limitation even when osteoinductive factors are preserved during processing. For specific indications, vascularized bone grafts from various locations, including fibula, scapula, iliac crest and others, are taken and transplanted into given bone defects (Ozaki et. al., 1997).

2.1.2 Synthetic bone substitutes

Degradable and non-degradable implant materials can be divided into synthetically produced metals and metallic alloys, ceramics, polymers and composites or modified natural materials (Mayer and Wiesmann, 2006). Whereas non-resorbable materials such as steel or titanium alloys are commonly used for prosthetic devices, resorbable substitute materials are currently being investigated for their use in bone and cartilage replacement therapies (Mayer and Wiesmann, 2006). Acrylate-based bone cements provide high mechanical stability after polymerization (Lewis, 1997). They are widely used for fixation of total joint prosthesis, vertebroplasty and for craniofacial bone defects. However, despite sophisticated modes of application, they do not possess osteogenic or osteoinductive properties and are slowly resorbed, if at all. Within the last two decades, many other biogenic and synthetic materials have been evaluated for their use as bone substitutes. Calcium phosphate- and apatite-based bone cements (porous composites and the most widespread ceramics used for bone reconstruction), as well as other types of biomaterials have been clinically applied for the treatment of fractures and bone defects (Jupiter *et al.*, 1997). Depending on their chemical composition and porosity, they are osteoconductive, biodegradable and are integrated into given bone defects (Kneser *et al.*, 2006).

2.2 Advanced approaches to treating bone defects

2.2.1 Osteoinductive substances

Although osteoinductive substances are clinically applied for the reconstruction of bone defects or for acceleration of fracture healing, only small numbers of patients have been treated and application modes and indications are not yet completely standardized. Platelet rich plasma contains, in addition to platelet derived growth factor (PDGF), a variety of different growth factors, depending on the processing and application modes andit enhances bone formation in experimental and clinical settings (Thornwarth *et al.*, 2006). Demineralized bone matrix (DBM) is prepared from allogeneic or xenogenic bone and is commercially available for clinical application in various formulations (Maddox *et al.*, 2000). Its osteoinductive potential is highly variable and depends not only on the donor but also on the processing protocols. DBM is commonly used in combination with other types of biomaterials. Bone morphogenetic proteins (BMPs) have been identified as the most relevant osteoinductive factor in demineralized bone matrix (Reddi *et al.*, 1998).

2.2.2 Allogenic bone tissue engineering products used in clinical practice and trials

The first registered tissue engineered bone product is called Osteocel, launched in the USA in July 2005 by Osiris Therapeutics Inc. It was the first product containing viable allogenic adult stem cells to be offered for the repair, replacement or reconstruction of bone defects. Osteocel promotes bone regeneration and is used to treat spinal defects or hard-to-heal fractures, in which the bone is shattered or pieces are missing. The producer declares this product to be the first bone matrix product to provide all three bone growth properties: osteoconduction, osteoinduction, and osteogenesis.

Osteocel is made from mesenchymal stem cells, which are mixed with spongy bone material obtained from human donors or cadavers. Because the cells are not manipulated (only harvested, processed and stored for later use, much like organs used for transplant), Osteocel is classified by the Food and Drug Administration (FDA) as a tissue transplant, not as a drug or a medical device. This product did not therefore have to go through the multiyear testing and approval process, which would most likely be be required for other stem cell products being developed.. Osteocel is an allogeneic tissue engineered product, exhibiting low immunogenicity and no activation of lymphocytes T in mixed leukocyte reaction testing *in vitro*. Osteocel grafts have been used since 2005 in over 30,000 procedures, with no reported adverse events.

Trinity™ by Blackstone Medical Inc. is another allograft substance that has recently begun to be used. Trinity BMSC are pre-immunodepleted and therefore do not stimulate local T-cell proliferation but instead are activated to act as osteoblasts and to stimulate bone formation. This local response can accelerate healing, earlier weight-bearing, healing and filling of bone voids in patients that have had excision of bony masses. In previous animal models, the use of BMSC has been shown to increase bone healing in critical sized defects. Trinity is currently approved by the FDA for use in trauma and bone defects within the spine, and has not shown any significant adverse effects compared with standard bone substitute products.

2.2.3 Autologous cell treatment approaches to bone defects

A widely accepted approach is the use of autologous cells for bone regeneration, which are frequently prepared as in-hospital procedures or produced only for the local market.

According to Chatterjea *et al.* (2010), 7 human clinical studies have so far been conducted based on the use of BMSC.

In 2001, BioTissue Technologies AG launched a product called BioSeed®-Oral Bone, using periosteum samples as a source of cells with osteogenic potential. BioSeed®-Oral Bone is a 3D jawbone graft used to reconstruct the jaw bone, for example in sinus lift operations or lower jaw augmentation.

Aastrom Biosciences in the USA has produced bone regeneration products for the treatment of osteonecrosis of the femoral head (called the ON-CORE trial) and a product for the treatment of severe non-union fractures (i.e., atrophic non-unions), both of which are in Clinical Phase III of development. Bone Repair Cells (BRCs) were derived from a small sample of the patient's bone marrow that is processed using Aastrom's Tissue Repair Cell (TRC) Technology to generate larger numbers of stem and early progenitor cells with enhanced therapeutic potential. In the study, patients underwent standard open reduction and internal fixation surgery, in which BRCs were applied directly to the fracture site, together with an allograft bone matrix, to promote local bone regeneration. After the treatment with BRCs, patients with non-union tibia, humerus or femur fractures that had previously failed to heal after one or more "standard" medical procedures showed an overall healing rate of 91% after one year. The positive results from this study, together with early clinical data reported from osteonecrosis patients, further support the broad application of the proprietary TRC Technology in the field of orthopedics.

There have been several clinical reports about the treatment of critical-sized long bone defects with tissue engineering products using BMSC and scaffolds (Quarto *et al.*, 2001, Orozco *et al.*, 2005).

Some ongoing clinical trials are testing the treatment of non-union fractures and bone cysts by autologous mesenchymal stem cell percutaneous grafting as a minimally invasive implantation procedure (ClinicalTrials.gov Identifier: NCT01429012; NCT01206179; NCT01207193; NCT00916981; NCT00916981).

3. Testing different approaches for the production of autologous tissue engineered bone constructs

Aspects that need to be considered in planning a cell therapy/tissue engineering approach are:

- biological: cells to express adequate cell phenotype to produce bone tissue
- tissue engineering: scaffold that allows cell survival, is biodegradable, non-immunogenic, possesses appropriate biomechanical properties and is easy to handle
- surgical: adaption to the size and shape of the injury, a good clinical outcome could be expected upon the appropriate selection of clinical indications

Since 2005, we have been working on several bone tissue engineering projects employing various osteogenic cells:

- Engineering bone grafts using AO and rotating bioreactor – *in vitro study*
- Engineering bone grafts using BMSC- *in vitro study*
- Engineering bone grafts using ASC and perfusion culture - *in vitro study*
- Vascularization of tissue engineered bone grafts - *in vitro study*
- BMSC based bone grafts for the repair of long bone defects - *clinical project*
- Treatment of paradontal diseases with AO - *clinical project*

3.1 In vitro investigation of osteogenic potential of different cell sources

Various cell source, namely AO, BMSC, and ASCwere investigated in relation to different targeted clinical indications. The basic proof of osteogenic activity is mineralization of the extracellular matrix, which was found in all three investigated cell types. Additionally, specific gene expression and alkaline phosphatase activity was analyzed.

While AO were investigated to treat small volume defects in periodontal intrabony defects, both BMSC and ASC can be obtained in sufficient number from bone marrow aspirate or liposuction and proliferated enough to treat high volume bone defects (up to 50 cm^3) and were studied in relation to the treatment of more extensive, e.g., long bone defects (pseudoarthrosis).

ASC and BMSC were tested for the expression of mesenchymal stem cell markers and for their capacity for mineralization after osteogenic differentiation. All tested cells expressed markers of mesenchymal stem cells DG73, CD90 and CD105 and were negative for CD34, which is a marker of hematopoietic cells (Table 1). All tested cell types were also positive for mineralization, which occurred in cultures of alveolar osteoblasts after 1-2 weeks of cultivation in osteogenic medium and in cultures of both mesenchymal stem cell types after 2-3 weeks of cultivation in osteogenic medium (Fig. 1). The intensity of matrix mineralization, however, significantly varied among cell cultures from different donors.

Sample \ Marker	CD 105	CD 90	CD 73	CD34
BMSC 30	+	+++	+++	---
BMSC 31	+	+++	+++	---
BMSC 35	+	+++	+++	---
ASC 01 (P3)	+	+++	+++	---
ASC 02 (P3)	++	+++	+++	---
ASC 03 (P3)	+	+++	+++	---
ASC 04 (P6)	+++	+++	+++	---

Table 1. Expression of mesenchymal stem cell markers by BMSC and ASC

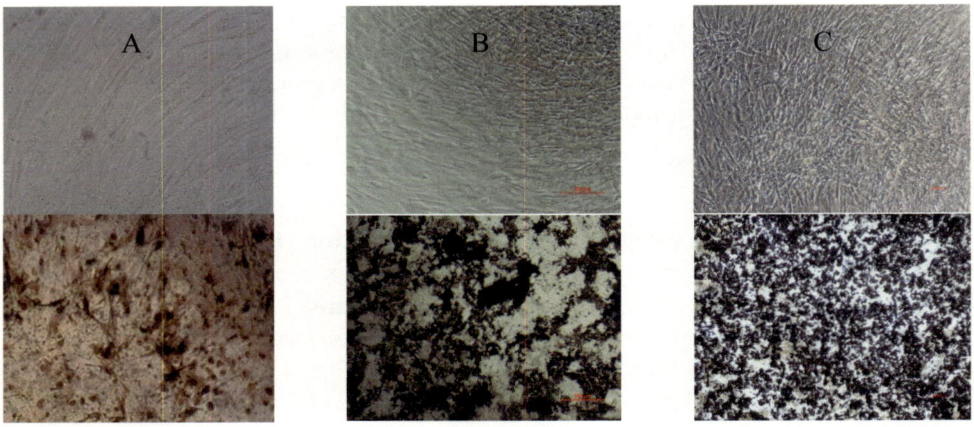

Fig. 1. Mineralization of the matrix is occurring in cultures of AO (A), BMSC (B) and ASC (C) after induction of osteogenic differentiation (upper line: culture in normal medium, bottom line: cultures in osteogenic medium). Von Kossa staining. (photos were taken at a magnification of 100x).

3.1.1 Engineering bone grafts using Alveolar Osteoblasts (AO) and a rotating bioreactor

AO can be isolated from alveolar bone tissue that is normally discarded prior to treatment of periodontal diseases. The use of alveolar bone tissue as a cell source for periodontal indications therefore represents no additional harm to the patient and is thus considered to be the optimal cell source for this application.

The aim of this study was to engineer bone grafts using AO for treating bone degeneration in periodontal diseases.

After harvesting a piece (approx. 40 mm^3) of maxillar or mandibular alveolar bone, primary explant culture and subsequently cell cultures of the first passage were established (Fig. 2).

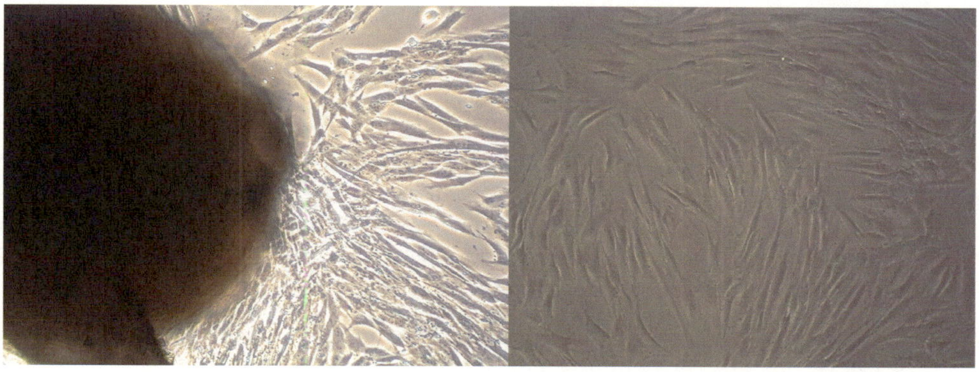

Fig. 2. Primary explant culture (left) and the first passage (right) of AO cells for the treatment of periodontitis (photos were taken at magnification 100x).

Expanded AO with proven osteogenic potential were loaded onto macroporous hydroxyapatite granules together with fibrin glue, which enabled the formation of solid grafts (Fig. 3), and cultured in medium supplemented with osteogenic differentiation factors for up to three weeks in a rotating bioreactor. Light and scanning electron microscopic examinations of the cell-seeded constructs showed a uniform cell distribution, as well as cell attachment and growth into the interior region of the hydroxyapatite granules (Fig. 4). Cells in tissue constructs exhibited growth patterns of enhanced proliferation during the first two weeks of cultivation, followed by a decrease in cell numbers.

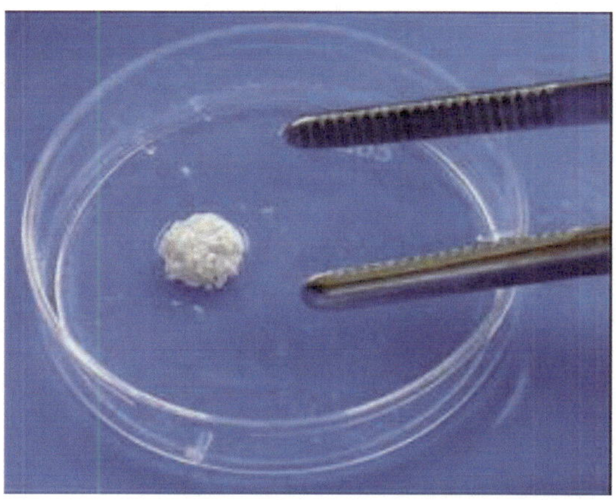

Fig. 3. Bone tissue engineered graft for the treatment of periodontal intrabony defects, macroscopic view of the graft.

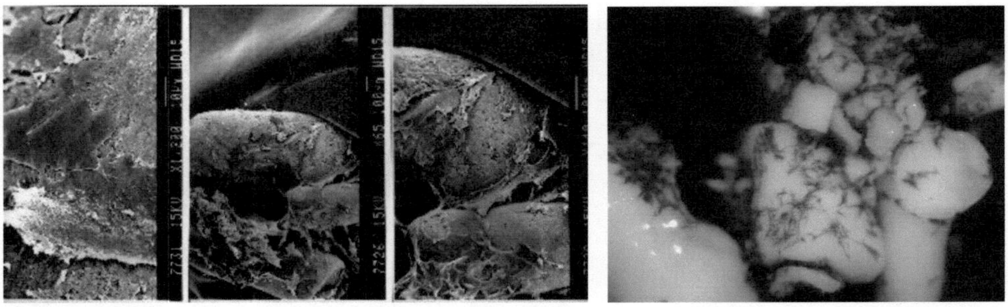

Fig. 4. AO growing over the hydroxyapatite granules of the scaffold one week after graft formation A: scanning electron microscopy, B: stereomicroscope (cells stained with MTT)

The osteogenic potential of the cells was tested by observation of the mineralisation capacity and analysis of gene expression of three important marker genes for osteogenesis: alkaline phoshatase, osteopontin and osteocalcin. Alkaline phosphatase activity was higher at three weeks in all cultures in osteogenic medium than in the control medium. Gene expression levels exhibited patterns of osteogenic differentiation (Maličev *et al.*, 2008).

We showed that bone-like constructs with viable cells exhibiting differentiated osteogenic phenotype can be prepared by cultivation of AO on hydroxyapatite granules.

3.1.2 Engineering bone grafts using Bone Marrow derived Stem Cells (BMSC)

Bone Marrow derived Stem Cells (BMSC) - also termed mesenchymal stem cells (MSC) or multipotent adult progenitor cells (MAPC) - are progenitors of skeletal tissue components such as bone, cartilage, muscle, the hematopoiesis-supporting stroma and adipocytes (Pittenger *et al.*, 1999; Flanagan *et al.*, 2001). The development of methods for isolation, expansion and controlled differentiation of BMSC offers possibilities of using these cells as an integral component of various clinical applications of tissue engineering, especially in reparative osteogenesis.

The aim of this study was to engineer bone grafts using BMSC for treating long bone defects in patients with pseudoartrosis.

After harvesting bone marrow from the iliac crest (approx. 30 ml), mononuclear cells were separated by gradient centrifugation and seeded in primary culture. Non-adherent cells were washed out after 24 hours and adherent cells were expanded and passaged to obtain a sufficient number of cells.

Osteogenic differentiation was carried out in confluent monolayer cultures of the second passage, which was confirmed by positive von Kossa staining (calcium deposits) and staining for the enhanced presence of alkaline phosphatase (Fig. 5). In addition, higher gene expression levels of bone sialoprotein II, osteopontin and BMP2 were determined in BMSC after osteogenic differentiation compared to control BMSC.

Porous TCP granules were used as a scaffold. The cells were seeded directly onto the granules to achieve an approximate total of 1×10^6 cells per 1 mL of the tissue engineering bone construct. The granules were "glued" by inducing fibrin clot formation with the addition of thrombin (Fig. 6). Cell viability in the tissue bone construct was confirmed by

MTT staining. Light microscopy examination of the cell-seeded constructs showed a uniform distribution of viable cells (Fig. 7).

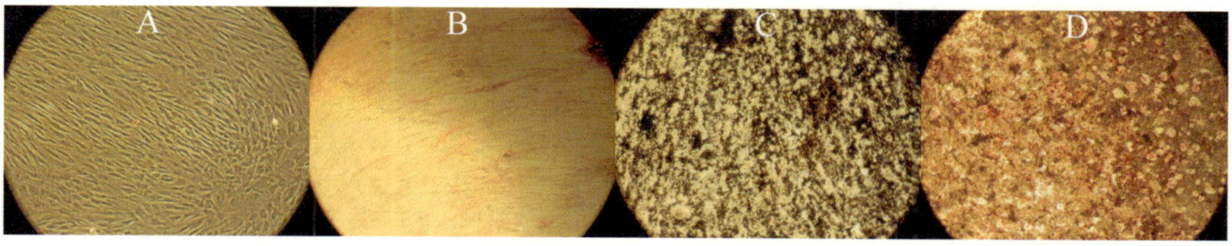

Fig. 5. Second passage of BMSC after 18 days in basal growth medium (A,B) and in osteogenic medium (C,D), respectively. The cells were subsequently stained for calcium deposits according to von Kossa (A,C) and for alkaline phosphatase (B,D) (photos were taken at magnification 100x).

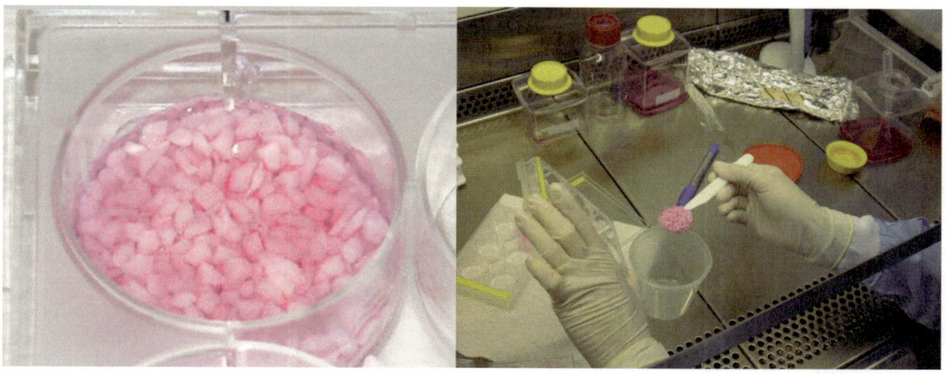

Fig. 6. Preparation of bone implant composed by BMSC after osteogenic differentiation, TCP granules and fibrin glue.

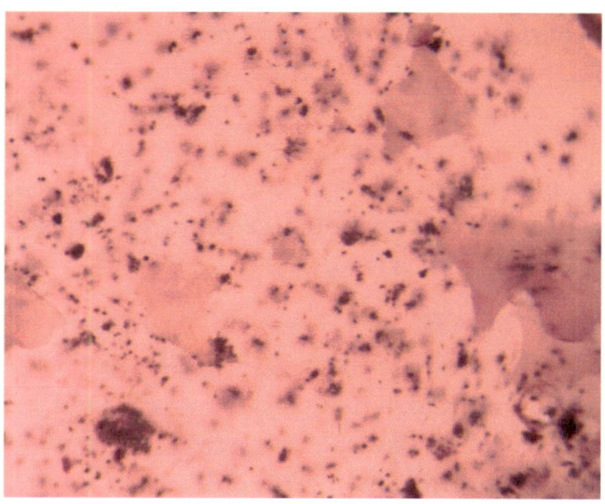

Fig. 7. BMSC after osteogenic differentiation and seeding onto the granules of TCP and staining with MTT, showing an equal distribution of cells in the graft (stereomicroscope, photos were taken at magnification 20x).

3.1.3 Engineering bone grafts using Adipose-derived Stem Cells (ASC) in a perfusion bioreactor

ASC are an attractive cell source for autologous bone tissue engineering, due to their easy accessibility and abundance, as well as their potential for osteogenic differentiation (Zuk *et al.*, 2001). In combination with scaffolds with mechanical properties similar to native bone, they could enable engineering of bone grafts for treating load-bearing sites.

The aim of this study was to engineer bone grafts using ASC on decellularized bone scaffolds and to evaluate the effects of long term perfusion culture conditions (enabling efficient cell nutrition and gas exchange) on the quality (cell distribution and bone matrix formation) of bone grafts. Perfusion culture has already been proved to be beneficial for BMSC based grafts in terms of cell distribution and bone matrix deposition (Gomes *et al.*, 2003; Grayson *et al.*, 2008).

Human ASC were isolated from lipoaspirates of three different donors, characterized and expanded up to the 3rd passage. The osteogenic potential of ASC was tested using von Kossa and Alizarin Red staining. For the perfusion study, cells were seeded on decellularized bovine trabecular bone scaffolds (4 mm ∅ × 4 mm) and subsequently cultured in two different medias (control and osteogenic), in static culture and perfusion bioreactors (Fig. 9). Four experimental groups were formed: (*i*) control-static, (*ii*) control-perfused, (*iii*) osteo-static and (*iv*) osteo-perfused. After 5 weeks, constructs were evaluated for cell viability (live/dead assay), DNA content (PicoGreen), cell distribution (4'-6-diamidino-2-phenylindole - DAPI), collagen (Trichrome), osteopontin and sialoprotein (immunohistochemistry).

ASC from three different donors showed that osteogenic culture conditions resulted in strong mineral deposition, as confirmed by von Kossa and Alizarin Red staining. Additionally, these data show a significant donor-to-donor variability in the osteogenesis of ASC (Fig 8).

During cultivation of ASC grafts, the DNA content increased in all experimental groups and was generally higher under osteogenic than under control conditions. Histological analysis demonstrated that grafts cultured in osteogenic medium contained more total collagen, bone sialoprotein and osteopontin than matching controls. Additionally, under static culture conditions, cell growth and matrix deposition were located mostly at the construct periphery, while perfused constructs exhibited a more even cell and matrix distribution throughout the scaffold volume (Figs. 10 and 11).

In summary, a combination of ASC as cell source, decellularized bone as scaffold and perfused culture conditions in combination with osteo-inductive supplements, provides a promising approach to obtaining high quality tissue engineered bone grafts. Furthermore, cultivation of ASC in a perfusion bioreactor improves cell and bone matrix distribution within the graft and therefore assures a superior cultivation environment to static culture, especially for larger grafts and for longer periods of time. However, for the successful application of ASC based bone grafts in clinical settings, the donor-to-donor variability in the osteogenic potential of ASC needs to be considered. (Fröhlich *et al.*, 2010)

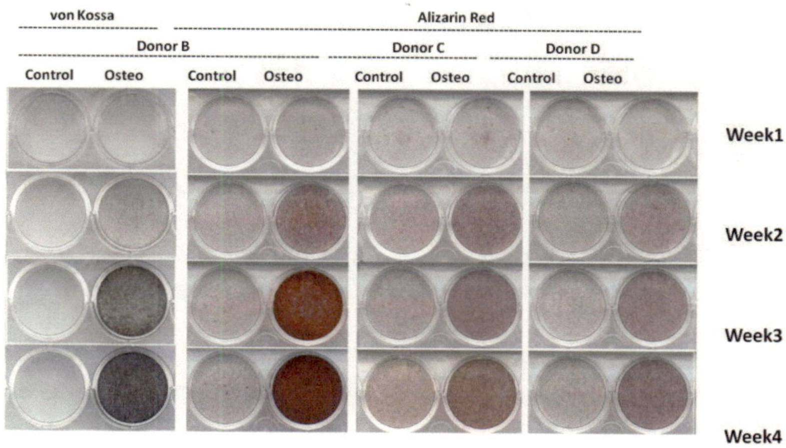

Fig. 8. Osteogenic potential of ASC. ASC of three donors (B, C, D) were cultured under either control or osteogenic medium for various time lengths and were stained with Alizarin Red (red) and von Kossa (black). (Fröhlich *et al.*, 2010)

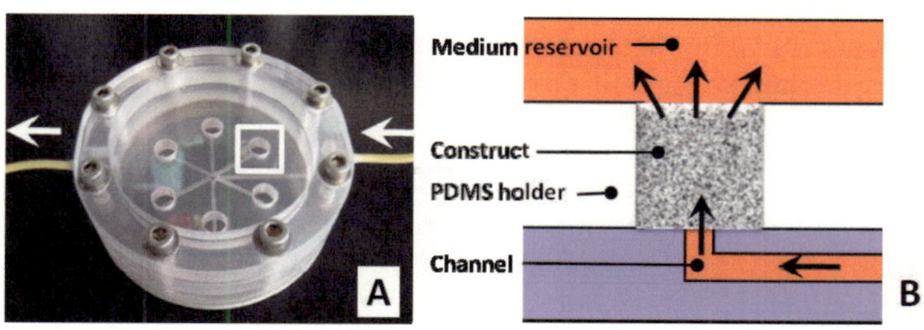

Fig. 9. Perfusion bioreactor used in the study. The region indicated by a white rectangle in A is shown schematically in panel B. Medium flows throughout the scaffold, as indicated by arrows (B). (Fröhlich *et al.*, 2010)

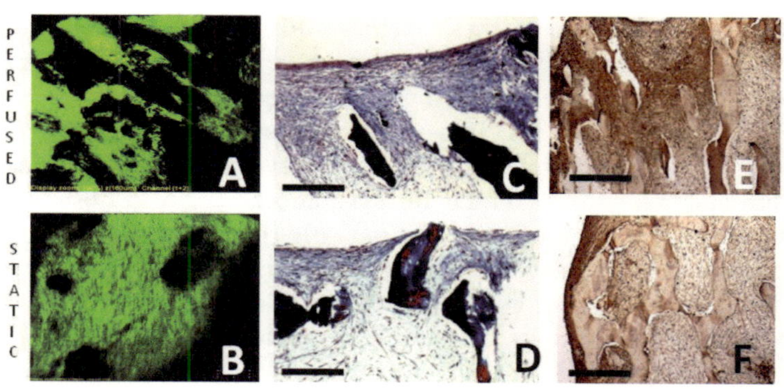

Fig. 10. Long term cultivation of ASC grafts in static and perfused culture. Live/dead staining of the central part of the cultured grafts under perfused (A) and static (B) conditions. Collagen (C, D) (blue) and osteopontin (E, F) (brown) deposition within the scaffold is more abundant and more uniformly distributed under perfused conditions (C, E) than under static culture (D, F). The scale bar is 0.5 mm. (Fröhlich *et al.*, 2010)

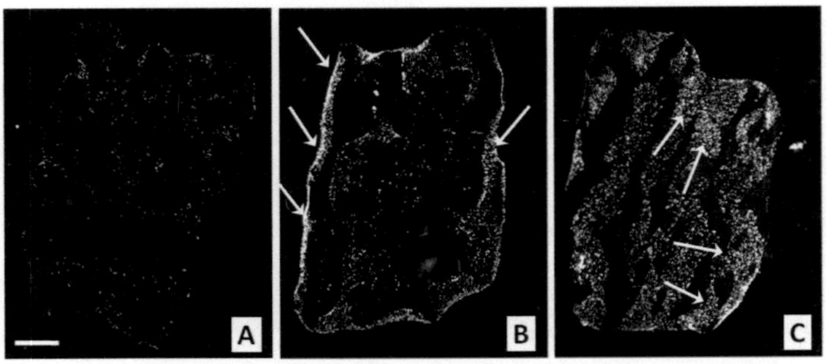

Fig. 11. Cell distribution within the ASC bone grafts. Constructs were stained with DAPI to visualize cell distribution (cell nuclei shown in white). Seeding resulted in an even initial distribution of cells throughout the scaffold (A). After 5 weeks of static culture, cells were found mostly in the outer regions of the constructs (indicated by arrows) (B). After 5 weeks of culture with medium perfusion, cells were more evenly distributed throughout the construct volume (indicated by arrows) (C). The scale bar is 0.5mm. (Fröhlich *et al.*, 2010)

3.2 Vascularization of tissue engineered bone grafts

Vascularization is of critical importance for the integration and survival of larger engineered bone grafts on implantation, since it ensures efficient gas and nutrition exchange with all cells within the tissue.

There are several approaches being utilized in order to vascularize bone grafts, and generally one or a combination of three major principles can be followed (Fig. 12). *In vivo* pre-vascularization employs the implantation of the bone grafts into environments rich in vascular supply (subcutaneous, intramuscular or intraperitoneal sites), where the constructs can be invaded with new vascular networks at their surfaces. However, transplantation to the site of interest is impossible without damaging the initial vascular network. Vascularization of an implanted graft can also be accelerated by the utilization of angiogenic factors. Growth factors, such as VEGF, PDGF and FGF, play a crucial role in angiogenesis (Jain *et al.*, 2003). Incorporation of these factors into scaffolds and control of their local release rate and delivery regime is one possibility for accelerating vascular in-growth *in vivo*. Another way of achieving vascularization of tissue engineered bone grafts is co-culturing endothelial and osteogenic cells into bone constructs engineered *in vitro* – the so called *in vitro* prevascularization approach. Endothelial cells have the potential to form new vessels within the scaffolds, with the potential to anastomose with the host vasculature when implanted *in vivo*. Moreover, endothelial cells not only contribute to forming the vasculature to deliver nutrients to the bone but are also important in terms of interaction with and differentiation of osteoprogenitor cells (Rouwkema et al., 2006; Unger *et al.*, 2007). Adult endothelial cells can be used as a source of endothelial cells, but recently, adult mesenchymal stem cells have also been shown to have the potential to differentiate toward the endothelial lineage (Miranville et al., 2004; Valarmathi et al., 2008). (Reviewed in Fröhlich *et al.*, 2008).

In addition to endothelial cells, smooth muscle cells or pericytes are also necessary for forming a functional vasculature. We exploited the vasculogenic potential of ASC and showed that ASC spontaneously, as well as in induced cultures, formed up to 1 mm long endothelial structures. In the same manner, ASC had the potential for smooth muscle phenotype (Fig. 13). (Fröhlich *et al.*, 2009) Since they have all the necessary types of cells - osteogenic (Fig. 8, Fig. 9) and vascular (Fig. 13), ASC seem to be an ideal source of cells for engineering autologous vascularized bone grafts. However, optimal culture conditions for the co-existence of various cell types still need to be determined.

BMSC have also been tested for their smooth muscle and endothelial phenotype. BMSC expressed α smooth- muscle actin characteristic of smooth muscle cells (Fig. 14), but did not form endothelial structures, as seen with ASC (data not shown).

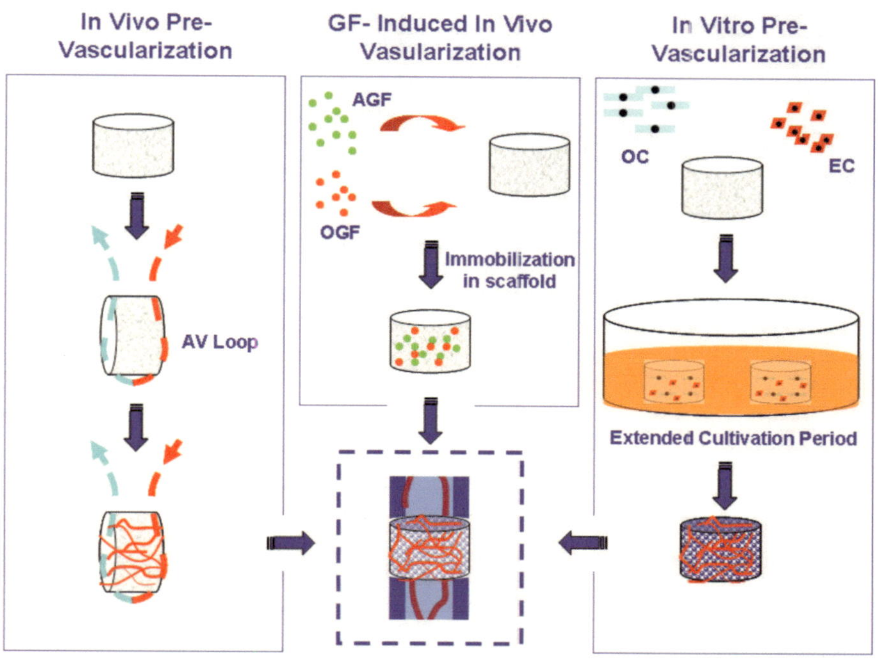

Fig. 12. Approaches to vascularizing engineered bone scaffolds. *Left:* The arterio-venous (AV) loop as an example of an *in vivo* approach for pre-vascularizing scaffolds. **Center:** One cell-free approach is to immobilize angiogenic growth factors (AGF) and osteogenic growth factors (OGF) in scaffolds and directly implant into the site of interest. In this method, the growth-factors induce migration of angiogenic and osteo-progenitor cells and provide them with the stimuli for neo-vessel formation and osteogenic differentiation. **Right:** The cell-based, tissue-engineering approach utilizes osteogenic cells (OC) and endothelial cells (EC) in a three-dimensional co-culture. (Fröhlich *et al.*, 2008)

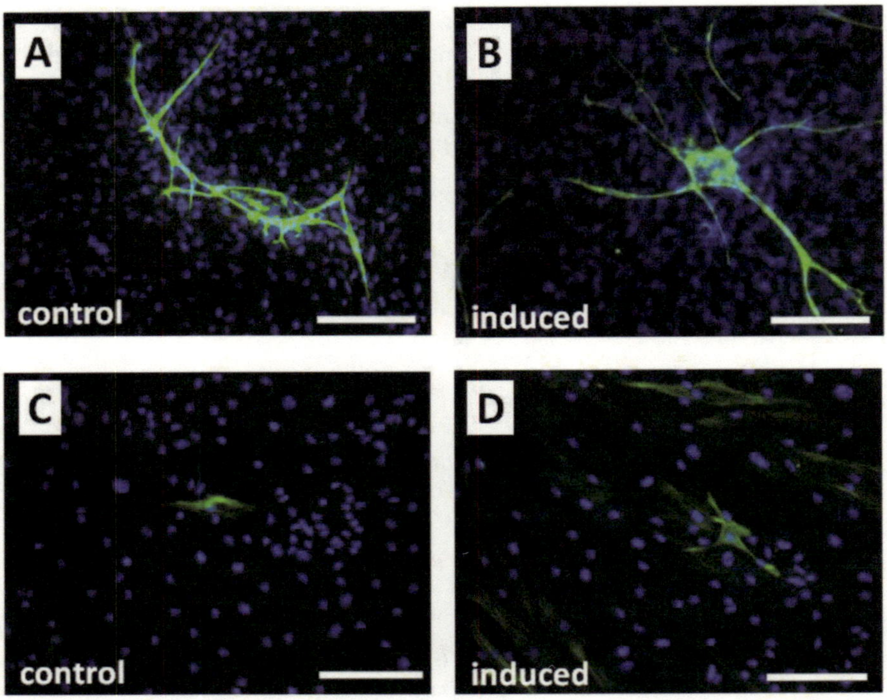

Fig. 13. ASC as a cell source for vascularization of bone constructs. ASC formed up to 1 mm long CD31 positive endothelial structures (green) with *the* close proximity of surrounding cells (blue stained nuclei) when cultured in stromal (control) medium (A). ASC also formed endothelial structures in endothelial medium *but* the structures were less numerous and without the specific pattern of surrounding cells (B). After induction with smooth muscle medium, *the* number of α smooth-muscle actin positive cells (green) increased (D) in comparison to *the* control medium (C). *The* scale bar is 200 μm. (Fröhlich et al., 2009)

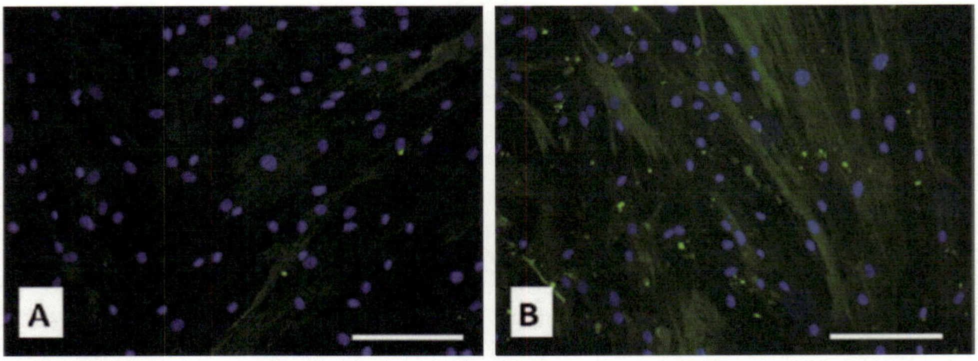

Fig. 14. BMSC of 3rd passage were induced to differentiate into the smooth muscle phenotype. Positive staining for α smooth-muscle actin was evident in the control medium (A) and was further enhanced by exposing the cells to induction medium (B). *The scale bar is 200 μm.*

3.3 Clinical projects for bone regeneration

The first concept (Section 3.1.1.) - employing osteoblasts of cancellous alveolar bone loaded onto the HA granules - has been developed to treat periodontal diseases (BoneArt™-A). The second approach (Section 3.1.2.) - employing bone marrow derived mesenchymal stem cells (BMSCs) differentiated into osteoblasts and loaded onto the TCP granules - has been developed to treat long bone defects (BoneArt™-S). In both cases, the principle is similar (Fig. 15): cells are isolated and proliferated from autologous tissue harvested from patients. Using a criopreservation step, we can adapt to the time of predicted implantation. When cells are proliferated to the desired number, they are seeded on scaffold material. Induction of differentiation can be added to the protocol before or after bone graft preparation. The bone graft can either be implanted immediately or submitted to conditions that stimulate osteogenesis prior to implantation. Grafts need to be tested according to the quality control (QC) protocol, ensuring the safety and efficiency of the product.

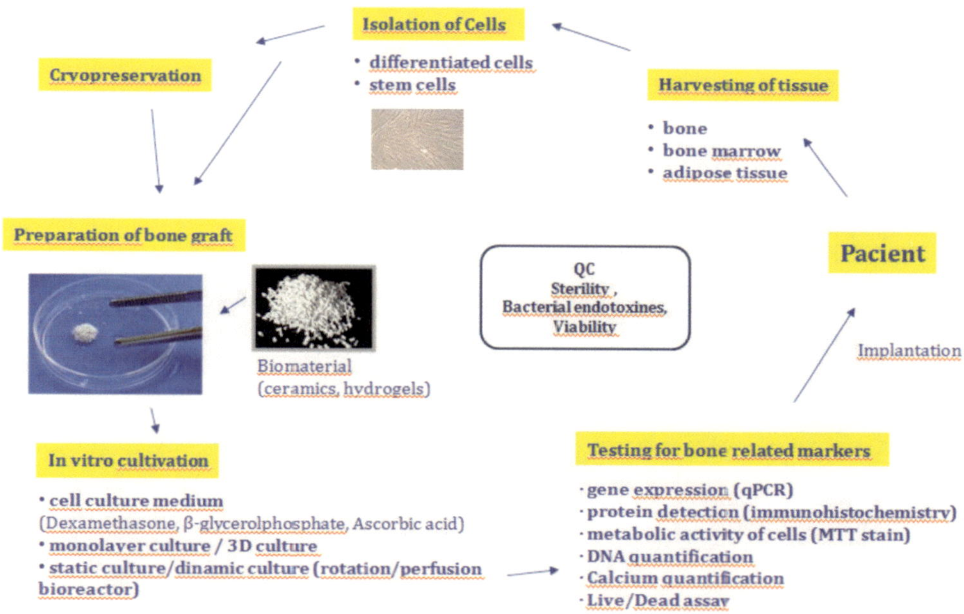

Fig. 15. Principle of bone tissue engineering for clinical application.

3.3.1 Autologous Alveolar Osteoblasts (AO) for the treatment of paradontosis

Periodontal diseases (periodontitis) is a chronic, infectious, inflammatory disease that affects the dental attachment apparatus - i.e. the tissues that support and anchor the teeth to the jaw; these include the cementum, periodontal ligament ans alveolar bone. If left untreated, periodontal disease may result in complete destruction of the alveolar bone as well as the othe supporting tissues. (Lin *et al.*, 2008).

The possibility of enhancing bone regeneration by implanting alveolar osteoblasts (AO) in combination with an appropriate scaffold is of clinical interest, particularly in reconstructive maxillofacial surgery and periodontology (Lin et al., 2008). In the research project, the concept was tested that artificial matrices, seeded with cells of osteogenic potential, may be implanted into sites where osseous damage has occurred, which could lead to significant osseous regeneration.

Firstly, the growth and differentiation of alveolar bone cells in tissue-engineered constructs and in monolayer cultures, as a basis for developing procedures for routine preparation of bone-like tissue constructs, were compared (Maličev *et al.*, 2008).

Autologous constructs as described above (Section 3.1.1) were prepared to treat six patients with aggressive periodontitis by an implantation of a cell-based alkaline phosphatase approach. The operative implantation procedures were carried out without any complications and no side effects were detected that could be assigned to the tissue engineered construct. The newly forming bone is clearly seen in X-rays 3 months after implantation (Fig. 16). Clinical evaluation at 6 months and 12 months after implantation showed a significantly higher gain of clinical attachment in cases in which cellularized grafts were implanted in comparison with the control group (implantation of the scaffold alone) (Fig. 17).

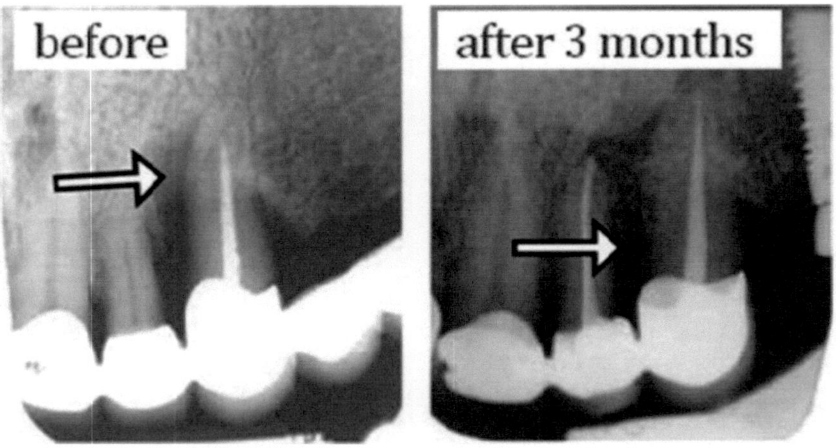

Fig. 16. X-ray of implanted site before and 3 months after treatment. Arrows indicate the limit of bone tissue.

In the first observation period after implantation, there was a significantly higher gain of clinical attachment in sites at which cells were added, compared to sites at which only material was implanted, while no difference is observed in the second period. Overall, in cases in which cells were implanted together with biomaterial, the bone regeneration process was faster and more efficient.

This clinical project confirmed the positive effect of autologous cell therapy for bone regeneration.

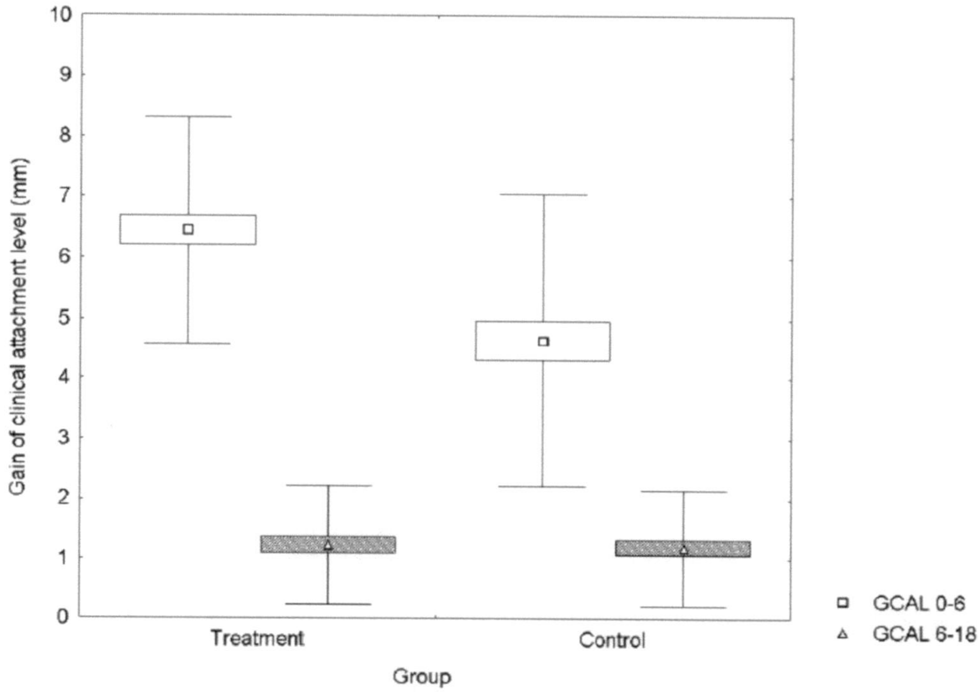

Fig. 17. Gain of clinical attachment measured up to 6 months after implantation (GCAL 0-6) and gain of clinical attachment in the period from first measurement to second measurement up to 18 months after implantation.

3.3.2 Autologous Bone Mesenchymal Stem Cells (BMSC) for the treatment of large long bone defects

A project to evaluate the concept of bone tissue engineering for the treatment of severe long bone defects was carried out using autologous BMSC differentiated into osteoblasts as a cell source and a TCP scaffold in combination with fibrin glue (Figs. 5-7). The bone marrow was harvested from the patient's posterior iliac crest. BMSC were isolated and expanded to the desired number according to Pittinger *et al.,* (1999) with some modifications as described in 3.1.2 (Krečič Stres *et al.,* 2007). Expanded cells with proven osteogenic potential were loaded onto macroporous TCP granules together with fibrin glue, which enabled the formation of solid grafts (Fig. 6). An outline of the procedure for the preparation of tissue engineered bone graft is shown in Figure 18.

The tissue engineered bone construct was surgically implanted to fill gaps in the long bone of patients, mainly for the treatment of pseudoarthrosis in the femur or tibia.

Six patients with a history of multiple failing surgical revisions were treated according to the described procedure (Fig. 19A). None of the patients had any side affects connected with the treatment procedure. Preliminary results were promising since they suggested ossification of the bone defects on X-ray (Fig. 19 B,C). Scintigraphy (^{99m}Tc DPD) also showed evident perfusion and osteoblast activity in the implanted site. At the intermediate observation (5-14

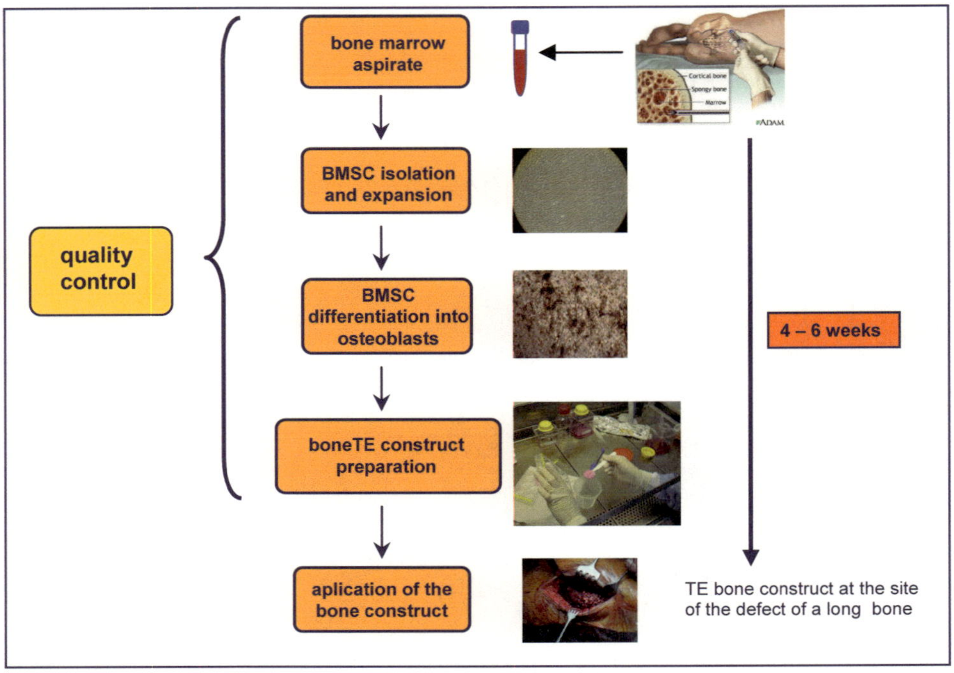

Fig. 18. An outline of the procedure for the preparation of tissue engineered bone graft BoneArt™ for long bone defect treatment from autologous BMSC as carried out in Educell Ltd.

months after implantation), bone bridging or callus formation was observed in 4 out of 6 patients and 3 patients were allowed full weight bearing of the treated leg.

However, the final evaluation of the clinical outcome did not show the expected results. Factors that probably contributed to the failure of these treatments were:

- the extensive volume of the missing tissue (up to 50 ml), which hindered perfusion of the graft as it was designed
- damaged/inadequate surrounding tissue (fibrotic tissue after burns…)
- septic events prior to cell implantation

Due to the small number of patients included in our study, as well as their clinical history, we cannot reach general conclusions about how useful a cell based treatment approach could be in the treatment of non-unions.

Several clinical reports do show successful results of implantation of tissue engineering bone tissues although, especially in large defects, in which a tissue engineering approach is expected to help after other treatments have failed, probably more advanced treatment concepts, considering also perfusion and vascularisation of the tissue, should be developed.

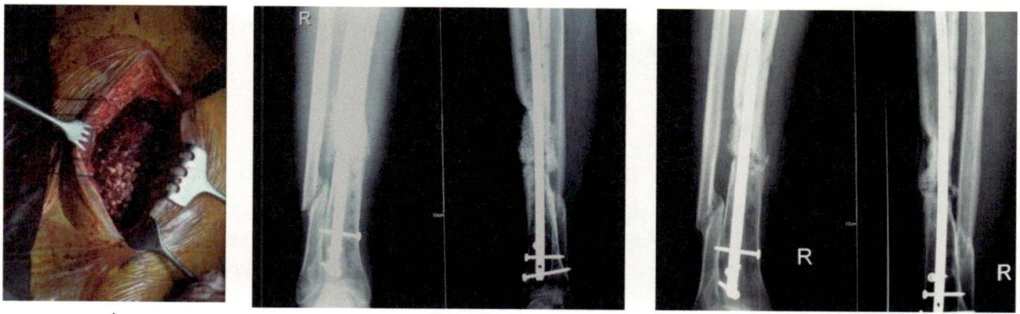

A B C

Fig. 19. Clinical application of tissue engineered bone graft. A: implantation of the graft, B: X-ray immediately after implantation, granulation of the TCP in a graft is visible, C: X-ray 6 months after implantation, the formation of new bone can be observed on the proximal part of the tibia, however complete bone filling was not achieved and a defect remained on the distal part of the non-union

4. Conclusion

Despite the high regeneration capacity of bone tissue, surgical procedures used in reparative osteogenesis do not consistantly result in structural and functional recovery. This state is associated with the disintegration or insufficiency of cambial cells in bone tissue and osteogenic deficiency. Cell based therapies are a new therapeutic approaches in regenerative medicine and using autologous cells is a promising strategy for bone regeneration.

We tested the three of the most studied and relevant sources of osteogenic cells: osteoblasts (from alveolar bone), bone marrow derived stem cells (BMSC) and adipose derived stem cells (ASC). We showed that all three cell sources posses adequate proliferation capacity for potential tissue engineering applications and their differentiation capacity was also proven by testing mineralisation of the extracellular matrix as well as gene expression, specific for osteogenic differentiation.

However, clinical application of a tissue engineering approach is not reflecting the enormous effort in research and preclinical development that has been invested so far - there is still a severe, unmet need for technologies that will facilitate bone tissue regeneration.

Our clinical projects indicate a positive effect of cell based therapies for the treatment of bone defects; in the case of alveolar bone tissue as well as in the case of long bone defects. However, there are limitations in the technology, especially in the treatment of large defects.

Extensive research on tissue vascularization might help cell and tissue engineering technologies become more prospective in bone regeneration. From this aspect, the vascular potential of mesenchymal stem cells seems to indicate a promising area for further bone tissue vascularization research.

Although basic research on osteogenic differentiation potentials of stem and other osteogenic cells is crucial for understanding the bone tissue engineering area, and promises great potential for its use in clinics, only experience from clinical applications will give relevant information and final answers regarding the usefulness of cell and tissue engineered products for various clinical indications.

5. Acknowledgements

Grants for research activities for the development of bone tissue engineered bone grafts were provided by the Ministry of defense of the Republic of Slovenia, grant no.: TP MIR 06/ RR/ 12 and TP MIR 07/RR/18 (coordinator N.Kregar Velikonja), by the Ministry of Higher Education, Science and Technology of the Republic of Slovenia or Slovenian research agency project grants no. 3311-04-824-6325 (project L4-6325, coordinator M.Knežević), no. 3311-04-823-6011 (project L3-6011, coordinator P. Rožman), no. 3211-05-000197 (coordinator E.Maličev), no. 3311-04-831828 (young research grant for M.Froelich) and program no. P3—0371 (coordinator P. Rožman).

Authors would like to thank all co-workers in these projects for their contribution in presented research.

6. References

Bianco, P.; Riminucci, M.; Gronthos, S. & Robey, PG. (2001). Bone Marrow Stromal Stem Cells: Nature, Biology, and Potential Applications. Stem Cells, 19(3):180-192.

Boyan, B.D.; Hummert, T.W.; Dean, D.D. & Schwartz, Z. (1996). Role of Material Surfaces in Regulating Bone and Cartilage Cell Response. Biomaterials., 17:137-146.

Braddock, M.; Houston, P.; Campbell, C. & Ashcroft, P. (2001). Born Again Bone: Tissue Engineering for Bone Repair. News Physiol. Sci., 16:208-213.

Cancedda, R.; Bianchi, G., Derubeis, A. & Quarto, R. (2003). Cell Therapy for Bone Disease: A Review of Current Status. Stem Cells, 21:610-619.

Carlson, N.E. & Roach, R.B. Jr. (2002). Platelet-rich Plasma: Clinical Applications in Dentistry. J. Am. Dent. Assoc. 133(10):1383–6.

Chatterjea, A., Meijer, G., Van Blitterswijk, C.& de Boer, J. (2010) Clinical Application of Human Mesenchymal Stromal Cells for Bone Tissue Engineering. Stem Cells International. Volume 2010 , Article ID 215625, 12 pages. doi:10.4061/2010/215625

Cowan, C.M.; Shi, Y.Y.,;Aalami, O.O.; Chou, Y.F.; Mari, C.; Thomas, R.; Quarto, N.; Contag, C.H.; Wu, B. & Longaker, M.T. (2004). Adipose-derived Adult Stromal Cells Heal Critical-Size Mouse Calvarial Defects. Nat Biotechnol 22, 560,

De Long, W.G.; Einhorn, T.A.; Koval, K.; McKee, M.; Smith, W. & Sanders, R. (2007). Bone Grafts and Bone Graft Substitutes in Orthopaedic Trauma Surgery. A Critical Analysis. J Bone Joint Surg Am, 89:649-658.

Ebraheim, N.A; Elgafy, H. & Xu, R. (2001). Bone-graft Harvesting From Iliac And Fibular Donor Sites: Techniques and Complications. J Am Acad Orthop Surg, 9:210–8.

Flanagan, N. (2001). Advances in Stem Cell Therapy. Genetic Engineering News, 21(9):1,29,61,66.

Frank, O.; Heim, M.; Jakob, M.; Barbero, A.; Schafer, D.; Bendik, I.; Dick, W.; Heberer, M. & Martin, I. (2002). Real-time Quantitative RT-PCR Analysis of Human Bone Marrow Stromal Cells During Osteogenic Different. in Vitro. J Cell Biochem, 85(4): 737-746.

Fröhlich, M.; Grayson, W.L.; Marolt, D.; Gimble, J.M.; Kregar-Velikonja, N. & Vunjak-Novakovic, G. (2010). Bone Grafts Engineered from Human Adipose-Derived Stem Cells in Perfusion Bioreactor Culture. Tissue Eng Part A: 16(1):179-89.

Fröhlich, M.; Grayson, W.L.; Wan, L.Q.; Marolt, D.; Drobnic, M. & Vunjak-Novakovic, G. 2008. Tissue Engineered Bone Grafts: Biological Requirements, Tissue Culture and Clinical Relevance. Curr Stem Cell Res Ther: 3(4):254-64.

Fröhlich, M.; Grayson, W.L.; Marolt, D.; Gimble, J.M.; Kregar-Velikonja, N. & Vunjak-Novakovic, G. Bone Grafts Engineered from Adipose-Derived Stem Cells in Perfusion Bioreactor Culture. Annual Meeting of the Orthopaedic Research Society, Las Vegas, Feb 2009.

Gazdag, A.R.; Lane, J.M.; Glaser, D. & Forster, R.A. (1995). Alternatives to Autogenous Bone Graft: Efficacy and Indications. J Am Acad Orthop Surg, 3:1–8.

Grayson, W.L.; Bhumiratana, S.; Cannizzaro, C.; Chao, P.H.; Lennon, D.P.; Caplan, A.I. & Vunjak-Novakovic, G. (2008). Effects of Initial Seeding Density and Fluid Perfusion Rate on Formation of Tissue-Engineered Bone. Tissue Eng Part A. Nov;14(11):1809-20.

Gomes, M.E.; Sikavitsas, V.I.; Behravesh, E.; Reis, R.L. & Mikos, A.G. Effect of Flow Perfusion on the Osteogenic Differentiation of Bone Marrow Stromal Cells Cultured on Starch-Based Three-Dimensional Scaffolds. J Biomed Mater Res A. 2003 Oct 1;67(1):87-95.

Hattori, H.; Masuoka, K.; Sato, M.; Ishihara, M.; Asazuma,T.; Takase, B.; Kikuchi, M. & Nemoto, K. (2006). Bone Formation Using Human Adipose Tissue-Derived Stromal Cells and a Biodegradable Scaffold. J Biomed Mater Res B Appl Biomater 76B, 230

Hipp, J. & Atala, A. (2004). Tissue Engineering, Stem Cells, Cloning, and Parthenogenesis: new Paradigms for Therapy. Journal of Experimental & Clinical Assisted Reproduction., 1:3.

Hutmacher, W.M. & Sittinger, M. (2003). Periosteal Cells in Bone Tissue Engineering. Tissue Eng, 9(Suppl 1): 45-64.

Jain, R.K. Molecular Regulation of Vessel Maturation. Nat Med 2003. 9:685-93.

Jorgensen, C.; Gordeladze, J. & Noel, D. (2004). Tissue Engineering Through Autologous Mesenchymal Stem Cells. Current Opinion in Biotechnology, 15(5):406-410.

Jupiter, J.B; Winters, S.; Sigman, S.; Lowe, C.; Pappas, C.; Ladd, A.L.; Van Wagoner, M. & Smith, S.T. (1997). Repair of Five Distal Radius Fractures with an Investigational Cancellous Bone Cement: a Preliminary Report. J Orthop Trauma, 11:110–6.

Kneser, U.; Schaefer, D.J.; Polykandriotis, E. & Horch, R.E.J. (2006). Tissue Engineering of Bone: the Reconstructive Surgeon's Point of View. J. Cell. Mol. Med, 10(1):7-19.

Krečič Stres, H.; Krkovič, M.; Koder, J.; Maličev, E.; Marolt, D.; Drobnič, M. & Kregar Velikonja, N. (2007). Mesenchymal Stem Cells: a Modern Approach to Treat Long Bones Defects. In T Jarm, P Kramar, A Zupanic (eds) 11th Mediterranean Conference on Medical and Biomedical Engineering and Computing. Berlin Heidelberg: Springer 2007, pp 253-256.

Langer, R. & Vacanti, J.P. Tissue Engineering. Science. 1993, 260: 920–6.

Lendeckel, S.M; Jödicke, A.; Christophis, P.; Heidinger, K.; Wolff, J.; Fraser, J.K.; Hedrick, M.H.; Berthold, L. & Howaldt, H.P. (2004). Autologous Stem Cells (Adipose) and

Fibrin Glue Used to Treat Widespread Traumatic Calvarial Defects: Case Report. Journal of Cranio-Maxillofacial Surgery, 32(6):370-373.

Lewis, G. (1997). Properties of Acrylic Bone Cement: State of the Art Review, J Biomed Mater Res, 2:155–82.

Lin, N.H.; Gronthos, S. & Bartold, P.M. (2008). Stem Cells and Periodontal Regeneration. J Tissue Eng Regen Med, 2(4):169-83.

Logeart-Avramoglou, D.; Anagnostou, F.; Bizios, R. & Petite, H. (2005). Engineering Bone: Challenges and Obstacles. J Cell Mol Med, 9(1):72-84.

Maddox, E.; Zhan, M.; Mundy, G.R.; Drohan, W.N. & Burgess, W.H. (2000). Optimizing Human Demineralized Bone Matrix for Clinical Application. Tissue Eng, 6: 441–8.

Malicev, E.; Marolt, D.; Kregar Velikonja, N.; Kreft, M.E.; Drobnic, M. & Rode, M. (2008) Growth and Differentiation of Alveolar Bone Cells in Tissue Engineered Constructs and Monolayer Cultures. Biotechnol Bioeng, 100(4):773-81.

Meinel, L.; Karageorgiou, V.; Hofmann, S.; Fajardo, R.; Snyder, B.; Chunmei, L.; Zichner, L.; Langer, R.; Vunjak-Novakovic, G. & Kaplan, D. (2004). Engineering Bone-Like Tissue in Vitro Using Human Bone Marrow Stem Cells and Silk Scaffolds. Journal of Biomedical Materials, 71A(1):25-34.

Meinel, L.; Fajardo, R.; Hofmann, S.; Langer, R.; Chen, J.; Snyder, B.; Vunjak-Novakovic, G. & Kaplan, D. (2005). Silk Implants for the Healing of Critical Size Bone Defects. Bone, 37(5):688-698.

Meyer, U. & Wiesmann, H.P. (2006). Bone and Cartilage Engineering. Berlin Heidelberg: Springer, pp 7-24.

Miranville, A.; Heeschen, C.; Sengenes, C.; Curat, C.A.; Busse, R. & Bouloumie, A. Improvement of Postnatal Neovascularization by Human Adipose Tissue-Derived Stem Cells. Circulation 2004. 110:349-55.

Muschler, G.F.; Chizu Nakamoto, C. & Griffith, L.G. (2004). Engineering Principles of Clinical Cell-Based Tissue Engineering, J Bone Joint Surg Am 86:1541-1558.

Nakahara, H.; Bruder, S.P.; Goldberg, V.M. & Caplan, A.I. (1990). In Vivo Osteochondrogenic Potential of Cultured Cells Derived from the Periosteum. Clin Orthop Relat Res, 259:223-232.

Ohgushi, H. & Caplan, A.I. (1999). Stem Cell Technology and Bioceramics: from Cell to Gene Engineering, J Biomed Mater Res, 48:913-927.

Orozco, L.; Rodriguez, L.; Torrico, C.; Douville, J.; Hock, J.M.; Armstrong, R.D.; Garcia, J. & Solano, C.(2005). Clinical Feasibility Study: the Use of Cultured Enriched Autologous Bone Marrow Cells to Treat Refractory Atrophic and Hypotrophic Nonunion Fractures. at http://scholar.google.com/scholar?hl=en&lr=&q=cache:ss358GFA5XcJ:www.aastrom.com/pdf/Whitepaper_Barcelona-051205.pdf+Orozco+clinical+feasibility

Ozaki, T.; Hillmann, A.; Wuisman, P. & Winkelmann, W. (1997). Reconstruction of Tibia by Ipsilateral Vascularized Fibula and Allograft. 12 Cases with Malignant Bone Tumors. Acta Orthop Scand, 68:298–301.

Peterson, B.; Zhang, J.; Iglesias, R.; Kabo, M.; Hedrick, M.; Benhaim, P. & Lieberman, J.R. (2005). Healing of Critically Sized Femoral Defects, Using Genetically Modified

Mesenchymal Stem Cells from Human Adipose Tissue. Tissue Engineering, 11(1-2): 120-129.

Pittenger, M.F.; Mackay, A.M.; Beck, S.C.; Jaiswal, R.K.; Douglas, R.; Mosca, J.D.; Moorman, M.A.; Simonetti, D.W.; Craig, S. & Marshak, D.R. (1999). Multilineage Potential of Mesenchymal Stem Cells. Science, 284:143-147.

Porter. J.R.; Ruckh, T.T.; & Popat, K.C. (2009). Bone Tissue Engineering: A Review in Bone Biomimetics and Drug Delivery Strategies. American Institute of Chemical Engineers Biotechnol. Prog., 25: 1539–1560.Praemer, A.; Furner, S. & Rice, D. (1992). Musculoskeletal Conditions in the United States. In: American Academy of Orthopaedic Surgeons, Park Ridge, Illinois, 85-124.

Ross, R.; Raines, E.W. & Bowen-Pope, D.F. (1986). The Biology of Platelet-Derived Growth Factor. Cell Review, Jul 18;46(2):155-69.

Rouwkema, J.; De Boer, J. & Van Blitterswijk, C.A. Endothelial Cells Assemble into a 3-Dimensional Prevascular Network in a Bone Tissue Engineering Construct. Tissue Eng 2006. 12:2685-93.

Quarto, R.; Mastrogiacomo, M.; Cancedda, R.; Ketepov, S.M.; Mukhachev, V.; Lavroukov, A.; Kon, E. & Marcacci, M. (2001). Repair of Long Bone Defects with the Use of Autologous Bone Marrow Stromal Cells. N Eng J Med, 344(5):385-386.

Reddi, A.H. (1998). Role of Morphogenetic Proteins in Skeletal Tissue Engineering and Regeneration. Nat Biotechnol, 16:247–52.

Sanchez, A.R.; Sheridan, P.J. & Kupp, L.I. (2003). Is Platelet-rich Plasma the Perfect Enhancement Factor? A current review. Int J Oral Maxillofac Implants, Jan-Feb;18(1):93-103.

Schimming, R. & Schmelzeisen, R. (2004). Tissue-engineered bone for Maxillary Sinus Augmentation. J Oral Maxillofacial Surg, 62(6): 724-729.

Thorwarth, M.; Wehrhan, F.; Schultze-Mosgau, S.; Wiltfang, J. & Schlegel, K.A. (2006). PRP Modulates Expression of Bone Matrix Proteins in Vivo Without Long-Term Effects on Bone Formation. Bone, 38:30–40.

Turhani, D.; Watzinger, E.; Weissenbock, M.; Yerit, K.; Cvikl, B.; Thurnher, D. & Ewers, R. (2005). Three-dimensional Composites Manufactured with Human Mesenchymal Cambial Layer Precursor Cells as an Alternative for Sinus Floor Augmentation: an in Vitro Study. Clin Oral Implants Res, 16(4): 417-424.

Unger, R.E.; Sartoris, A.; Peters, K. et al. Tissue-like Self-assembly in Cocultures of Endothelial Cells and Osteoblasts and the Formation of Microcapillary-like Structures on Three-dimension Porous Biomaterials. Biomaterials 2007. 28:3965-76.

Valarmathi, M.T.; Yost, M.J.; Goodwin, R.L. & Potts, J.D. A Three-dimensional Tubular Scaffold that Modulates the Osteogenic and Vasculogenic Differentiation of Rat Bone Marrow Stromal Cells. Tissue Eng Part A 2008. 14:491-504.

Xiao, Y.; Qian, H.; Young, W.G. & Bartold, P.M. (2003). Tissue Engineering for Bone Regeneration Using Differentiated Alveolar Bone Cells in Collagen Scaffolds. Tissue Eng, 9(6): 1167-1177.

Zhu, S.J.; Choi, B.H.; Huh, J.Y.; Jung, J.H.; Kim, B.Y. & Lee, S.H. (2006). A Comparative Qualitative Histological Analysis of Tissue-Engineered Bone Using Bone Marrow

Mesenchymal Stem Cells, Alveolar Bone Cells, and Periosteal Cells. Oral Surgery, Oral Medicine, Oral Pathology, Oral Radiology, and Endodontology, 101(2):164 – 169.

Zuk, P.A.; Zhu, M.; Mizuno, H.; Huang, J.; Futrell, J.W.; Katz, A.J.; Benhaim, P.; Lorenz, H.P. & Hedrick, M.H. (2001). Multilineage Cells From Human Adipose Tissue: Implications for Cell-based Therapies. Tissue Eng, 7(2):211-228.

3

Signals Between Cells and Matrix Mediate Bone Regeneration

Ron Zohar
Private Practice, Toronto, Canada

1. Introduction

Bone regeneration procedures aim at recapitulating optimal wound healing where tissue components are restored to the form and function required for tissue and organ homeostasis (Zohar &Tenenbaum 2005; Bueno &Glowacki 2009; Dimitriou et al. 2011). Examples of ideal bone regeneration include the healing of a healthy tooth extraction socket or a simple bone fracture. This is not the case in non-union fractures, or extensive damage as a result of tumour removal or bone subjected to chemotherapy, where the overall wound healing ability may be compromised (Dimitriou et al. 2011). Bone is a specialized connective tissue consisting of osteoblasts, osteocytes and osteoclasts embedded in a mineralized matrix capable of remodelling, renewing and load bearing. Optimal bone regenerative therapy will enhance mineralized tissue wound healing through enrichment of the wound/bone defect with a matrix scaffold to support the wound, cells that will give rise to osteoprogenitors and inducer molecules, such as growth factors to amplify activity of cells or events responsible for bone formation. New regenerative approaches may include a combination of these factors in part or as a whole. The temporal, spatial activity and maturation of these three components (i.e. cells, matrix and inducer molecules) during bone regeneration has to be a coordinated and integrative process. Delayed, reduced or lack of activity of any of these components may result in repair and not regeneration of a remodelling functional bone. Cell therapy is compared to the gold standard of autogenous bone marrow grafting, which is considered to be enriched with mesenchymal stem cells, osteoprogenitors and inducer molecules; marrow grafting usually offers predicative regenerative approach. Matrix grafting has to offer mechanical support for the regenerative process to interact with the differentiating osteoprogenitor cells and provide the conditions for the cells to deposit host bone matrix. Grafted inducer molecules need to interact with both the developed matrix and differentiating osteoprogenitors to assure bone matrix deposition and mineralization (Figure 1).

Our earlier studies focused on the isolation and differentiation of bone stem cells, osteoinductive cytokines and matrix development and maturation. The spatial and temporal sequence of matrix molecules expression used to sort stem-like cells population, single application of bone morphogenetic protein-7 (BMP7) induced differentiation of these cells to osteoblasts (Zohar et al. 1997a; Zohar et al. 1998; Zohar et al. 1997b). For bone cells to differentiate or for the bone matrix to mature and mineralize, cross talk between matrix and cells is required to activate bone transcription factors associated with signaling pathways and osteogenic protein expression. Communication between matrix and osteoprogenitor cells is crucial to form a mature, weight-bearing bone. This communication is mediated through secreted growth factors, matrix or matrix associated molecules and activated receptors on the differentiating bone cells.

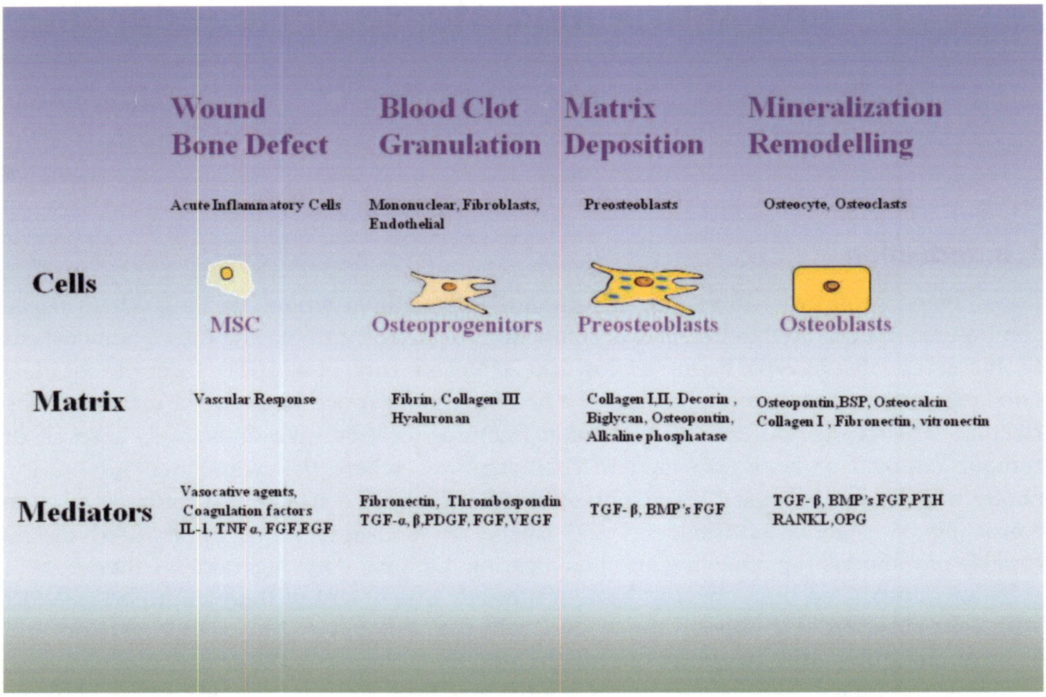

Fig. 1. Wound healing in bone regeneration follows a temporal sequence of ideal healing where a clean wound start healing through bleeding, clot formation and recruitment of mesenchymal stem cells which will differentiated to bone forming cells. The successful differentiation of mesenchymal stem cells to osteoblasts dependent of the temporal and spatial recruitment and expression of cells, matrix and bone related mediators. Matrix would form through adequate blood supply, stable clot formation and deposition of bone matrix that will mineralize. Osteoblasts and osteocytes will differentiate with matrix maturation and will secret mediators and bone specific proteins.

Various animal wound models in number of animal species are used to asses regenerative approaches include rodents, rabbits, sheep, goats, cats, dogs and primates (Gomes &Fernandes 2011; Intini et al. 2007; Kim et al. 2007; Artzi et. al 2003 a,b; Meinig 2002 ; Lemperle et al. 1998). New experimental approaches attempt to regenerate critical-size defects in the affected bone that won't heal without therapeutic intervention. Comparing results between animal models is challenging due to different wound models, different bones used, healing rate, unique animal physiology, whether or not the bone is weight bearing and a variety of protocols. Mice are the animals of choice for transgenic analyses for the significance of the permanent present or absence of one or two molecules (Kim et al.

2007; Masaki &Ide 2007). Large animal models on the other hand are preferred for a slower healing process resembling human physiology; however due to the high cost, control of animals and lower sample number, their use is more limited. The tibia or femur are usually used for the fracture model in a load-bearing area and the calvaria may be used for critical size bony defects in a non-loaded area (Alberius &Gordh 1996; Au et al. 2007; Landry et al. 1996).

Regenerative regimens usually focuses on one of the main components of the missing mineralized tissue: matrix, cells or inducer molecules. While expression patterns were identified for cell differentiation and matrix maturation, ongoing interactions during healing through receptors and signal molecules determines whether the outcome is repair or full regeneration. Thus, evaluating these interactions and the ability of the host wound area to support the process is a major determinant of regeneration. This chapter will focus on the importance of signaling between matrix and bone cells and how growth factors or inducer molecules can mediate this interaction and lead to the regeneration of bone tissue.

2. Bone wound healing

Bone wound healing in primates may involve formation of cartilaginous template, leading to endochondral ossification and/or intramembranous ossification (Dimitriou et al. 2011 ; Javed et al. 2011). Both processes require the commitment of adult stem cells toward bone-forming or osteoprogenitor cells (Figure 1). It is well recognized that adult bone contains a reservoir of mesenchymal stem cells responsible for physiological remodelling of bone and reconstruction during wound healing (Awad et al. 1999; Pittenger et al. 1999). Notably, mesenchymal stem cells are multipotential and capable of differentiation not only to bone forming cells but also to chondrocytes, adipocytes or fibroblasts, as shown in vitro and in vivo studies (Ghilzon et al. 1999; Owen 1988). Commitment of mesenchymal stem cells is thought to be irreversible, and thus signals during the early stages of the wound healing where mesenchymal stem cells differentiation to osteoprogenitors occurs is crucial for bone regenerative process. The ability to induce mesenchymal stem cells to express osteoblastic markers is dependent on transcription of bone-related genes activated by specific signalling, such as wingless-type MMTV integration site (Wnt) family which control osteoblasts differentiation (Hoeppner et al. 2009; Secreto et al. 2009). Important mediators in these pathways activated by Wnt will be the Runx2 (Cbfa1) and Osterix transcription factors. These proteins control expression and repression of genes that will direct the commitment of mesenchymal stem cells toward osteoblasts (Liu, W. et al. 2001). Runx2-deficient mice exhibit neonatal lethality due to absence of bone. In the absence of Runx2 there will be no osteoblast differentiation or ossification. Haploinsufficiency of Runx2 in humans results in cleidocranial dysplasia, a disease characterized by abnormal bone development, formation and decreased bone density (Notoya et al. 2004; Post et al. 2008; Xiao et al. 2004). Cytokines derived from the TGFβ superfamily, such as BMP-4, induce the expression of these transcription factors and thus bone-specific proteins such as alkaline phosphatase (AP), collagen I, bone sialoprotein (BSP), osteocalcin(OCN), osteopontin (OPN), integrin and TGFβ receptors. The expression of these markers serves to ascertain osteoblastic differentiation and evaluate the progression of bone formation. Unfortunately, at present, clear markers to identify and isolate mesenchymal stem cells or osteoprogenitors are not available and the lack of hematopoietic stem cells markers, as well as cellular morphological characteristics, such as undeveloped cytoplasmic structure, are the only reliable criteria for osteoprogenitors (Belmokhtar et al. 2011; Bernardo et al. 2011; Vater et al. 2011).

Following differentiation of osteoprogenitor cells, a stage of the committed cells proliferation and growth cell cycle changes, accompanied by regulation of proliferation-related genes, such as histones, c-myc and c-fos being upregulated; secretion of matrix proteins, such as collagen I, II, III; alkaline phosphatase; fibronectin (Figure 1); as well as cytokines like FGF-2, TGF and BMBs members (Augello &De Bari 2010). Osteoprogenitors mature to secretory osteoblasts with a reduction in mitotic activity and formation of collagenous extracellular matrix (ECM) enriched with bone-specific proteins such as AP, OCN, BSP and OPN. Osteoblasts also secrete osteoprotegerin (OPG) a member of the TNF superfamily to reduce osteoclastic bone resorption by binding with the receptor activator of NF-kappaB ligand (RANKL) (Takahashi et al. 1999). Osteoblasts express receptors to mediate connections between ECM and cells; this connection is primarily mediated through integrins, which will attach to the ECM and intracellular will activate the actin cytoskeleton, initiating cellular signal transduction of proteins such as mitogen-activated protein kinase (MAP kinase) and the SMAD pathway (Blair et al. 2008; Komori 2011). Decrease in matrix formation precedes deposition of hydroxyapatite crystals in the mature collagen organized in a quarter-staggered pattern with 68nm gaps to house hydroxyapatite crystals, which accumulate on the collagen fibers within them and flattening of the active osteoblasts, which may undergo apoptosis or become trapped in the mineralized matrix as osteocytes (Kogianni &Noble 2007).

3. Bone regenerative therapy - Present approaches

There are multiple approaches and various grafting materials available for bone regenerative therapy. The noble regenerative objective is the same for all suggested approaches: living, functional, remodelling bone! Different studies evaluating the success of fracture regeneration or repair estimate the failure rate as 10% or more. Common factors in failure are: lack of vascularity, improper correction, delayed union, non-union and revision surgery (Jones et al. 2000; Lee et al. 2004; Osti et al. ; Parker et al. 2011; Smith, T. O. et al. 2009). The tibia is the most common bone to fracture in children and adults. Corrections that exhibit non-union complications present greater challenge to regenerative therapy (Garrison et al. 2011; Mashru et al. 2005). Other than fixation of fructure, there are also non-invasive approaches used to improve healing, such as electromagnetic field or ultrasound stimulations (Griffin et al. 2011). Distraction osteogenesis is another approach which encourages bone formation through gradual distraction of defect surfaces, requires long treatment, sensitive technique and prolonged healing for the patient; it also serves as a burden to the health system (Heo et al. 2008). Autologous bone marrow grafting is the most predictable approach to achieve regeneration. Bone can be harvested from the iliac crest of the pelvis, or alternatively, reamers can be used to harvest the intramedullary canal of long bones (Hak &Pittman 2010; Valimaki &Aro 2006). If a larger volume of grafts required, allograft or biomaterials are sometimes used in conjunction with autograft.

Present descriptors of grafting materials other than their source (i.e. allograft, autogenous, alloplast, cancellous, cortical), refer to grafts as being capable of osteoconduction, osteoinduction, mechanical support, cell exclusion, cement and filler. Regeneration of bone is a very clear outcome, and unless osteoblast differentiation taking place, new bone matrix deposition and interaction between the two during de novo bone formation and remodelling, no real regeneration could occur. The assumption that placing an allograft that may contain BMP's, collagen matrix or an even high number of mesenchymal stem cells will result in regeneration in every case cannot be true. Without receptive wound environment where osteoprogenitors have signals for differentiation and deposition of new bone matrix, healing by fibrous or cartilage or adipose tissues may occur. Thus, using terminology like osteoconduction and osteoinduction would only suggest of the potential of regenerative approach or material, but it is not necessarily predictive of the desired outcome in specific host, specific wound, and specific surgical approach. The clinical results suggest variability of wound healing (Garrison et al. 2011).

Since most new bone graft or regenerative product is first tested for its biological activity, rather than focusing on osteoinduction and osteoconduction, this chapter classifies present grafts by their contribution to one of the major missing components of the missing bone: cells, matrix, and mediators (Figure 1). To evaluate the present state of bone regenerative therapy, it is worthwhile to see how each approach can contribute to the restoration of one of these three components.

4. Matrix grafting

Matrix serves as an organized framework for bone as a tissue and organ, offering mechanical support, and facilitate preservation of form and adaptive protection to internal organs through ongoing remodelling (Grabowski 2009; Scott et al. 2008). Osteogenic cells, like most other matrix-associated cells, cannot survive or differentiate without adhesion to their matrix (Popov et al. 2011). Thus, the importance of bone matrix in addition to acting as mechanical scaffold, is to mediate the biological activities of bone cells and signals that maintain homeostasis, remodelling and ability for wound healing. The mature mineralized bone matrix is composed of ~20% organic components, primarily collagens I, III and V and less than 5% noncollagenous proteins. The latter consists of proteoglycans, such as versican, decorin, and hyaluronan, adhesions molecules such as fibronectin and vitronectin, and specialized proteins like OCN, BSP, OPN and cytokines (Nagata et al. 1991). The collagen fibrils structure house the hydroxyapatite crystals which tend to be oriented in the same direction as the collagen fibrils. The collagen network also mediates adhesion to cells primarily through integrin receptors connected to collagen or the associated non-collagenous proteins. A mineralized bone matrix not only increases the mechanical strength of the bone but also act as reservoir for specialized proteins and cytokines, such as BMP's.

Matrix proteins mediate not only maturation and mineralization of bone matrix, but also bone cell differentiation and signalling. Bone cell differentiation is detected through the differential expression of matrix molecules such as collagen OPN, BSP, AP and OCN. Expression of AP, collagen I and OPN are considered an early markers, while BSP, OCN and a second peak of OPN are considered a late mineralization associated marker (Aronow et al. 1990; Binderman et al. 2011; Lynch, M. P. et al. 1995).Our studies of OPN expression, which is not restricted to bone, but can be used as a useful marker for early and late differentiation of osteogenic cells. We isolated a population of small cells that do not express OPN, AP, collagen I and that are enriched with stromal stem cells capable of generating bone, fat and cartilage (Zohar et al. 1998; Zohar et al. 1997b). We have isolated BMP-responsive cells, which will undergo chondrogenic differentiation with continuous stimulation of BMP-7 or osteogenic differentiation with single dose (Zohar et al. 1998). Thus, evaluating the expression of matrix proteins can help determine the status of mesenchymal stem cells differentiation.

4.1 Matrix-based grafts can be autologous, allogeneic or biomaterials

Autologous bone grafts, such as the marrow graft, marrow aspirates, will contain cancellous and/or cortical or blocks such as vascularised Graft and will carry cells, matrix and potentially inductive molecules (Friedrich et al. 2009; Sotereanos et al. 1997). A vascularised graft will carry blood vessels to enrich the wound with nutrients and soluble mediators, which may support or inhibit bone formation and carry periosteum enriched with osteoprogenitor cells. There is less necrosis of grafted material during healing and vascularised grafts are thought to be a very reliable option for reconstructing non-union or osteonecrosis defects (Friedrich et al. 2009; Gaskill et al. 2009; Sotereanos et al. 1997). The difficulty with all autogenous grafts is the quantity and morbidity, such as non-stress fracture for donor sites (Friedrich et al. 2009). The cancellous or cortical block graft may carry cells and cytokines, and their quantity and effectiveness is related to the age and state of the donor area. Cortical block graft will contain the least amount of cells and mediators and considered to function primarily as scaffolding which is more susceptible to infection and necrosis.

In allogeneic bone matrix grafts, cadaver bone is a common source of allograft. To generate a safe allograft, the bone is subjected to irradiation or freeze-drying and is thus devoid of any cellular components (Nguyen et al. 2007). Allografts are prepared as particulate, morselized or block, with mineralized or demineralized bone particles that are easy to shape and mold. Demineralized bone matrix serves as a natural matrix as well as decellularized matrices that could derive from dura or intestine of various animals (Costain &Crawford 2009; Kligman et al. 2003; Mroz et al. 2006). Allografts have very limited, if any, biological activity and serve primarily as osteoconductive and mechanical support. The main advantage is ample supply (Hamer et al. 1996). Reports of infection transfer, matrix alteration during processing and limited remodelling of the grafted bone reduce the likelihood of full regeneration (Nguyen et al. 2007) unless combined with autologous bone (Matejovsky et al. 2006) to add osteoprogenitors and mediators that can append biological activity to the dead bone particles.

Matrix proteins-based polymers are very popular, as are collagen, fibrin, hyaluronic acid, fibronectin and BSP. These proteins are delivered as membranes, sponges, gels, demineralized bone particles, small intestinal submucosa, dura or even urinary bladder (Chajra et al. 2008; Smith, I. O. et al. 2009; von der Mark et al. 2010). The problem with generating these polymers is fairly low solubility; the organic purified polymers is costly and hard to extract, purify and stabilize; risk of immunogenicity; and variations based on the batch.

Biomaterials and synthetic bone substitutes are currently used as fillers and/or scaffolds for the missing bone structure (Gosain et al. 2009; Healy et al. 1999; Shekaran &Garcia 2010 ; Wojtowicz et al. 2010). The design and fabrication of matrix-based regenerative materials is aimed at restoring the natural bone matrix properties as a whole or in part. Reconstruction of missing bone using matrices involves the planning of macrostructures as well as microstructure of the engineered matrix (Cholewa-Kowalska et al. 2009; Huang &Miao 2007; Vater et al. 2009). Macrostructures to fill and adapt to the space to assure sufficient quantity and/or provide mechanical support for the surrounding tissue or cells carried. Microstructures of micron or nanotechnology designs of particles or pores are used to encourage cell adhesion, colonization and absorption of proteins or required molecules. An ideal scaffold will have highly interconnected macroporosity to allow host bone tissue and blood vessels to grow into the scaffold (Healy et al. 1999). Popular building blocks are hydroxyapatite (HA), calcium phosphates (CP), tricalcium phosphate (TCP) and bioactive glasses (Behnamghader et al. 2008; Muschler et al. 1996; Valimaki &Aro 2006). They form a carbonated apatite layer when grafted, which is very similar to bone mineral; this will attract attachment of collagen fibres and eventually should be replaced by host tissue, mineralized matrix and cells. Other scaffolds consist of combinations of poly (lactic-co-glycolic acid) (PLGA), alginate and chitosan (Huang &Miao 2007; Jose et al. 2009; Liu, X. et al. 2009; Mishra et al. 2009; Renghini et al. 2009). These polymers can also be used to carry cytokines for controlled release at the wound and/or to carry mesenchymal stem cells. Different studies use different mixtures of these materials or different preparation protocols. The requirement for most preparations is to offer bioactivity and mechanical support. Bioactivity of the scaffold is measured by the number of host cells attached to its surface and interaction with the material to transform them into functional osteoblasts. The mineralized bone matrix will appose directly onto the surface of the material which ideally will have the ability to degrade over time (Holy et al. 2000). It is important that the material will degrade at a rate that allows the newly formed tissue to gradually replace the scaffold, both as a mechanical structure and in terms of space occupied. Finally, and this is where most current materials fail, the material needs mechanical properties that allow the device to be implanted without losing mechanical properties, still allowing sufficient loading of the newly formed tissue (Au et al. 2007; Smit et al. 2010). As of yet, no one has reported a material that fulfills all these requirements. The new scaffolds, usually termed composite scaffolds, maybe coated with proteins to increase cell adhesion, carry cells or cytokines with sustained release (Ameer et al. 2002; Bueno &Glowacki 2009; Gupta et al. 2011 ; Nie et al. 2008).

Bioactivity of biomaterial can be modified through chemical and physical alterations. Nanotechnology approaches try to mimic cell surface properties through approaches such as controlling space between ligands connected to biomaterials (Smith, I. O. et al. 2009). Using the proper spacing will enable, for example, integrin receptor clustering to enable propagation of signals through ligation. Another line of research focuses on molecules that work in synergy with receptors to promote cell adhesion and differentiation; for example, fibronectin, laminin and BSP contain heparan sulphate binding domains that interact with molecules on the cell surface in conjunction to integrin binding. Thus, using cell membrane molecule, such as syndecan which has three sites of heparan sulphate, would augment ligation of fibronectin or RGD sequence by integrin receptors (Whiteford et al. 2007; Yamada et al. 2010).

5. Cellular grafting

Cellular grafting for bone regeneration is a rapidly developing area. This approach had been used for many years through autologous bone grafting, which contains high numbers of bone-committed cells in marrow aspirates, or in bone particles or blocks containing cells embedded in their own matrix (Hak &Pittman 2010; Papakostidis et al. 2008; Tiedeman et al. 1995). The objective of new approaches is to obtain an unlimited amount of adult stem cells, comparing new cellular sources to the gold standard of autogenous bone marrow stromal cell, which are considered to be enriched with osteoprogenitors. Notably, the frequency of osteoprogenitors in young rodent marrow is about 0.0005% (Falla et al. 1993) and up to 0.3% in fetal periosteal tissues. Adult marrow shows a reduction of these precursor cells in number and quality (Stolzing et al. 2008). We have used single cell flow cytometric sorting to isolate osteoprogenitors from fetal rodent periosteal tissues. These cells when plated and stimulated exhibit high proliferative capacity and enhanced osteogenic potential. Notably, these cells consisted of only a very small fraction of the fetal bone tissues. Thus, even in young fetal tissues osteoprogenitors consist of only a very small fraction of bone tissue and usually reside in a well-protected niche. Moreover, during seeding, grafting and transfer of cells to the wounded area there is loss of cells through apoptosis or cytotoxic effects of mediators in the wound area (Giannoni et al. 2009). Regeneration efforts focus on the ability to deliver mesenchymal stem cells to the wound, which will differentiate to the osteoblastic lineage. Differentiation requires the commitment of mesenchymal stem cells to osteoblasts, exhibiting bone-specific gene expression. Osteoblast-specific gene expression is a fairly clear analysis of proteins like AP, OPN, BSP, OCN that are selectively expressed in bone. For the mesenchymal stem cells to form new bone and regenerate the wound, cells need to attach, proliferate, differentiate and survive. Mesenchymal stem cells from marrow seem to be the most predictable source for osteoprogenitor cells and a safe autologous grafting. Unfortunately, bone marrow stromal cell consists of heterogenous population that are subject to age changes; not only does their number deplete, but also their quality and ability to generate new bone is reduced (Benayahu 2000; Stolzing et al. 2008; Zhou et al. 2008). Thus, in the aging population where bone wound healing is compromised, harvesting autologous sufficient number of mesenchymal stem cells from marrow may not be that predictable.

Other sources for bone forming cells could be the umbilical cord, peripheral blood, adipose tissue, dental pulp or periodontium (Goodwin et al. 2001; Honda et al. 2011; Rhee et al. 2010; Yamamoto et al. 2007). Human embryonic stem (hES) cells also being considered as an option due to their fast growth and the fact that these cells, if kept as undifferentiated cell lines, are pluripotential and capable of differentiating to many tissue types under the right conditions (Bahadur et al. 2011; Lerou &Daley 2005). The hES has the advantage of unlimited supply, minimal immune response and no need for a second surgical site (Watt &Hogan 2000). Ethical dilemmas, as well as work needed to control their growth in the targeted tissue, seem to be the main concerns limiting their use. Animal experimentations results are inconsistent and complexed by grafted cell death, formation of teratomas and tumours have been observed (Blum &Benvenisty 2008; Brederlau et al. 2006).

Autologous mesenchymal stem cells derived from bone marrow is still the preferred cellular source and iliac crest harvesting is the most common source. The simple approach could be through bone marrow aspirates or the harvest of cancellous bone enriched in osteoprogenitors. These cells can sometimes go through in vitro expansion before being loaded onto a scaffold or other carrier (Bernardo et al. 2011; Caplan &Correa 2011; Kuo et al.

2011). Gene therapy for insertion or activation of selected genes through transfection or electroporation is often attempted on mesenchymal stem cells (Stender et al. 2007). Due to the morbidity associated with marrow mesenchymal stem cells harvesting, need for a second surgical site, limited amounts of grafting material and lack of mechanical stability in extensive defects composites of mesenchymal stem cells with non-autologous grafting materials are frequently used (Caplan et al. 1997; Dimitriou et al. 2011).

The question is if delivery of bone marrow stromal cell containing stem cells to different wounds will assure a predictable and consistent outcome. Mesenchymal stem cells differentiation, proliferation and survival is dependent on their surrounding matrix, signals to express receptors and secrete signaling molecules. Large size defects with a potentially compromised host may offer a local environment that is not supportive or even inhibitory for bone formation. For example, it has been shown that disruption of integrin activity in mesenchymal stem cells will result in cell death and lack of differentiation (Popov et al. 2011). Various combinations have been prepared in an attempt to find a predictable and consistent graft (Schofer et al. 2011). Notably, at present, even if the number of mesenchymal stem cells is high, without the right matrix and cytokine's support bone differentiation and maturation may not occur.

6. Inducer molecules

The ability of demineralized bone matrix to induce bone formation in the subcutaneous sites of rodents, as reported by Dr. Urist, revolutionized our approach to bone therapy and studies of bone regeneration (Urist 1965; Urist et al. 1967). These studies demonstrated that the non-mineralized fraction of the bone stores molecules that can derive osteogenic differentiation and initiate bone formation in ectopic sites. Factors such as BMP's consist of only a very small fraction of the bone matrix and cannot be purified from bone for scientific or clinical use; however, these factors were cloned and prepared as recombinant molecules or peptides with very potent biological activity (Reddi &Cunningham 1993; Sampath et al. 1992). Inducer molecules can be delivered in a carrier or integrated into expression vehicles through ex vivo transfer to grafted cells, or infected through viruses that will target the tissues; these approaches fall under the category of gene therapy (Table 1)(Franceschi et al. 2000; Mason et al. 1998). Transient transfection and conditional expression approaches achieved in mice and other animals, thorough adeno and lentiviral, as well as non-viral, approaches such as electroporation (Franceschi et al. 2000; Holstein et al. 2009; Kawai et al. 2006). Gene delivery approaches being used in an ex vivo and in vivo gene delivery can also be utilized in humans to deliver genes to marrow stromal cells (Belmokhtar et al. 2011; Chen et al. 2011). Expression control modifications at embryo through transgenic animals or conditional modifications which, dependent on the initiator or temporary gene alteration in adult animals, assist in determining the relative importance of cell, matrix or inducer molecules to mineralized tissue healing. Gene therapy is still not available for regular clinical use, due to inability to assure target of specific cells only and adequate control over the gene transfer transcription, translation and expression in a temporal and spatial manner that will support bone regeneration. Other issues limiting clinical use are concern of viral vectors, control on the expression, immune response and potential for other non-controlled mutations. Moreover, The applications of gene transfer and control in human is not always as efficient or predictable as shown in rodents or primates in animal experiment models (Gomes &Fernandes 2011; Sharma et al. 2011).

Matrix grafting	Cellular grafting	Inducer molecules	Techniques
Vascularised Graft	mesenchymal stem cells -bone marrow aspirate	(bone morphogenetic proteins (BMPs	Gene therapy- transfection, Transduction
Matrix molecules- Collagen, fibrin, hyaluronic acid, BSP,OPN	Cancellous graft- iliac, distal femur, proximal or distal tibia	platelet-derived growth factor-PDGF, Fibroblast growth factor, Vascular Endothelial growth factor,	Recombinant proteins
Mineral- Hydroxyapatite, β- Tricalcium phosphate(TCP),	Other sources of adult stem cells- peripheral blood adipose	Transforming growth factor's	Peptides
Polymers- poly (lactic-co-glycolic acid) (PLGA), alginate and chitosan	ES-Embryonic stem cells	insulin-like growth factor-I,II	Nanotechnology
Calcium phosphate or sulphate, glass ceramics	Umbilical cord	endothelial growth factor	Cellular in-vitro expansion, differentiation induction,
DBM- Demineralised bone matrix	Dental – follicle, pulp, periodontal	Hormones- parathyroid hormone, Growth Hormone	Scaffolds- Three-dimensional porous scaffolds, coated, biodegradable
cancellous bone allograft		Pepetides- FHRRIKA, FNIII 7-10,P15, DGEA (Asp-Gly-Glu-Ala), RGD, PTH 1-34, and PTH 1-84	Morcellized bone grafting, freeze-drying
Cortical		Denosumab-antibody to RANKL	Purified proteins Membranes, Mesh
Block graft		agonists of the prostaglandin receptors EP2 and EP4	Distraction osteogenesis

Table 1. Classification of Grafting

At present, about 20 BMPs have been identified, with about eight having osteogenic effects: BMP-2, 3, 4, 6, 7, 8, 12, 14. BMP-7 or OP-1 is the subject of many studies approved for clinical use and exhibits a very potent osteoinductive effect in vivo and in vitro. BMP-7 effects on mesenchymal stem cells include increased migration, differentiation and induction of bone formation through endochondral as well as intramembranous ossification (Giannoudis et al. 2009). Many other cytokines are the subject of ongoing investigations and use such as platelet-derived growth factor (PDGF), transforming growth factor-β (TGF β), insulin-like growth factor-I and II (IGF), vascular endothelial growth factor (VEGF), endothelial growth factor (EGF), parathyroid hormone (PTH), growth hormone (GH) and fibroblast growth factor (FGF). Some are prepared as synthetic peptides where only the active sequence is synthesized; often the peptide will be more potent that the whole molecule. Examples of these peptides include PTH [PTH(1-34); Forteo (or teriparitide) and PTH 1-84, P24 is a 24-amino acid peptide derived from BMP2 capable of induction of ectopic bone (Lin et al. 2010; Wu et al. 2008). The growth factors that are approved for clinical use in human and received the Food and Drug Administration (FDA) approval for bone regeneration are BMP-2, 7 and PDGF-BB (Caplan &Correa 2011; Kanakaris et al. 2008; Lynch, S. E. et al. 2006; Mulconrey et al. 2008). These growth factors will predictably stimulate bone formation, and when compared to the gold standard of autologous bone grafting, these growth factors meet the expectations of inducing bone regeneration in a high percentage of the clinical cases (Garrison et al. 2011). Advantages include ample supply, convenient grafting carriers, osteoinduction, no need for a second surgical site and no significant immune responses. The reported concerns are no cellular component, no osteoconduction support, lower mechanical strength of the newly formed bone, expensive and variability in induction. These growth factors are carried or released by various materials that may alter their effects and potency (Nauth et al. 2011). It is beyond the scope of this chapter to describe the molecular mechanism known for each of these growth factors or the expression of their receptors and associated signaling pathways. Each of these growth factors is a subject of numerous clinical trials and reports and suggestions on its most potent use for bone regeneration. Their effects are dependent on the availability of cells, the expression of the appropriate receptors and biological half-life at the bone defect.

Bone formation can also be induced by non-growth factor molecules, such as matrix components or proteins that will encourage mesenchymal stem cells cell adhesion, migration, proliferation, differentiation and survival (Popov et al. 2011). Matrix components like collagen will not only induce bone cells directly but also their ability to bind other potent molecules, such as growth factors, thrombospondin, decorin, biglycan, OPN, OCN, BSP, fibronectin, vitronectin and hydroxyapatite (Bentley &Tralka 1983; Ber et al. 1991; Bergmann et al. 1990). Control of expression of receptors to mediate bone matrix adhesion would be another approach, through antibodies or fragments that will induce their expression; for example, the Denosumab human monoclonal antibody that inhibits osteoclastic activity through binding to RANKL and safe even for systemic use (Miller 2009).

Matrix proteins can be used as purified proteins or synthetic peptides. Purified collagen is one source of a primary matrix molecule derived from human, bovine or porcine sources as purified fibrillar collagen or composite with other minerals that can be use to fill defects or mixed with other grafts (Gleeson et al. 2011; Muschler et al. 1996; Thula et al. 2011). Proteoglycans, such as hyaluronan, can be purified from the human umbilical cord, cultures of cells or bacteria. The non-collagenous proteins of the bone, such as OPN, BSP and OCN, can also be purified and used to coat biomaterials, or mixed with grafting materials. The use of synthetic peptides as a whole molecule or just active sequence is a more accurate approach, as it may may be missing post-translational modifications found on the native purified protein. It would be a cleaner and safer product as far as immune reactions or carrying impurities for clinical use. Recombinant molecules and synthetic peptide technologies are becoming more popular as well as more accurate, pure and have reduced variability in mediating osteogenic cell adhesion and bone formation. RGD (arginine - glycine - aspartate) is a well-characterized sequence in number of matrix proteins including fibronectin, OPN, BSP and vitronectin that mediate attachment of osteogenic cells to integrin receptors (Hsiong et al. 2009; Pallu et al. 2009). RGD will usually ligate αVβ3-integrin, but also αvβ1, α8β1, αvβ8, αvβ6, αvβ5, and αIIbβ3. RGD being synthesised as linear as well as cyclic peptide as some studies also suggest that the cyclic form may offer better presentation that is more potent in inducing osteoblastic differentiation (Hsiong et al. 2009). Collagen I adheres to bone cells via α2β1 integrin receptor (Mizuno et al. 2000) through DGEA (Asp-Gly-Glu-Ala) motif. Its recognition sequences and competition for this association with DGEA peptide could inhibit osteoblastic differentiation (Takeuchi et al. 1996). Fibronectin fragments FNIII 7-10, α5β1 integrin specific enhanced osteoblastic differentiation in bone marrow stromal cells and can upregulate adherence to titanium implants (Petrie et al. 2008). P15 is a 15-amino acid sequence derived from Collagen I, α1 chain and in clinical use (Gomar et al. 2007; Pettinicchio et al.). P15 enhances osteoblastic cell adhesion and differentiation to osteoblasts. Other peptides will be FHRRIKA, derived from the heparin binding site of BSP, human vitronectin peptide HVP (351-359) and osteopontin-derived peptides (Healy et al. 1999; von der Mark et al. 2010).

Most of these peptides and growth factors show great promise in in vitro studies and great potential in human trials and therapy (Bosetti et al. 2007; Nauth et al. 2011; Rose et al. 2004). Unfortunately, the animal and human analyses seem to exhibit wide variability (Faour et al. 2011; Giannoudis &Dinopoulos 2010; Papakostidis et al. 2008; Shekaran &Garcia 2011). An important factor in the application of these peptides and growth factors is the delivery system, as are the biochemical properties of the surrounding matrix and accessibility of the cells and the relevant receptors for their signaling.

The nature of the biomaterial, the surface to be coated or the carrying polymer, scaffold or gel will have an impact on the availability of the inducer or the ligand used to attach the differentiating bone cells. A common problem will be the hydrophobic surfaces of biomaterials, which will be covered by plasma and absorb abundant proteins such as albumin. This will make any ligand attached to the biomaterial less accessible, while more hydrophilic surfaces, such as culture dishes coated with ECM proteins, will encourage cell adhesions. Nanotechnology used to space ligands, such as RGD, affects cells adhesion, clustering and increases affinity between ligand and the receptors through both chemical and physical modifications. These approaches will enable osteoprogenitors to differentiate and migrate in the desired direction (Hirschfeld-Warneken et al. 2008). Designs aimed at creating the right topography of the biomaterial, as well as chemical alteration of serine residues or energy molecules such as purines that will change the availability of the inducer, will have impact on the ability of osteoprogenitors to differentiate (Costa et al. 2011; Mager et al. 2011; Vater et al. 2009).

7. Concluding remarks

The number of bone regeneration tools is growing every day, some but not all of which are listed above (Table 1). Unfortunately, there is no single tool available that can predictably match the gold standard of autologous marrow bone grafting. To restore the missing bone matrix, cells and inducer molecules need to act in a synergistic manner. Indeed, the new regenerative approaches are based on composite grafting, including matrix replacement, mesenchymal stem cells and inducer molecules. Most composites grafts focus on merging osteoconductive scaffolds with osteoinductive agents, such as BMP, or with cells (Bueno &Glowacki 2009; Lin et al.2010). Nanotechnology improves matrix characteristics for cell adherence, survival and differentiation, delivery vehicle for cell, proteins or gene carriers also improve macro mechanical properties (Shekaran &Garcia 2010; Smith, I. O. et al. 2009; Zhang et al. 2007). The research of forming a scaffold with organic and non-organic parts, which is mechanically strong, bioresorbable, carries inducer molecules and cells, and will adhere to the newly forming bone and still be affordable, is challenging. These are hard objectives to achieve. At present, a composite graft that can match the success of autologous marrow bone grafting does not exist.

The question is whether our quest for an ideal composite graft that will fit and regenerate most, if not all, bone wounds in every host is a realistic one. This chapter classified the three main components needed to restore missing bone tissue and outlined some of the tools and techniques(Figure 1). It is unlikely that composite grafts will be successful as autogenous grafting without having individual "custom made composite graft". We can mix autogenous marrow aspirates with the scaffold, but still most of the grafted components will not derive from the host. Host factor variables should dictate our regenerative approach for supplementing either matrix, cellular and inductive molecules at the right composition to increase our success. Bony defects are rarely uniform and healing patterns may vary, especially in human subjects. Other than local factors, host factors such as age, medications and chronic conditions may impact wound healing in general. Our future ability to design and adapt our regenerative tools may aid in boosting critical wound healing factors required in a compromised site or individual.

A different approach is suggested, in which the clinical team will be able to identify the difficulties associated with particular wounds, such as size, mechanics, blood supply and whether or not the bone is load bearing. Host factors to be considered include age, medications and other systemic conditions that may compromise wound healing. Based on these analyses of the available tools (Table 1), a list will be presented to the lab with physical, chemical and inductive requirements. An individual composite graft will be constructed for the wound that will meet and boost the particular requirements of the specific wound. With advancement of clinical diagnosis and scientific and biotechnological tools, this approach may be more predictable in achieving bone regeneration.

8. References

Alberius, P. and M. Gordh (1996). Failure of onlay bone grafts to integrate over the calvarial suture: observations in adult isogeneic rats. *J Craniomaxillofac Surg* 24(4): 251-5

Ameer, G. A., T. A. Mahmood and R. Langer (2002). A biodegradable composite scaffold for cell transplantation. *J Orthop Res* 20(1): 16-9.

Aronow, M. A., L. C. Gerstenfeld, T. A. Owen, M. S. Tassinari, G. S. Stein and J. B. Lian (1990). Factors that promote progressive development of the osteoblast phenotype in cultured fetal rat calvaria cells. *J Cell Physiol* 143(2): 213-21

Artzi, Z., Givol, N., Rohrer, M.D., Nemcovsky, C.E., Prasad, H.S. & Tal, H. (2003) Qualitative and quantitative expression of bovine bone mineral in experimental bone defects. Part 1: Description of a dog model and histological observations. Journal of Periodontology 74, 1143-1152.

Artzi, Z., Givol, N., Rohrer, M.D., Nemcovsky, C.E., Prasad, H.S. & Tal, H. (2003) Qualitative and quantitative expression of bovine bone mineral in experimental bone defects. Part 2. Morphometric analysis. Journal of Periodontology 74, 1159-1160

Au, A. G., V. James Raso, A. B. Liggins and A. Amirfazli (2007). Contribution of loading conditions and material properties to stress shielding near the tibial component of total knee replacements. *J Biomech* 40(6): 1410-6.0021-9290 (Print) 0021-9290 (Linking)

Augello, A. and C. De Bari (2010). The regulation of differentiation in mesenchymal stem cells. *Hum Gene Ther* 21(10): 1226-38.1557-7422 (Electronic) 1043-0342 (Linking)

Awad, H. A., D. L. Butler, G. P. Boivin, F. N. Smith, P. Malaviya, B. Huibregtse and A. I. Caplan (1999). Autologous mesenchymal stem cell-mediated repair of tendon. *Tissue Eng* 5(3): 267-77

Bahadur, G., M. Morrison and L. Machin (2011). Beyond the 'embryo question': human embryonic stem cell ethics in the context of biomaterial donation in the UK. *Reprod Biomed Online* 21(7): 868-74.1472-6491 (Electronic)1472-6483 (Linking)

Behnamghader, A., N. Bagheri, B. Raissi and F. Moztarzadeh (2008). Phase development and sintering behaviour of biphasic HA-TCP calcium phosphate materials prepared from hydroxyapatite and bioactive glass. *J Mater Sci Mater Med* 19(1): 197-201.0957-4530 (Print) 0957-4530 (Linking)

Belmokhtar, K., T. Bourguignon, M. E. Worou, G. Khamis, P. Bonnet, J. Domenech and V. Eder (2011). Regeneration of Three Layers Vascular Wall by using BMP2-Treated MSC Involving HIF-1alpha and Id1 Expressions Through JAK/STAT Pathways. *Stem Cell Rev*.1558-6804 (Electronic)1550-8943 (Linking)

Benayahu, D. (2000). The Hematopoietic Microenvironment: The Osteogenic Compartment of Bone Marrow: Cell Biology and Clinical Application. *Hematology* 4(5): 427-435.1607-8454 (Electronic)1024-5332 (Linking)

Bentley, S. A. and T. S. Tralka (1983). Fibronectin-mediated attachment of hematopoietic cells to stromal elements in continuous bone marrow culture. *Exp Hematol* 11(2): 129-38.0301-472X (Print)0301-472X (Linking)

Ber, R., T. Kubota, J. Sodek and J. E. Aubin (1991). Effects of transforming growth factor-beta on normal clonal bone cell populations. *Biochem Cell Biol* 69(2-3): 132-40

Bergmann, P., N. Nijs-De Wolf, T. Pepersack and J. Corvilain (1990). Release of parathyroid hormonelike peptides by fetal rat long bones in culture. *J Bone Miner Res* 5(7): 741-53

Bernardo, M. E., A. M. Cometa, D. Pagliara, L. Vinti, F. Rossi, R. Cristantielli, G. Palumbo and F. Locatelli (2011). Ex vivo expansion of mesenchymal stromal cells. *Best Pract Res Clin Haematol* 24(1): 73-81.1532-1924 (Electronic) 1521-6926 (Linking)

Binderman, I., A. Yaffe, R. Zohar, D. Benayahu and H. Bahar (2011). Tissue engineering of bone: an ectopic rat model. *Front Biosci (Schol Ed)* 3: 61-8.1945-0524 (Electronic) 1945-0516 (Linking)

Blair, H. C., M. Zaidi, C. L. Huang and L. Sun (2008). The developmental basis of skeletal cell differentiation and the molecular basis of major skeletal defects. *Biol Rev Camb Philos Soc* 83(4): 401-15.1469-185X (Electronic)0006-3231 (Linking)

Blum, B. and N. Benvenisty (2008). The tumorigenicity of human embryonic stem cells. *Adv Cancer Res* 100: 133-58.0065-230X (Print) 0065-230X (Linking)

Bosetti, M., F. Boccafoschi, M. Leigheb and M. F. Cannas (2007). Effect of different growth factors on human osteoblasts activities: a possible application in bone regeneration for tissue engineering. *Biomol Eng* 24(6): 613-8.1389-0344 (Print)1389-0344 (Linking)

Brederlau, A., A. S. Correia, S. V. Anisimov, M. Elmi, G. Paul, L. Roybon, A. Morizane, F. Bergquist, I. Riebe, U. Nannmark, M. Carta, E. Hanse, J. Takahashi, Y. Sasai, K. Funa, P. Brundin, P. S. Eriksson and J. Y. Li (2006). Transplantation of human embryonic stem cell-derived cells to a rat model of Parkinson's disease: effect of in vitro differentiation on graft survival and teratoma formation. *Stem Cells* 24(6): 1433-40.1066-5099 (Print)1066-5099 (Linking)

Bueno, E. M. and J. Glowacki (2009). Cell-free and cell-based approaches for bone regeneration. *Nat Rev Rheumatol* 5(12): 685-97.1759-4804 (Electronic)1759-4790 (Linking)

Caplan, A. I. and D. Correa (2011). PDGF in bone formation and regeneration: New insights into a novel mechanism involving MSCs. *J Orthop Res*.1554-527X (Electronic)0736-0266 (Linking)

Caplan, A. I., M. Elyaderani, Y. Mochizuki, S. Wakitani and V. M. Goldberg (1997). Principles of cartilage repair and regeneration. *Clin Orthop*(342): 254-69

Chajra, H., C. F. Rousseau, D. Cortial, M. C. Ronziere, D. Herbage, F. Mallein-Gerin and A. M. Freyria (2008). Collagen-based biomaterials and cartilage engineering. Application to osteochondral defects. *Biomed Mater Eng* 18(1 Suppl): S33-45.0959-2989 (Print)0959-2989 (Linking)

Chen, J., H. Chen, P. Li, H. Diao, S. Zhu, L. Dong, R. Wang, T. Guo, J. Zhao and J. Zhang (2011). Simultaneous regeneration of articular cartilage and subchondral bone in vivo using MSCs induced by a spatially controlled gene delivery system in bilayered integrated scaffolds. *Biomaterials* 32(21): 4793-805.1878-5905 (Electronic)0142-9612 (Linking)

Cholewa-Kowalska, K., J. Kokoszka, M. Laczka, L. Niedzwiedzki, W. Madej and A. M. Osyczka (2009). Gel-derived bioglass as a compound of hydroxyapatite composites. *Biomed Mater* 4(5): 055007.1748-605X (Electronic)1748-6041 (Linking)

Costa, M. A., A. Barbosa, E. Neto, A. Sa-e-Sousa, R. Freitas, J. M. Neves, T. Magalhaes-Cardoso, F. Ferreirinha and P. Correia-de-Sa (2011). On the role of subtype selective adenosine receptor agonists during proliferation and osteogenic differentiation of human primary bone marrow stromal cells. *J Cell Physiol* 226(5): 1353-66.1097-4652 (Electronic)0021-9541 (Linking)

Costain, D. J. and R. W. Crawford (2009). Fresh-frozen vs. irradiated allograft bone in orthopaedic reconstructive surgery. *Injury* 40(12): 1260-4.1879-0267 (Electronic)0020-1383 (Linking)

Derwin, K. A., A. R. Baker, J. P. Iannotti and J. A. McCarron (2009). Preclinical models for translating regenerative medicine therapies for rotator cuff repair. *Tissue Eng Part B Rev* 16(1): 21-30.1937-3376 (Electronic)1937-3368 (Linking)

Dimitriou, R., E. Jones, D. McGonagle and P. V. Giannoudis (2011). Bone regeneration: current concepts and future directions. *BMC Med* 9: 66.1741-7015 (Electronic)1741-7015 (Linking)

Falla, N., V. Van, J. Bierkens, B. Borremans, G. Schoeters and U. Van Gorp (1993). Characterization of a 5-fluorouracil-enriched osteoprogenitor population of the murine bone marrow. *Blood* 82(12): 3580-91

Faour, O., R. Dimitriou, C. A. Cousins and P. V. Giannoudis (2011). The use of bone graft substitutes in large cancellous voids: Any specific needs? *Injury*.1879-0267 (Electronic)0020-1383 (Linking)

Franceschi, R. T., D. Wang, P. H. Krebsbach and R. B. Rutherford (2000). Gene therapy for bone formation: in vitro and in vivo osteogenic activity of an adenovirus expressing BMP7. *J Cell Biochem* 78(3): 476-86.0730-2312 (Print)0730-2312 (Linking)

Friedrich, J. B., S. L. Moran, A. T. Bishop and A. Y. Shin (2009). Free vascularized fibula grafts for salvage of failed oncologic long bone reconstruction and pathologic fractures. *Microsurgery* 29(5): 385-92.1098-2752 (Electronic)0738-1085 (Linking)

Garrison, K. R., I. Shemilt, S. Donell, J. J. Ryder, M. Mugford, I. Harvey, F. Song and V. Alt (2011). Bone morphogenetic protein (BMP) for fracture healing in adults. *Cochrane Database Syst Rev*(6): CD006950.1469-493X (Electronic)1361-6137 (Linking)

Gaskill, T. R., J. R. Urbaniak and J. M. Aldridge, 3rd (2009). Free vascularized fibular transfer for femoral head osteonecrosis: donor and graft site morbidity. *J Bone Joint Surg Am* 91(8): 1861-7.1535-1386

Ghilzon, R., C. A. McCulloch and R. Zohar (1999). Stromal mesenchymal progenitor cells. *Leuk Lymphoma* 32(3-4): 211-21 the above report in

Giannoni, P., S. Scaglione, A. Daga, C. Ilengo, M. Cilli and R. Quarto (2009). Short-time survival and engraftment of bone marrow stromal cells in an ectopic model of bone regeneration. *Tissue Eng Part A* 16(2): 489-99.1937-335X (Electronic)1937-3341 (Linking)

Giannoudis, P. V. and H. T. Dinopoulos (2010). Autologous bone graft: when shall we add growth factors? *Orthop Clin North Am* 41(1): 85-94; table of contents.1558-1373 (Electronic)0030-5898 (Linking)

Giannoudis, P. V., N. K. Kanakaris, R. Dimitriou, I. Gill, V. Kolimarala and R. J. Montgomery (2009). The synergistic effect of autograft and BMP-7 in the treatment of atrophic nonunions. *Clin Orthop Relat Res* 467(12): 3239-48.1528-1132 (Electronic)0009-921X (Linking)

Gleeson, J. P., N. A. Plunkett and F. J. O'Brien (2011). Addition of hydroxyapatite improves stiffness, interconnectivity and osteogenic potential of a highly porous collagen-based scaffold for bone tissue regeneration. *Eur Cell Mater* 20: 218-30.1473-2262 (Electronic)1473-2262 (Linking)

Gomar, F., R. Orozco, J. L. Villar and F. Arrizabalaga (2007). P-15 small peptide bone graft substitute in the treatment of non-unions and delayed union. A pilot clinical trial. *Int Orthop* 31(1): 93-9.0341-2695 (Print)0341-2695 (Linking)

Gomes, P. S. and M. H. Fernandes (2011). Rodent models in bone-related research: the relevance of calvarial defects in the assessment of bone regeneration strategies. *Lab Anim* 45(1): 14-24.1758-1117 (Electronic)0023-6772 (Linking)

Goodwin, H. S., A. R. Bicknese, S. N. Chien, B. D. Bogucki, C. O. Quinn and D. A. Wall (2001). Multilineage differentiation activity by cells isolated from umbilical cord blood: expression of bone, fat, and neural markers. *Biol Blood Marrow Transplant* 7(11): 581-8.

Gosain, A. K., H. Chim and J. S. Arneja (2009). Application-specific selection of biomaterials for pediatric craniofacial reconstruction: developing a rational approach to guide clinical use. *Plast Reconstr Surg* 123(1): 319-30.1529-4242 (Electronic)

Grabowski, P. (2009). Physiology of bone. *Endocr Dev* 16: 32-48.1662-2979 (Electronic)1421-7082 (Linking)

Griffin, X. L., M. L. Costa, N. Parsons and N. Smith Electromagnetic field stimulation for treating delayed union or non-union of long bone fractures in adults. *Cochrane Database Syst Rev*(4): CD008471.1469-493X (Electronic)1361-6137 (Linking)

Gupta, V., G. H. Mun, B. Choi, A. Aseh, L. Mildred, A. Patel, Q. Zhang, J. E. Price, D. Chang, G. Robb and A. B. Mathur (2011). Repair and Reconstruction of a Resected Tumor Defect Using a Composite of Tissue Flap-Nanotherapeutic-Silk Fibroin and Chitosan Scaffold. *Ann Biomed Eng*.1521-6047 (Electronic) 0090-6964 (Linking)

Hak, D. J. and J. L. Pittman (2010). Biological rationale for the intramedullary canal as a source of autograft material. *Orthop Clin North Am* 41(1): 57-61; table of contents.1558-1373 (Electronic)0030-5898 (Linking)

Hamer, A. J., J. R. Strachan, M. M. Black, C. J. Ibbotson, I. Stockley and R. A. Elson (1996). Biochemical properties of cortical allograft bone using a new method of bone strength measurement. A comparison of fresh, fresh-frozen and irradiated bone. *J Bone Joint Surg Br* 78(3): 363-8.0301-620X (Print)

Healy, K. E., A. Rezania and R. A. Stile (1999). Designing biomaterials to direct biological responses. *Ann N Y Acad Sci* 875: 24-35.0077-8923 (Print)0077-8923 (Linking)

Heo, C. Y., S. Kwon, G. H. Back and M. S. Chung (2008). Complications of distraction lengthening in the hand. *J Hand Surg Eur Vol* 33(5): 609-15.1753-1934 (Print)0266-7681 (Linking)

Hirschfeld-Warneken, V. C., M. Arnold, A. Cavalcanti-Adam, M. Lopez-Garcia, H. Kessler and J. P. Spatz (2008). Cell adhesion and polarisation on molecularly defined spacing gradient surfaces of cyclic RGDfK peptide patches. *Eur J Cell Biol* 87(8-9): 743-50.0171-9335 (Print)0171-9335 (Linking)

Hoeppner, L. H., F. J. Secreto and J. J. Westendorf (2009). Wnt signaling as a therapeutic target for bone diseases. *Expert Opin Ther Targets* 13(4): 485-96.1744-7631 (Electronic)1472-8222 (Linking)

Holstein, J. H., P. Garcia, T. Histing, A. Kristen, C. Scheuer, M. D. Menger and T. Pohlemann (2009). Advances in the establishment of defined mouse models for the study of fracture healing and bone regeneration. *J Orthop Trauma* 23(5 Suppl): S31-8.1531-2291 (Electronic)0890-5339 (Linking)

Holy, C. E., M. S. Shoichet and J. E. Davies (2000). Engineering three-dimensional bone tissue in vitro using biodegradable scaffolds: investigating initial cell-seeding density and culture period. *J Biomed Mater Res* 51(3): 376-82.

Honda, M. J., M. Imaizumi, S. Tsuchiya and C. Morsczeck (2011). Dental follicle stem cells and tissue engineering. *J Oral Sci* 52(4): 541-52.1880-4926 (Electronic)1343-4934 (Linking)

Hsiong, S. X., T. Boontheekul, N. Huebsch and D. J. Mooney (2009). Cyclic arginine-glycine-aspartate peptides enhance three-dimensional stem cell osteogenic differentiation. *Tissue Eng Part A* 15(2): 263-72.1937-3341 (Print)1937-3341 (Linking)

Huang, X. and X. Miao (2007). Novel porous hydroxyapatite prepared by combining H2O2 foaming with PU sponge and modified with PLGA and bioactive glass. *J Biomater Appl* 21(4): 351-74.0885-3282 (Print)0885-3282 (Linking)

Intini, G., S. Andreana, F. E. Intini, R. J. Buhite and L. A. Bobek (2007). Calcium sulfate and platelet-rich plasma make a novel osteoinductive biomaterial for bone regeneration. *J Transl Med* 5: 13.1479-5876 (Electronic)1479-5876 (Linking)

Javed, A., H. Chen and F. Y. Ghori Genetic and transcriptional control of bone formation. *Oral Maxillofac Surg Clin North Am* 22(3): 283-93, v.1558-1365 (Electronic)1042-3699 (Linking)

Jones, G. L., G. M. McCluskey, 3rd and D. T. Curd (2000). Nonunion of the fractured clavicle: evaluation, etiology, and treatment. *J South Orthop Assoc* 9(1): 43-54.1059-1052 (Print)1059-1052 (Linking)

Jose, M. V., V. Thomas, K. T. Johnson, D. R. Dean and E. Nyairo (2009). Aligned PLGA/HA nanofibrous nanocomposite scaffolds for bone tissue engineering. *Acta Biomater* 5(1): 305-15.1878-7568 (Electronic)1742-7061 (Linking)

Kanakaris, N. K., G. M. Calori, R. Verdonk, P. Burssens, P. De Biase, R. Capanna, L. B. Vangosa, P. Cherubino, F. Baldo, J. Ristiniemi, G. Kontakis and P. V. Giannoudis (2008). Application of BMP-7 to tibial non-unions: a 3-year multicenter experience. *Injury* 39 Suppl 2: S83-90.1879-0267 (Electronic)0020-1383 (Linking)

Kawai, M., K. Bessho, H. Maruyama, J. Miyazaki and T. Yamamoto (2006). Simultaneous gene transfer of bone morphogenetic protein (BMP) -2 and BMP-7 by in vivo electroporation induces rapid bone formation and BMP-4 expression. *BMC Musculoskelet Disord* 7: 62.1471-2474 (Electronic)1471-2474 (Linking)

Kim, J. B., P. Leucht, K. Lam, C. Luppen, D. Ten Berge, R. Nusse and J. A. Helms (2007). Bone regeneration is regulated by wnt signaling. *J Bone Miner Res* 22(12): 1913-23.0884-0431 (Print)0884-0431 (Linking)

Kligman, M., A. Rotem and M. Roffman (2003). Cancellous and cortical morselized allograft in revision total hip replacement: A biomechanical study of implant stability. *J Biomech* 36(6): 797-802.0021-9290 (Print)0021-9290 (Linking)

Kogianni, G. and B. S. Noble (2007). The biology of osteocytes. *Curr Osteoporos Rep* 5(2): 81-6.1544-1873 (Print)1544-1873 (Linking)

Komori, T. (2011) Signaling networks in RUNX2-dependent bone development. *J Cell Biochem* 112(3): 750-5.1097-4644 (Electronic)0730-2312 (Linking)

Kuo, H. C., C. C. Chiu, W. C. Chang, J. M. Sheen, C. Y. Ou, R. F. Chen, T. Y. Hsu, J. C. Chang, C. C. Hsaio, F. S. Wang, C. C. Huang, H. Y. Huang and K. D. Yang (2011). Use of proteomic differential displays to assess functional discrepancies and adjustments of human bone marrow- and Wharton jelly-derived mesenchymal stem cells. *J Proteome Res* 10(3): 1305-15.1535-3907 (Electronic)1535-3893 (Linking)

Landry, P. S., A. A. Marino, K. K. Sadasivan and J. A. Albright (1996). Bone injury response. An animal model for testing theories of regulation. *Clin Orthop Relat Res*(332): 260-73.0009-921X (Print)0009-921X (Linking)

Lee, C., J. Dorcil and T. E. Radomisli (2004). Nonunion of the spine: a review. *Clin Orthop Relat Res*(419): 71-5.0009-921X (Print)0009-921X (Linking)

Lemperle, S. M., C. J. Calhoun, R. W. Curran and R. E. Holmes (1998). Bony healing of large cranial and mandibular defects protected from soft-tissue interposition: A comparative study of spontaneous bone regeneration, osteoconduction, and cancellous autografting in dogs [In Process Citation]. *Plast Reconstr Surg* 101(3): 660-72

Lerou, P. H. and G. Q. Daley (2005). Therapeutic potential of embryonic stem cells. *Blood Rev* 19(6): 321-31.0268-960X (Print)0268-960X (Linking)

Lin, Z. Y., Z. X. Duan, X. D. Guo, J. F. Li, H. W. Lu, Q. X. Zheng, D. P. Quan and S. H. Yang (2010). Bone induction by biomimetic PLGA-(PEG-ASP)n copolymer loaded with a novel synthetic BMP-2-related peptide in vitro and in vivo. *J Control Release* 144(2): 190-5.1873-4995 (Electronic)0168-3659 (Linking)

Liu, W., S. Toyosawa, T. Furuichi, N. Kanatani, C. Yoshida, Y. Liu, M. Himeno, S. Narai, A. Yamaguchi and T. Komori (2001). Overexpression of Cbfa1 in osteoblasts inhibits osteoblast maturation and causes osteopenia with multiple fractures. *J Cell Biol* 155(1): 157-66.

Liu, X., W. Huang, H. Fu, A. Yao, D. Wang, H. Pan, W. W. Lu, X. Jiang and X. Zhang (2009). Bioactive borosilicate glass scaffolds: in vitro degradation and bioactivity behaviors. *J Mater Sci Mater Med* 20(6): 1237-43.1573-4838 (Electronic)0957-4530 (Linking)

Lynch, M. P., J. L. Stein, G. S. Stein and J. B. Lian (1995). The influence of type I collagen on the development and maintenance of the osteoblast phenotype in primary and passaged rat calvarial osteoblasts: modification of expression of genes supporting cell growth, adhesion, and extracellular matrix mineralization. *Exp Cell Res* 216(1): 35-45

Lynch, S. E., L. Wisner-Lynch, M. Nevins and M. L. Nevins (2006). A new era in periodontal and periimplant regeneration: use of growth-factor enhanced matrices incorporating rhPDGF. *Compend Contin Educ Dent* 27(12): 672-8; quiz 679-80.1548-8578 (Print)

Mager, M. D., V. Lapointe and M. M. Stevens (2011). Exploring and exploiting chemistry at the cell surface. *Nat Chem* 3(8): 582-9.1755-4349 (Electronic)1755-4330 (Linking)

Masaki, H. and H. Ide (2007). Regeneration potency of mouse limbs. *Dev Growth Differ* 49(2): 89-98.0012-1592 (Print)0012-1592 (Linking)

Mashru, R. P., M. J. Herman and P. D. Pizzutillo (2005). Tibial shaft fractures in children and adolescents. *J Am Acad Orthop Surg* 13(5): 345-52.1067-151X (Print)1067-151X (Linking)

Mason, J. M., D. A. Grande, M. Barcia, R. Grant, R. G. Pergolizzi and A. S. Breitbart (1998). Expression of human bone morphogenic protein 7 in primary rabbit periosteal cells: potential utility in gene therapy for osteochondral repair. *Gene Ther* 5(8): 1098-104 the above report in

Matejovsky, Z., Jr., Z. Matejovsky and I. Kofranek (2006). Massive allografts in tumour surgery. *Int Orthop* 30(6): 478-83.0341-2695 (Print)0341-2695 (Linking)

Meinig, R. P. (2002). Polylactide membranes in the treatment of segmental diaphyseal defects: animal model experiments in the rabbit radius, sheep tibia, Yucatan minipig radius, and goat tibia. *Injury* 33 Suppl 2: B58-65.0020-1383 (Print)0020-1383 (Linking)

Miller, P. D. (2009). Denosumab: anti-RANKL antibody. *Curr Osteoporos Rep* 7(1): 18-22.1544-2241 (Electronic)1544-1873 (Linking)

Mishra, R., B. Basu and A. Kumar (2009). Physical and cytocompatibility properties of bioactive glass-polyvinyl alcohol-sodium alginate biocomposite foams prepared via sol-gel processing for trabecular bone regeneration. *J Mater Sci Mater Med* 20(12): 2493-500.1573-4838 (Electronic)0957-4530 (Linking)

Mizuno, M., R. Fujisawa and Y. Kuboki (2000). Type I collagen-induced osteoblastic differentiation of bone-marrow cells mediated by collagen-alpha2beta1 integrin interaction. *J Cell Physiol* 184(2): 207-13.0021-9541 (Print)0021-9541 (Linking)

Mroz, T. E., E. L. Lin, M. C. Summit, J. R. Bianchi, J. E. Keesling, Jr., M. Roberts, C. T. Vangsness, Jr. and J. C. Wang (2006). Biomechanical analysis of allograft bone treated with a novel tissue sterilization process. *Spine J* 6(1): 34-9.1529-9430 (Print)1529-9430 (Linking)

Mulconrey, D. S., K. H. Bridwell, J. Flynn, G. A. Cronen and P. S. Rose (2008). Bone morphogenetic protein (RhBMP-2) as a substitute for iliac crest bone graft in multilevel adult spinal deformity surgery: minimum two-year evaluation of fusion. *Spine (Phila Pa 1976)* 33(20): 2153-9.1528-1159 (Electronic)0362-2436 (Linking)

Muschler, G. F., S. Negami, A. Hyodo, D. Gaisser, K. Easley and H. Kambic (1996). Evaluation of collagen ceramic composite graft materials in a spinal fusion model. *Clin Orthop Relat Res*(328): 250-60.0009-921X (Print)0009-921X (Linking)

Nagata, T., H. A. Goldberg, Q. Zhang, C. Domenicucci and J. Sodek (1991). Biosynthesis of bone proteins by fetal porcine calvariae in vitro. Rapid association of sulfated sialoproteins (secreted phosphoprotein-1 and bone sialoprotein) and chondroitin sulfate proteoglycan (CS-PGIII) with bone mineral. *Matrix* 11(2): 86-100

Nauth, A., B. Ristevski, R. Li and E. H. Schemitsch (2011). Growth factors and bone regeneration: how much bone can we expect? *Injury* 42(6): 574-9.1879-0267 (Electronic)0020-1383 (Linking)

Nguyen, H., D. A. Morgan and M. R. Forwood (2007). Sterilization of allograft bone: effects of gamma irradiation on allograft biology and biomechanics. *Cell Tissue Bank* 8(2): 93-105.1389-9333 (Print)1389-9333 (Linking)

Nie, H., B. W. Soh, Y. C. Fu and C. H. Wang (2008). Three-dimensional fibrous PLGA/HAp composite scaffold for BMP-2 delivery. *Biotechnol Bioeng* 99(1): 223-34.1097-0290 (Electronic)0006-3592 (Linking)

Notoya, M., E. Otsuka, A. Yamaguchi and H. Hagiwara (2004). Runx-2 is not essential for the vitamin D-regulated expression of RANKL and osteoprotegerin in osteoblastic cells. *Biochem Biophys Res Commun* 324(2): 655-60.0006-291X (Print)0006-291X (Linking)

Osti, M., H. Philipp, B. Meusburger and K. P. Benedetto Analysis of failure following anterior screw fixation of Type II odontoid fractures in geriatric patients. *Eur Spine J*.1432-0932 (Electronic)0940-6719 (Linking)

Owen, M. (1988). Marrow stromal stem cells. *J Cell Sci Suppl* 10: 63-76

Pallu, S., J. C. Fricain, R. Bareille, C. Bourget, M. Dard, A. Sewing and J. Amedee (2009). Cyclo-DfKRG peptide modulates in vitro and in vivo behavior of human osteoprogenitor cells on titanium alloys. *Acta Biomater* 5(9): 3581-92.1878-7568 (Electronic)1742-7061 (Linking)

Papakostidis, C., G. Kontakis, M. Bhandari and P. V. Giannoudis (2008). Efficacy of autologous iliac crest bone graft and bone morphogenetic proteins for posterolateral fusion of lumbar spine: a meta-analysis of the results. *Spine (Phila Pa 1976)* 33(19): E680-92.1528-1159 (Electronic)0362-2436 (Linking)

Parker, M. J., J. Kendrew and K. Gurusamy (2011). Radiological predictive factors in the healing of displaced intracapsular hip fractures. A clinical study of 404 cases. *Hip Int* 21(4): 393-8.1724-6067 (Electronic)1120-7000 (Linking)

Petrie, T. A., J. E. Raynor, C. D. Reyes, K. L. Burns, D. M. Collard and A. J. Garcia (2008). The effect of integrin-specific bioactive coatings on tissue healing and implant osseointegration. *Biomaterials* 29(19): 2849-57.0142-9612 (Print)0142-9612 (Linking)

Pettinicchio, M., T. Traini, G. Murmura, S. Caputi, M. Degidi, C. Mangano and A. Piattelli Histologic and histomorphometric results of three bone graft substitutes after sinus augmentation in humans. *Clin Oral Investig*.1436-3771 (Electronic)1432-6981 (Linking)

Pittenger, M. F., A. M. Mackay, S. C. Beck, R. K. Jaiswal, R. Douglas, J. D. Mosca, M. A. Moorman, D. W. Simonetti, S. Craig and D. R. Marshak (1999). Multilineage potential of adult human mesenchymal stem cells. *Science* 284(5411): 143-7 the above report in

Popov, C., T. Radic, F. Haasters, W. C. Prall, A. Aszodi, D. Gullberg, M. Schieker and D. Docheva (2011). Integrins alpha2beta1 and alpha11beta1 regulate the survival of mesenchymal stem cells on collagen I. *Cell Death Dis* 2: e186.2041-4889 (Electronic)

Post, S., B. M. Abdallah, J. F. Bentzon and M. Kassem (2008). Demonstration of the presence of independent pre-osteoblastic and pre-adipocytic cell populations in bone marrow-derived mesenchymal stem cells. *Bone* 43(1): 32-9.8756-3282 (Print)1873-2763 (Linking)

Reddi, A. H. and N. S. Cunningham (1993). Initiation and promotion of bone differentiation by bone morphogenetic proteins. *J Bone Miner Res* 8(Suppl 2): S499-502

Renghini, C., V. Komlev, F. Fiori, E. Verne, F. Baino and C. Vitale-Brovarone (2009). Micro-CT studies on 3-D bioactive glass-ceramic scaffolds for bone regeneration. *Acta Biomater* 5(4): 1328-37.1878-7568 (Electronic)1742-7061 (Linking)

Rhee, S. C., Y. H. Ji, N. A. Gharibjanian, E. S. Dhong, S. H. Park and E. S. Yoon (2010). In vivo evaluation of mixtures of uncultured freshly isolated adipose-derived stem cells and demineralized bone matrix for bone regeneration in a rat critically sized calvarial defect model. *Stem Cells Dev* 20(2): 233-42.1557-8534 (Electronic)1547-3287 (Linking)

Rose, F. R., Q. Hou and R. O. Oreffo (2004). Delivery systems for bone growth factors - the new players in skeletal regeneration. *J Pharm Pharmacol* 56(4): 415-27.0022-3573 (Print)0022-3573 (Linking)

Sampath, T. K., J. C. Maliakal, P. V. Hauschka, W. K. Jones, H. Sasak, R. F. Tucker, K. H. White, J. E. Coughlin, M. M. Tucker, R. H. Pang, C. Corbett, E. Ozkaynak, H. Oppermann and D. C. Rueger (1992). Recombinant human osteogenic protein-1 (hOP-1) induces new bone formation in vivo with a specific activity comparable with natural bovine osteogenic protein and stimulates osteoblast proliferation and differentiation in vitro. *J Biol Chem* 267(28): 20352-62

Schofer, M. D., A. Veltum, C. Theisen, F. Chen, S. Agarwal, S. Fuchs-Winkelmann and J. R. Paletta (2011). Functionalisation of PLLA nanofiber scaffolds using a possible cooperative effect between collagen type I and BMP-2: impact on growth and osteogenic differentiation of human mesenchymal stem cells. *J Mater Sci Mater Med* 22(7): 1753-62.1573-4838 (Electronic)0957-4530 (Linking)

Scott, A., K. M. Khan, V. Duronio and D. A. Hart (2008). Mechanotransduction in human bone: in vitro cellular physiology that underpins bone changes with exercise. *Sports Med* 38(2): 139-60.0112-1642 (Print)0112-1642 (Linking)

Secreto, F. J., L. H. Hoeppner and J. J. Westendorf (2009). Wnt signaling during fracture repair. *Curr Osteoporos Rep* 7(2): 64-9.1544-2241 (Electronic)1544-1873 (Linking)

Sharma, A. K., M. I. Bury, A. J. Marks, N. J. Fuller, J. W. Meisner, N. Tapaskar, L. C. Halliday, D. J. Matoka and E. Y. Cheng (2011). A nonhuman primate model for urinary bladder regeneration using autologous sources of bone marrow-derived mesenchymal stem cells. *Stem Cells* 29(2): 241-50.1549-4918 (Electronic)1066-5099 (Linking)

Shekaran, A. and A. J. Garcia (2010). Nanoscale engineering of extracellular matrix-mimetic bioadhesive surfaces and implants for tissue engineering. *Biochim Biophys Acta* 1810(3): 350-60.0006-3002 (Print)0006-3002 (Linking)

Shekaran, A. and A. J. Garcia (2011). Extracellular matrix-mimetic adhesive biomaterials for bone repair. *J Biomed Mater Res A* 96(1): 261-72.1552-4965 (Electronic)1549-3296 (Linking)

Smit, T. H., T. A. Engels, S. H. Sontjens and L. E. Govaert (2010). Time-dependent failure in load-bearing polymers: a potential hazard in structural applications of polylactides. *J Mater Sci Mater Med* 21(3): 871-8.1573-4838 (Electronic)0957-4530 (Linking)

Smith, I. O., X. H. Liu, L. A. Smith and P. X. Ma (2009). Nanostructured polymer scaffolds for tissue engineering and regenerative medicine. *Wiley Interdiscip Rev Nanomed Nanobiotechnol* 1(2): 226-36.1939-0041 (Electronic)1939-0041 (Linking)

Smith, T. O., C. Hedges, R. MacNair, K. Schankat and J. A. Wimhurst (2009). The clinical and radiological outcomes of the LISS plate for distal femoral fractures: a systematic review. *Injury* 40(10): 1049-63.1879-0267 (Electronic)0020-1383 (Linking)

Sotereanos, D. G., A. Y. Plakseychuk and H. E. Rubash (1997). Free vascularized fibula grafting for the treatment of osteonecrosis of the femoral head. *Clin Orthop Relat Res*(344): 243-56.0009-921X (Print)0009-921X (Linking)

Stender, S., M. Murphy, T. O'Brien, C. Stengaard, M. Ulrich-Vinther, K. Soballe and F. Barry (2007). Adeno-associated viral vector transduction of human mesenchymal stem cells. *Eur Cell Mater* 13: 93-9; discussion 99.1473-2262 (Electronic)1473-2262 (Linking)

Stolzing, A., E. Jones, D. McGonagle and A. Scutt (2008). Age-related changes in human bone marrow-derived mesenchymal stem cells: consequences for cell therapies. *Mech Ageing Dev* 129(3): 163-73.0047-6374 (Print)0047-6374 (Linking)

Takahashi, N., N. Udagawa and T. Suda (1999). A new member of tumor necrosis factor ligand family, ODF/OPGL/TRANCE/RANKL, regulates osteoclast differentiation and function. *Biochem Biophys Res Commun* 256(3): 449-55

Takeuchi, Y., K. Nakayama and T. Matsumoto (1996). Differentiation and cell surface expression of transforming growth factor-beta receptors are regulated by interaction with matrix collagen in murine osteoblastic cells. *J Biol Chem* 271(7): 3938-44.0021-9258 (Print)0021-9258 (Linking)

Thula, T. T., D. E. Rodriguez, M. H. Lee, L. Pendi, J. Podschun and L. B. Gower (2011). In vitro mineralization of dense collagen substrates: a biomimetic approach toward the development of bone-graft materials. *Acta Biomater* 7(8): 3158-69.1878-7568 (Electronic)1742-7061 (Linking)

Tiedeman, J. J., K. L. Garvin, T. A. Kile and J. F. Connolly (1995). The role of a composite, demineralized bone matrix and bone marrow in the treatment of osseous defects. *Orthopedics* 18(12): 1153-8

Urist, M. R. (1965). Bone: formation by autoinduction. *Science* 150(698): 893-9.0036-8075 (Print)0036-8075 (Linking)

Urist, M. R., B. F. Silverman, K. Buring, F. L. Dubuc and J. M. Rosenberg (1967). The bone induction principle. *Clin Orthop Relat Res* 53: 243-83.0009-921X (Print)0009-921X (Linking)

Valimaki, V. V. and H. T. Aro (2006). Molecular basis for action of bioactive glasses as bone graft substitute. *Scand J Surg* 95(2): 95-102.1457-4969 (Print)1457-4969 (Linking)

Vater, C., A. Lode, A. Bernhardt, A. Reinstorf, C. Heinemann and M. Gelinsky (2009). Influence of different modifications of a calcium phosphate bone cement on adhesion, proliferation, and osteogenic differentiation of human bone marrow stromal cells. *J Biomed Mater Res A* 92(4): 1452-60.1552-4965 (Electronic)1549-3296 (Linking)

von der Mark, K., J. Park, S. Bauer and P. Schmuki (2010). Nanoscale engineering of biomimetic surfaces: cues from the extracellular matrix. *Cell Tissue Res* 339(1): 131-53.1432-0878 (Electronic)0302-766X (Linking)

Watt, F. M. and B. L. Hogan (2000). Out of Eden: stem cells and their niches. *Science* 287(5457): 1427-30

Whiteford, J. R., V. Behrends, H. Kirby, M. Kusche-Gullberg, T. Muramatsu and J. R. Couchman (2007). Syndecans promote integrin-mediated adhesion of mesenchymal cells in two distinct pathways. *Exp Cell Res* 313(18): 3902-13.0014-4827 (Print)0014-4827 (Linking)

Wojtowicz, A. M., A. Shekaran, M. E. Oest, K. M. Dupont, K. L. Templeman, D. W. Hutmacher, R. E. Guldberg and A. J. Garcia (2010). Coating of biomaterial scaffolds with the collagen-mimetic peptide GFOGER for bone defect repair. *Biomaterials* 31(9): 2574-82.1878-5905 (Electronic)0142-9612 (Linking)

Wu, B., Q. Zheng, X. Guo, Y. Wu, Y. Wang and F. Cui (2008). Preparation and ectopic osteogenesis in vivo of scaffold based on mineralized recombinant human-like collagen loaded with synthetic BMP-2-derived peptide. *Biomed Mater* 3(4): 044111.1748-605X (Electronic)1748-6041 (Linking)

Xiao, Z. S., A. B. Hjelmeland and L. D. Quarles (2004). Selective deficiency of the "bone-related" Runx2-II unexpectedly preserves osteoblast-mediated skeletogenesis. *J Biol Chem* 279(19): 20307-13.0021-9258 (Print)0021-9258 (Linking)

Yamada, Y., K. Hozumi, F. Katagiri, Y. Kikkawa and M. Nomizu (2010). Biological activity of laminin peptide-conjugated alginate and chitosan matrices. *Biopolymers* 94(6): 711-20.0006-3525 (Print)0006-3525 (Linking)

Yamamoto, N., H. Akamatsu, S. Hasegawa, T. Yamada, S. Nakata, M. Ohkuma, E. Miyachi, T. Marunouchi and K. Matsunaga (2007). Isolation of multipotent stem cells from mouse adipose tissue. *J Dermatol Sci* 48(1): 43-52.0923-1811 (Print)0923-1811 (Linking)

Zhang, Y., J. Song, B. Shi, Y. Wang, X. Chen, C. Huang, X. Yang, D. Xu and X. Cheng (2007). Combination of scaffold and adenovirus vectors expressing bone morphogenetic protein-7 for alveolar bone regeneration at dental implant defects. *Biomaterials* 28(31): 4635-42.0142-9612 (Print)0142-9612 (Linking)

Zhou, S., J. S. Greenberger, M. W. Epperly, J. P. Goff, C. Adler, M. S. Leboff and J. Glowacki (2008). Age-related intrinsic changes in human bone-marrow-derived mesenchymal stem cells and their differentiation to osteoblasts. *Aging Cell* 7(3): 335-43.1474-9726 (Electronic)1474-9718 (Linking)

Zohar, R., W. Lee, P. Arora, S. Cheifetz, C. McCulloch and J. Sodek (1997a). Single cell analysis of intracellular osteopontin in osteogenic cultures of fetal rat calvarial cells. *J Cell Physiol* 170(1): 88-100

Zohar, R., C. A. McCulloch, K. Sampath and J. Sodek (1998). Flow cytometric analysis of recombinant human osteogenic protein-1 (BMP-7) responsive subpopulations from fetal rat calvaria based on intracellular osteopontin content. *Matrix Biol* 16(6): 295-306.

Zohar, R., J. Sodek and C. McCulloch (1997b). Characterization of Stromal Progenitor cells enriched by flow cytometry. *Blood* 90(9): 3471-81

Zohar, R. and H. C. Tenenbaum (2005). How predictable are periodontal regenerative procedures? *J Can Dent Assoc* 71(9): 675-80

4

Tissue Engineering of Bone: Critical Evaluation of Scaffold Selection

Itzhak Binderman[1], Avinoam Yaffe[2], Yuval Samuni[3], Hila Bahar[1], Joseph Choukroun[4] and Philippe Russe[5]
[1]*Department of Oral Biology, Maurice and Gabriela Goldschleger, School of Dental Medicine, Sackler Faculty of Medicine, Tel Aviv University,*
[2]*Department of Prosthodontics, Haddasah School of Dental Medicine, Hebrew University of Jerusalem,*
[3]*Department of Oral and Maxillofacial Surgery, Barzilai Medical Center, Ashkelon,*
[4]*Syfac, Nice,*
[5]*Private Practice, Reims,*
[1,2,3]*Israel*
[4,5]*France*

1. Introduction

The first report on tissue engineering (TE) dates back to the book of Genesis where "...the Lord God sent a deep sleep on the man, and took one of the bones from his side while he was sleeping, joining up the flesh against its place" (Genesis). Interestingly and perhaps inadvertently, the importance of bone as a scaffold in the process of tissue engineering was acknowledged even in the scriptures. The ultimate goal of TE is to regenerate and replace structural and functional deficits of tissue, beyond its natural healing capacity. For that purpose, external regenerative resources including scaffolds, cells and growth/trophic factors (GF) either alone or in combination are employed (Place et al., 2009; Tanner, 2010;, Rokn et al., 2011). The general strategy of TE uses undifferentiated cells seeded within a scaffold which defines the geometry of the replacement tissues, and provides environmental cues to promote the development of new tissues (Zuk, 2008;,Place et al., 2009; Binderman et al., 2011.). It is now well understood that the cell-scaffold interaction is a crucial part of TE and should mimic the interaction between cell surface receptors and the extracellular matrix (ECM). The ECM, composed of various macromolecules such as proteoglycans, collagens, laminins, fibronectins and sequestered growth factors, is responsible for regulating cellular functions including survival, adhesion, proliferation, migration, differentiation, and matrix deposition (Binderman et al., 2011). Furthermore, it is now widely accepted that bone matrix and its cellular environment constitute one of the best known niche of adult stem cells both for hematopoiesis and mesenchymal tissues (Ferrer, et al., 2010). Given the complexity of living tissue, current approach for TE does not support attempts to recreate tissue *ex vivo*. Instead, one should develop synthetic materials that will establish key interactions with cells and unlock the body's innate powers of organization and self-repair (Place et al., 2009;

Binderman et al., 2011). This principal of using the body as a "bioreactor" guides our development strategy for TE of bone (Stevens, et al., 2005).

Autografts and allogenic grafts are routinely used in the clinic. Nevertheless, the morbidity associated with harvesting of autografts and their limited availability, and the inferior mechanical properties of allogenic grafts have spurred the search for the optimal artificial bone substitute material. Numerous artificial bone grafts are commercially available. These include macroporous bioactive ceramic granules made from calcium sulphate, tricalcium phosphate (ß-TCP), synthetic hydroxyapatite (HA), and biphasic calcium phosphate (a mixture of TCP and HA) (Hing, 2004; Jones et al., 2010). Although widely used, calcium sulphate and to a lesser extent TCP, dissolve very rapidly, often resulting in a new defect. HA on the other hand degrades very slowly, thus impeding the apposition of new bone. In fact, osteoprogenitor cells in conjunction with scaffold and osteogenic factors were used to create bone tissue both in vitro and in vivo (Binderman,et al., 2011). These engineered bone grafts have been shown to posses the capacity for osteogenesis, but also for osteoconduction and even bioactivity. Ideally, the engineered bone should form a structural and functional connection with the host bone, also termed as physical connectivity. Unfortunately, vascularization of engineered bone tissue remains a major obstacle in achieving a clinically sized bone grafts.

While the physical and chemical requirements for scaffold composition and design for TE of bone are well defined, our ability to produce them is still limited. Scaffold should be manufactured from bioactive material that allows attachment of cells to its surface and their transformation to functional osteoblasts. Its design should contain macropores, 200-500 μm across, to allow in-growth of bone tissue and blood vessels, and apposition of mineralized bone matrix directly on the surface of the material (Hing et al.,2004; Zuk, 2008). On the other hand, biocompatible scaffold are less desirable since they allow formation of bone arbitrarily. Additionally, scaffold should be constructed from degradable material thus enabling the newly formed tissue to gradually replace it. Finally, scaffolds should maintain their mechanical stability and allow loading of the newly formed composite tissue. To the best of our knowledge, as of yet no such material has been reported to have all these characteristics.

This paper focuses on a specific rodent model which provides a remarkable tool for the study of TE of bone in a non-bone ectopic site. A comparative analysis of commercially available scaffolds is presented. To complement the analysis, clinical biopsies of grafted sinuses are also shown.

2. Animal studies

Dark Agouti (DA) inbred rats were used to study the bioactive properties of four bone graft materials (BGM), ranging from weak biocompatability to high bioactivity, namely (a) Cerabone, inorganic bovine bone treated, manufactured by aap Implantate AG, Dieburg, Germany, (b) Bio-Oss, mineral of bovine bone, manufactured by Geistlich Pharma AG, Wolhusen, Germany, (c) NanoBone, synthetic silicium rich hydroxyapatite, manufactured by Artoss GMBH, Rostock,Germany, and (d) ReproBone, synthetic tricalcium phosphate and hydroxyapatite (40:60, %) mineral, manufactured by Ceramisys, Sheffield, England; all in particulate form. Our model is highly reproducible and has two major advantages: (a) bone is formed under the osteogenic environment of bone marrow and BGM, in a subcutaneos site; this way it excludes the direct effects of host bone tissue surroundings, and (b) new bone is formed within 3 weeks after implantation (Yaffe et al.,2003; Bahar et al., 2003; Bahar et al., 2010; Binderman et al., 2011). In brief, the BGM is mixed with freshly harvested femural marrow (3:1, v/v) of 6-8 weeks old DA rats. The mixture is immediately implanted into a subcutaneous space prepared by blunt dissection in the thoracic region of other native DA rat (Figure 1). Three weeks later, animals were euthanized and the subcutaneous implant was harvested for microradiography and histology.

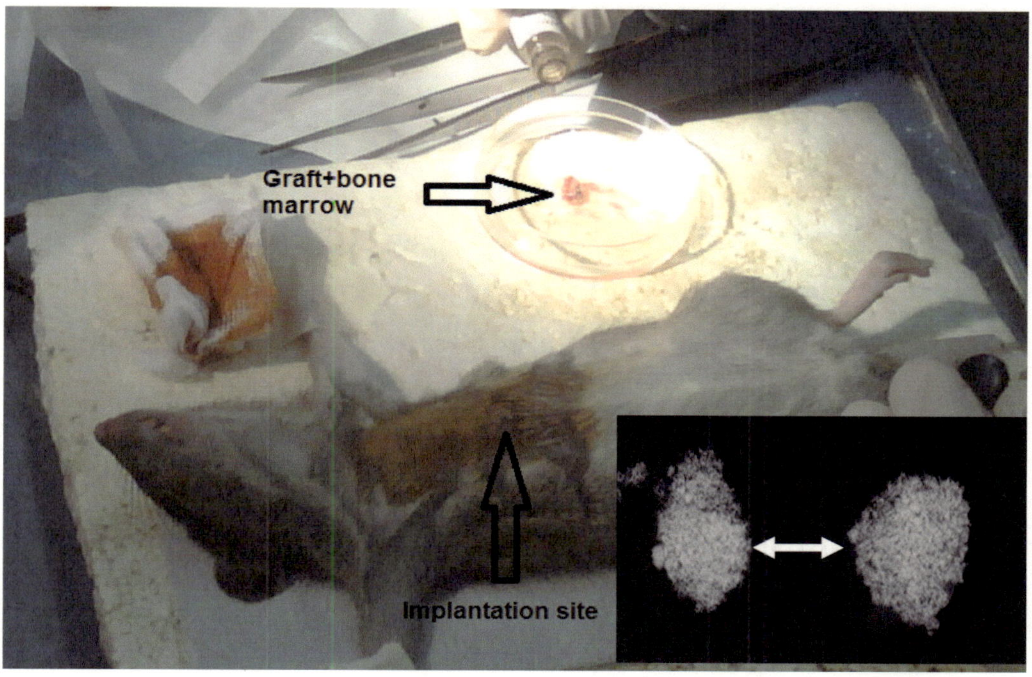

Fig. 1. Surgical implantation of BGM mixed with fresh marrow at the thoracic site of DA rats. Three weeks later the BGM implant was removed for microradiography (white arrows, x3).

Our histological evaluation included the following aspects: (i) the ability of the BGM to recruit osteogenic cells onto its surface, triggering bone deposition directly on the BGM structure, (ii) the ability of new bone to allow ingrowth of blood vessels and formation of new marrow, (iii) the recruitment of osteoclasts to resorb BGM. In this manner, a comparative analysis of commercially available BGM's was performed. Previously, we have shown histologically that fresh marrow interacting with demineralized bone matrix (DBM) of DA rats produced an ossicle consisting of a thin cortical bone surrounding numerous trabecullii which occupied new active marrow tissue (Yaffe et al.,2003; Bahar et al., 2003; Bahar et al., 2010; Binderman, 2011). Here, we compared the interaction of BGM's with fresh marrow that leads to osteogenesis in the thoracic subcutaneous site of DA rats. Moreover, we evaluated the reaction between the same BGM's in the osteogenic environment of the human maxillary sinus, on growth and deposition of bone.

In the present study, microradiography of all implanted BGM's revealed a similar view of composite including BGM material and mineralized new bone (Fig. 1). In contrast, histological analysis demonstrated differences between the biocompatible and the bioactive BGMs as demonstrated in the following figures. Deposition of cell-rich new bone can be seen in close proximity to the particles of processed bovine bone matrix of both Cerabone (Figs. 2a and 2b) and Bio-Oss (Figs. 2c and 2d).

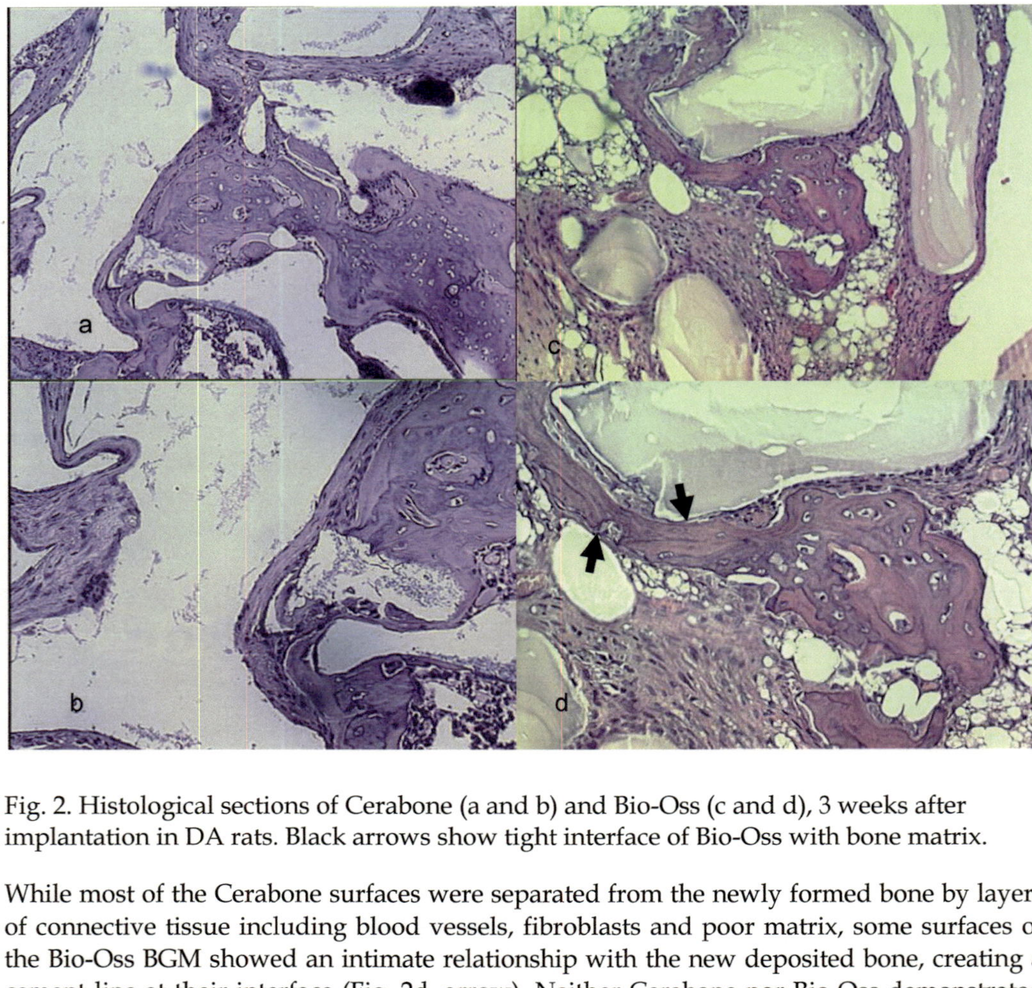

Fig. 2. Histological sections of Cerabone (a and b) and Bio-Oss (c and d), 3 weeks after implantation in DA rats. Black arrows show tight interface of Bio-Oss with bone matrix.

While most of the Cerabone surfaces were separated from the newly formed bone by layers of connective tissue including blood vessels, fibroblasts and poor matrix, some surfaces of the Bio-Oss BGM showed an intimate relationship with the new deposited bone, creating a cement line at their interface (Fig. 2d, arrow). Neither Cerabone nor Bio-Oss demonstrated active bone marrow or osteoclasts, suggesting poor resorptive properties. Although both Bio-Oss and Cerabone are composed of the mineral portion of bovine bone (no collagen was expected to be present) we found residual collagen in demineralized histological sections of Bio-Oss but not of Cerabone. The possibility exists that bone mineral may protect the organic material during the process of Bio-Oss preparation. Whether the improved biocompatability of Bio-Oss in comparison to that of Cerabone, could be attributed to the presence of bone matrix should be further investigated (Rokn et al.,2011). Nevertheless, these findings suggest that Bio-Oss and Cerabone are biocompatible rather than bioactive and that they have poor resorbable qualities.

NanoBone which is composed of hydroxyapatite enriched by silicium (24%) is considered to be bioactive (Gotz et al., 2008, Jones, et al.,2010). In our DA rat system, deposition of bone was seen in intimate association and engulfing the surfaces of the Nanobone particles, indicating high bioactivity (Fig. 3).

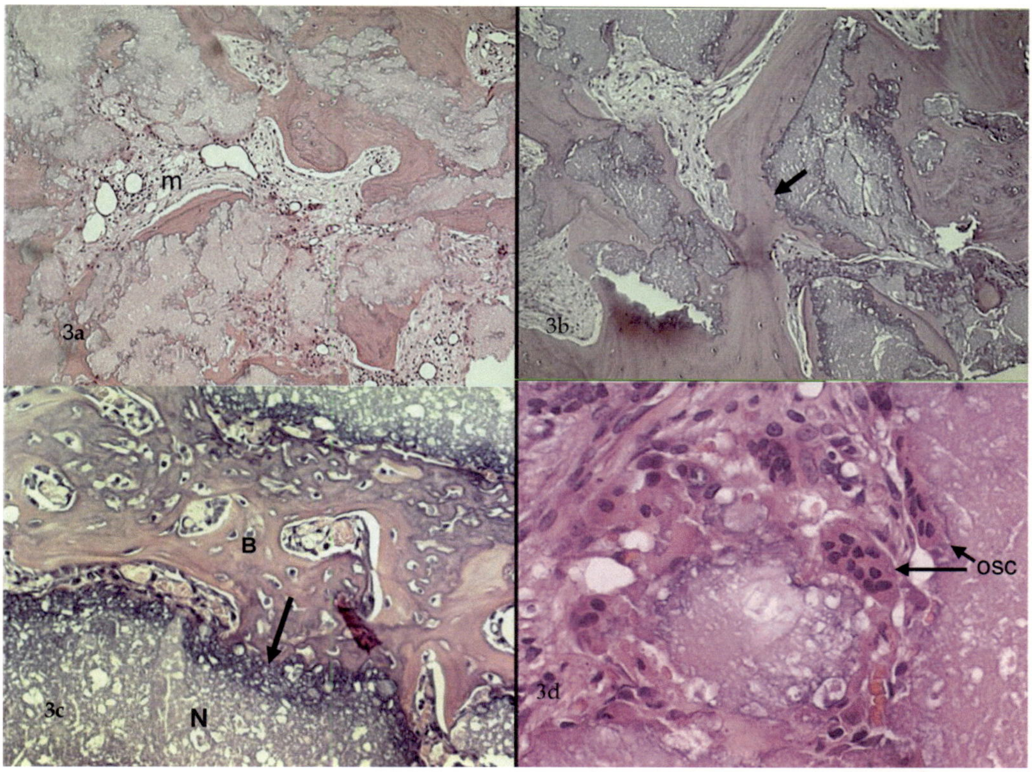

Fig. 3. Histological sections of NanoBone BGM 3 weeks after implantation. 3a shows connectivity of interfaces between bone and NanoBone; m=marrow spaces, including vessels and cells. 3b and 3c shows the bone interfaces with NanoBone by cement line (Black arrows). B= bone, N= NanoBone BGM,. 3d shows several osteoclasts (OSC) , indicated by arrows on NanoBone surfaces.

Active bone marrow and many blood vessels surrounded by new bone were also seen in these histological sections. A structure of bone-BGM-bone continuity and tight connectivity of mineralized matrices that occured can provide an optimal BGM for implant anchorage and function. In the demineralized histological sections an organic residue basic material was present where the NanoBone particles reside. Because NanoBone is strictly mineral, it seems that the organic material is composed of blood proteins that are absorbed by this BGM. It was already suggested that blood proteins are absorbed mainly by silicium, thus allowing attachment of osteognic cells onto them (Jones et al.,2010). Surprisingly, in the case of ReproBone that is composed of 40% beta-tricalcium phosphate (ß-TCP) and 60% hydroxyapatite (HA) bone was deposited directly on its surfaces, in a similar fashion observed in NanoBone bioactive BGM.

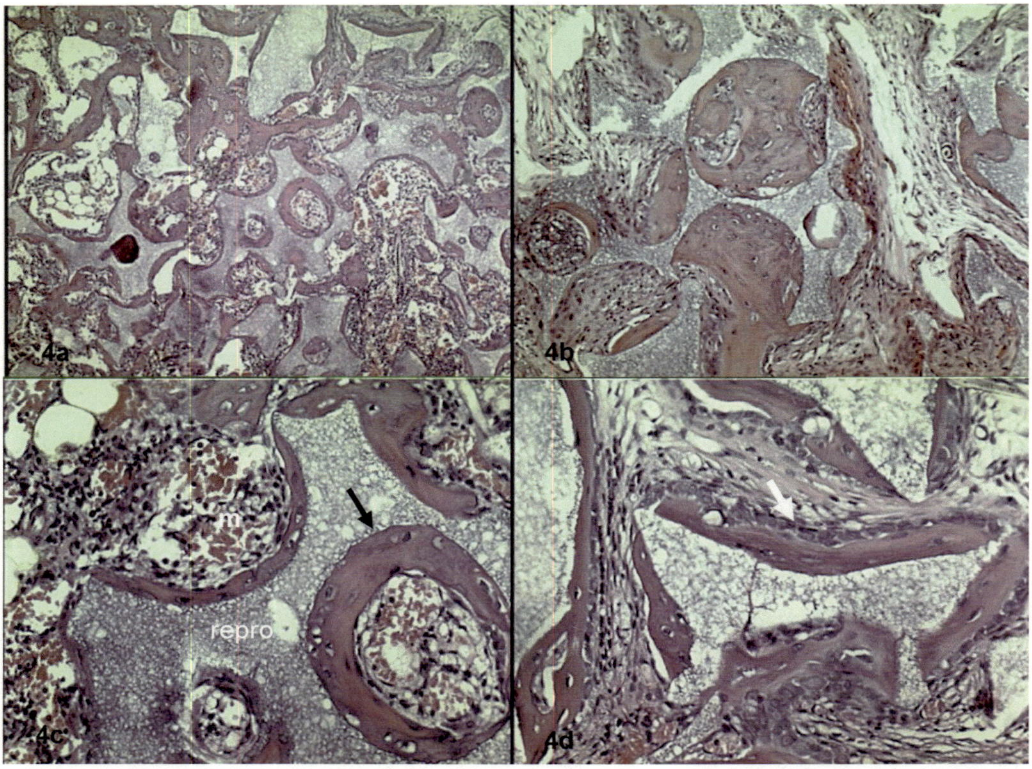

Fig. 4. Histological sections of ReproBone, 3 weeks after implantation in DA rats. 4a and 4b, show connectivity of bone and ReproBone througout the section. Newly formed bone is filling many of the macropores. 4c emphasizes the bone ReproBone interface (arrow), the highly active marrow and blood vessls, m=marrow, repro=ReproBone residual organic material. 4d shows very active osteoblasts linning new bone interfacing the BGM (white arrow).

Our observation that organic basic residue is seen in demineralized histological sections of Reprobone, similar to that seen in NanoBone sections, may suggest that also here glycoproteins from the blood are strongly absorbed throughout the Reprobone material. Furthermore, Reprobone but not NanoBone allowed blood vessels ingrowth into macropores, and new bone-surrounded marrow cells to be deposited on the pores (Fig. 4, arrow). This ingrowth into macropores is reminiscent of Howship lacunae in normal bone, and was not seen in other BGM's tested in this study. The presence of osteoclasts on the surfaces of the NanoBone (Figure 3d, osc) and Reprobone indicate active remodeling and therefore high degree of bioactivity. These results support the use of Reprobone since it fulfills the criteria for an excellent BGM, namely, highly bioactive, allows blood vessels and bone ingrowth into its pores, contains active marrow and undergoes active remodeling (Figure 4).

3. Clinical studies

To complement the animal data we present histological sections of BGM-grafted sinus biopsies taken from patients who underwent sinus lift procedures. All surgical procedures and biopsies were performed by Dr. Philipe Russe. In brief, under local anesthesia full thickness flap was elevated to access the anterior wall of the maxillary sinus. Using piezosugery (piezosurge III by Mectron) a bony lid was detached and the Schneiderian membrane was elevated. BGM was then placed into the sinus and rehydrated with the plated-rich-fibrin (PRF) exsudate and metronidazole (Fig. 5a). The bony window was placed back, covered with PRF membranes and the flap was sutured (Fig. 5c). 4-6 months later, bone core biopsies were taken at the implant site using a trephine (Dentsply Frios 51-4091) with external and internal diameters of 3.1 and 2 mm, respectively (Figure 5d). Biopsies were then pushed out gently of the trephine, and taken for histology.

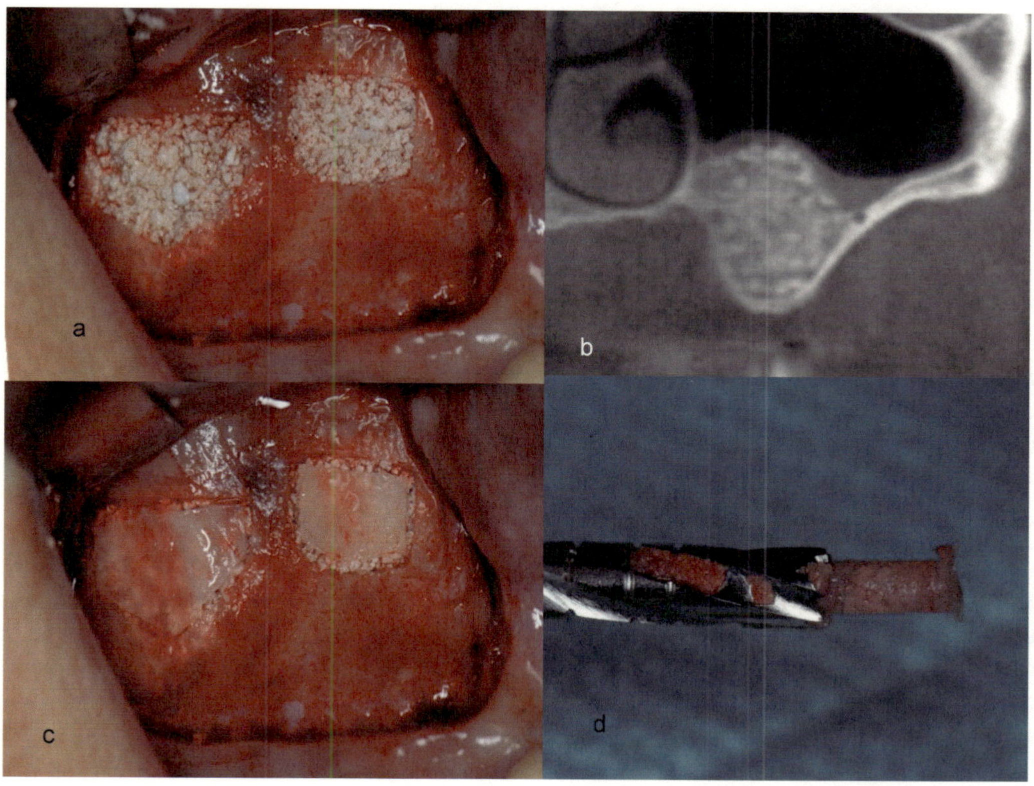

Fig. 5. (a) BGM it seen in sinus immediately after filling (b) Conen beam of sinus filled with BGM. (c) Opening of Sinus is covered by bone and PRF, before suturing back the mucosa. (d) The bone core biopsy before implant insertion (4-6 month after grafting sinus with BGM).

The osteogenic potential of the Schneiderian membrane has been previously described (Srouji, et al., 2010; Srouji, et al., 2009; Kim, et al., 2009). Histological sections from sinus grafted with Cerabone (Fig. 6c and 6d) reveal that this although considered biocompatible, the Cerabone is separated from the newly formed bone by a layers of soft connective tissue.

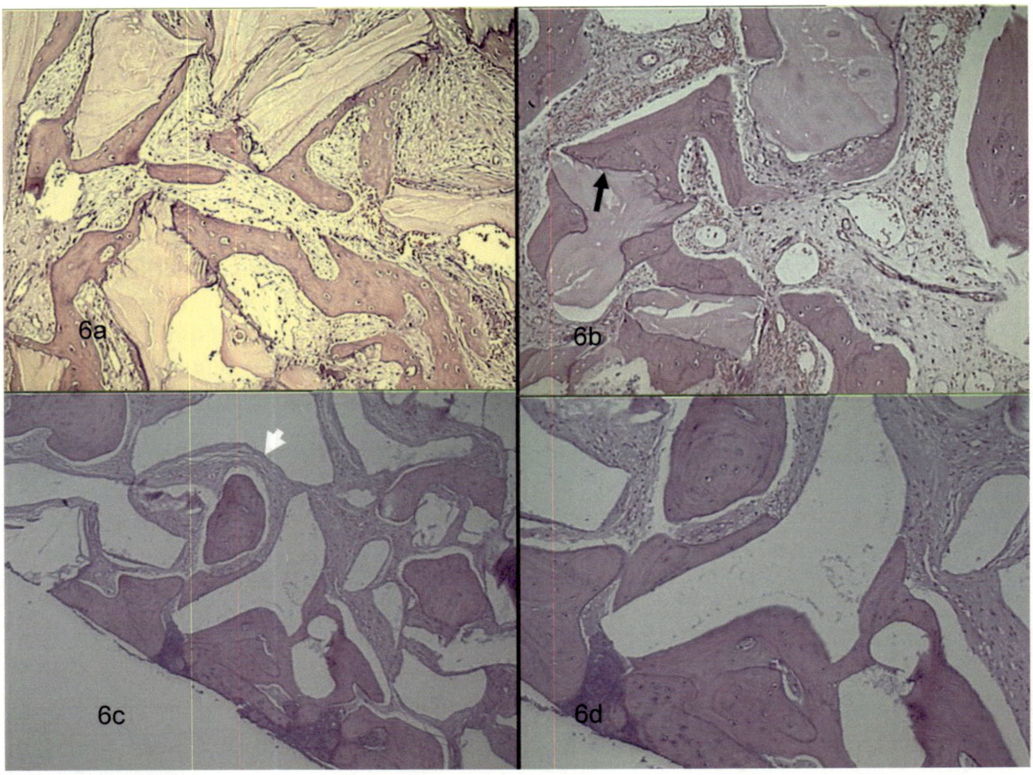

Fig. 6. Histological sections of biopsies of Bio-Oss (a and b) and Cerabone (c and d). Here we can clearly see that the interface between Bio-Oss and new bone is tight in many surfaces (black arrow in 6b), while Cerabone is separated from bone by soft connective tissue (white arrow in 6c).

Thus, the biopsy from the grafted Cerabone showed islands of grafted BGM and islands of mineralized bone, surrounded by soft connective tissue. In contrast, histological sections from biopsies of sinus grafted with Bio-Oss demonstrated new osteocyte-rich bone surrounding the Bio-Oss particles. The particles were separated from bone by few layers of connective tissue, cells and matrix and many of the Bio-Oss surfaces interphased with bone, creating a cement line of physical bond between bone and Bio-Oss (Fig. 6d, arrow). Bone could not be seen in any of the pores of this BGM. In the marrow spaces fat cells and sparse fibrous tissue, and no osteoclasts could be seen on the surfaces of Bio-Oss. Still, since it is made of cortical bone, clinicians feel that its bio-mechanical qualities are such that implants are well anchored and stabilized in the bone and Bio-Oss composite. These findings suggest that Bio-Oss is highly biocompatible and to some extent bioactive, in agreement with our animal observations. Biopsies taken from sinuses grafted with NanoBone or ReproBone presented an integrated mosaic of these BGM's and bone. A physical connectivity of BGM and bone was seen throughout the sections of both NanoBone and ReproBone. The marrow spaces were usually rich in blood cells and blood vessels, and both osteoclastic activity and matrix-producing active osteoblasts were present (Fig. 7).

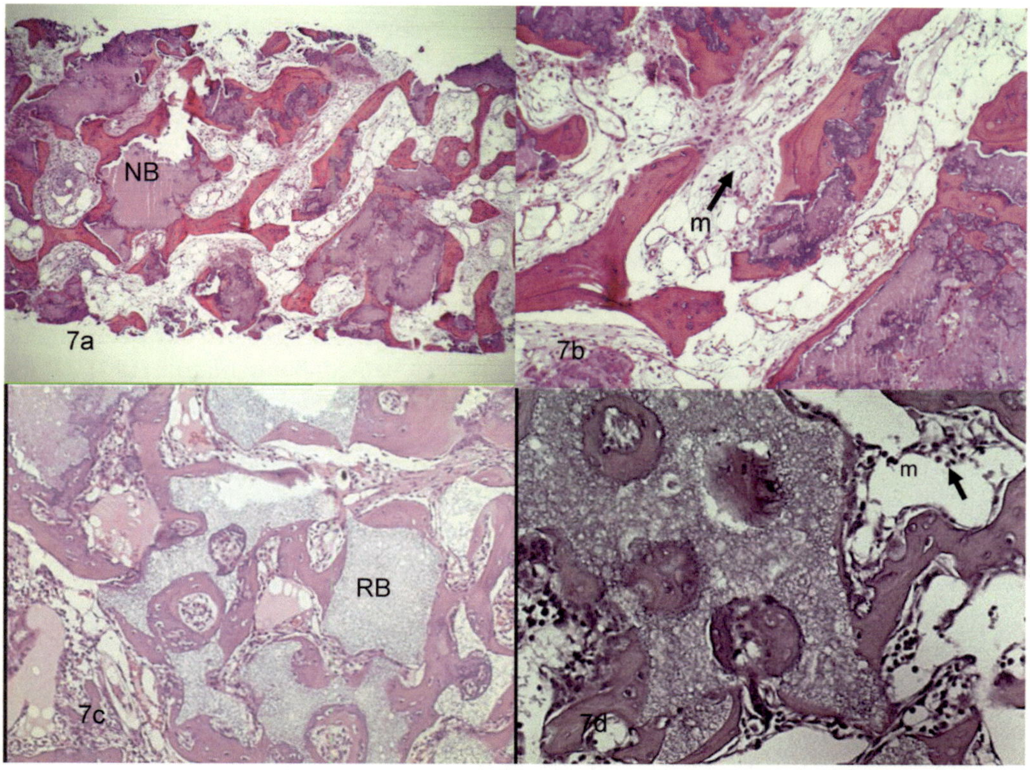

Fig. 7. Histological sections of biopsies 4-6 month after grafting in sinuses of humans. 7a and 7b the BGM is NanoBone (NB). 7c and 7d sections from ReproBone (RP).

In both NB and RP sections connectivity of bone and BGM could be seen. Bone marrow (m) was viable and blood cells were dispersed. Macropores were filled with new bone ,blood vessel and marrow. Interestingly, bone ingrowth into large pores was seen mostly in Reprobone sections but not in those of NanoBone. These results are in accord with the animal data and demonstrate the uniqueness of Reprobone in attracting ingrowth of blood vessels and bone into many of macropores. Viable marrow was seen in both NanoBone and ReproBone biopsies, indicating active bone surfaces (Ferrer, et al., 2010). This group (Ferrer,et al.,2010) proposed that hematopoietic and mesenchymal cells in marrow are much dependent on active osteoblasts. It seems that both ReproBone and NanoBone support bone formation that active in producing viable marrow and is can also undergo remodeling by osteoclasts. The data presented demonstrate the resemblance of Cerabone, Bio-Oss, NanoBone and ReproBone characteristics in the DA rat model and in human grafted sinuses.

4. Conclusions

In this chapter we presented a remarkable animal model for tissue engineering of bone in a non-osseous site. This model allows for bone generation in a very efficient and reproducible manner. Furthermore, it provides an in vivo measuring tool for assessing the biocompatibility and bioactivity of BGM's. Hydroxyapatite bone mineral (HA) and its calcium phosphate derivatives that lack the organic components of bone are considered biocompatible BGM's. They are routinely employed in bone grafting procedures to restore or fill bone defects. However, if structured to absorb serum components that can attract and bind active cells from their near environment, then these BGM's will be converted into bioactive BGM's. Consequently, the attached cells are expected to express their osteogenic phenotype and deposit bone directly on BGM's surfaces, and induce new, viable and active hematopoietic marrow in the new TE bone. Our animal studies and also clinical biopsies demonstrate that ReproBone is a highly bioactive BGM.

Of interest is our observation that Reprobone when processed as a wet moldable product to be delivered through syringe (Reprobone Novo), inhibited osteogenesis in our DA rat model (Fig. 8c and Fig. 8d). Interestingly, another product named Bonit Matrix (DOT Gmbh, Rostock , Germany) that was shown to be bioactive in our model, but when processed for delivery by syringe (Ossa Nova), demonstrated only granulation tissue surrounding the particles but no bone formation (Fig. 8a and 8b). Furthermore, multi-nucleated giant cells were also seen (Fig. 8d). We assume that these (and other) products that are intended for syringe delivery have a smaller particle size of less than 50 μm. Condensation of small particles may change their ability to interact with cells properly.

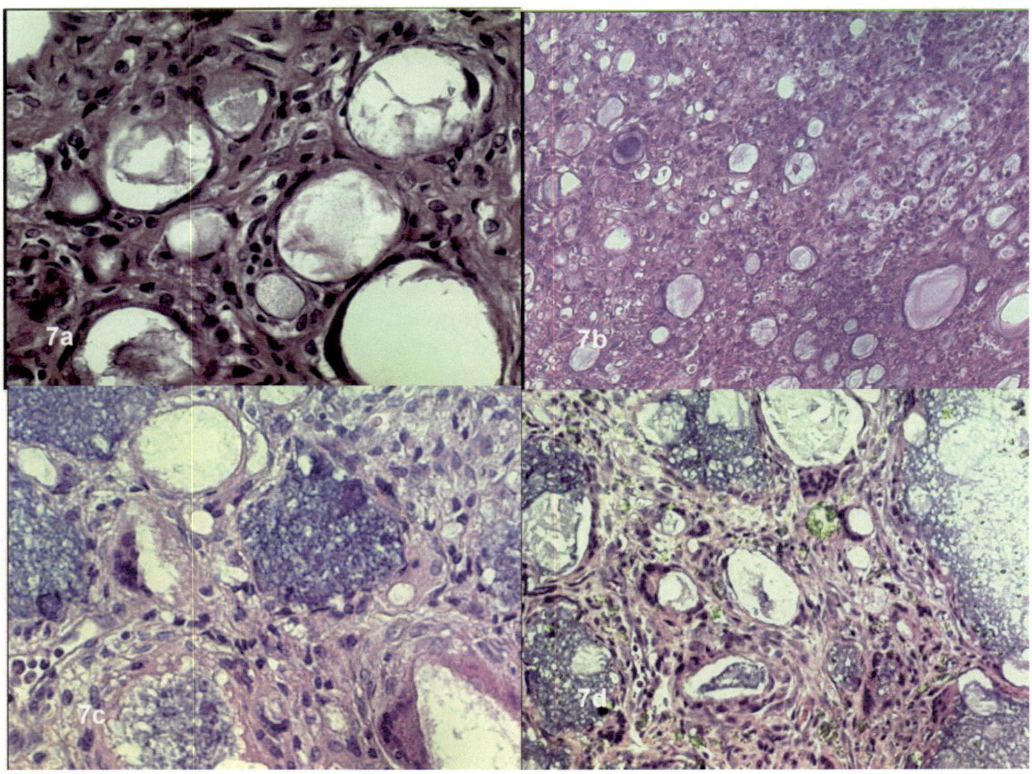

Fig. 8. Histological sections of BGM's delivered by syringe including Ossa Nova (8a and 8b) and Repro-Novo (8c and 8d). Both were mixed with bone marrow and implanted in DA rats. See in both BGM's no bone was visible.

The process of bone engineering depends on the normal cascade of wound healing which begins with the inflammatory response. At the same time, the process requires an immediate interplay between the progenitor cells and the BGM surface. This reaction is followed by the recruitment, proliferation and differentiation of osteoprogenitor cells, synthesis of ECM proteins, and angiogenesis.

We have recently demonstrated that as soon as 3-4 days after implantation of DBM and fresh marrow in DA rats, a significant upregulation of osteoblast and of angiogenic genetic profile was measured (Bahar, et al., 2007). These findings support our view that very early after grafting, accurate and tuned interactions between the unique surface of BGM and the extracellular and cellular environment are crucial in leading the pathway for de-novo engineering of bone.

It is well accepted that continuity and connectivity of bone trabeculli is essential to the transmission of functional forces in our body. For example, in osteoporotic patients spontaneous fractures occur when trabecullii in long bones are resorbed and connectivity is disrupted. Consequently, forces are transmitted through alternative and vulnerable pathways that are unable to absorb them. We therefore propose that a clinician should use BGM's that produce physical connectivity with bone. If the mechanical properties, like strength and stiffness of the BGM are similar to that of bone, such connectivity of bone-BGM will provide an excellent biomaterial for implant function. Whether NanoBone or ReproBone physical properties are in the range of compact bone is plausible.

5. References

Bahar, H.; Yaffe, A.; Binderman, I.(2003). The influence of nacre surface and its modification on bone apposition: a bone development model in rats. *Journal of Periodontology*, 74, 366–371.

Bahar H, Benayahu D, Yaffe A, Binderman I.(2007) Molecular signaling in bone regeneration. *Crit Rev Eukaryot Gene Expr.*, 17,(2), 87-101. Review.

Bahar, H.; Yaffe, A.; Boskey, A.; Binderman I.(2010). Influence of bone derived matrices on generation of bone in an ectopic rat model. *Journal of Orthopedic Research*, 28,(5), : 664-700.

Binderman, I.; Yaffe, A.; Zohar, R.; Benayahu, D.; Bahar, H. (2011). Tissue engineering of bone: an ectopic rat model. *Frontiers in Biosciences (Scholar Edu)*, 3, 61-68.

Ferrer,S.M.; Michurina,T.V.; Ferraro,F.; Mazloom, A.R.; MacArthur, B.D.; Lira,S.A.; Scadden, D.T.; Maayan, A.; Enikolopov, G.N.; Frennete, P.S. (2010). Mesenchymal and Haematopoietic stem cells form a unique bone marrow niche. *Nature*, 466, 829-834.

Genesis. "And the LORD God caused a deep sleep to fall upon the man, and he slept; and He took one of his ribs, and closed up the place with flesh instead thereof. 22, And the rib, which the LORD God had taken from the man, made He a woman, and brought her unto the man". *Bible, Genesis*, chapter 2, sentence 21.

Gotz, W.; Gerber, T.; Lossdorfer, S.; Henkel, K.O.; Heinemann, F. (2008) Immunohistochemical characterization of nanocrystalline hydroxyapatite silica gel (Nanobone) osteogenesis: a study on biopsies from human jaws. *Clin Oral Implants Res* 19, 1016-1026.

Hing, K. A.; Best, S. M.; Tanner, K. E.; Bonfield,W.; Revell, P. A. (2004). Mediation of bone ingrowth in porous hydroxyapatite bone graft substitue. *J. Biomed. Mater. Res.* A, , 68, 187-200.

Jones, J.R.; Lin,S.; Yue, S.; Lee,P.D.; Hanna, J.V.; Smith,M.E.; Newport, R.J. (2010). Bioactive glass scaffolds for bone regeneration and their hierarchical characterisation. *Journal of Engineering in Medicine*, 224, 1373-1387.

Kim, S.W.; Lee, I.K.; Yun, K.I.; Kim, C.H.; Park, J.U. (2009). Adult stem cells derived from maxillary sinus membrane and their osteogenic differentiation. *Int. J Oral Maxillofac Implants,* 24(6):991-998.

Place, E.S.; Evans, N.D.; Stevens, M.M. (2009) Complexity in biomaterials for tissue engineering. *Nature Materials*, 8, 457-470.

Rokn, A.R.; Khodadoostan, M.A.; Reza, A.A.; Ghahroudi, R.; Motahhary, P.; Fard, M.J.K.; De Bruyn, H.; Afzalifar, R.; Soolar, E.; Soolari A. (2011). Bone Formation with Two Types of Grafting Materials: A Histologic and Histomorphometric *The Open Dentistry Journal*, 5, 96-104.

Srouji, S.; Ben David D.; Lotan, R.; Riminucci, M.; Livne, E.; Bianco,P. (2010). The innate osteogenic potential of the maxillary sinus (Schneiderian) membrane: an ectopic tissue transplant model simulating sinus lifting. *Int. J Oral Maxillofac Surgery*, 39(8): 793-801.

Srouji, S.; Kizhner, Y.; Ben David D.; Riminucci, M.;,Bianco,P.; Livne, E. (2009). The Schneiderian membrane contains osteoprogenitor cells: in vivo and in vitro study. *Calcif Tissue Int.*, 84(2):138-145.

Stevens, M. M.; Marini, R.P.; Schaefer, D.; Aronson, J.; Langer, B.; Shastri, V.P. (2005) In vivo engineering of organs: the bone bioreactor. *Proc. Natl Acad. Sci. USA* 102, 11450–11455.

Tanner, K. (2010). Bioactive composites for bone tissue engineering. *Journal of Engineering in Medicine*, 224, 1359-1372.

Yaffe, A.; Kollerman. R; Bahar, H.; Binderman, I.(2003) The influence of alendronate on bone formation and resorption in a rat ectopic bone development model. J *J of Periodontol* 74, 44–50.

Zuk, P.A. (2008). Tissue Engineering Craniofacial Defects With Adult Stem Cells? Are We Ready Yet? *Pediatric Research*, vol 63,(No5), 478-486.

5

Tissuue Engineering in Maxillar Sinus Lifting: A Comparation of Differents Grafts and Confocal Laser Scanning Microscopic Evaluation

Andrea Ballini, Michele Scivetti, Stefania Cantore, Biagio Rapone, Gianfranco Favia and Felice Roberto Grassi
Department of Dental Sciences and Surgery-University of Bari "Aldo Moro"
Italy

1. Introduction

1.1 Tissue engineering

The term tissue engineering was originally coined to denote the construction in the laboratory of a device containing viable cells and biologic mediators (e.g., growth factors and adhesins) in a synthetic or biologic matrix, which could be implanted in patients to facilitate regeneration of particular tissues. The role of tissue oxygenation in wound healing became the focal point in the 1980s. Tissue oxygenation enhances phagocytic and bactericidal ability of host immune cells and supports collagen as well as other protein synthetic events. The importance of growth factors in enhancing wound healing has become the focus of research in the present day. In addition, a link has been established between tissue oxygenation and growth factors. Macrophage stimulation causes the release of angiogenic and other growth factors that support wound healing and resist infection [1]. In general, tissue engineering combines three key elements, namely scaffolds (collagen, bone mineral), signaling molecules (growth factors), and cells (osteoblasts, fibroblasts). Tissue engineering has been redefined presently as the relatively new, highly promising field of reconstructive biology, which draws on the recent advances in medicine and surgery, molecular and cellular biology, polymer chemistry, and physiology. These principles of tissue engineering have found widespread application in several branches of dentistry, such as periodontics, oral and maxillofacial surgery, and oral implantology. In the field of implant dentistry, the most frequently encountered problems at the implantation site are lack of adequate bone and proximity to anatomic structures, such as the maxillary sinus and the inferior alveolar nerve canal. Advanced surgical procedures that act as an adjunct in dental implants consist of sinus grafting and guided bone regeneration. These procedures are quite predictable when proper surgical protocols are established and followed. In this preliminary study we performed histological analisys using confocal laser scanning microscopy (CLSM), with the aim of evaluating the differences between graft and newly formed bone.

2. Confocal laser scanning microscopy

CLSM is a recent technique that has proved to be a valuable tool in the study of tissue and cell biology, as well as in the in vivo and in vitro detection of fluorescent markers. CLSM analysis shows a different fluorescence of different tissues. Specifically, the technique consists of illuminating the sample with a mono- and/or bi-chromatic punctiform laser source. The resulting emission energy is detected by a spatially filtered optical system, the pinhole, which eliminates light signals arising from out-of-focus planes. The possibility of visualizing each sample at different focal planes (optical sectioning) allows the acquisition of a series of images which can be then assembled for three-dimensional reconstruction and analysis [2]. In this study CLSM has been used because is a simple technique and also an easy way to underline differences between graft and human bone (either pre-existing and new-formed). The following picture shows how CLSM enhances graft matherial (bovine hydroxyapatite) compared with native bone. (Fig CLSM_1 and _2)

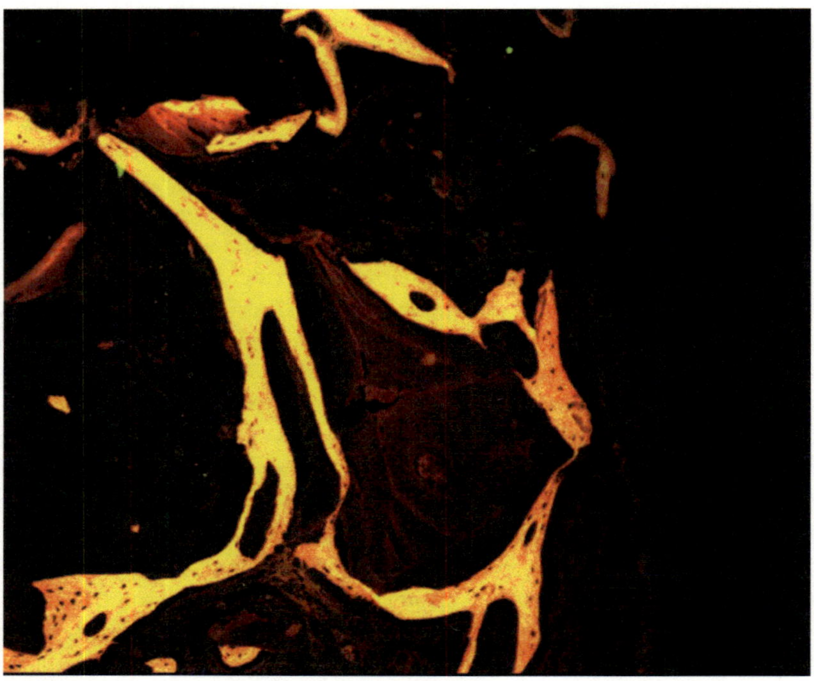

Fig. CLSM_1. Well-defined difference between graft matherial (dark brown) and newly-formed bone (yellow) in a case of sinus-lift bone regeneration (CLSM, original magnification 100X).

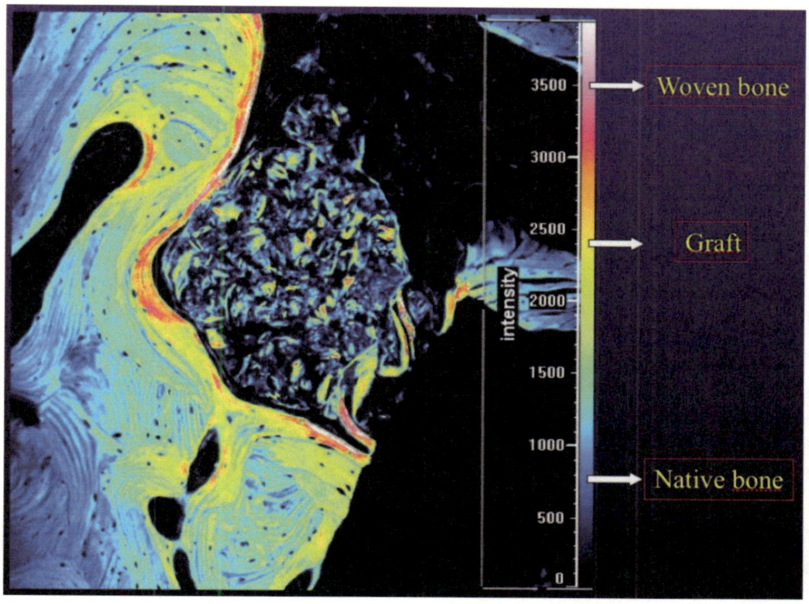

Fig. CLSM_2. Specimen of regenerated bone observed by CLSM: the color scale allows to define several tissues, on the base of different emitting autofluorescence (CLSM; original magnification 200X).

3. Demineralized Freeze-Dried Bone Allograft (DFDBA) and Freeze-Dried Bone Allograft (FDBA)

Bone graft materials commonly used for these procedures are demineralized freeze-dried bone allograft (DFDBA) and freeze-dried bone allograft (FDBA). The osteoinductive properties of DFDBA have made it the grafting material of choice as compared to FDBA, xenografts, and alloplasts. However, the osteoinductive potential of DFDBA procured from different bone banks or from different batches of the same bank may vary highly. The bioactivity of DFDBA seems to be dependent on the age of the donor; the younger the donor the more osteoinductive the graft material [3]. This controversy as well as concerns about disease transmission has pushed clinicians toward using xenografts and alloplastic materials. Although these materials are biocompatible and are osteoconductive in nature, clinical outcomes are unpredictable. The problem that arises next is how to improve clinical outcomes by improving the properties of these grafts.

4. Platelet concentrates

In general, platelet concentrates are blood-derived products used for the prevention and treatment of hemorrhages due to serious thrombopenia of the central origin. The development of platelet concentrates as bioactive surgical additives that are applied locally to promote wound healing stems from the use of fibrin adhesives. Since 1990, medical science has recognized several components in blood, which are a part of the natural healing process; when added to wounded tissues or surgical sites, they have the potential to accelerate healing. Fibrin glue was originally described in 1970 and is formed by polymerizing fibrinogen with thrombin and calcium. It was originally prepared using donor plasma; however, because of the low concentration of fibrinogen in plasma, the stability and quality of fibrin glue were low. These adhesives can be obtained autologously from the patient or can be obtained commercially (Tisseel, Baxter Healthcare). These products are heat-treated, thus immensely reducing, but not entirely eliminating, the risk of disease transmission. Therefore, the commercially available adhesives constitute an infinitely small risk of disease transmission. PRP is an autologous modification of fibrin glue, which has been described and used in various applications with apparent clinical success. PRP obtained from autologous blood is used to deliver growth factors in high concentrations to the site of bone defect or a region requiring augmentation. [4]

Platelet-rich plasma (PRP) is a platelet concentrate that has been used widely to accelerate soft-tissue and hard-tissue healing. The preparation of PRP has been described by several authors. Platelet-rich fibrin (PRF) was first described by Choukroun *et al.* in France. It has been referred to as a second-generation platelet concentrate, which has been shown to have several advantages over traditionally prepared PRP. Its chief advantages include ease of preparation and lack of biochemical handling of blood, which makes this preparation strictly autologous.

Platelets isolated from peripheral blood are an autologous source of growth factors. When platelets in a concentrated form are added to graft materials, a more predictable outcome is derived. Platelet-rich plasma (PRP) is an easily accessible source of growth factors to support bone- and soft-tissue healing. It is derived by methods that concentrate autologous platelets and is added to surgical wounds or grafts and to other injuries in need of supported or accelerated healing. A blood clot is the center focus of initiating any soft-tissue healing and bone regeneration. In all natural wounds, a blood clot forms and starts the healing process. PRP is a simple strategy to concentrate platelets or enrich natural blood clot, which forms in normal surgical wounds, to initiate a more rapid and complete healing process. A natural blood clot contains 95% red blood cells, 5% platelets, less than 1% white blood cells, and numerous amounts of fibrin strands.

A PRP blood clot contains 4% red blood cells, 95% platelets, and 1% white blood cells. PRF is in the form of a platelet gel and can be used in conjunction with bone grafts, which offers several advantages including promoting wound healing, bone growth and maturation, graft stabilization, wound sealing and hemostasis, and improving the handling properties of graft materials[5]. PRF can also be used as a membrane.

Clinical trials suggest that the combination of bone grafts and growth factors contained in PRP and PRF may be suitable to enhance bone density. In an experimental trial, the growth factor content in PRP and PRF aliquots was measured using Elisa kits. The results suggest that the growth factor content (PDGF and TGF-β) was comparable in both. Another experimental study used osteoblast cell cultures to investigate the influence of PRP and PRF on proliferation and differentiation of osteoblasts. In this study, the affinity of osteoblasts to the PRF membrane appeared to be superior. [6] PRF has many advantages over PRP. It eliminates the redundant process of adding anticoagulant as well as the need to neutralize it. The addition of bovine-derived thrombin to promote conversion of fibrinogen to fibrin in PRP is also eliminated. The elimination of these steps considerably reduces biochemical handling of blood as well as risks associated with the use of bovine-derived thrombin. The conversion of fibrinogen into fibrin takes place slowly with small quantities of physiologically available thrombin present in the blood sample itself. Thus, a physiologic architecture that is very favorable to the healing process is obtained due to this slow polymerization process. Literature pertaining to PRF is found in French, and the material is being used widely in France. The popularity of this material should increase considering its many advantages. The findings of Wiltfang *et al.* [5] from a series of clinical trials are encouraging, in that they show improved properties of PRF as compared with PRP. In future, more histologic evaluations from other parts of the world are required to understand the benefits of this second-generation platelet concentrate.

5. The Antral Membrane Balloon Elevation (AMBE) technique

Many edentulous posterior maxilla are found to be encumbered by alveolar resorption and increased pneumatization of the sinus. These factors limit the quantity and quality of bone necessary for successful implant placement in these areas. One solution is to use shorter implants, but this often results in an unfavorable crown-root ratio. To create an improved environment in such regions, the classic sinus floor elevation with bone augmentation is a well-accepted technique. However, when the edentulous area is limited to a zone between 1 and 2 teeth, lifting the membrane becomes difficult and may subject it to iatrogenic injury. The antral membrane balloon elevation technique, which is introduced in this preliminary report, is a modification of the currently used sinus lift. It elevates the membrane easily and makes the antral floor accessible for augmentation with grafting materials. The edentulous posterior maxilla presents special challenges to the implant surgeon that are unique compared with other areas in the mouth. After tooth extraction, the initial decrease in bone width is secondary to resorption of the buccal bone plate. As the edentulous area continues to atrophy, there is a continuing loss of bone height and density and an increase in antral pneumatisation [6,7]. The maxilla is primarily trabecular or spongy bone enclosed within thin cortical layers. In addition, the posterior maxilla contains the least dense bone in the oral environment. In some cases, the alveolus may be 2 mm high or shorter and of poor quality. Even if an individual were to have a modicum of bone, the resulting short endosseous implants would lead to insufficient anchorage, questionable integration, and unfavourable crown-root ratios. It has been written that as much as a minimum of 10 mm of bone height is necessary for successful implant stabilization and integration. The antral membrane balloon elevation (AMBE) technique lifts the sinus membrane with minimal trauma and is particularly useful in areas that are difficult to reach. It is beneficial when teeth are adjacent to the edentulous area that requires augmentation. The AMBE technique is accomplished with a limited incision, minimal mucoperiosteal flap reflection, and a small window. The membrane is elevated to the medial wall of the sinus cavity avoiding sharp dissection around the roots of adjacent teeth. Thus, morbidity, blood loss, operative time, and postoperative pain and complications are reduced when compared with the conventional procedure. Sinus lift surgery is predictable and is usually not technically demanding. However, it is a more difficult surgical technique when teeth are adjacent to the edentulous area. It presents a far lesser challenge in the totally edentulous posterior maxilla [8]. The AMBE is a modification and combination of surgical techniques that adds sufficient bone height to allow placement of longer implants of up to 16 mm. The balloon operation and graft procedure described herein can be used to augment a severely atrophic ridge and does not depend on the accessible ridge height, as does the crestal approach, which uses trephines and osteotomes. Numerous approaches to the management of the deficient posterior maxilla have been described. Among them is the 2-stage classic approach with a 4- to 6-month interregnum required before the placement of implants. Fifteen millimeters or more of bone height can be achieved by this operation. A single-stage procedure following the same protocol as the 2-stage approach also is frequently used. The requirement here is for a beginning bone height of at least 4 to 6 mm. No lateral osteotomy into the antral cavity is required when using the Summers technique. It uses a number of concave-tipped, tapered osteotomes that are used to both enlarge and deepen the osteotomies while pushing the garnered bone apically beneath the tented membrane [9,10]. This is not unlike the green stick fracture method, which adds 2 to 3 mm of bone height beneath the elevated but unsullied sinus membrane. A recently published trephine bonecore sinus elevation graft is a new technique that permits autogenous bone grafting from 5 to 10 mm [11].

5.1 Advantages of the AMBE technique

The use of the AMBE allows the surgeon to elevate the sinus membrane with minimal risk of tearing and with a conservative, tissue-sparing surgical approach. This reduces postoperative pain, bleeding, possibilities of infection, and the other morbid symptoms often associated with sinus lift procedures. The technique introduced in this article is often completed within 30 minutes. It is especially beneficial when access is difficult and when adjacent teeth are present next to the edentulous area.

5.2 Disadvantages of the AMBE technique

Unlike some of the currently used techniques, which are performed from a crestal approach, AMBE requires a buccal fenestration and a larger incision than do other alternative operations.

The AMBE technique with augmentation has been a highly successful and predictable procedure. It facilitates lifting the sinus membrane gently and displacing it upward. The graft material is deposited into the space thus created. Implants may be placed simultaneously with the graft.

5.3 Aim

The aim of our preliminary work is to evalutate the role of PRF in enhancing bone regeneration in maxillary sinus grafting, performed for implant placement and prosthetic rehabilitation.

6. Preliminary study

6.1 Patients and methods

This study included 10 (ten) patients that underwent to bilateral sinus lifting surgical procedure for implant placement and prosthetic rehabilitation. Sinus grafting has been performed in both sides in the same surgical procedure, using AMBE technique, following this prothocol: right sinus received only bovine hydroxyapatite (Bio-Oss®) as grafting matherial, while left sinus received Bio-Oss® and PRF previously obtained by using Choukroun technique [12]. During the stage-two surgery, performer after four months, ten trephine cores were taken, using a trephine bur instead pilot(s) drill, one for each one for each graft site, and sent for histologic examination. Each specimen was first fixed in a buffered solution of 10% formalin and then immersed in an appropriate fixing and decalcifying solution (Mielodec, Bio Optica, Milan, Italy) for 90 min, rinsed in 70% ethyl alcohol for 30 min, and then conventionally processed for histopathologic paraffin embedding, thin sectioning at 5 mm perpendicular to the long axis, and staining with hematoxylin–eosin. Histological examination was carried out using a Nikon Eclipse E600 microscope (Nikon Corporation, Tokyo, Japan), equipped with argon–ion and helium–neon lasers, emitting at 488- and 543-nm wavelengths respectively, which allows both optical and confocal laser scanning microscopic analyses. The Nikon EZ C1 software (Nikon Corporation, version 2.10, Coord Automatisering) was used for image processing.

Statistical analyses were performed using the SPSS statistical software package for Microsoft Windows (version 14.0, SPSS Inc, Chicago, IL, USA).

6.2 Results

When analyzed by CLSM, the samples showed a strong natural autofluorescence pattern that was not homogenous, either quantitatively or qualitatively. We have inferred that such an intense fluorescence is due to the collagen component of the organic matrix, different between equine and human, in agreement with a number of studies reporting high autofluorescence intensity of collagen [13].

In fact, Figure 1 exhibited variable degrees of autofluorescence intensity, where higher scale values are related to higher fluorescence intensity. Different layers of bone showed different fluorescence extent and the autofluorescence variability was also qualitative, according to different stages of bone growt. Autofluorescence appears homogeneous in graft, were collagen fibrils are linear; otherwise, newly formed bone shows different autofluorescence degrees patterns as can be seen from Figure 2 due to the different hardness between woven and mature bone.

After statistical and histomorphometrical analisis, even if on a little cohort with few specimen, preliminary interesting results should be undelined. PRF seems to significatively enhance bone re generation: in fact, in all specimens obtained from left graft side (Bio-Oss® with PRF) we observed large amounts of mature bone than contralateral side. At 4 months, histology showed a rich connective and vascular tissues surrounding graft particles with a low quantity of newly-formed woven bone (Figures 3(a), 3(b)) in right samples (only Bio-Oss®) while large amounts dense mature bone with marrow and blood vessels are observed in left samples

(Fig. 1). At higher magnification, it was observed that the graft remained engaged with the bone with a few number of resorption lacunae and no inflammatory infiltrate (Fig. 5). New bone was also functionally organized with osteons and Havers channels.

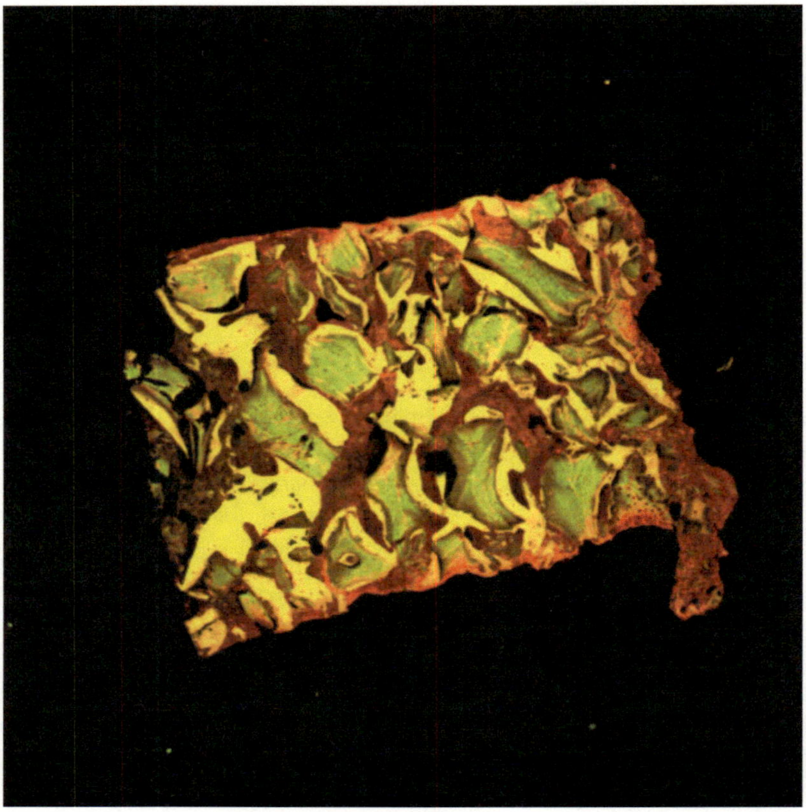

Fig. 1. Histology of one of the biopsy cores at 4 months. A large amount of new and well-characterized bone is present. (CLSM, original magnification 100X).

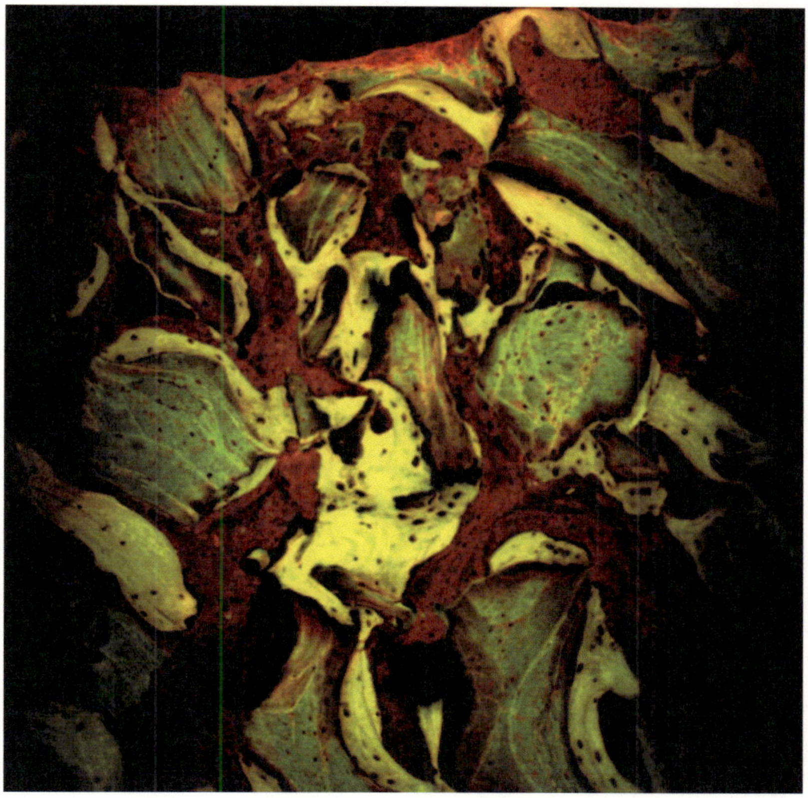

Fig. 2. Particular of Fig. 1, newly formed bone shows different autofluorescence degrees patterns. (CLSM, original magnification 200X).

The principles of tissue engineering have been applied to dental surgery. PRF is a natural medical device that seems to enchances wound healing and bone–soft tissue regeneration. The purpose of this preliminary studyis to demonstrate the possibility of obtaining mature bone regeneration in humans with the combination of a biomaterial graft and PRF in a short period of time. Similar results are also been obtained with others proteins, such as rh-PDGF [14] and in our opinion all these result shuold be compared.

The limited conclusions of this study are that this technique provides a very important new amount of bone in a short period of time, and this bone already has all the characteristics of mature bone around the scaffold. Since there is a lack of available literature on this topic, this case study should encourage clinicians to use this technique, although further studies are needed to provide evidence-based conclusions.

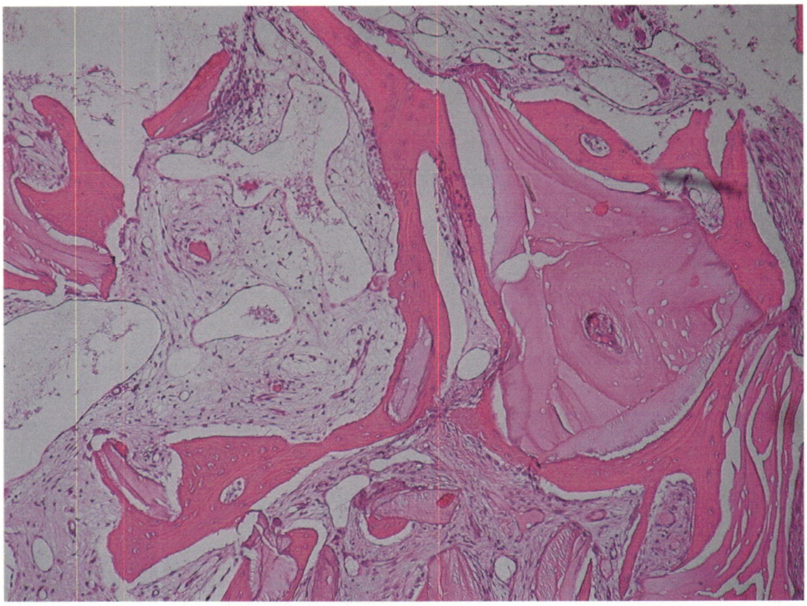

(a)

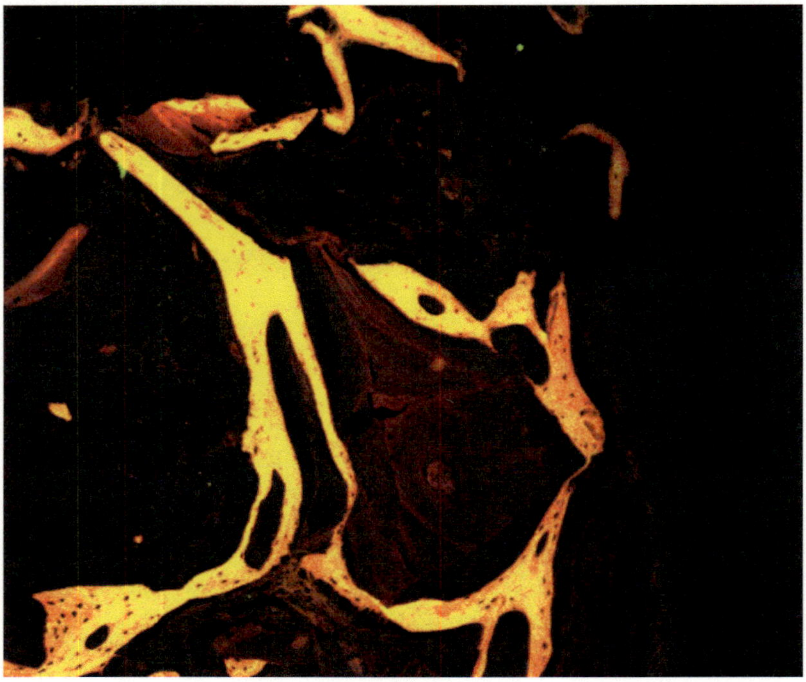

(a)

Fig. 3. Right sample (only Bio-Oss®) showing connective and vascular tissues surrounding graft particles with a low quantity of newly-formed woven bone (Optical and CLSM, original magnification 200X)

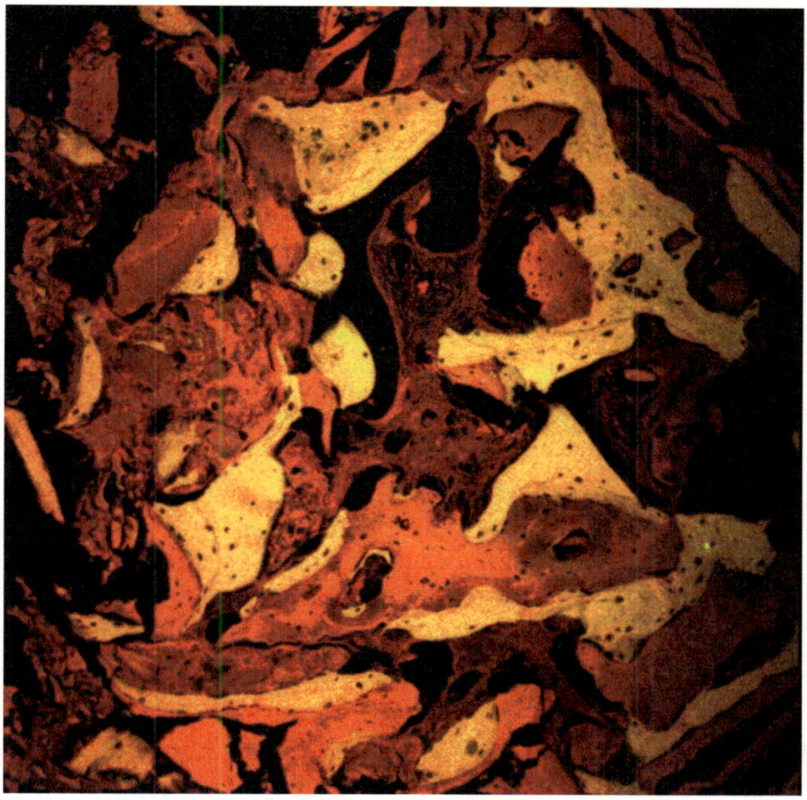

Fig. 4. Graft particles strictly closed with the bone, resorption lacunae and no inflammatory infiltrate. (CLSM, original magnification 100X).

7. Tissue engineering in the future

Bone tissue engineering is the general term for a number of ways by which bony tissue lost as a result of trauma and disease might be restored. It is possible to use cells alone (as in the case of bone marrow transplantation), however for most applications in regenerative medicine, cells in combination with appropriate scaffolds and carriers are more commonly used [15].

The design of scaffolds for bone tissue engineering includes a large number of factors related to structural integrity, superficial properties, incubating and cell growth conditions, and cell/biomaterial interactions [15]. One of the main factors that have an influence on the cellular and molecular mechanisms is the biophysical stimulus transmitted to the mesenchymal tissue [15]. This stimulus is linked to the architecture and the material properties of the scaffold that will serve as a host to receive external stimuli for matrix production. The ideal scaffold is capable of transferring the most favourable stimulus on the re-generating tissue hence allowing the times for the regeneration to be reduced and the optimal me-chanical properties of the regenerate to be obtained.

Mechano-regulation models can be utilized in bone tissue engineering to optimize the morphology, the porosity, the mechanical properties etc of scaffolds as well as the environment conditions. Such an issue has been recently investigated in different studies [15].

Stem cells are object of a growing interest from the international scientific community for their potentialities in regenerative medicine for their interactions with scaffolds.

At the moment, many laboratories working in this field are designing and setting up biocompatible and biodegradable scaffolds that might facilitate cell differentiation and *in vitro* reconstruction tissues of interest, before grafting them to patients.

Stem cells are defined as clonogenic cells capable of both self-renewal and multi-lineage differentiation. A population of these cells have been identified in human dental pulp.

Dental Pulp Stem Cells (DPSCs) were found in adults teeth and have been shown to differentiate, under particular conditions, into various cell types including osteoblat-like cells.

The stem cells in these locations lie dormant in a non-proliferating state until they are required to participate in local repair and regeneration. The stem cells in these locations lie dormant in a non-proliferating state until they are required to participate in local repair and regeneration.

In a few our previous experiences we studied the osteoblastic phenotype developed by DPSCs cultured in osteogenic medium.

In particular, the tipical osteoblast markers such as alkaline phosphatase (ALP), collagen type-I (COLL-1) and mineralized matrix production were analyzed. [16-18].

Results are promising and with DPSCs researchers can offer the possibility of realizing scaffolds for bone tissue engineering with a customized external shape as well as with structures to increase the mass transport of oxygen and nutrients.

8. Conclusions

A more robust integration is required, in future, between biology, mechanics and materials science. This should lead to the development of mechano-regulation models that more accurately describe physiological processes such as the fracture healing, the tissue genesis etc. Future perspectives of numerical simulations of biomaterial scaffolds for bone tissue engineering rely also on the development of new methods to account for the multi-scale dimension of the problems.

As a conclusion, bone tissue engineering is an emerging multidisciplinary field that can revolutionize the ways we improve the health and quality of life for millions of people worldwide. The future of computational models integrating the finite element method and mechano-regulation algorithms appears promising. More realistic models of biological/physiological processes need to be simulated; however, in order to make the implementation of these algorithms, affordable for a clinical use, more efforts need to be put into the development of powerful computational tools.

9. Acknowledgments

Funding for Stem Cell work was awarded from Ministero dell'Istruzione, dell'Università e della Ricerca -PRIN 2009 (Progetto di Ricerca d'Interesse Nazionale- Grant 2009)

10. References

[1] Marx ER. Platelet-rich plasma: A source of multiple autologous growth factors for bone grafts. In : Lynch SE, Genco RJ, Marx RE, editors. Tissue engineering, applications in maxillofacial surgery and periodontics, Quintessence Publishing Co, Inc: Illinois; 1999. p. 71-82.

[2] Scivetti M, Pilolli GP, Corsalini M, Lucchese A, Favia G. Confocal laser scanning microscopy of human cementocytes: analysis of three-dimensional image reconstruction. Ann Anat. 2007;189(2):169-74

[3] Schwartz Z, Somers A, Mellonig TJ, Carnes DL Jr, Dean DD, Cochran DL, et al. Ability of commercial demineralized freeze-dried bone allograft to induce new bone formation is dependent on donor age but not gender. J Periodontol 1998;69:470-8.

[4] Carlson ER. Bone grafting the jaws in the 21 st century: The use of platelet-rich plasma and bone morphogenetic protein. Alpha Omegan 2000;93:26-30.

[5] Par Wiltfang J, Terheyden H, Gassling V, Acyl A. Platelet rich plasma (PRP) vs. platelet rich fibrin (PRF): Comparison of growth factor content and osteoblast proliferation and differentiation in the cell culture. *In:* Report of the 2nd International Symposium on growth Factors (SyFac 2005).

[6] Sanchez AR, Sheridan PJ, Kupp LI. Is platelet-rich plasma the perfect enhancement factor? A current review. Int J Oral Maxillofac Implants 2003;18:93-103.

[7] Thomas GJ. Sinus lifts: a possible solution to the atrophic maxilla. J Macomb Dent Soc. 1990;29:9–11.

[8] Garg A. Augmentation grafting of the maxillary sinus for the placement of dental implants: anatomy, physiology, and procedure. Implant Dent. 1994;8:36–45

[9] Woo I, Le BT. Maxillary sinus floor elevation: a review of anatomy and two techniques. Implant Dent. 2004;13:28–30.

[10] Rosen P, Summers RB, Mellado JR, Salkin LM, Shanaman RH, Manuel HM. The bone-added osteotome sinus floor elevation technique: multicenter retrospective report of consecutively treated patients. Int J Oral Maxillofac Implants. 1999;14:853–858.

[11] Soltan M, Smiler DG. Trephine bone core sinus elevation graft. Implant Dent. 2004;13:148–152.

[12] Dohan DM, Choukroun J, Diss A, Dohan SL, Dohan AJ, Mouhyi J, Gogly B. Platelet-rich fibrin (PRF): a second-generation platelet concentrate. Part I: technological concepts and evolution. Oral Surg Oral Med Oral Pathol Oral Radiol Endod. 2006 Mar;101(3):e37-44. Epub 2006 Jan 19.

[13] Prentice AID. Autofluorescence of bone tissues. J Clin Pathol. 1967;20:717–719.

[14] De Angelis N, Scivetti M., Lateral Ridge Augmentation Using an Equine Flex Bone Block Infused with Recombinant Human Platelet-Derived Growth Factor BB: A Clinical and Histologic Study. Int J Periodontics Restorative Dent. 2011 Jul-Aug;31(4):383-8.

[15] Ballini A, De Frenza G, Cantore S, et al. In vitro stem cell cul-tures from human dental pulp and periodontal ligament: new prospects in dentistry. Int J Immunopathol Pharmacol. 2007; 20: 9-16.

[16] Ballini A, Capodiferro S, Cantore S, Crincoli V, Grassi FR. Dental pulp stem cells curriculum vitae. Journal of osteology and biomaterials. 2010; 1: 23-27.

[17] Mori G, Centonze M, Brunetti G, Ballini A, Colucci S, Grano M, Grassi FR. Osteogenic properties of human dental pulp stem cells. J Biol Regul Homeost Agents. 2010; 24: 167-75.

[18] Boccaccio A, Ballini A, Pappalettere C, Tullo D, Cantore S, Desiate A Finite Element Method (FEM), Mechanobiology and Biomimetic Scaffolds in Bone Tissue Engineering. Int J Biol Sci. 2011; 7(1):112-132

Part 2

Regenerative Bone Therapy

6

Bioresorbable Collagen Membranes for Guided Bone Regeneration

Haim Tal, Ofer Moses, Avital Kozlovsky
and Carlos Nemcovsky
Department of Periodontology and Implantology,
Tel Aviv University
Israel

1. Introduction

Localized lack of bone volume in the jaws may be due to congenital, post-traumatic, postsurgical defects or different disease processes. Increasing the bone volume has long been an attractive field of basic and clinical research. The introduction of implant therapy, and the proven relationship between long-term prognosis of dental implants and adequate bone volume at the implant site (Lekholm *et al.* 1986), dramatically increased the interest of both clinicians and scientists in this field, making augmentation procedures an important part of contemporary implant therapy. Basically, four methods have been described to augment bone volume: a. *osteoinduction*, using appropriate growth factors (Reddi 1981; Urist 1965); b. *osteoconduction*, using grafting materials that serve as scaffolds for new bone growth (Buch *et al.* 1986; Reddi *et al.* 1987); c. *distraction osteogenesis*, by which a surgically induced bone fracture enables slow controlled pulling apart of the separated bone fragments (Ilizarov 1989a,b); d. *guided bone regeneration*, which allows selective bone tissue growth into a space maintained by tissue barriers (Dahlin *et al.* 1988, 1991a; Kostopoulos & Karring 1994; Nyman & Lang 1994). Among the different methods, guided bone regeneration (GBR) is the most popular and best documented for the treatment of localized bone defects in the jaws, probably due to its relative simplicity of use while allowing the placement of endosseous implants in areas of the jaw with bony defects and/or insufficient bone volume. Highly predictable success rates can be achieved using GBR; in fact, it has been shown that success rates of implants placed at GBR treated sites and sites without bone augmentation are comparable (Hammerle *et al.* 2002).

Guided Tissue Regeneration (GTR) has been introduced into clinical dental practice over 30 years ago, soon after Melcher (1970,1976), followed by Karring *et al.* 1980 and Nyman *et al.* 1980, presented its basic principles to the professional community. The mechanisms of GBR generally followed the same principles i.e. that under certain conditions, cells that originate from tissues adjacent to an exclusively provided space are able to form their parent tissue. In order to allow exclusive repopulation of cells from desired tissues into that space, preference must be given by preventing access of cells from neighboring undesired tissues using tissue barriers, commonly referred to as membranes (Figs. 1,2).

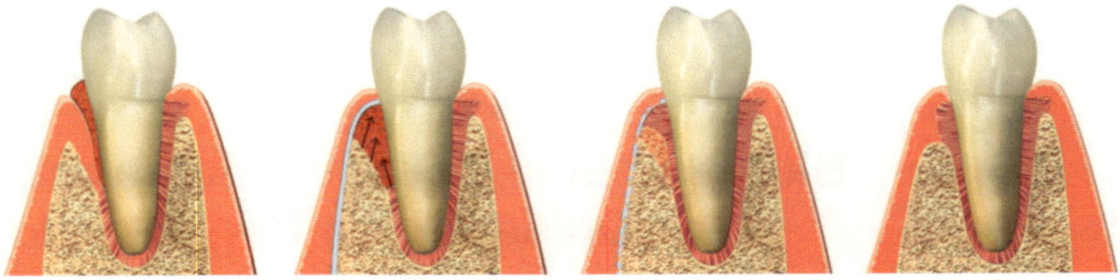

Fig. 1. Guided tissue regeneration scheme describing the use of a resorbable barrier membrane. a. vertical periodontal lesion in the mandible b. resorbable membrane is stabilized over the debrided lesion and covered by the mucosal flap c. after 3- 5 months the membrane starts to resorb; new bone, new periodontal ligament and new cementum are visible d. reestablishment of most of the periodontal attachment apparatus is completed.

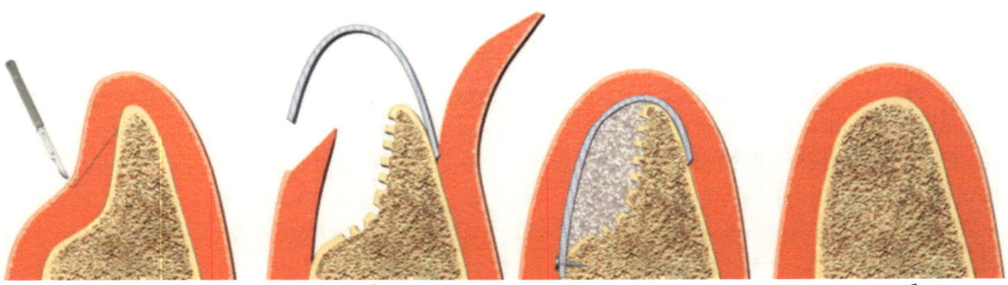

Fig. 2. Guided bone regeneration scheme describing the use of a resorbable barrier membrane. a. bony defect is diagnosed b. the defect is debrided, bone cortex is perforated, membrane supporting scaffold material and a membrane are placed. c. the membrane is stabilized and shaped to dictate the desired bone contours d. a few months later bone regeneration is observed restoring the desired shape of the jaw.

2. Tissue barriers

A wide range of tissue barrier materials has been used in experimental and clinical studies in GTR/GBR procedures, including polytetrafluoroethylene (PTFE), expanded PTFE (e-PTFE), polyglactin 910, polylactic acid, polyglycolic acid, polyorthoester, polyurethane, polyhydroxybutyrate, calcium sulfate, freeze-dried fascia lata,freeze-dried dura mater allografts, native and/or synthetic collagen, micro titanium mesh, and titanium foils. Among the requirements set for GTR/GBR devices are safety and effectiveness. Based on the strict criteria applied to cosmetic/plastic medicine, where no life threatening conditions are involved, documentation of the effectiveness of the procedures and materials should be available while at the same time adverse effects emerging from the implanted devices should be kept to a minimum. In 1994, Hardwick *et al.* formulated a list of criteria for tissue barriers used for GTR; these may also be applied for barriers used for GBR. Briefly, the main essential qualities expected are: a. biocompatibility, b. cell occlusiveness, c. integration by the host tissues, d. clinical manageability and ease to apply, e. space making ability. Since bioabsorbable and biodegradable barriers are degraded in vivo and are absorbed by the body, another additional two requirements need to be fulfilled for these, i.e. that tissue reactions resulting from the resorption of the barriers and it's by-products should be minimal and reversible and that the regenerative process should not be negatively affected (Gottlow 1993). There are a number of factors which are critical for the success or failure of GTR/GBR; these include membrane stability, duration of barrier function, ample blood fill of the area for regeneration, enhanced access of bone and bone marrow-derived cells to that space, and prevention of soft tissue dehiscences over the membrane.

Tissue barriers may be classified according to several criteria. For the purpose of this chapter, however, tissue barriers are divided into two main groups: non-resorbable and resorbable. Briefly, non-resorbable barriers are made of thin sheets of materials, mainly polymers. They are stable, non-degradable and biocompatible. The earliest commercial and most popular non-resorbable membranes were expanded polytetrafluoroethylene (ePTFE) membranes; these became a standard for bone regeneration shortly after GBR has been recognized as accepted dental therapy. Expanded PTFE is a polymer with high stability in biologic systems. It resists breakdown by host tissues and by microbes and does not elicit immunologic reactions (Fig. 3). The main limitation of ePTFE and other non

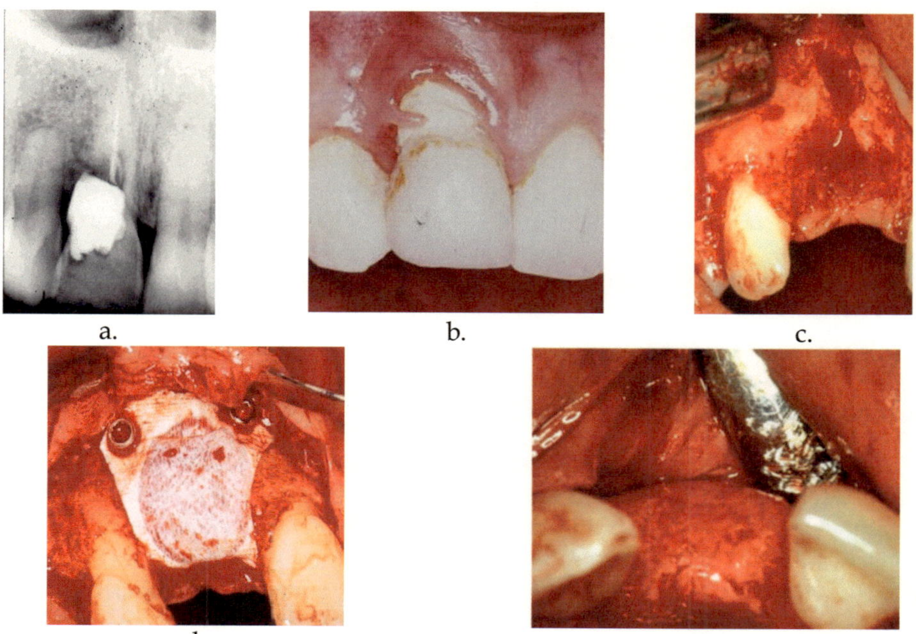

a. b. c. d. e.

Fig. 3. Successful GBR procedure aimed for implant site development using ePTFE membrane. a. radiographic view showing secondary caries and root resorbtion indicate extraction of the upper right central incisor b. clinical view and examination support the radiographic findings c. extraction of the tooth reveals significant bone loss at the esthetic zone d. guided bone regeneration using e-PTFE membrane stabilized by two buccal and one palatal titanium screws, and by demineralized freezed dried bone allograft supporting the sub-membraneous space. e. six months following the procedure the membrane is removed presenting adequate Bone volume and proper contours for an implant supported fixed restoration.

resorbable membranes is the need for an additional surgery for membrane retrieval. Non-resorbable membranes have to be removed during a second surgical intervention imposing significant additional morbidity on the patient and additional risk for tissue damage.

A frequent complication associated with non-resorbable membranes is soft tissue dehiscense and membrane exposure; these may be followed by infection. Animal experiments (Gotfredsen *et al.* 1993; Kohal *et al.* 1999a) and clinical studies (Gher *et al.* 1994; Simion *et al.*1994; Hämmerle *et al.* 1998) have shown that premature membrane exposure and wound dehiscence impair the amount of bone regeneration. Figure 4 presents the exposure and removal of an infected e-PTFE membrane.

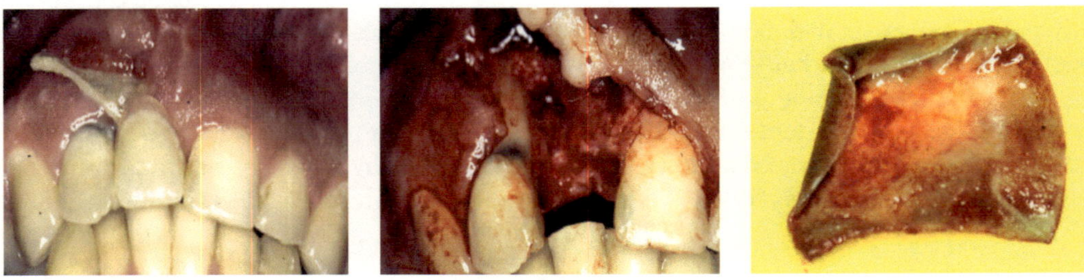

a. b. c.

Fig. 4. Failure of a GBR procedure due to premature exposure and infection of an e-PTFE membrane three weeks following membrane placement. a. soft tissue dehiscence expose the membrane to the oral environment. The membrane is contaminated with dental plaque. b. The membrane is surgically removed and the defect debrided c. clinical view of the e-PTFE membrane after removal.

With the development of bioresorbable membranes, and the increasing evidence regarding their effectiveness, the use of non-resorbable membranes has become limited to specific indications.

3. Bioresorbable membranes

Bioresorbable membranes offer many advantages compared with non-resorbable materials. Apart from the fact that there is no need for a second surgical intervention for removal of the membrane, they present improved soft tissue healing (Lekovic *et al.* 1997, 1998; Zitzmann *et al.* 1997); the incorporation of the membranes by the host tissues and rapid resorbtion if exposed eliminate open microstructures prone to bacterial contamination (Zitzmann *et al.* 1997). Bioresorbable materials that may be used for the fabrication of resorbable membranes belong to the groups of natural or synthetic polymers. The best known groups of polymers used for medical purposes are *aliphatic polyesters* and *collagen*. Over the past decade membranes are made of polyglycolide, polylactide or copolymers thereof or of collagen (Hutmacher & Hürzeler 1995, Tal *et al.* 1991, 1996; Tal & Pitaru 1992; Moses et al 2005, , Friedmann *et al.* 2002, von Arx *et al.* 2001,2005, Rothamel 2005). The qualities of a wide variety of bioresorbable membranes have been investigated experimentally and clinically (Lundgren *et al.* 1994; Mayfield *et al.* 1997; Simion *et al.* 1997; Zitzmann *et al.* 1997; Buser et al 1999, Tal et al 2008a; Tal et al 2008b). Most researchers and clinicians agree that in comparison to non-resorbable membranes, sites treated with bioresorbable membranes present a lower rate of complications, and have thus largely replaced the non-resorbable e-PTFE membranes becoming the standard for most clinical situations. In a few well designed studies which compared bioresorbable and non-resorbable membranes, (Zitzmann *et al.* 1997, 2001; Christensen *et al.* 2003) no significant difference was found between the two treatment modalities. It should be noted however, that Chiapasco *et al.* (2006), who carried out a systematic review of comparitive studies between bioresorbable and non-resorbable membranes, concluded that drawing definite conclusions was impossible due to a lack of sufficient well designed studies.

The development of bioabsorbable membranes for GTR started in the late 80's, soon after the concept had been considered *accepted dental medicine*; in the early 90's however, with the emerging usage of endosseous implants and the growing need for implants site development, regenerative barriers, especially bioabsorbable membranes, received more attention. Therapeutic success, complications and failures were reported, most of which dealing with inflamatory reactions in the tissues adjacent to some bioresorbable membranes, (Sandberg *et al.* 1993; Piatelli *et al.* 1995;Aaboe *et al.* 1998; Kohal *et al.* 1999; Schliephake *and Krachtl.* 1997).

4. Developing collagen membranes

Among the different materials which were experimented and clinically examined for their potential application as regenerative tissue barriers, collagen appeared to be an optimal choice and was considered to meet most requirements expected from bioabsorbable membranes. Collagens are a family of proteins with a well determined triple helical configuration. Among these proteins, collagen Type I is most prevalent, constituting approximately 25% of the body's proteins and about 80% of the connective tissue proteins. Collagen Type I polymerizes to form aggregates of fibers and bundles. Collagens are continuously remodeled in the body by degradation and synthesis. Type I collagen is degraded only by a specific enzyme - collagenase, and is resistant to any non-specific proteolytic degradation. Collagen biocompatibility, biodegradability and low immunogenicity render it advantageous for extensive application in pharmaceutical or biotechnological disciplines.

Collagen is a weak antigen and most of its antigenicity resides in the tellopeptides non-helical terminals of the molecule. These terminals may be removed by enzymes such as pepsin, producing atellocolagen. Atellocolagen's weak antigenicity and weak immunogenicity (Cooperman and Michaeli 1984; Schlegel et al 1997). As well as its relative resistance to degradation, make collagen an optimal choice for implantable devices. Additional advantages that it possess are hemostasis, chemotaxix of periodontal ligament and gingival fibroblasts (Postlethwaite et al 1978, Yaffe et al 1984, Locci et al 1997), easy manipulation and ability to augment tissue thickness (Tal et al 1996). To the best of the author's knowledge, cross-linked medical collagen was first introduced for the purpose of guided tissue barriers in 1984 (Chu 1985, 1987). Using type I and type III dermal bovine collagen, cross linked with Glutaraldehyde (Collagen Inc. Palo Alto, CA) Becker and Tal (1984 - unpublished) have successfully examined the application of collagen membranes on class II furcation lesions in the lower mandibular molar (Fig. 5). It is noteworthy however that extreme care should be taken to avoid any remnants of amids or aldehydes in the final products since if left behind, these may induce a severe inflammation and damage to the peripheral tissues (Fig.6). Further laboratory improvements, turned this gluteraldehyde cross-linked collagen membranes into a safe and effective device (Pfeifer et al 1989; van Swol et al 1993). The main limitation of this membrane (Periogen, Collagen Inc. Palo Alto, CA) was a relative rapid absorbtion time varying from 4 – 8 weeks.

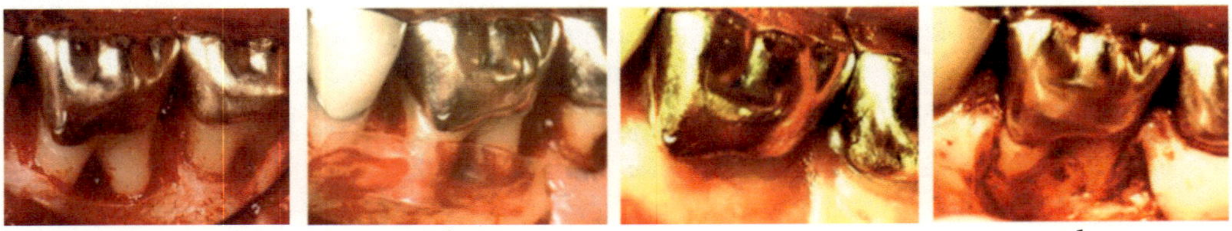

a. b. c. d.

Fig. 5. Clinical presentation of a class II Furcation lesion in a lower first molar tooth treated with gluteraldehyde cross-linked type I bovine collagen membrane. a. furcal defect is carefully debrided b. membrane is placed to seal the class II lesion c. clinical view 1 year after surgery d. one year re-entry presents impressive regeneration of the lesion. (Becker and Tal, 1984; unpublished).

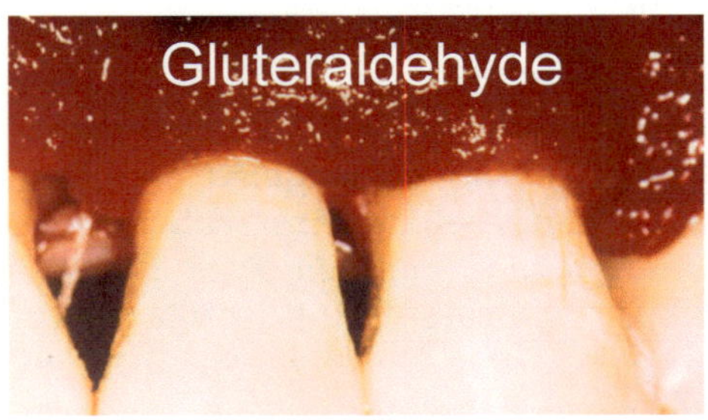

Fig. 6. Severe inflammatory gingival reaction in a patient in whom a prototype glutaraldehyde cross-linked bovine collagen membrane was placed beneath the buccal gingiva of the upper front segment, in an attempt to treat gingival recession. The clinical assumption was that gluteraldehyde remnants left after the cross-linking process were the cause for this severe tissue reaction.

4.1 Ribose induced cross linking of medical collagen

Since the resistance of collagen fibrils towards degradation directly relates to the density of the intermolecular cross-links, various collagen cross-linking methods have been used including aldehyde fixatives and imides and treatments such as hydration and radiations. The main drawbacks of such treatments were toxicity and inability to accurately control the degree of cross-linking. In 1988 Tanaka et al. described a novel technique to cross-link collagen that is safe and clinically effective. Briefly, their invention was based on the fact that a glycated collagen-based matrix may be prepared by non-enzymatic glycosylation of native collagen fibrils to irreversibly cross-link collagen polypeptide chains. The sugars employed as cross-linking agents, especially D-ribose, are typically non-toxic and non-immunogenic. Since the degree of cross-linking correlates well with mechanical and biodegradation characteristics of the collagen matrix, the matrix can be readily bio-stabilized as desired by monitoring the degree of cross-linking during the procedure (Fig. 7). It has been suggested that ribose sugar, practically allows an unlimited degree of cross-linking providing a collagen barrier which is extremely resistant to enzymatic degradation. Clinically, cross-linked collagen membranes retain integrity for longer periods of time (Paul *et al.* 1992).

Fig. 7. Synthetic production of Enzymatic cross-linked medical collagen (modified from Colbar Life Science LTD. Ramat Hasharon, Israel.

4.2 Bilayered collagen membrane with an internal polylactide layer

Experimental investigations showed that degradation of collagen membranes may start within 4 days to 4 weeks after membrane placement (Zhao *et al.* 2000; Owens & Yukna 2001). It was therefore questionable whether rapidly degraded membranes can successfully serve as tissue barriers for GBR. In search for different techniques to delay membrane degradation, von Arx et al (2002) hypothesized that combining a bilayer collagen membrane with an internal polylactide layer may prolong its barrier function, as well as enhance the mechanical properties of this hybrid membrane. They evaluated a prototype biodegradable bioresorbable collagen-polylactide hybrid membrane and compared it to ePTFE membrane placed at similar sites in the canine. Histological and histomorphometrical analysis of 2 months specimens, and percentage calculations for areas showing bone regeneration showed that ePTFE sites healed significantly better than the experimental membrane sites with 53% - 96.9% bone regeneration compared to 57% - 85% respectively. The experimental membrane induced moderate infiltration of lymphocytes and plasma cells adjacent to empty spaces corresponding to polylactide fragments. In addition, these reactions appeared to provoke subsequent resorption of newly formed bone. No such findings were seen in ePTFE sites. The authors concluded that the prototype membrane could not be recommended for clinical use.

4.3 Chemical cross-linked porcine type I and III collagens

In an ongoing search for an improved collagen membrane which combines optimal biodegradation over time, high vascularization, favorable tissue integration, and low foreign body reaction, Rothamel et al (2005) compared the biodegradation of differently cross-linked collagen membranes in rats applying the above mentioned parameters. Five commercially available and three experimental membranes (VN) were included: (1) BioGide (BG) (non-cross-linked porcine type I and III collagens), (2) BioMend (BM), (3) BioMendExtend (BME) (glutaraldehyde cross-linked bovine type I collagen), (4) Ossix (OS) (enzymatic-cross-linked bovine type I collagen), (5) TutoDent (TD) (non-cross-linked bovine type I collagen, and (6-8) VN(1-3) (chemical cross-linked porcine type I and III collagens). Specimens were randomly allocated in unconnected subcutaneous pouches on the back of Wistar rats. Histological examination and histometrical evaluation of explanted specimens representing 2, 4, 8, 16, and 24 weeks showed that the highest vascularization and tissue integration was noted for BG followed by BM, BME, VN and TD, while OS exhibited none. Subsequently, biodegradation of BG, BM, BME and VN was faster than TD. Ossix showed only a minute amount of superficial biodegradation 24 weeks following implantation. Biodegradation of TD, BM, BME, and VN was associated with the presence of inflammatory cells. Within the limits of this study, it was concluded that cross-linking of bovine and porcine-derived collagen types I and III was associated with prolonged biodegradation, decreased tissue integration and vascularization, and in case of TD, BM, BME and VN foreign body reactions (Fig 8).

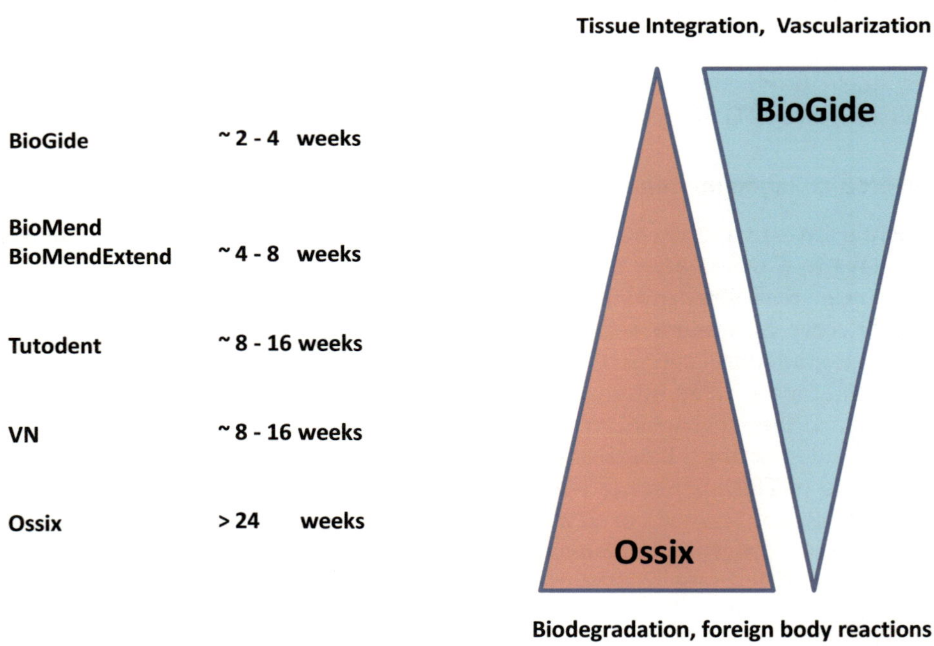

Fig. 8. Prolonged biodegradation seems to be associated with tissue integration, vascularization, foreign body reactios and frequency of tissue dehiscence. Partially modified from Rothamel et.al.(2005).

5. Collagen barrier membranes exposed to the oral environment by spontaneous mucosal dehiscence

Most commercially available collagen barriers have extensively been investigated in both animal models, (Pfeifer *et al.* 1989; Tal *et al.* 1991, 1996; Hyder *et al.* 1992; Tal & Pitaru 1992; Crigger *et al.* 1996) and human studies (Van Swol et al 1993; Al Arrayed *et al.* 1995; Zitzmann *et al.* 1997; Tal 1998, 2004) presenting comparable clinical results to non-resorbable membranes (Cortellini *et al.* 1996; Caffesse *et al.* 1997). It is generally agreed that collagen membranes show a lower incidence of spontaneous exposure to the oral environment compared with non-resorbable membranes, and unlike non-resorable membranes, soft tissue healing following exposure of collagen membranes involves no infection (Friedmann *et al.* 2002; Moses *et al.* 2005). In a comparative study between prematurely exposed non-resorbable membranes (ePTFE), non-cross-linked collagen membranes (BioGide) and cross-linked collagen membranes (Ossix), the latter were claimed to be superior, and capable of supporting healing (Moses *et al.* 2005). This finding was partially explained by the authors' impression that in prematurely exposed cases, the cross-linked collagen membranes has the capacity to withstand bacterial collagenolytic degradation while facilitating soft tissue healing over the exposed membranes (Moses *et al.* 2005). In an attempt to better understand collagen barrier bio-durability and integrity in sites treated with cross-linked and non-cross-linked collagen barrier membranes, Tal et al (2008a) have clinically and histologically investigated GBR treated sites showing spontaneous mucosal perforations over barrier membranes in humans. In 52 patients, bony defects were grafted with xenograft and covered with collagen resorbable barrier membranes: 26 with cross-linked collagen membranes (*Ossix*™) and 26 with non-cross-linked collagen membranes (BioGide®) (Fig.9). Post-surgical spontaneous mucosal perforations and membrane exposures were recorded. Surgical sites were monitored for 6 months before implant placement.

During implant placement procedure, full thickness soft tissue discs, 3 mm in diameter, were punched out from 18 *Ossix*™ and 18 BioGuide® sites for histologic examination. Of the 52 sites, 33 were intact and 19 (36.5%) were associated with spontaneous membrane exposure: 13 (50%) were in *Ossix*™ sites and 6 (23.1%) in *BioGuide*® sites ($p < 0.05$). Perforation sites clinically healed within 2-4 weeks over both CL and NCL membranes. Histologically, *Ossix* membranes were intact in all 9 non-perforated sites, but in the perforated *Ossix*™ sites, membranes were interrupted in 5 and undetected in four. *BioGuide*® membranes were undetected in all 18 specimens examined. In 3 non-perforated *Ossix* membrane sites, ossification was associated with or within the membrane. It was concluded that *Ossix*™ membranes were more resistant to tissue degradation than *BioGide*®. At non-perforated sites, *Ossix*™ membranes maintained integrity during the 6 months study period of time while *BioGide*® membranes disintegrated (Fig.10).

Neither CLM (*Ossix*™) nor NCLM (*BioGide*®) were resistant to degradation when exposed to the oral environment. *Ossix*™ membranes were associated with a significantly higher incidence of tissue perforations compared with *BioGide®e* membranes (Fig. 11).

There is no satisfactory explanation for the increased incidence of spontaneous exposure incidence at CLM treated sites compared with NCLM sites. Based on the observation that

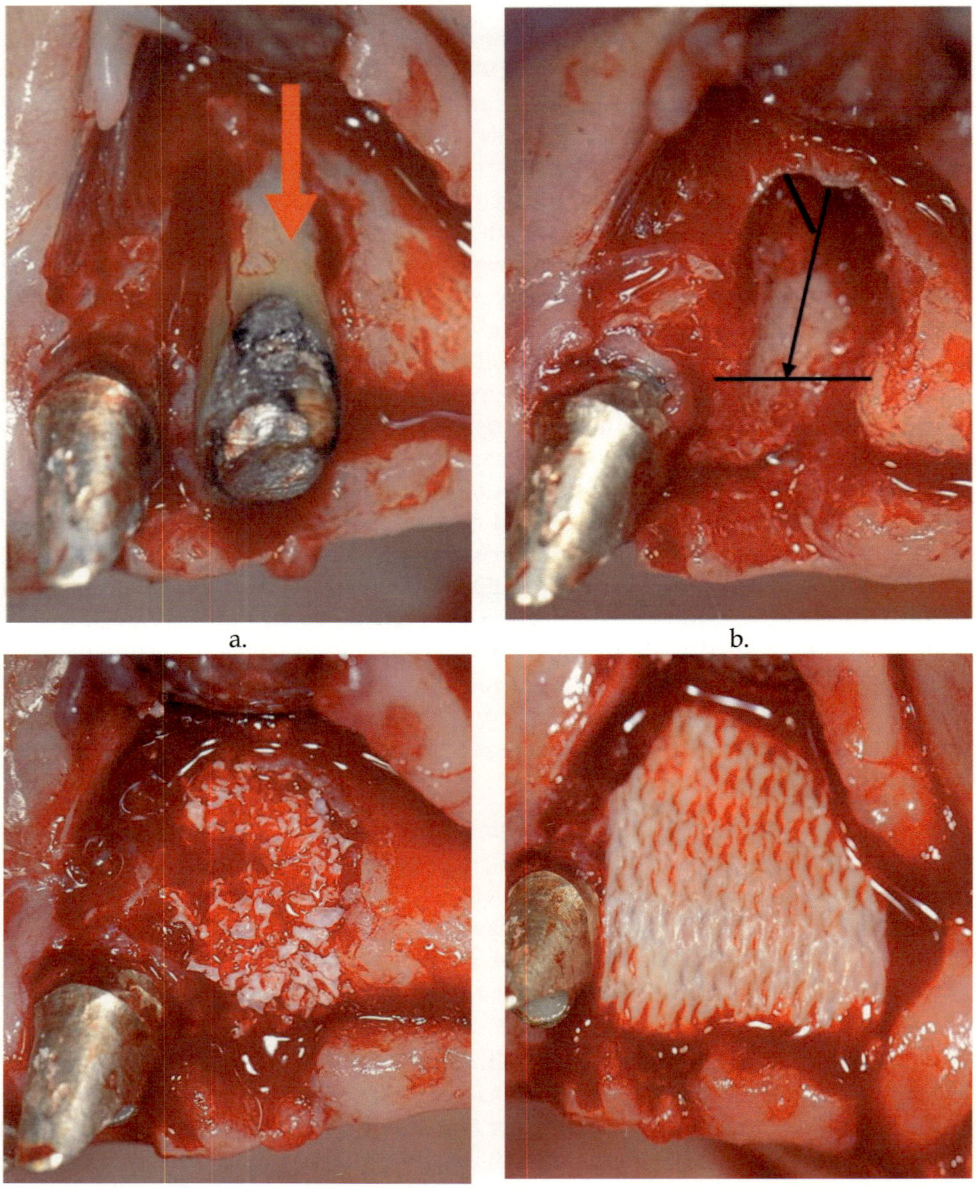

a. b. c. d.

Fig. 9. Clinical view of a ridge preservation procedure a. severe bone loss and complete resorbtion of the buccal plate of the alveolus associated with a longitudinal root fracture. b. the fresh extraction socket is debrided and measured c. the socket and bony defect are grafted with xenograft d. the grafted socket and defect are covered with a cross-linked collagen resorbable membrane.

the degree of cross-linking is negatively correlated with the attachment and proliferation of PDL fibroblasts and human osteoblast-like cells observed in in-vitro cultures (Rothamel *et al.* 2004), the tendency of the CLM to split from the adjacent connective tissue may be due to lack of attachment between the two, and lack of membrane vascularization in the early healing phase. This phenomenon has been described by Rothamel et al. (2005) who examined the biodegadation of differently cross-linked collagen membranes in the rat and found that the Ossix™ membrane exhibited only slight superficial vascularization and minimal surface resorption after 24 weeks compared with Bio-Gide® - a native dermal porcine collagen which showed early vascularization and over 80% resorption after 8 weeks. It should be stressed that in their animal study Rothamel et al (2005) made complete tissue closure in extra-oral subcutaneous pouches, while in our human study (Tal *et. al.* 2008a) a significant number of sites were spontaneously exposed to the oral environment. However, bone growth patterns shown in this study, included membrane ossification beneath the membrane leaflets and new bone growth adherent to the membrane (Figs. 12,13), suggesting high bio-compatibility between the membrane and osteoblasts (Rothamel *et al.* 2004). Although this study was limited to evaluation of membrane bio-durability rather than sub-membraneous bone regeneration, it is noteworthy that a recent histological evaluation of changes during ossification and cellular events at GBR sites, has shown that in the rat maxillae, NCLM (BioGide®) associated with defect-derived bone had integrated to such a degree that it was difficult to distinguish the membrane-derived new bone from that generated in the cavity (Taguchi *et al.* 2005). Based on these observations, these authors suggested that NCLM participated in the osteogenic differentiation.

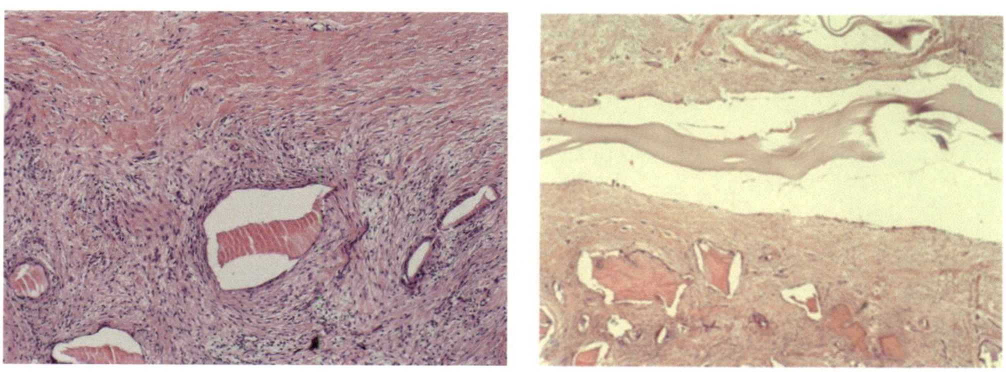

a. b.

Fig. 10. Histological examination of non perforated sites 6 months after membranes placement. a. the native collgen membrane (*BioGide®*) completely disintegrated, and its location can be identified by the abrupt transition from mature supramambraneous connective tissue to young cellular submembraneous connective tissue b. the cross-linked membrane (*Ossix™*) maintained integrity and separates between the mature supramambraneous connective tissue and the newly established submembraneous connective tissue.

The findings described by Tal et al (2008a,b) are in agreement with the report of von Arx *et al.* (2005) who claimed that in the rabbit calvarium cross-linked collagen membranes clearly displayed prolonged membrane integrity compared with non-crossed-linked collagen

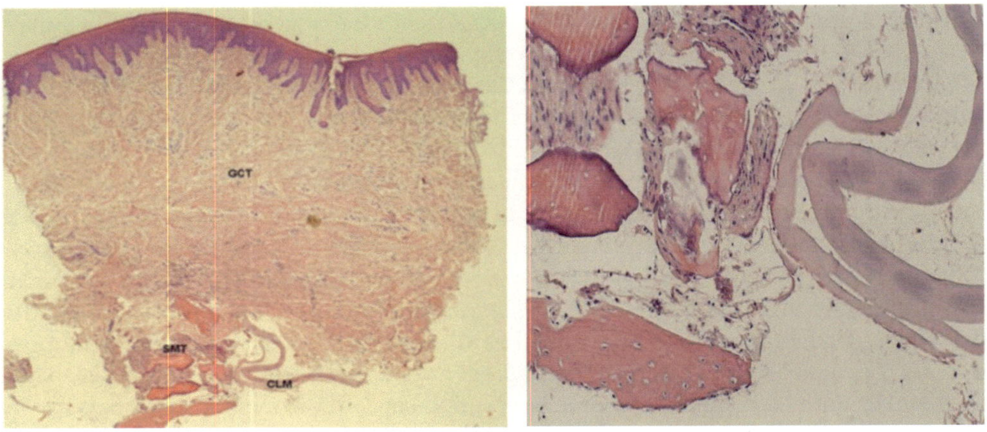

a. b.

Fig. 11. a. Histological aspect of CLM remnants in a perforated site. The membrane (CLM) is interrupted under the perforation and identified at the peripheral borders of the specimen (H&E, original magnification x20). b. Higher magnification (x40) from a. showing membrane remnants (right) associated with graft particles and new bone growth.

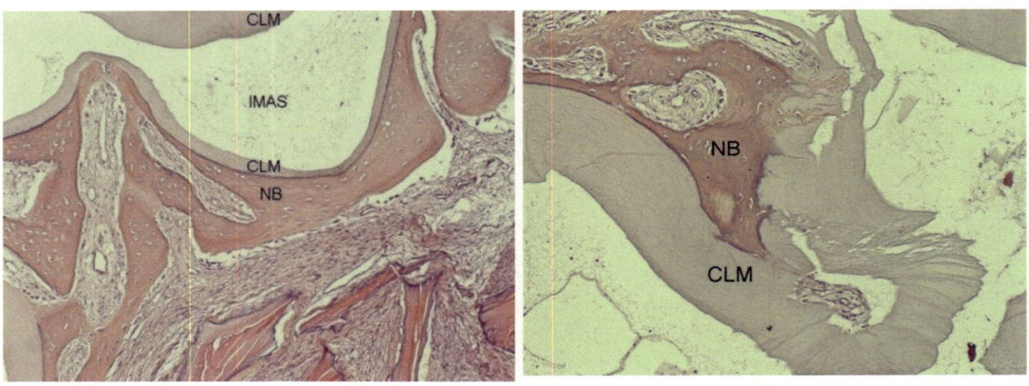

a. b.

Fig. 12. Histologic view of CLM remnants in a non-perforated site showing a. new bone adhered to the membrane, (H&E, original magnification x20). b. Occasionally bone growth is observed between the membrane leaflets penetrating the membrane (CLM) and perhaps replacing it (H&E, original magnification x40).

membrane. However, since their experiment was performed in a close wound, no tissue inflammatory response was involved. The studies of Tal et al (2008a,b) disagree with the interpretation that in prematurely exposed cases crossed-linked collagen membranes have "the capacity to withstand bacterial collagenolytic degradation" when exposed by soft tissue dehiscence (Moses *et al.* 2005). It is noteworthy that long epithelial ridges were observed in several of the healed perforation lesions associated with the crossed-linked group (FIG 14). Similar epithelial ridges have been previously described by Tal and Dayan (2000) and Tal *et al.*(2001) in human specimens retrieved from mucosal early perforations over submerged implants. While no scientific proof is available, it may be logical to assume that these epithelial ridges are remnants of epithelial down-growth that separates the exposed membrane from the perforated gingival connective tissue during early healing and before the exposed membrane resorbs.

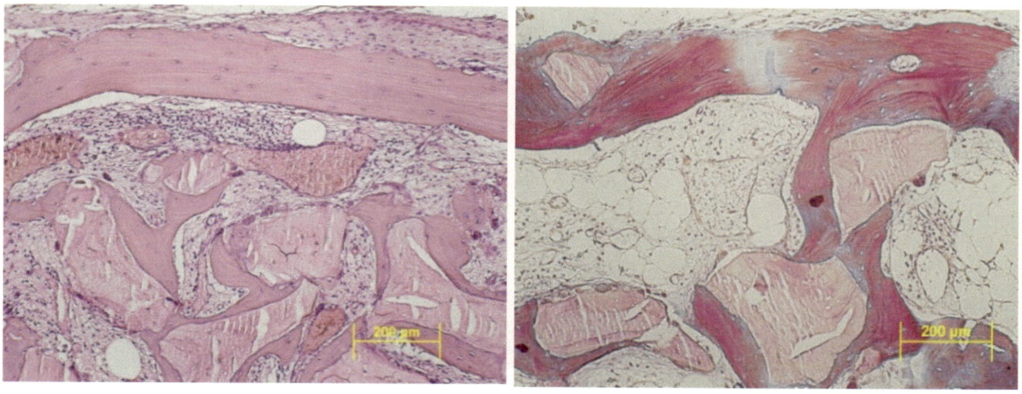

a. b.

Fig. 13. Histologic view of new bone adhered to an ossified CLM in a non-perforated GBR site; new bone in intimate contact with graft particles and the resorbed membrane is observed. a. H&E, original magnification (x100). b. (x100).

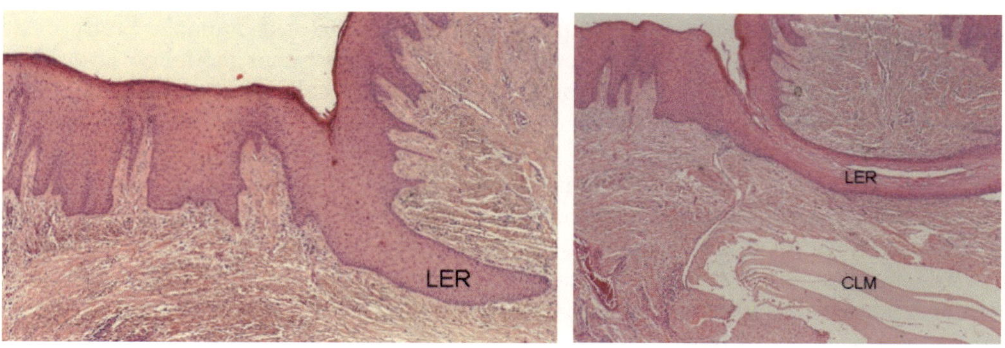

a. b.

Fig. 14. Histological aspect of a CLM specimen from a perforated site. Long epithelial ridge (LER) is shown at the periphery of the healed perforation. b. Long epithelial ridge (LER) is shown along the cross-linked membrane remnants (CLM), peripheral to the perforation site. A thin layer of connective tissue separates it from the membrane (H&E, original magnification x20).

To further understand the bio-degradation of cross-linked and non-cross-linked collagen membranes, an animal study was undertaken in which membranes were experimentally exposed to the oral environment (Tal et al 2008b). In 8 cats, 48 surgical procedures were performed, 3 along each side of the palate: 32 soft tissue perforations were made, and 16 full thickness mini-flaps were raised. Cross-linked and non-cross-linked collagen membranes discs were placed either under the perforations and peripheral mucosa and left exposed (experimental), or covered by the flaps (controls) (FIG.15).

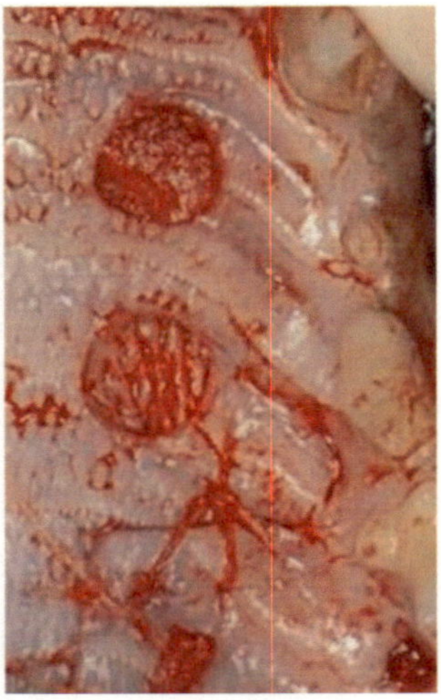

Fig. 15. Clinical view of two experimental perforation sites (top/middle) and one control site (bottom) in the cat palate. Top - Cross-linked collagen membrane disc placed under the surgically perforated mucosa; Middle – non cross-linked membrane disk placed under the surgically perforated mucosa; Bottom - membrane disk placed under surgically elevated mucosal flap and then covered (control).

The four treatment modalities were equally distributed among 8 animals. Study design provided 7 and 28 days histological specimens for each treatment modality. Histological observations revealed that cross-linked and non-cross-linked collagen membranes remained intact in the control sites during the 28 days study period. At 7 and 28 days, Cross-linked membranes appeared interrupted in 3 and 2 experimental sites, respectively, and were undetected in the remaining experimental sites. There was no statistical difference between control specimens and between CLM and NCLM of the different treatment modalities (Fig. 16). It was concluded that if covered, both cross-linked and non-cross-linked membranes were resistant to tissue degradation and maintained continuity throughout the study. However, none of the membranes was resistant to degradation when exposed to the oral environment, even after 7 days.

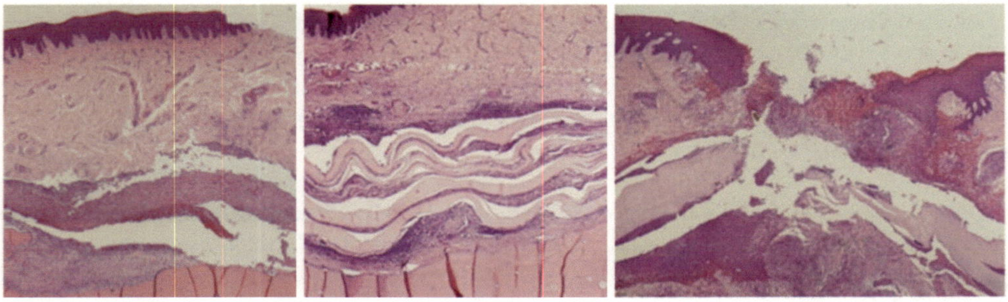

a. b. c.

Fig. 16. Histologic view of a. 28-day control NCLM specimen. The NCLM presents intimate contact with the palatal bone on the one side and with the soft connective tissue on the other side. b. 28-day control CLM specimen. The CLM presents intimate contact with the palatal bone on the one side and with the soft connective tissue on the other side. Connective tissue penetration between the membrane and the bone is present and membrane separation to leaflets is observed. c. 7-day experimental CLM specimen showing disintegration of the membrane under the perforation. Submembraneous granulation tissue growth and peripheral epithelial migration are observed. Inflammatory infiltrate is abundant around and between the membrane layers.

6. Increasing the longevity of non cross-linked collagen membranes

6.1 Double layer native non cross-linked collagen preparation

In search for formulas that will extend the functional activity of collagen membranes *in vivo,* Kozlovsky et al (2009) have evaluated the bio-degradation of a two layer preparation of a native Type I and Type III porcine collagen membrane (Bio-Gide® (BG) (Geistlich, Wolhusen, Switzerland)) and compared it with that of the commercially available single layer membrane. It has been shown that four weeks after subcutaneous implantation of BG membrane in the rat clavaria, reduction in membrane thickness combined with nearly complete biodegradation was reported (Zhao et al 2000, Rothamel et al 2005). When applied in surgical pouches in the palate of mongrel dogs (Owens & Yukna 2001) moderate to complete degradation was reported 4-8 weeks following implantation. It has been suggested that application of a second layer of Bio-Gide® (BG) membrane (double layer technique) may reduce micro movement and improve its stabilization, thus enhancing its protective effect in the sub-membranous augmented area (von Arx & Buser 2006). Kozlovsky et al (2009) investigated the influence of the additional layer of the barrier membrane on its durability by measuring quantitatively the in vivo degradation of barriers composed of one vs. two layers of Bio-Gide® The degree of membrane disintegration was histochemically measured based on the amount of residual membrane collagen labeled with biotin (Zohar *et al.* 2004) . Two circular bony defects, 5-mm in diameter, were made in the calvaria of Wistar rats. Bio-Gide® membrane 5-mm diameter discs, labeled with biotin, were placed in these defects either as a mono layer (MLM) or as a double layer (DLM) (Fig 17).

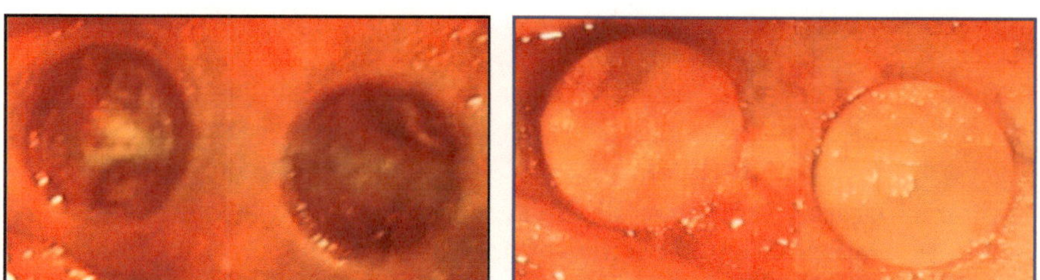

a. b.

Fig. 17. Photographs of the surgical procedure. (a) on the mid-line of the parietal bone, (along the sagittal suture) two similar, 5-mm diameters, approximately 1 mm deep) bony defects are prepared. (b) The labeled membrane discs are laid over one defect as a mono layer membrane (MLM) (right) and over the other defect as double layer membrane (DLM) (left)

Rats were sacrificed after 4 and 9 weeks and histology performed. Membranes were stained with Horseradish peroxidase-conjugated Streptavidin and Aminoethyl Carbazole as substrate for detection of biotinylated collagen. Collagen degradation in the mono layer BG ranked 60% during 4 weeks post application. The use of two layers of membranes resulted in a similar (60%) resorption rate of the initial collagen content however, since the initial total collagen content in the double layer membrane was roughly twice the amount of the monolayer one, the amount of residual collagen at 4 weeks was significantly greater in the double layer sites. The rate of collagen degradation for mono layer and double layer membrane sites at 9 weeks was similar (~80%), with twofold amount of residual collagen content at the DLM sites. In addition, the residual thickness of the DLM compared with the MLM was 81% and 74% after 4 and 9 weeks respectively. Thus, it has been shown that despite of similar collagen degradation rates of both MLM and DLM, the application of a second layer of Bio-Gide® (BG) membrane (double layer technique) results in a significantly greater residual amount of collagen, at least up to 9 weeks post surgery in rats. Since membrane thickness was reduced by ~30% between 4 and 9 weeks while the collagen surface area was reduced by 50-60% during that period, it may be suggested that the main pattern of resorption is an internal rather than an external one. Since the Bio-Gide® membrane is structured like an interconnected porous system with large interstices, it is most suitable for transmembranous formation of blood vessels (Schwarz et al 2006), facilitating its resorption (Rothamel et al 2004, Schwarz et al 2006). Indeed the transmembranous vascularization of the membrane was manifested histologically already 4 weeks following implantation and become well-defined through all the layers of the membrane 9 weeks following implantation (Fig 18).

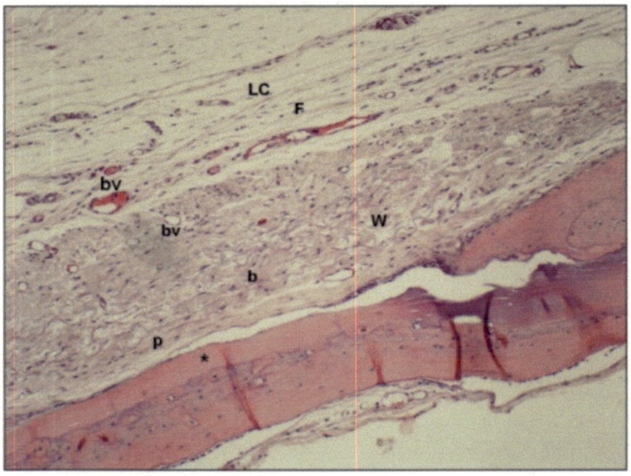

Fig. 18. Photomicrographs of hematoxylin and eosin-stained mono-layer membrane (MLM), 9-weekpost-implantation. Original magnification × 100. bv- blood vessel; p- periosteum-like structure; b- eosinophilic collagen bundles; w- wavy short fibers; LC- loose connective tissue;.

Schwarz et al (2006) proposed that the vascularization process may also contribute to membrane degradation since the monocytes penetrating through the blood vessel wall may differentiate into macrophages.Kozlovsky et al (2009) reported that in spite of the difference in the thickness of the 2 membrane preparations, similar degradation rate of 80% for both membranes was measured at 9 weeks. Since the trans-membranous formation of blood vessels is essential for collagen resorption (Schwarz et al 2006), it seems that the vascularization of the double layer membrane was not impaired by its increased thickness.

It has been claimed that increasing the density of cross links between collagen molecules has a negative effect on membrane biocompatibility (Schwarz et al 2006, Rothamel et al 2004), membrane to tissue integration and vascularization, and inhibits attachment and proliferation of PDL fibroblasts and osteoblasts (Rothamel *et al.* 2004, 2005). Using a second layer of resorbable, cross-linked membrane avoids these disadvantages, while extending membrane longevity. In the double layer 9 weeks membrane specimens, central intra - membrane neo-ossification was clearly identified with collagen fibers embedded in the osteoid (Fig 19), confirming a phenomena previously described by others (Taguchi et al 2005, Tal et al 2008).

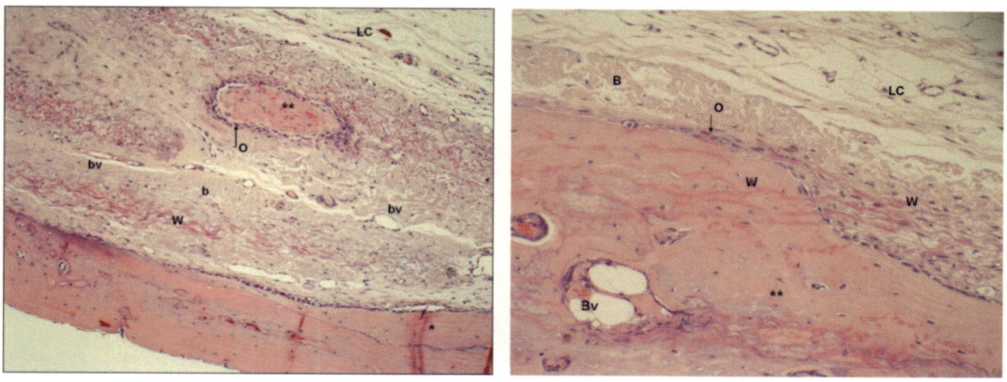

a. b.

Fig. 19. Photomicrographs of a double-layer membrane (DLM) membranes 9-week post-implantation. a. P-periosteum- like structure; B - eosinophilic collagen bundles; W - wavy short fibers; LC - loose connective tissue ;BV-blood vessel; O - osteoblasts. *Newly formed bone at the floor of the defect; **central ossification. (Hematoxylin and eosin Original magnification × 100). b. Newly formed bone (**) adjacent to the membrane collagen bundles (B) with osteoblasts (O) arranged on the surface. Strongly eosinophilic wave-like long collagen fibers (W) of the membrane penetrate the osteoblastic layers and incorporate into the bone matrix of the centrally neo-ossified membrane. BV, blood vessel; LC, loose connective tissue (Hematoxylin and eosin; Original magnification × 200).

While this finding has never been fully understood, it may be speculated that the significant increase in membrane thickness and longevity result in increasing angiogenesis and cellular population of the collagen matrix, leading to cell proliferation, differentiation and ossification.

6.2 Increasing collagen structural stability by collagen membranes immersion and systemic tetracycline administration

In vivo breakdown of medical collagen has thoroughly been investigated. Collagenase, the enzyme responsible for collagen biodegradation, belongs to the matrix metalloproteinases (MMP's) family of enzymes which are normally present in Mammalian tissues. Collagenase which is produced by a number of body tissues and cells, degrades collagen as part of the physiological connective tissue remodeling. Neutrophils, monocytes and fibroblasts are believed to play a major role in collagen degradation mechanisms in wounds healing by releasing MMPs (Reynolds 1994, Armstrong 2002). In vivo biodegradation of collagen membranes depends on collagenolytic activity; it has, therefore, been suggested that in addition to increasing the collagen structural stability by cross-linking, inhibition of MMP's may further slow down the collagen degradation (Golub *et al* 1998). It has been shown that Tetracycline, an antibiotic with anti-collagenolytic properties, inhibits MMPs activity (Golub *at al* 1997, Greenwald *at al* 1998). It has also been shown that the use of tetracycline-coated expanded polytetrafluoroethylene (ePTFE) barrier membranes results in additional gain of clinical periodontal attachment, most likely due to its antimicrobial activity (Zarkesh *et al.* 1999). Modulation of wound-healing has been successful applying sub dosage formulation of TTC, lacking antimicrobial activity while retaining the MMP inhibitory capacity. This phenomenon probably occurs due to TTC inhibition of extra-cellular MMP, likely due to its chelating activity on Ca^{++} and Zn^{++} ions (Golub *et al.* 1987). Furthermore, inflammatory cytokines including TNF-alpha, IL-1 beta, and IL-6 are markedly down-regulated in patients during treatment with tetracyclines. This phenomenon also reduces the amount of MMP's present in inflamed tissues, contributing to a reduction of the collagenolytic activity (Chung *et al* 1997). The effect of immersing collagen membranes in varying TTC concentration solutions on the in vitro degradation rate was evaluated (Moses *et al.* 2001) . Membranes were incubated in either phosphate buffered saline or with TTC-HCl dissolved in concentrations of 5 mg/ml, 50 mg/ml or 100 mg/ml. The TTC impregnated and non-impregnated membranes were incubated with either bacterial collagenase or cultures of human bone lineage cells. Membrane degradation was examined on days 2, 4, 7, and 14. It was concluded that collagen membranes immersed in 50 mg/ml TTC solution exhibited the longest degradation time, both in the clostridial collagenase and the human bone cell lineage assays. Immersion in a 50 mg/ml TTC solution before implantation was, therefore, considered most effective in delaying collagen membranes degradation (Moses *et al.* 2001) [FIG 20].

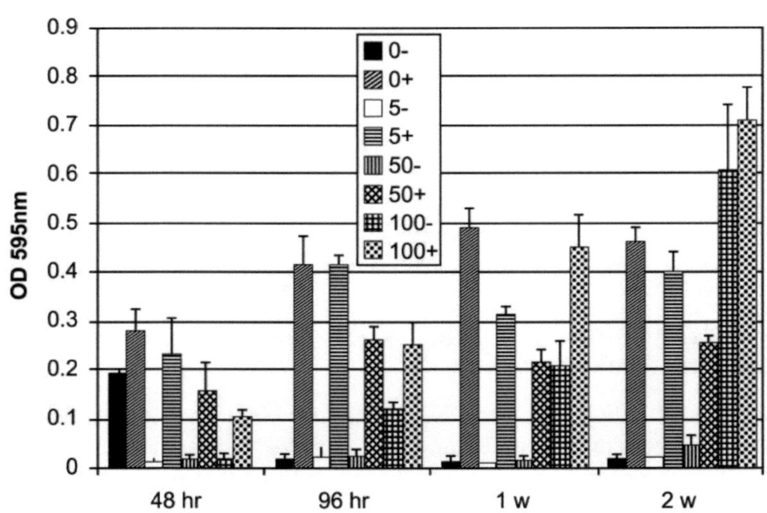

Fig. 20. Degradation of collagen membrane (measured as optical density (OD) at the 595 nm wave length with (+) and without (-) collagenase in different Tetracycline concentration.

These findings suggested the hypothesis that immersion of collagen in TTC prior to implantation could delay collagen degradation also in-vivo. The effect of soaking collagen membranes in different concentrations of tetracycline hydrochloride solutions (TTC) on membranes degradation rate in the rat calvaria (Zohar *et al.* 2004) was evaluated. Prior to implantation in the rat calvaria membranes were labeled with Biotin. The histological slides were stained with Avidin and horseradish peroxidase (HRP) to detect remnants of biotinylated collagen. Staining intensity was correlated to the amount of the remaining collagen and analyzed by image-analysis software. It was found that the staining intensity of membranes that were soaked in 50 mg/ml of TTC exhibited > 11-fold higher intensity than PBS soaked membranes (control) (Figs. 21-23). It was concluded that soaking collagen membranes in 50 mg/ml TTC solution is an effective, practical, and simple tool to reduce membrane degradation rate in vivo.

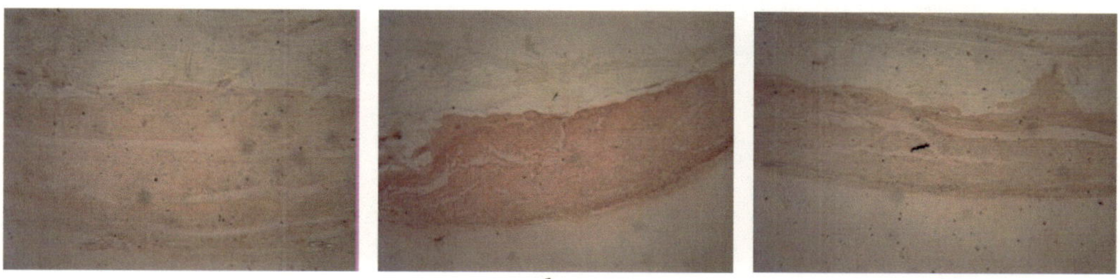

a. b. c.

Fig. 21. Histological view (x40) of the membranes 7 days after implantation with different tetracycline (TTC) concentrations. a. Phosphate-buffered saline alone (0 mg/mL TTC); b. 50 mg/mL TTC. c. 100 mg/mL TTC. Collagen stained in red/brown with Avidin-Biotin-HRP reaction.

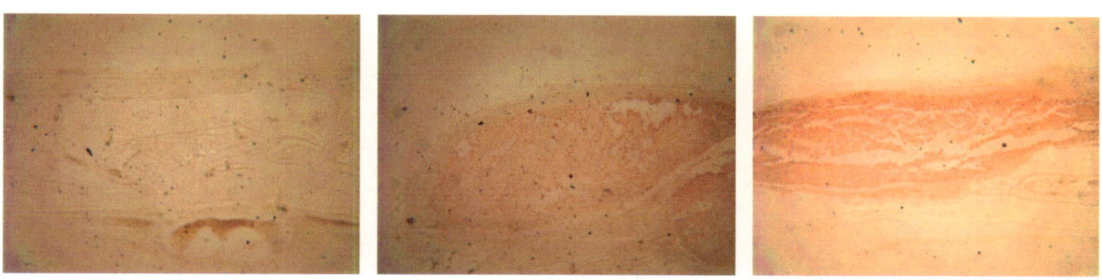

a. b. c.

Fig. 22. Histological view (x40) of the membranes 21 days after implantation with different tetracycline (TTC) concentrations. a. Phosphate-buffered saline alone (0 mg/mL TTC); b. - 50 mg/mL TTC. c. 100 mg/mL TTC. Collagen stained in red/brown with Avidin-Biotin-HRP reaction.

The question of whether administration of subantimicrobial dose of TTC may have an effect on the degradation of collagen membranes in vivo still remained. Evaluation of the in vivo degradation of collagen membranes treated by combined TTC immersion and systemic administration in the rat have showed that immersion of collagen membranes in TTC solution prior to their implantation and systemic administration of TTC significantly decreased membranes' degradation (Figs. 24, 25). It was therefore concluded that this technique may offer a treatment alternative to reduce bio-degradation and enhance bio-durability of certain collagen membranes.(Moses *et al.* 2010).

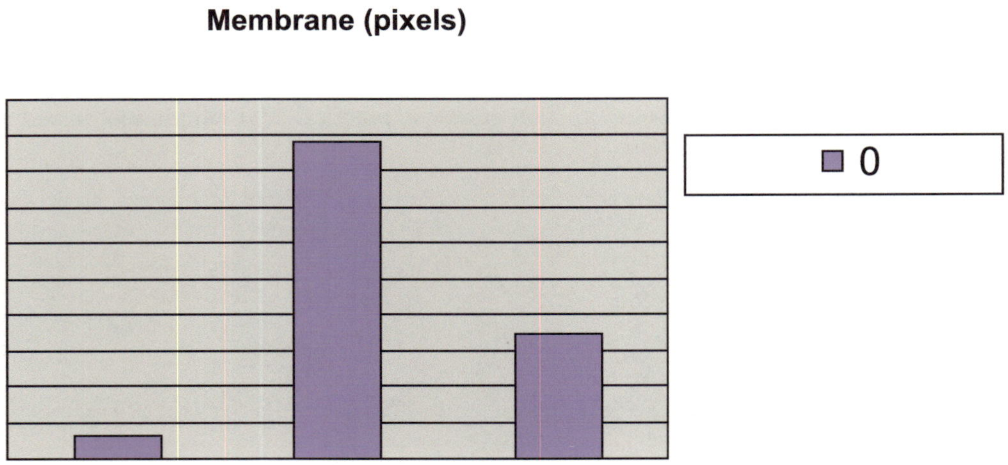

Fig. 23. Color intensity of biotinylated material as measured by number of colored pixels within collagen membrane discs from all 3 groups, 3 weeks.

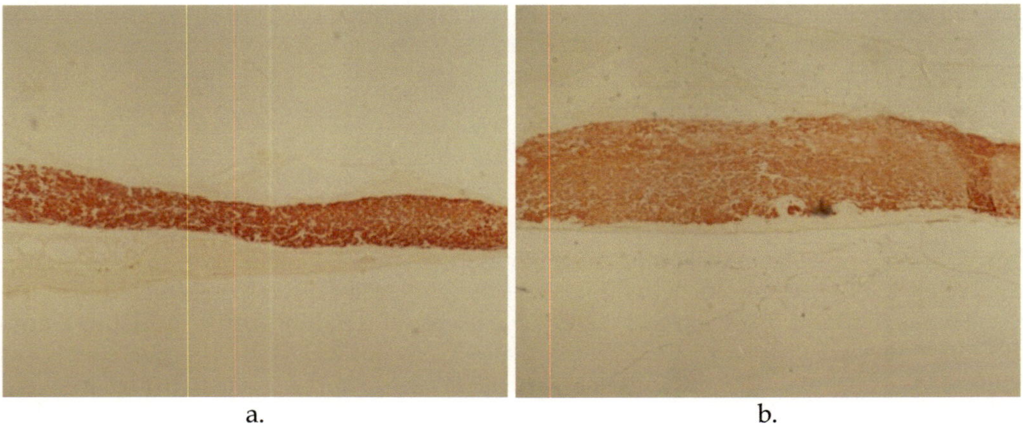

Fig. 24. Histological view (x40) of the membranes 21 days after implantation with different tetracycline (TTC) concentrations with systemic administration of TTC. a. Phosphate-buffered saline alone (0 mg/mL TTC); b. 50 mg/mL TTC. Collagen stained in red/brown with Avidin-Biotin-HRP reaction.

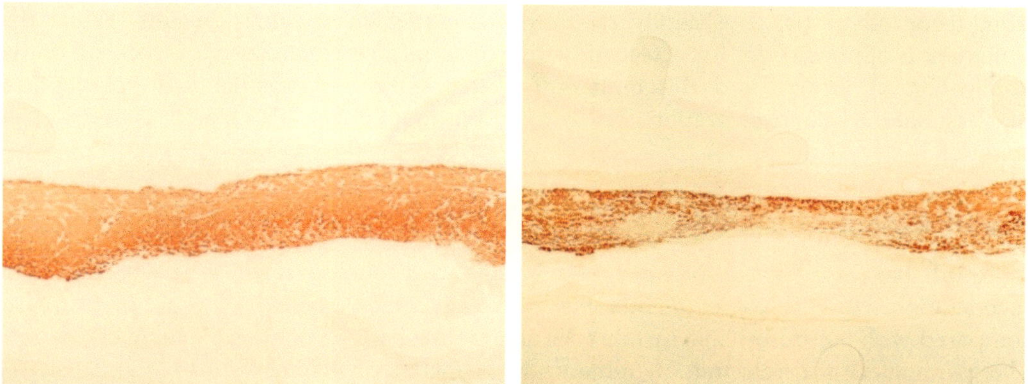

a. b.

Fig. 25. Histological view (x40) of the membranes 21 days after implantation with different tetracycline (TTC) concentrations without systemic administration of TTC: a. Phosphate-buffered saline alone (0 mg/mL TTC); 2. 50 mg/mL TTC. b. Collagen stained in red/brown with Avidin-Biotin-HRP reaction.

7. Membrane supporting materials

Publishd reports have shown that a wide range of membrane supporting materials are suitable fulfilling the requirements set, making harvesting of autogenous bone unnecessary for most procedures (See Chapters 11-14). It has been shown that in self membrane-supporting defects, where there is no need to support the membrane in order to provide space, tissue regeneration is superior to that achieved by using membrane supporting agents; in fact, the optimal osteoconductive membrane supporting material is one that interferes the least with bone regeneration within the membrane occluded space i.e. blood clot (Fig. 26). Among many materials, deproteinized bovine bone mineral, a fully synthesized homogenous hydroxyapatite and beta tricalcium phosphate, and demineralized and mineralized freezed

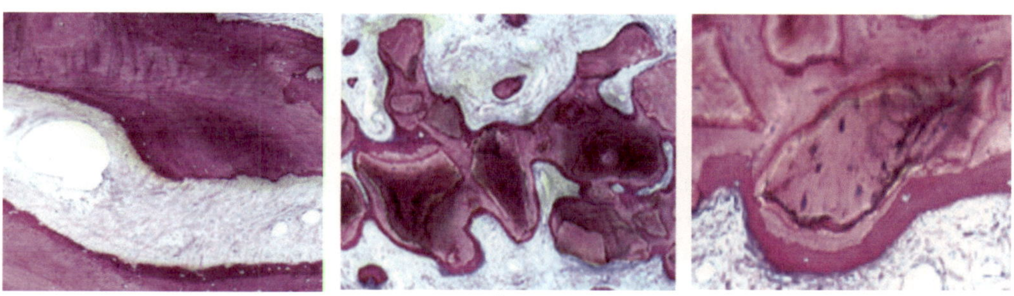

a. b. c.

Fig. 26. Experimental guided periodontal regeneration in the dog. a. new bone growth (above) in a membrane provided space filled with natural coagulum. New cementum (below) and separating periodontal ligament were also regenerated. b. GTR of the same space filled with deproteinized Bovine Bone Mineral. Paticles of the xenograft intimately surrounded by bone are observed c. higher magnification from b. showing a particle of deproteinized Bovine Bone Mineral intimately surrounded by new bone.

dried bone allografts, consistently demonstrated satisfactory clinical results. While this statement is applicable for for osteoconductive materials, current research is focused on the application of growth and differentiation factors - natural proteins and polypeptides that regulate tissue regeneration. To learn more about these, the reader is referred to chapters 1-5.

8. Conclusions

Bioresorbable and non-resorbable membranes can successfully be used for bone regeneration. Bioresorbable membranes generally show better clinical performance compared with non-resorbable membranes and are the barriers of choice wherever possible. The parameters for selecting membranes also include mechanical properties, risk for spontaneous early exposure and ease of clinical handling. The search for improved barriers is underway; presently, collagen's biocompatibility, biodegradability and low immunogenicity render it advantageous for application in pharmaceutical or biotechnological disciplines in general, and in guided bone regeneration in particular. Among the membranes examined by our group *Ossix*™* (heavily cross-linked collagen) was found to be the most resistant to biodegradeation but showed the least tissue integration and vascularization, while *BioGide* (native, non-cross-linked collagen) presented high tissue integration and vascularization. Clinically, *Ossix*™* was associated with a higher rate of soft tissue dehiscence followed by rapid resorbtion following membrane exposure to the oral environment.

9. Acknowledgement

Production of this chapter was supported by The Gerald A Niznick Chair of Implantology.

10. References

Aaboe, M., Pinholt, E.M., Schou, S. & Hjørting-Hansen, E. (1998). Incomplete bone regeneration of rabbit calvarial defects using different membranes. *Clinical Oral Implants Research* 9, 313–320.

Al Arrayed, F., Adam, S., Moran, J. & Dowell, P. (1995) Clinical trial of cross-linked human type I collagen as a barrier material in surgical periodontal treatment. *Journal of Clinical Periodontology* 22, 371-379.

Armstrong DG, Jude EB. (2002). The role of matrix metalloproteinases in wound healing. J Am Podiatr Med Assoc: 92:12-8.

Buch, F., Albrektsson, T. & Herbst, E. (1986). The bone growth chamber for quantifi cation of electrically induced osteogenesis. *Journal of Orthopedic Research* 4, 194–203.

Buser D., Dula, K., Hess, D., Hirt, H. P. & Belser, U. C. (1999) Localized ridge augmentation with autografts and barrier membranes. *Periodontology 2000* 19, 151-163.

Caffesse, R. G., Mota, L. F., Quinones, C. R. & Morrison, E. C. (1997) Clinical comparison of resorbable and non-resorbable barriers for guided periodontal tissue regeneration. *Journal of Clinical Periodontology* 24, 747-752.

* Ossix-P ™ Colbar, Ramat Hasharon, Israel

Chiapasco, M., Zaniboni, M. & Boisco, M. (2006). Augmentation procedures for the rehabilitation of defi cient edentulous ridges with oral implants. *Clinical Oral Implants Research* 17 (Suppl 2), 136–159.

Christensen, D.K., Karoussis, I.K., Joss, A., Hammerle, C.H. &Lang, N.P. (2003). Simultaneous or staged installation with guided bone augmentation of transmucosal titanium implants. A 3-year prospective cohort study. *Clinical Oral Implants Research* 14, 680–686.

Chu G. (1985). Process for preparing malleable collagen and the product thereof. United States Patent No. 4,557,764

Chu G. (1987). Collagen membranes for medical use. United States Patent No. 4,689,399

Chung CP, Kim DK, Park YJ, Nam KH, Lee SJ. (1997) Biological effects of drug-loaded biodegradable membranes for guided bone regeneration.bJ Periodontal Res;32:172-5

Cooperman L, Michaeli D. (1984). The immunogenicity of injectable collagen. I. A 1-year prospective study. J Am Acad Dermatol. 10:638-46

Cortellini, P., Pini Prato, G. & Tonetti, M. S. (1996) Periodontal regeneration of human intrabony defects with bioresorbable membranes. A controlled clinical trial. *Journal of Periodontology* 67, 217-223.

Crigger, M., Bogle, G. C., Garrett, S. & Gantes, B. G. (1996) Repair following treatment of circumferential periodontaldefects in dogs with collagen and expanded polytetrafluoroethylene barrier membranes. *Journal of Periodontology* 67, 403-413.

Dahlin, C., Linde, A., Gottlow, J. & Nyman, S. (1988). Healing of bone defects by guided tissue regeneration. *Plastic and Reconstructive Surgery* 81, 672–677.

Dahlin, C., Alberius, P. & Linde, A. (1991a). Osteopromotion for cranioplasty. An experimental study in rats using a membrane technique. *Journal of Neurosurgery* 74 (3), 487–491.

Friedmann, A., Strietzel, F. P., Maretzki, B., Pitaru, S. & Bernimoulin, J. P. (2002) Histological assessment of augmented jaw bone utilizing a new collagen barrier membrane compared to a standard barrier membrane to protect a granular bone substitute material. *Clinical Oral Implants Research* 13, 587-594.

Gher, M.E., Quintero, G., Assad, D., Monaco, E. & Richardson, A.C. (1994). Bone grafting and guided bone regeneration for immediate dental implants in humans. *Journal of Periodontology* 65, 881–991.

Golub LM, McNamara TF, D'Angelo G, Greenwald RA, Ramamurthy NS. (1987). A non-antibacterial chemically-modified tetracycline inhibits mammalian collagenase activity. J Dent Res. 7;66:1310-4

Golub LM, Lee HM, Greenwald RA, Ryan ME, Sorsa T, Salo T, Giannobile WV.(1997). A matrix metalloproteinase inhibitor reduces bone-type collagen degradation fragments and specific collagenases in gingival crevicular fluid during adult periodontitis. Inflamm Res.;46:310-9

Golub LM, Lee HM, Ryan ME, Giannobile WV, Payne J, Sorsa T. (1998). Tetracyclines inhibit connective tissue breakdown by multiple non-antimicrobial mechanisms. Adv Dent Res.8;12(2):12-26.

Gotfredsen, K., Nimb, L., Buser, D. & Hjørting-Hansen, E. (1993). Evaluation of guided bone regeneration around implants placed into fresh extraction sockets. An experimental study in dogs. *Journal of Oral and Maxillofacial Surgery* 51, 879–884.

Gottlow, J. (1993). Guided tissue regeneration using bioresorbable and non-resorbable devices: Initial healing and longterm results. *Journal of Periodontology* 64, 1157–1165.

Greenwald RA, Golub LM, Ramamurthy NS, Chowdhury M, Moak SA, Sorsa T. (1998). In vitro sensitivity of the three mammalian collagenases to tetracycline inhibition: relationship to bone and cartilage degradation. Bone.; 22:33-8

Hämmerle, C.H.F., Brägger, U., Schmid, B. & Lang, N.P. (1998). Successful bone formation at immediate transmucosal implants. *International Journal of Oral & Maxillofacial Implants* 13, 522–530.

Hammerle, C.H., Jung, R.E. & Feloutzis, A. (2002). A systematic review of the survival of implants in bone sites augmented with barrier membranes (guided bone regeneration) in partially edentulous patients. *Journal of Clinical Periodontology* 29 (Suppl 3), 226–231; discussion 232–223.

Hardwick, R., Scantlebury, T.V., Sanchez, R., Whitley, N. & Ambruster, J. (1994). Membrane design criteria for guided bone regeneration of the alveolar ridge In: Buser, D., Dahlin, C. & Schenk, R.K., eds. *Guided Bone Regeneration in Implant Dentistry.* Chicago, Berlin: Quintessence, pp. 101–136.

Hutmacher, D. & Hürzeler, M.B. (1995). Biologisch abbaubare Polymere und Membranen für die gesteurte Gewebe- und knochenregeneration. *Implantologie* 1, 21–37.

Hyder PR, Dowell P, Dolby AE. (1992) Freeze-dried, cross-linked bovine type I collagen: Analysis of properties. *J Periodontol*;63:182-186.

Ilizarov, G.A. (1989a). The tension-stress effect on the genesis and growth of tissues: Part I. The infl uence of stability of fi xation and soft tissue reservation. *Clinical Orthopaedics* 238, 249–281.

Ilizarov, G.A. (1989b). The tension-stress effect on the genesis and growth of tissues: Part II. The infl uence of the rate and frequency of distraction. *Clinical Orthopaedics* 239, 263–285.

Karring, T., Nyman, S. & Lindhe, J. (1980). Healing following implantation of periodontitis affected roots into bone tissue. *Journal of Clinical Periodontology* 7, 96–105.

Kohal, R.J., Trejo, P.M., Wirsching, C., Hürzeler, M.B. & Caffesse, R.G. (1999a). Comparison of bioabsorbable and bioinert membranes for guided bone regeneration around non-submerged implants. An experimental study in the mongrel dog. *Clinical Oral Implants Research* 10, 226–237.

Kostopoulos, L. & Karring, T. (1994). Augmentation of the rat mandible using guided tissue regeneration. *Clinical Oral Implants Research* 5, 75–82.

Kozlovsky A, Aboodi G, Moses O, Tal H, Artzi Z, Weinreb M, Nemcovsky CE. (2009).Bio-degradation of a resorbable collagen membrane (Bio-Gide) applied in a double-layer technique in rats.Clin Oral Implants Res. 2009 Oct;20(10):1116-23

Lekholm, U., Adell, R. & Lindhe, J. (1986). Marginal tissue reactions at osseointegrated titanium fi xtures. (II) A crosssectional retrospective study. *International Journal of Oral & Maxillofacial Surgery* 15, 53–61.

Lekovic, V., Kenney, E.B., Weinlaender, M., Han, T., Klokkevold, P.R., Nedic, M. & Orsini, M. (1997). A bone regenerative approach to alveolar ridge maintenance following tooth extraction. Report of 10 cases. *Journal of Periodontology* 68, 563–570.

Lekovic, V., Camargo, P.M., Klokkevold, P.R., Weinlaender, M., Kenney, E.B., Dimitrijevic, B. & Nedic, M. (1998). Preservation of alveolar bone in extraction sockets using bioabsorbable membranes. *Journal of Periodontology* 69, 1044–1049.

Locci P, Calvitti M, Belcastro S, Pugliese M, Guerra M, Marinucci L, Staffolani N, Becchetti E (1997).Phenotype expression of gingival fibroblasts cultured on membranes used in guided tissue regeneration. J Periodontol. 68:857-63.

Lundgren, D., Mathisen, T. & Gottlow, J. (1994a). The development of a bioresorbable barrier for guided tissue regeneration. *The Journal of the Swedish Dental Association* 86, 741–756.

Mayfield, L., Nobréus, N., Attström, R. & Linde, A. (1997). Guided bone regeneration in dental implant treatment using a bioabsorbable membrane. *Clinical Oral Implants Research* 8, 10–17.

Melcher AH. (1970). Repair of wounds in the periodontium of the rat. Influence of periodontal ligament on osteogenesis. Arch Oral Biol. 15,1183-204.

Melcher AH. (1976). On the repair potential of periodontal tissues. J Periodontol. 47, 256-60.

Moses, O., Pitaru, S., Artzi, Z. & Nemcovsky, C. (2005). Healing of dehiscence-type defects in implants placed together with different barrier membranes: a comparative clinical study. *Clinical Oral Implants Research* 16, 210–219.

Moses, O, Nemcovsky CE, Tal H, Zohar R. (2001) Tetracycline modulates collagen membrane degradation in vitro. J Periodontol.;72:1588-93

Moses, O, Frenkel T, Tal H, Weinreb M, Bornstein MM, Nemcovsky CE. (2010) Effect of systemic tetracycline on the degradation of tetracycline-impregnated bilayered collagen membranes: an animal study. Clin Implant Dent Relat. Res. 12(4):331- 337.

Nyman, S., Karring, T., Lindhe, J. & Planten, S. (1980). Healing following implantation of periodontitis affected roots into gingival connective tissue. *Journal of Clinical Periodontology* 7, 394–401.

Nyman SR, Lang NP.;Guided tissue regeneration and dental implants. Periodontol 2000. 1994 Feb;4:109-18. Review. No abstract available

Owens, K. W. & Yukna, R. A. (2001) Collagen membrane resorption in dogs: Acomparative study. *Implant Dentistry* 10, 49-56.

Paul, B. F., Mellonig, J. T., Towle, H. J. 3rd & Gray, J. L. (1992) Use of a collagenbarrier to enhance healing in human periodontal furcation defects. *International Journal of Periodontics & Restorative Dentistry* 12, 123-131.

Pfeifer, J., Van Swool, L. M. & Ellinger, R. (1989) Epithelial exclusion and tissueregeneration using a collagen barrier in chronic periodontal defects: Histologicalstudy. *International Journal of Periodontics & Restorative Dentistry* 9, 263-273.

Piatelli, A., Scarano, A., Russo, P. & Matarasso, S. (1995). Evaluation of guided bone regeneration in rabbit tibia using bioresorbable and non-resorbable membranes. *Biomaterials* 17, 791–796.

Postlethwaite AE, Seyer JM, Kang AH. (1978) Chemotactic attraction of human fibroblasts to type I, II, and III collagens and collagen-derived peptides. Proc Natl Acad Sci U S A. 75:871-5

Reddi, A.H. (1981). Cell biology and biochemistry of endochondral bone development. *Collagen Related Research* 1, 209–226.

Reddi, A.H., Weintroub, S. & Muthukumaran, N. (1987). Biologic principles of bone induction. *Orthopedic Clinics of North America* 18, 207–212.

Reynolds JJ, Hembry RM, Meikle MC. (1994) Connective tissue degradation in health and periodontal disease and the roles of matrix metalloproteinases and their natural inhibitors. Adv Dent Res.;8(2):312-9.

Rothamel, D., Schwarz, F., Sculean, A., Monika, H., Scherbaum, W. & Becker, J. (2004) Biocompatibility of various collagen membranes in cultures of human PDL fibroblasts and human osteoblast-like cells. *Clinical Oral Implants Research* 15, 443-449

Rothamel, D., Schwarz, F., Sager, M., Herten, M., Sculean, A. & Becker, J. (2005) Biodegradation of differently cross-linked collagen membranes: an experimentalstudy in the rat. *Clinical Oral Implants Research* 16, 369-378.

Sandberg, E., Dahlin, C. & Linde, A. (1993). Bone regeneration by the osteopromotion technique using bioabsorbable membranes. An experimental study in rats. *Journal of Oral and Maxillofacial Surgery* 51, 1106–1114.

Schlegel AK, Möhler H, Busch F, Mehl A. (1997) *Preclinical and clinical studies of a collagen membrane (Bio-Gide).* Biomaterials. 18:535-8.

Schliephake, H. & Kracht, D. (1997). Vertical ridge augmentation using polylactic membranes in conjunction with immediate implants in periodontally compromized extraction sites: an experimental study in dogs. *International Journal of Oral & Maxillofacial Implants* 12, 325–334.

Schwarz F, Rothamel D, Herten M, Sager M, Becker J. (2006) Angiogenesis pattern of native and cross-linked collagen membranes: an immunohistochemical study in the rat. Clinical Oral Implants Research 17:403-9.

Simion, M., Baldoni, M., Rossi, P. & Zaffe, D. (1994a). A comparative study of the effectiveness of e-PTFE membranes with and without early exposure during the healing period. *International Journal of Periodontics and Restorative Dentistry* 14, 167–180.

Simion, M., Misitano, U., Gionso, L. & Salvato, A. (1997). Treatment of dehiscences and fenestrations around dental implants using resorbable and nonresorbable membranes associated with bone autografts: a comparative clinical study. *International Journal of Oral & Maxillofacial Implants* 12, 159–167.

Taguchi, Y., Amizuka, N., Nakadate, M., Ohnishi, H., Oda, K., Nomura, S. & MaedaT. (2005) A histological evaluation of guided bone regeneration induced by collagenous membrane. *Biomaterials* 26, 6158-6166.

Tal, H. (1998) Ridge preservation and gingival augmentation. *Independent Dentistry* 3, 96-100.

Tal, H. (2004) Healing of osseous defects by guided bone regeneration using ribosecross linked collagen membranes. *Journal of the Israel Dental Association* 21, 32- 40

Tal, H., Kozlovsky A, Pitaru S, Noff M. (1991): Monolayer versus bilayer-gel coated collagen membranes for periodontal guided tissue regeneration in the canine. *J Periodontology* 1991; abst#61:317.

Tal, H. & Pitaru, S. (1992) Formation of new periodontal attachment apparatus after experimental root isolation with collagen membranes in the dog. *International Journal of Periodontics & Restorative Dentistry* 3, 231-242.

Tal, H., Pitaru, S., Moses, O. & Kozlovsky, A. (1996) Collagen gel and membrane in guided tissue regeneration in periodontal fenestration defects in dogs. *Journal of Clinical Periodontology* 23, 1-6.

Tal, H. & Dayan, D. (2000) Spontaneous early exposure of submerged implants. III.Histopathology of perforated mucosa covering submerged implants. *Journal of Periodontology* 72, 1231-1235. Not in text

Tal, H., Artzi, Z., Moses, O., Nemcovsky, C. E. & Kozlovsky, A. (2001) Spontaneousearly exposure of submerged endosseous implants resulting in crestal one loss: A clinical evaluation between stage I and stage II surgery. *International Journal of Oral & Maxillofacial Implants* 16, 514-521.

Tal, H., Kozlovsky A, Artzi Z., Nemcovsky, C.E. , Moses O. (*2008*a). Long term bio-degradation of cross-linked and non-cross-linked collagen barriers in human guided bone regeneration *Clinical Oral Implants Res. 19):295-302.*

Tal, H., A. Kozlovsky, Z. Artzi, C.E. Nemcovsky, O. Moses (*2008b*). Cross-linked Cross-linked and non-cross-linked collagen barrier membranes disintegrate following surgical exposure to the oral environment: Histological study in the cat. *Clinical Oral Implants19:760-6.*

Tanaka S, Avigad G, Eikenberry EF, Brodsky B.(1988). Isolation and partial characterization of cross-links collagen chains dimerized by sugar-derived. *J Biol Chem*;263:17650-17657.

Urist, M.R. (1965). Bone: formation by autoinduction. *Science* 150, 893–899.

Van Swol RL, Ellinger R, Pfeifer J, Barton NE, Blumenthal N.Collagen membrane barrier therapy to guide regeneration in Class II furcations in humans. J Periodontol. 1993 Jul;64(7):622-9.

von Arx, T., Cochran, D.L., Hermann, J.S., Schenk, R.K., Higginbottom, F.L. & Buser, D. (2001). Lateral ridge augmentation and implant placement: An experimental study evaluating implant osseointegration in different augmentation materials in the canine mandible. *International Journal of Oral & Maxillofacial Implants* 16, 343–354.

von Arx, T., Cochran, D. L., Schenk, R. K. & Buser, D. (2002) Evaluation of a prototype trilayer membrane (PTLM) for lateral ridge augmentation: An experimental study in the canine mandible. *International Journal of Oral & Maxillofacial Surgery* 31, 190-199.

von Arx T, Buser D. (2006). Horizontal ridge augmentation using autogenous block grafts and the guided bone regeneration technique with collagen membranes: a clinical study with 42 patients. Clinical Oral Implants Research 17:359-66.

von Arx, T., Hafl iger, J. & Chappuis, V. (2005). Neurosensory disturbances following bone harvesting in the symphysis: a prospective clinical study. *Clinical Oral Implants Research* 16, 432–439.

Yaffe A, Ehrlich J, Shoshan S. (1984). Restoration of periodontal attachment employing enriched collagen solution in the dog. J Periodontol. 55:623-8.

Zarkesh N, Nowzari H, Morrison JL, Slots J (1999) Tetracycline-coated polytetrafluoroethylene barrier membranes in the treatment of intraosseous periodontal lesions. J Periodontol.;70:1008-16.

Zhao, S., Pinholt, E. M., Madsen, J. E. & Donath, K. (2000) Histological evaluation of different biodegradable and nonbiodegradable membranes implanted subcutaneouslyin rats. *Journal of Craniomaxillofacial Surgery* 28, 116-122.

Zitzmann, N.U., Naef, R. & Schärer, P. (1997). Resorbable versus nonresorbable membranes in combination with Bio- Oss for guided bone regeneration. *International Journal of Oral & Maxillofacial Implants* 12, 844–852.

Zitzmann, N.U., Schärer, P. & Marinello, C.P. (2001). Longterm results of implants treated with guided bone regeneration: A 5-year prospective study. *International Journal of Oral & Maxillofacial Surgery* 16, 355–366

Zohar R, Nemcovsky CE, Kebudi E, Artzi Z, Tal H, Moses O. (2004) Tetracycline impregnation delays collagen membrane degradation in vivo. J Periodontol. 75:1096-101.

7

Augmentation and Preservation of the Alveolar Process and Alveolar Ridge of Bone

Haim Tal, Zvi Artzi, Roni Kolerman, Ilan Beitlitum and Gal Goshen
Department of Periodontology and Dental Implantology, Tel Aviv University
Israel

1. Introduction

1.1 The alveolar process and alveolar ridge of bone in health and disease

The alveolar process is the part of the maxilla and the mandible that house and support the alveoli of the teeth. It develops in conjunction with the development and eruption of the teeth, over the basal bone and coronal to it. The alveolar process consists of an outer layer of *cortical bone*, an inner *cancellous bone*, and a special layer - alveolar bone proper - which together with the root, cementum and the periodontal membrane constitutes the dental attachment apparatus. The attachment apparatus supports the tooth in the jaw, on the one hand, and distribute forces generated by the teeth to the alveolus and bone peripheral to it, on the other hand. The forces transferred to the jaw due to teeth activities, influence the structure, architecture, size and density of the cancellous bone trabeculae. Fig. 1a-c shows a cross section through dentate sites in the mandible at a level corresponding to the roots and through an edentulous site (d). In health, the bone lining the wall of the socket (alveolar bone proper) is continuous with the cortical bone at the lingual and buccal aspects of the alveolar process (Fig.1a-c,e) however, if the buccal plate of bone is extremely thin (Fig.1. a,c) the buccal cortical plate and the alveolar bone proper unite having no cancellous bone between them. The different structures of the alveolar process, i.e. cortical and cancellous bone, are constantly undergoing remodeling in response to functional forces acting on the teeth. Once teeth are lost, the attachment apparatus is destroyed, and the alveolar process, mainly the alveolar ridge, undergoes significant structural changes; these are referred to as "disuse atrophy" (Fig. 1a,c,f).

Immediately after extraction the bony walls of the alveolus present significant resorption, the central part of the socket is partly filled up with woven bone and the extraction site becomes markedly reduced in size. Pietrokovski & Massler (1967) and Schropp *et al.* (2003) have shown that the edentulous site diminishes in all dimensions i.e. bucco-lingual, bucco-palatal and apico-coronal. At the same time, the soft tissues in the extraction site undergo adaptive changes that clinically may appear as deformations of the jaw(Fig. 2).

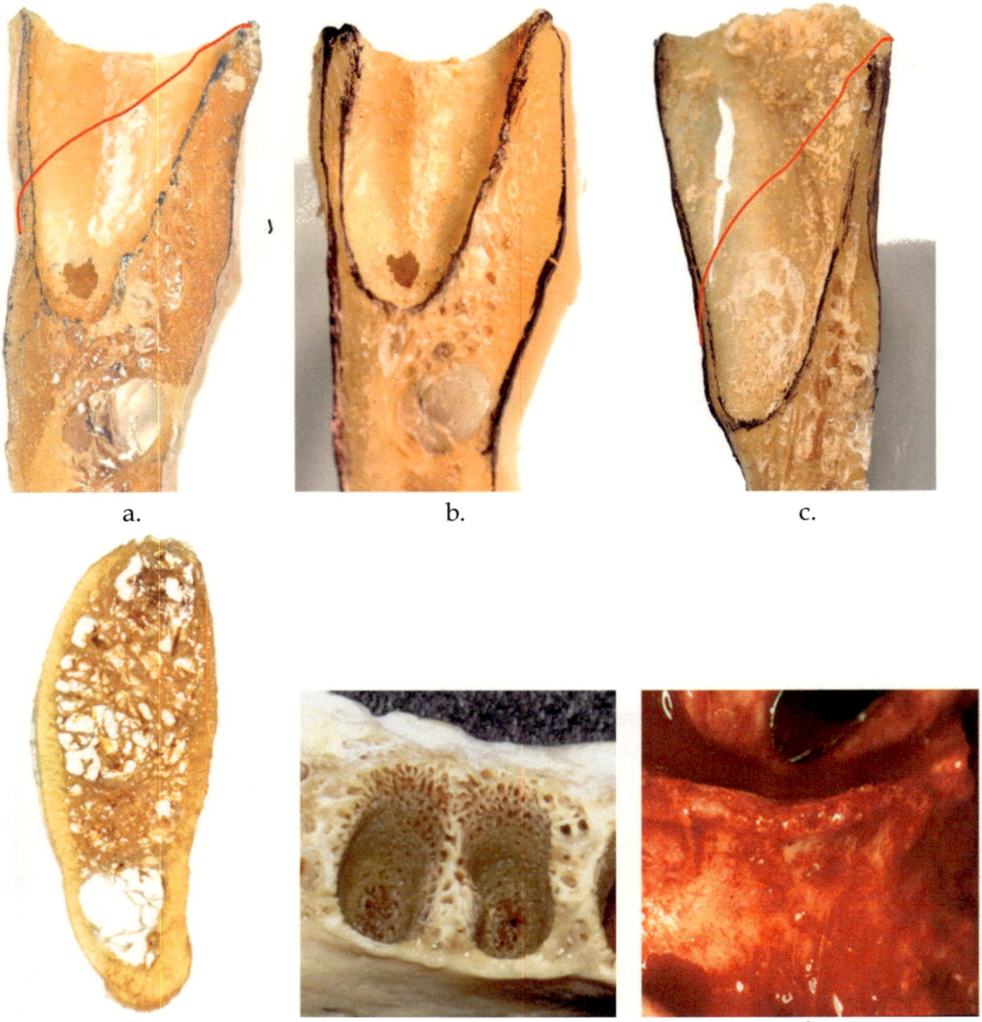

a. b. c. d. e. f.

Fig. 1. Different views of sites of a dry mandible. a. cross section through an empty alveolar socket of a mandibular canine tooth; the red line represents the expected bone contour that would be established had the tooth been removed; note that the buccal wall contains exclusively cortical bone. b. cross section through an empty alveolar socket of a mandibular premolar tooth; note that the buccal wall contains exclusively cortical bone in spite of its being relatively thick c. cross section through an empty alveolar socket of a mandibular canine tooth; the red line represents the expected bone contour that would be established had the tooth been removed; note that the buccal wall is extremely thin ("paper thin") and contains exclusively cortical bone d. cross section of an edentulous inter-radicular site a few months after tooth loss; there is less bone loss in this area compared with extraction socket sites. e. upper view of an empty socket of the lower second molar showing the cribriform alveolar bone proper f. clinical view of the anterior segment of an edentulous mandible 1 year after extraction; severe disuse atrophy is noted.

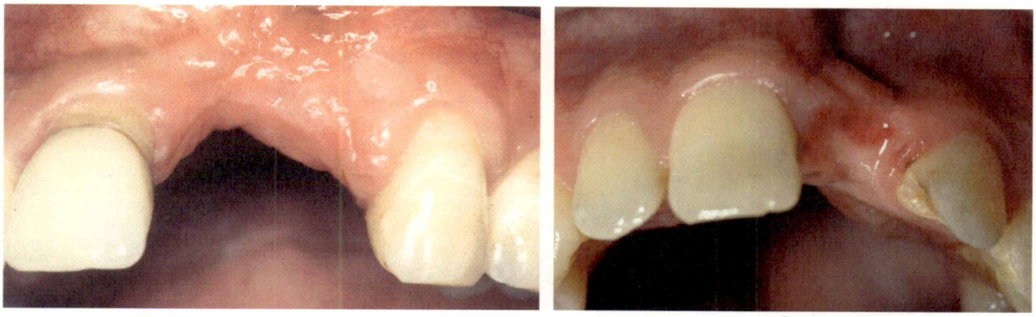

a. b.

Fig. 2. Deformations of the jaw due to severe bone loss and soft tissues adaptation in extraction sites associated with a. implant failure and b. traumatic injury; in the esthetic zone this deformations needs regenerative and/or plastic therapy.

1.2 The alveolar process status in relation to implant placement

Re-establishment of the natural dimensions of the alveolar process is essential for both functional rehabilitation and esthetic restoration; if missing teeth are to be restored with implant supported prostheses, restoring these dimensions is of crucial importance. It is agreed that endosseous implants should be completely embedded in bone and preferably surrounded by not less than 2 mm of bone in all aspects. In view of the changes in bone dimensions after tooth extraction, the issue relating to the "optimal" timing of implant placement has received much attention (Hammerle *et al.* 2004). Attempts made to identify the advantages and disadvantages of **early**, **delayed**, and **late** implant placement, led to incorporation of the knowledge in this field into a classification relating the timing of implant placement to the condition of soft and hard tissue healing as follows (Hammerle *et al.* 2004):

- **Type 1**: the implant is placed immediately following tooth extraction (Fig. 3a)
- **Type 2**: the implant is placed after soft tissues have healed and a mucosa is covering the socket entrance (Fig. 3b)
- **Type 3**: the implant is placed after substantial amounts of new bone have formed in the extraction socket (Fig. 3c)
- **Type 4**: the implant is placed in a fully healed ridge. (Fig. 3d)

Preservation of the alveolar process is dependent on the presence of teeth; after the teeth are lost the alveolar process posses gradual regression. The loss of teeth, and the loss of function within and peripheral to the socket results in adaptive alterations of the edentulous portion of the ridge; the alveolar ridge becomes markedly reduced in all dimension. The magnitude of this change was described by Pietrokovski and Massler (1967) who studied anthropometrically dry jaws, and Schropp *et al.* (2003) who clinically studied bone and soft tissue volume changes following the extraction of single premolars and molars. The later concluded that the buccal-lingual/palatal dimension during the first 3 months was reduced about 30%, and after 12 months the edentulous site had lost at least 50% of its original width. Furthermore, after 12 months of healing the buccal prominence was reduced to a level 1.2 mm apical of its lingual/palatal counterpart. It is noteworthy that frequently, the alveolar process has undergone pathologic changes prior to tooth loss due to traumatic injuries, chronic or aggressive periodontitis, periapical lesions, root fractures and resorption as well as severe periimplantitis. (Fig. 4).

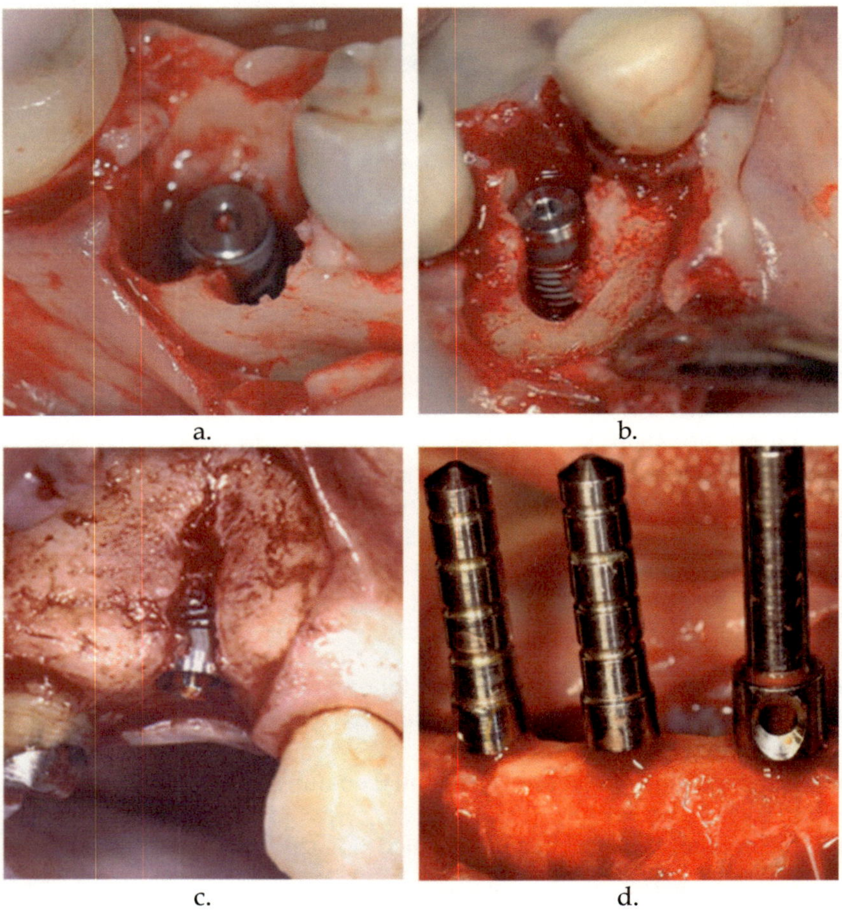

a. b.

c. d.

Fig. 3. Classification relating the timing of implant placement a.Type 1: an implant is placed immediately following a molar tooth extraction. b. Type 2: an implant is placed 2 months after implant removal. The soft tissues have healed and the mucosa covering the socket entrance was intact c. Type 3: an implant is placed 4 months after extraction of the upper left first premolar. Substantial amounts of new bone have formed in the extraction socket; a buccal dehiscence defect is associated with the buccal aspect of the implant. d. delayed implant placement in a fully healed (type 4) ridge, 1 year after extractions. Both, vertical and horizontal bone loss is evident.

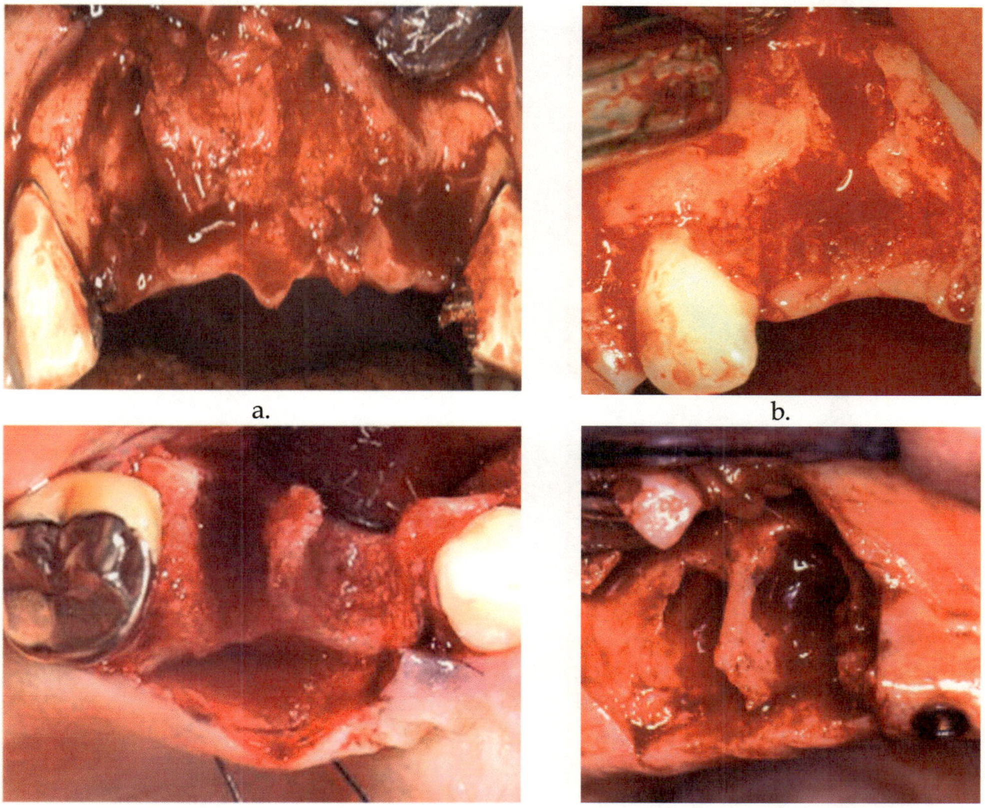

a. b. c. d.

Fig. 4. Clinical view of bone destruction in the alveolar process a. immediately after tooth removal due to advanced periodontitis b. immediately after removal of tooth remnants due to root resorbtion c. complete destruction of the alveolar buccal plates observed after removal of the first and second maxillary bicuspids d. immediately after removal of dental implants due to advanced peri-implantitis.

Radiographic studies dealing with the atrophy of the alveolar process have shown that in the first few months bone loss is obvious in the alveolar crest region, simultaneous to bone gain in the socket. Gain of bone in the socket continued until 6 months following extraction, being replaced by bone remodeling during the next 6 months to follow. Based on the volume of remaining bone, the edentulous sites were classified by Lekhom and Zarb (1985) into five different categories: A and B groups represent sites in which substantial amounts of the alveolar process still remain, whereas in groups C, D, and E, there are only minute remnants of the alveolar process present (Fig. 4)

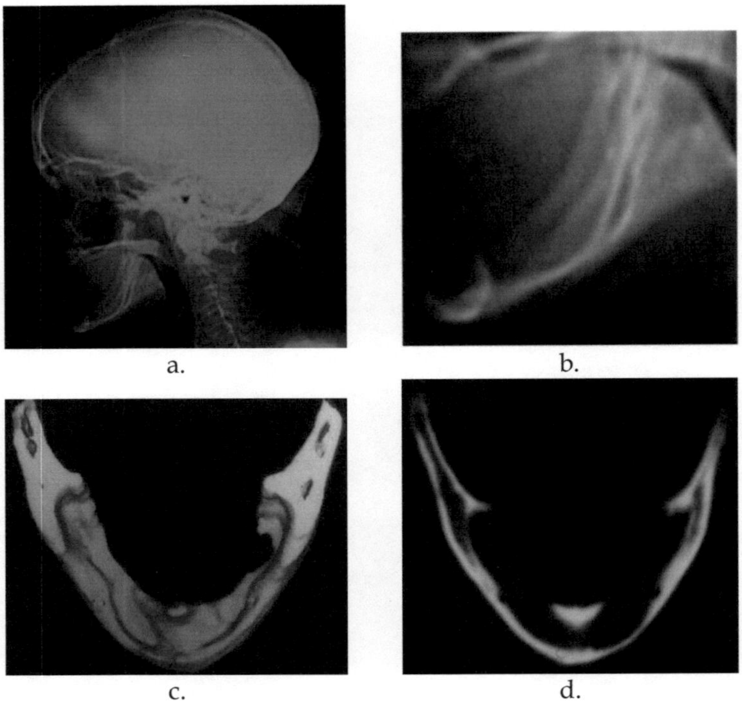

Fig. 5. CT scan demonstrating very severe (group E) bone loss due to implant failure followed by usage of a full removable denture over a 20 years period of time. Bone loss include the whole alveolar process and most of the basal bone. a. mid-sugittal section view through of the maxilla and mandible. b. higher magnification of the mid-sugittal section view through the mandible c. 3D view of the mandible from above; vertical bone loss beyond the level of the mental foramina is evident d. coronal section of the mandible at the level of the mental foramina.

Lekholm and Zarb (1985) further classified the residual bone according to "quality": Class 1 and class 2 relate to residual alveolar process presenting thick cortical plates and relatively small volume of bone marrow, while sites belonging to class 3 and class 4, present relatively thin walls of cortical bone and large amount of cancellous bone including trabeculae of lamellar bone and marrow. While the definition of the alveolar process is clear, there seems to be no distinct boundary between the alveolar process and the basal bone of the jaws. However, by definition the alterations, modifications and adjustments occurring in the alveolar process and ridge following tooth extraction include *intra-alveolar processes* and *extra-alveolar Processes;* these were described in details by Amler (1969), and later by Evian (1982).

Understanding the changes occurring to the alveolar process after extraction is of utmost importance when planning the rehabilitation of the edentulous jaw. Araújo & Lindhe (2005) studied histologically the processes alterations following tooth extraction in the dog at 1, 2, 4, and 8 weeks of healing. At 1 week after tooth extraction the socket was occupied by a coagulum. The presence of osteoclasts on the inner surface of the socket walls indicated that the bundle bone was being resorbed. At 2 weeks newly formed immature (woven) bone resided in the apical and lateral parts of the socket. In several parts of the socket walls the bundle bone has been replaced with woven bone. At 4 weeks after extraction the entire socket was occupied with woven bone and at some areas the newly formed woven bone was being replaced with a more mature type of bone. At 8 weeks a layer of cortical bone covered the entrance to the extraction site. The woven bone had been replaced with bone marrow and some trabeculae of lamellar bone. Signs of ongoing hard tissue resorption were observed on the outside and on the top of the buccal and lingual bone wall and the buccal bone wall was located apical of its lingual counterpart. Araújo & Lindhe (2005) concluded that the process of modeling and remodeling that occur following tooth extraction results in pronounced resorption of the various components of the alveolar ridge. The resorption of the buccal bone wall is more pronounced than that of the lingual/palatal wall and hence the center of the ridge will move in lingual/palatal direction. In the extreme case, the entire alveolar process may be lost and in such situations only the bone of the base of the mandible or the maxilla remains. In fact, with time, depending on functional and parafunctional activities, significant parts of the basal bone may be lost leaving but the cortical envelop in situ (Fig 5). Since this subject is beyond the scope of this book, for a systematic description of the histological and morphometrical changes in the alveolar process following tooth extraction the reader is referred to a detailed long-term experiment in the dog carried out by Cardaropoli *et al.* (2003).

1.3 The role of dental implants in bone healing and bone regeneration

Augmentation and regeneration procedures of the alveolar process and alveolar ridge have received special attention soon after the introduction of modern implant therapy (1970's). Successful restoration of health, function and esthetic appearance using dental implants require the establishment of conditions that promote bone and soft tissue integration to the implant. In addition, in a growing number of cases, treatment must also satisfy esthetic demands. After tooth removal it takes about 4–8 weeks before granulation tissue and provisional connective tissue/woven bone fill the extraction socket and its surface becomes covered with epithelium (Amler 1969; Zitzmann *et al.* 1999; Nemcovsky & Artzi 2002). The maturation of the soft tissue may require an even longer healing time before the soft tissue quality allows for precise management of a mucosal flap. This timing however, must be matched against the hard tissue reduction that results in by the socket walls resorption, especially that of the buccal plate of bone. Special care should be taken with flap elevation at sites where the mucosa adheres to the underlying bone or underlying scar tissue; in such cases flap separation from the bone may rupture the soft tissue resulting in soft tissue dehiscence, local infection, and compromised healing (Zitzmann *et al.* 1997). Thus, if bone height apical to the tip of the root is less than 3 mm, and obtaining primary implant stability in the bone is impossible, a more delayed approach is preferable and it is advised to wait until substantial bone fill has occurred, i.e. 10–16 weeks (Evian *et al.*1982). At that time newly formed woven bone occupies the socket area, however, by that time the walls of the socket are frequently severely resorbed. At this stage of healing it is possible to place the implant in a position that facilitates the prosthetic phase of the treatment. Six to 12 months after tooth extraction, the alveolar ridge is characterized by dense cortical bone that is lined by a mature keratinized mucosa. The advantage of placing implants into the mature edentulous ridge is since at that delayed stage of healing, further changes of the ridge morphology may be minimal and very slaw. The main disadvantages of such delayed implant placement is that the overall reduction of the ridge volume is significant, and its external contours may be deformed.

Depending on the pre-extraction bone loss, and the time elapsed from extraction to implant placement, the loss of ridge volume and changes in contours may require bone augmentation varying from minimal ridge preservation in the fresh extraction cases, to more complicated bone augmentation procedures in the very pronounced ones. Although each case requires a "custom made" treatment planning, in most cases, whenever possible, tooth replacement should be done as early as possible; the final decision regarding the timing for implant placement must however be based on a thorough understanding of the structural changes that occur in the alveolar process following tooth extraction, with and without implant placement.(Hämmerle *et al.* 2006)

1.4 Ridge correction in conjunction with implant placement

Implants may be placed immediately after the removal of teeth. Many claims have been made regarding the advantages of immediate implant placement (Chen *et al.* 2004) including implant positioning and bone preservation at the site of implantation, (Werbitt & Goldberg 1992; Barzilay 1993; Schwartz-Arad & Chaushu 1997a; Hammerle *et al.* 2004). It was proposed that placement of an implant in a fresh extraction socket may allow the preservation of bone tissue of the socket and the surrounding jaw by stimulating bone formation and osseointegration and hence counteract the adaptive alterations that occur to bone tissue following tooth loss (e.g. Denissen *et al.* 1993; Watzek *et al.* 1995;for review see Chen *et al.* 2004). Human clinical studies (Botticelli *et al.* 2004; Covani *et al.* 2004) and dog experiments (Araujo & Lindhe 2005; Araujo *et al.* 2006a,b) have shown that after 4 months of healing post extraction and immediate implant placement, the marginal gap between implants and socket bony walls had completely resolved, however, the thickness of the buccal as well as the palatal bone walls had become markedly reduced so that the implant surface could be seen through the very thin remaining buccal bone wall. The alveolar process next to implants placed in the palatal socket of the fresh extraction sites of extracted first maxillary premolars and next to implants placed in healed edentulous ridge at similar positions have been entirely resolved and the distance between the implant and the outer surface of the buccal bone plate had become markedly reduced. Based on clinical measurements, Botticelli *et al.* (2004) reported that during 4 months of healing following tooth extraction and implant placement the reduction of the buccal dimension was 56% (1.9 mm) while the reduction of the lingual dimension was 27% (0.8 mm). These findings which were based on measurements at 21 sites in 18 subjects, show that after implant placement all marginal gaps had practically become resolved and suggest that that implant placement in a fresh extraction socket may, in fact, not prevent the physiologic modeling/remodeling that occurs in the ridge following tooth removal. These findings further demonstrate that the bone(woven bone)-to-implant contact that was established during the early phase of socket healing following implant installation, was in part lost when the buccal bone wall underwent continued atrophy.

In summary, It is obvious that the alveolar process following tooth extraction will adapt to the altered functional demands by atrophy, and that an implant, in this respect, is unable to substitute for the tooth. The clinical problem associated with immediate implant placement may be that unless the implant is placed palatal or lingual to the natural position of the root, bone loss frequently cause the buccal portion of the implant to gradually lose its hard tissue coverage, and the metal surface may become visible through a thin peri-implant mucosa or even be exposed and cause functional and/or esthetic concerns (Fig. 6).

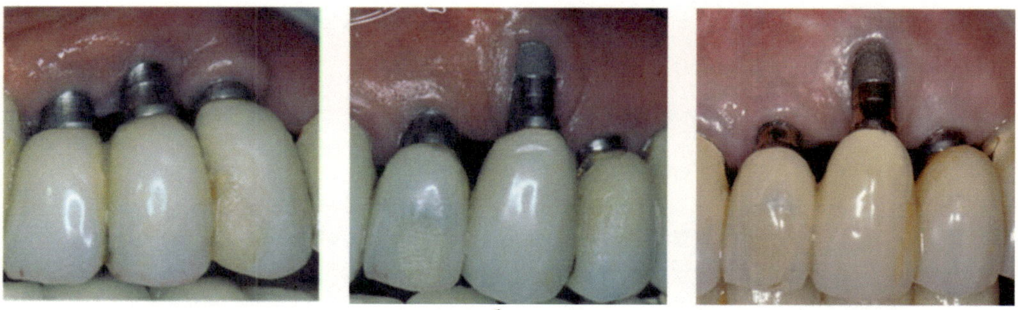

a. b. c.

Fig. 6. Gradual bone loss associated with the buccal portion of an implant placed in the natural position of the extracted root. Bone loss results in exposure of the metal surface and causing esthetic concerns. a. one year after implant loading b. three years after implant loading. c. eight years after implant loading.

These findings were supported by a recent clinical study (Grunder, 2011) who have shown that following implant placement into fresh extraction sockets the average horizontal loss of hard anf soft tissue measured 1.06 mm. Placing a subepithelial connective tissue graft using the tunnel technique in the labial area resulted in a slight (0.3mm) increase in the horizontal dimension of the ridge. Figure 7 presents a clinical case in which an implant is immediately placed after extraction of the mandibular first right molar tooth. An osteotomy at the center of the socket, through the septum, results in the establishment of an intimate contact between the surface of the implant body and the buccal and lingual base of the septum, the walls of the socket and the bone apical to it. Implant is stabilized at 45N. Osseoconductive graft material* and a coagulum resides in the void between the contact regions and peripheral bony walls of the socket. Six month later the implant is fully loaded. The final outcome after 1 year clearly shows that the buccal profile of the ridge is reduced in width, in spite of the atraumatic extraction, and immediately placed implant.

Atrophy of the edentulous ridge following tooth loss seems to be a biologic principle resulting in reduction of the width and the height of both the buccal and lingual bone plates; it is unavoidable and cannot be prevented by placing implants into the fresh extraction socket. This pathological phenomenon may be reduced by anchoring the implant deeper into the fresh socket, apical to it, and in a more lingual/palatal portion to that of the extracted tooth. In that case, bone regeneration procedures may be required to improve or retain bone volume and the buccal contour at a fresh extraction site. Adding a subepithelial connective tissue graft in the labial area may be favorable in the esthetic zone (Grunder 2011).

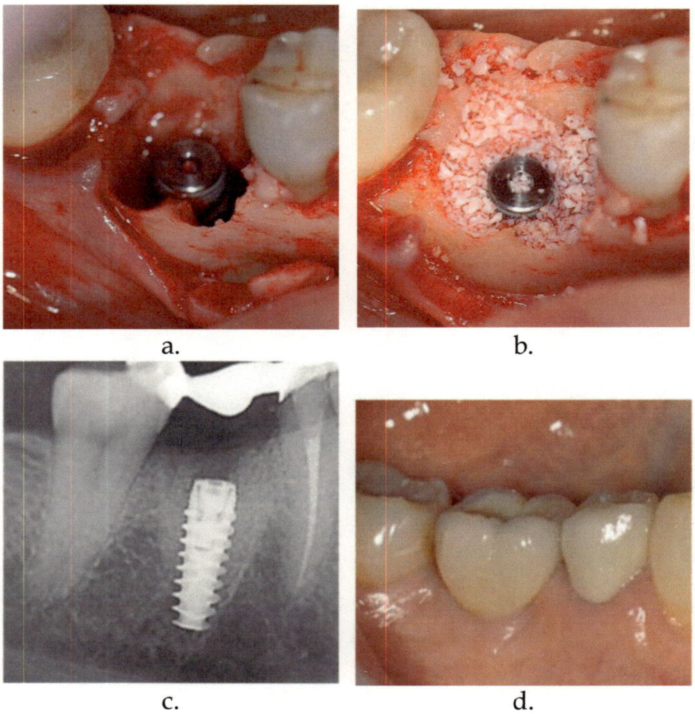

Fig. 7. Immediate placement of an endosseous implant in a lower molar fresh extraction socket. a. clinical view immediately after extraction and implant placement; implant surface is engaging the buccal and lingual remnants of the interadicular septum. b. the gaps between the implant surface and the socket walls is grafted with xenograft material*. c. periapical radiograph 6 months after implant placement. d. clinical view one year after extraction and implant placement shows a mild buccal deformation in the jaw owing to buccal bone loss and soft tissue adaptation.

2. Bone regeneration in the alveolar process of the jaw

Successful oral rehabilitation following tooth loss requires replacement of the missing roots, and satisfactory restoration of an adequate volume of bone; this is influenced mainly by health necessities, functional requirements, implant placement (Lekholm *et al.* 1986), and esthetic demands. Four methods have been described to achieve these goals: **osteoinduction** using growth factors (Urist 1965; Reddi 1981); **osteoconduction** using grafting material as a scaffold for new bone growth (Buch *et al.* 1986; Reddi *et al.* 1987); **distraction osteogenesis,** by which the two fragments of a surgically induced fracture are slowly pulled apart (e.g. Ilizarov 1989a,b); **guided bone regeneration (GBR),** which allows spaces maintained by barrier membranes to be filled with new bone (Dahlin *et al.* 1988, 1991a; Nyman & Lang 1994).

* Bio-Oss®, Geistlich Biomaterials, Wolhusen, Switzerland) alloplast 4Bone ™ SBS: BioMATLANE SARL, France. Particles size of 0.25-1 mm

2.1 Guided bone 1 regeneration in the alveolar process of bone

Guided bone regeneration (GBR) which is better documented than the other methods for the treatment of localized bone defects in the jaws will be dealt in this chapter in depth; for other techniques the reader is referred to chapters (8-10). GBR allows the placement of endosseous implants in areas of the jaw with insufficient bone volume. Lack of bone volume may be due to congenital, post-traumatic or postsurgical defects or results in from disease processes (Figs. 4,5). It has been claimed that the predictability and success which can be achieved with GBR procedures enable the clinician to obtain similar rates of treatment success at sites with bone defects compared to sites without defects (Hammerle *et al.* 2002). (see also chapter 6).

Guided bone regeneration frequently forms a part of complex treatments, but this chapter focuses on the aspects of bone augmentation at localized defects in the alveolar process. More than two decades have passed since the introduction of GBR into clinical practice. Today, general understanding of the mechanisms leading to regeneration of desired tissues still agrees with the initially published statements regarding guided tissue regeneration (Karring *et al.* 1980; Nyman *et al.* 1980, 1989). In brief, when a space is formed, cells from the adjacent tissues grow into this space to form their parent tissue, i.e. the tissue they migrated in from. In order to give preference to cells from desired tissues, tissue barriers, most commonly membranes, are placed to prevent cells from undesired tissues having access to the space. (Figure 8)

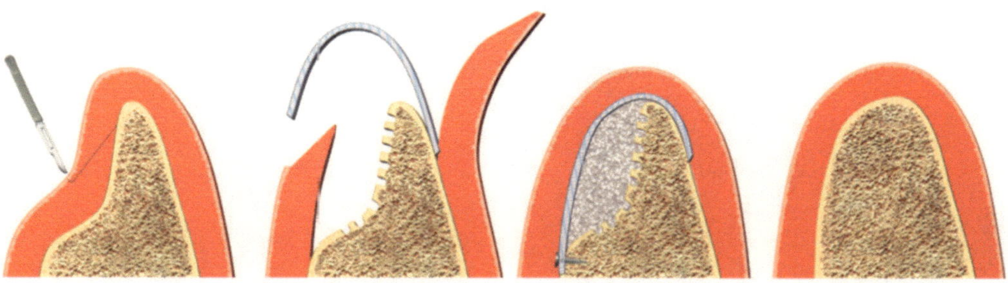

a. b. c. d.

Fig. 8. Guided bone regeneration scheme describing the use of a resorbable barrier membrane. a. bony defect is diagnosed b. the defect is debrided, bone cortex is perforated, and a membrane supporting scaffold material is placed.
c. the membrane is stabilized and shaped to dictate the desired bone contours
d. a few months later bone regeneration is observed restoring the desired shape of the jaw.

Experimental research regarding ridge augmentation using GBR has shown that in large surgically created defects in the alveolar ridge, treatment with membranes with or without the addition of grafts, entirely filled the space between the membrane and the bone with bone; in the absence of membranes, bone formation was lacking(Seibert & Nyman 1990). These findings received further support by later investigators who reported that GBR procedures can successfully be employed in the regeneration of alveolar ridge defects (Seibert & Nyman 1990; Schenk *et al.* 1994; Smukler *et al.* 1999). While intrabony alveolar ridge defects and lateral ridge augmentation has been shown to be predictable (Nyman *et al.* 1990; Dahlin *et al.* 1991b; Becker *et al.* 1994b; Buser *et al.* 1996; von Arx *et al.* 2005), vertical bone gain was initially less promising. Intensive efforts in GBR therapy were focused on vertical ridge augmentation, due to the great demands of this procedure. Clinical experiments have shown promising results when placing of autogenous bone grafts or bone substitute materials in combination with e-PTFE membranes of various configurations (Simion *et al.* 1994b, 1998 Tinti *et al.* 1996; Tinti & Parma-Benfenati 1998; Chiapasco *et al.* 2004). The membranes were supported either by the graft alone or additionally by implants protruding vertically from the host bone for various lengths. Employing vertical GBR, it was possible to achieve bone gain above the external borders of the jaw (Lundgren *et al.* 1995; Hämmerle *et al.* 1996, 1999; Schliephake & Kracht 1997; Schmid *et al.* 1997; Lorenzoni *et al.* 1998). Although in some experiments vertical bone formation reached up to 4 mm above the previous border of the alveolar crest, clinical attempts to regenerate vertical bone was not predictable, and bone growth to the top of the membrane was not consistently achieved. (Simion *et al.* 1994a).

Advances in guided bone regeneration had become possible thanks to a series of biomaterials including tissue barriers, bone grafts and bone graft substitutes. Recently, growth and differentiation factors and tissue engineering means have been added to the available stock of such materials. These biomaterials are briefly mentioned at the end of this chapter and are dealt with more thoroughly in chapters 11-14 Successful GBR depends on the ability of the different materials to provide a space into which bone originated granulation tissue can proliferate exclusively, partly due to peripheral cell oclusiveness achieved

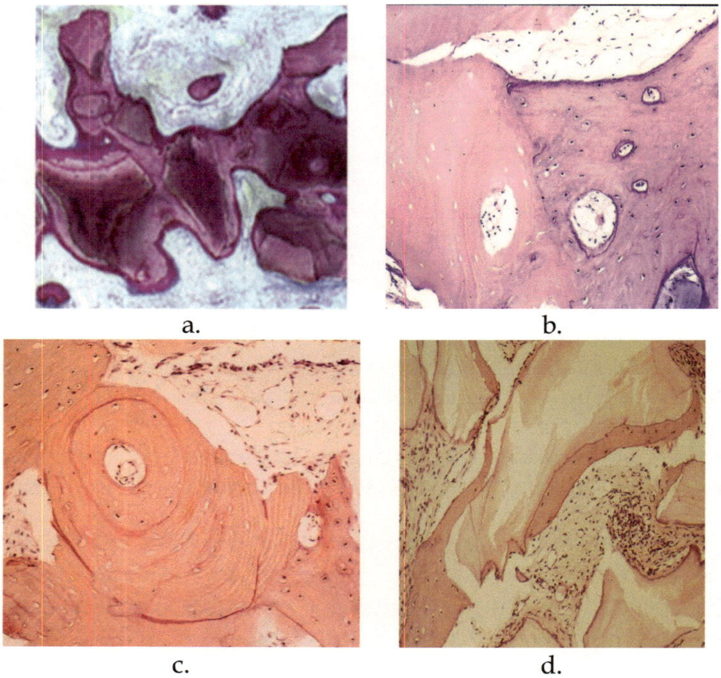

Fig. 9. Biopsy specimens demonstrating bone growth in the presence of osseoinductive and osseoconductive bone scaffold grafts. a. deproteinized bovine bone mineral completely surrounded with intimately integrated new bone in a GTR procedure b. demineralized freezed dried bone allograft (left) in intimate contact with new bone (right) c. demineralized freezed dried bone allograft in intimate contact with new bone. Notice a centrally developing osteon. d. deproteinized bovine bone mineral surrounded with new bone in a GBR procedure.

by tissue barriers. Success rates of the GBR procedure further depends on the stability of the healing site, minimal or no tissue reactions resulting from the presence and/or resorption of the occluding barriers, bone substitutes (Gottlow 1993), and osseoinductivity/conductivity of the bone scaffold in use (Figs. 9,10).

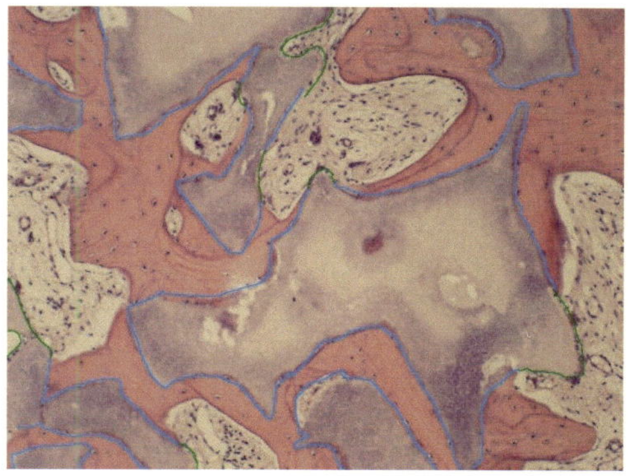

a.

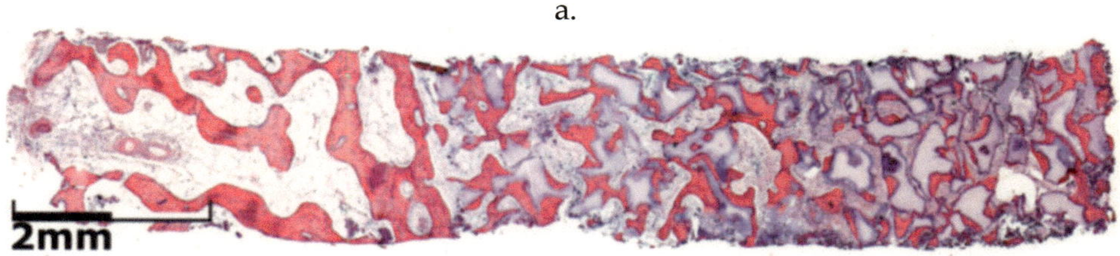

b.

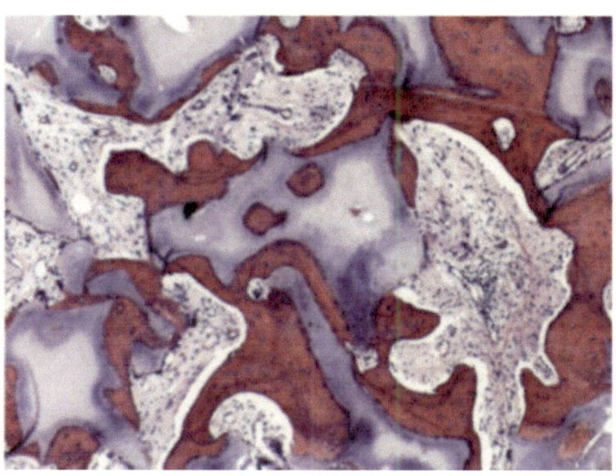

c.

d.

Fig. 10. a. A 9 months biopsy sample demonstrating alloplast graft particles** surrounded by vital bone and connective tissue. Bone to graft contact areas are marked with a blue line. (Hematoxylin & Eosin, original magnification x400)Histological view of a specimen obtained from a regenerating tissue in the sinus demonstrating alloplast graft particles** surrounded by vital bone and connective tissue.
b. (H&E original magnification x10). c. higher magnification from fig 10b. d. computer analysis of surface areas of new bone (red), graft material (blue), and connective tissue (yellow). (x40)

** A fully synthesized homogenous hydroxyapatite and beta tricalcium phosphate (HA : β-TCP) 60 : 40 alloplast 4Bone ™ SBS: BioMATLANE SARL, France. Particles size of 0.25-1 mm

2.2 Biomaterials for guided bone regeneration in the jaws – Animal and human studies

Extensive research has been conducted in search for the ideal enhancing bone repair and regeneration substance. Among the many available materials, bovine bone mineral (BBM) (Bio-Oss®, Geistlich Biomaterials, Wolhusen, Switzerland) is perhaps the most extensively researched one, presenting very favorable biocompatibility and osteoconductive qualities (Spector 1994; Jensen et al. 1996; Berglundh & Lindhe 1997; Boyne 1997; Hämmerle et al. 1997; Skoglund et al. 1997; Artzi & Nemcovsky 1998 Artzi et al 2000, 2001a,b,c, 2002). Based on the authors experience as well as other investigators, it has proved to be an appropriate scaffold in ridge deficiencies, peri-implant destruction, and sinus augmentation procedures (Smiler et al. 1992; Wetzel et al. 1995; Dies et al. 1996; Hürzeler et al. 1997; Valentini & Abensur 1997; Piattelli et al. 1999; Artzi et al. 2000, 2001a,b, 2002, 2003a,b 2005; Hallman et al. 2001b, 2002a).

In a 24 months comparative study (Artzi et al 2003a), the healing of surgical experimental defects grafted with bovine bone mineral was studied in the dog mandible, with and without tissue barrier membranes. Average bone area fraction at the bovine bone mineral uncovered sites was 23.1%, 44%, 63.4%, and 58.8% at 3, 6, 12, and 24 months, respectively. Differences were statistically significant between 3 to 6 and 6 to 12 months ($P<0.001$). At the membrane-protected sites, average bone area fraction was 26.4%, 51.7%, 61.2%, and 52.4%, at the respective periods. Differences were statistically significant between 3 to 6 months ($P<0.05$). However, Differences between the two sites with regard to the newly formed bone and particle presence were insignificant. At 3 and 6 months, newly formed bone, woven in nature, was incorporated with the grafted particles. High cellular bone with occasional osteoclasts was noted towards the surface of the mineral particles. Osteons were established in direct contact to particle configuration (Fig. 11.)

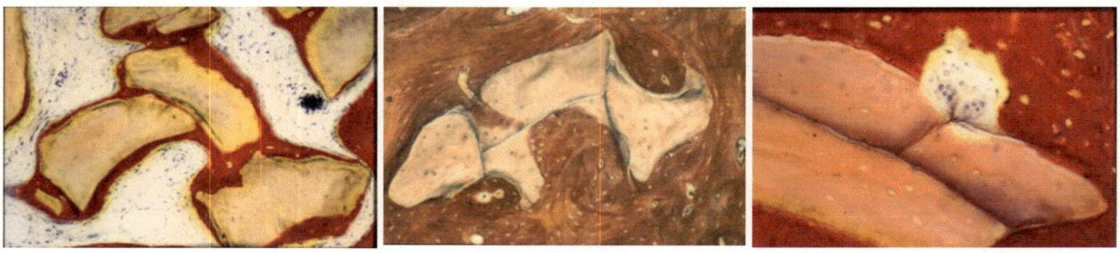

a. b. c.

Fig. 11. Photomicrograph of a bony defect grafted with bovine bone mineral (BBM). a. At 3 months, newly formed bone primarily surrounds the grafted BBM particles (Stevenel's blue and Van Gieson's picro fuchsin staining; original magnification x100). b. At 12 months, Haversian canal system; i.e., osteons, is established in proximity to the BBM particle and in accordance to its configuration (Stevenel's blue and Van Gieson's picro fuchsin staining; original magnification ×100). c. On higher magnification, note presence of multinucleated cells i.e., osteoclasts in proximity to the particle (Stevenel's blue and Van Gieson's picro fuchsin staining; original magnification×400).

At 1 and 2 years, the grafted sites showed complete bone healing configuration, however, the grafted particles - completely surrounded by the newly formed bone - were still predominant (Fig. 12). Osteons and lamellar bone arrangement were established but the bone was still highly cellular and osteoclasts could still be identified. The biomaterial did not show any substantial resorption within 2 years observation period of time.

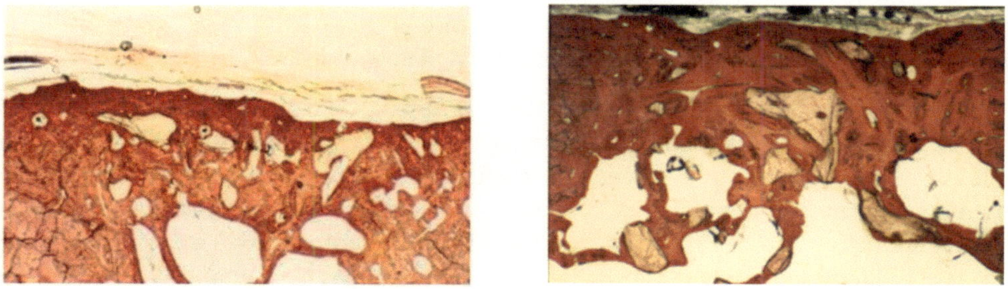

a. b.

Fig. 12. Photomicrograph of a bony defect grafted with bovine bone mineral particles a. At 12 months, protected by the membrane, the entire defect is filled with bone (Stevenel's blue and Van Gieson's picro fuchsin staining; original magnification ×20) b. At 24 months, the bovine bone mineral grafted site is filled with newly formed bone surrounding a substantial amount of grafted particles (Stevenel's blue and Van Gieson's picro fuchsin staining; original magnification ×20).

When bovine bone mineral was grafted under a configured titanium mesh serving as a contained stabilized vehicle to restore a deficient alveolar ridge (Artzi et al 2003), average bone fill of 81.2% ±7.98 was measured with remarkable height gain of 5.2 ± 0.79mm. Picrosirius red stained sections examined under polarized illumination, showed a gradual increase in new lamellar bone from the coronal to the most apical sections, reaching highest bone density near the most apical zone. (Fig 13 a,b).

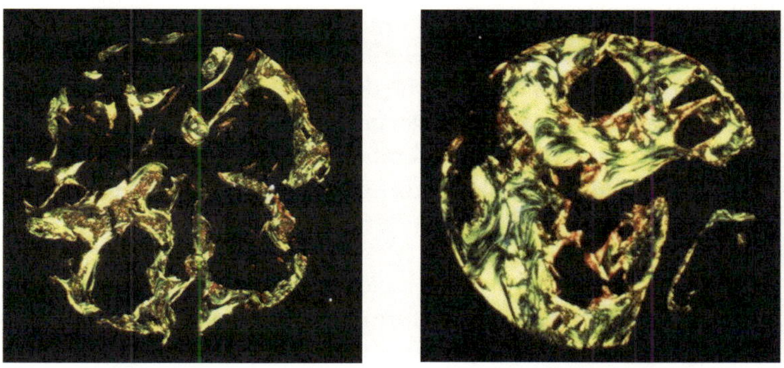

a. b.

Fig. 13. a. Photomicrograph of coronal section, mainly woven bone, containing both unorganized thin and thick collagen fibers with polarization colors of green to greenish-yellow (Picrosirius red staining with polarizing microscopy, x20 magnification). b. Apical section presenting a higher percentage of lamellar bone at 9 months consisting mainly of fibers with greenish-yellow and yellow polarization colors (Picrosirius red staining with polarizing microscopy, x20 original magnification).

Beta tricalcium phosphate (β-TCP) (Cerasorb®, Curasan, Kleinostheim, Germany), a ceramic alloplast, is another popular graft material, extensively researched with pleasing results

(Breitbart et al. 1995; Gao et al. 1997; Buser et al. 1998; Ohsawa et al. 2000; Szabo et al. 2001). β-TCP is biocompatible (Rosa et al 1995, Hossain et al 1996, Ohsawa, et al 2000) , ,and achieves favorable volumetric maintenance. (Breitbart et al 1995, ; Gao et al. 1997) In a comparative histomorphometric study in miniature pigs (Merten et al 2001), β-TCP has shown very favorable qualities of biodegradation and substitution.

Unlike deproteinized bovine bone mineral xenograft, β-TCP has shown extensive resorption within 12 to 84 months after grafting (Yamada et al. 1997; Wiltfang et al. 2002), raising the question on the relationship between the material resorption rate and amount of newly formed bone in the augmented sites. Looking into this query we have undertaken to explore the osteoconductive and resorbability expressed by both deproteinized bovine bone mineral and β-TCP in identical defects performed in the dog mandible (Artzi et al 2004). At the bovine bone mineral sites, newly formed bone was incorporated and primarily established near the native defect walls and around the grafted particles at 3 months. At 6 months most of the defect (51.7%± 2.5) was filled with bone. At 12 and 24 months, complete bone regeneration was evident, but the grafted mineral particles still dominated the grafted sites (Fig 14).

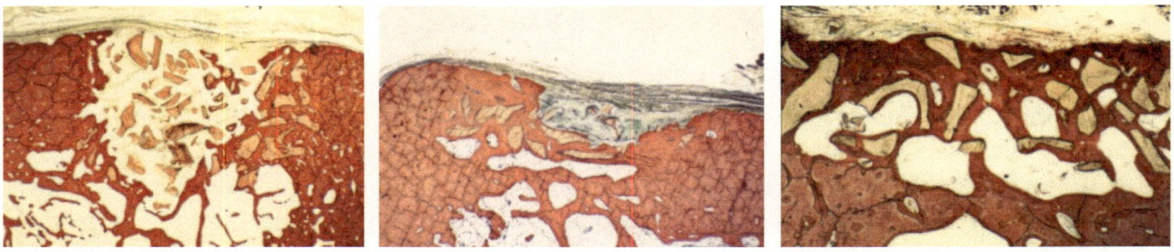

a. b. c.

Fig. 14. Photomicrograph of a bony defect grafted with bovine bone mineral particles. a. At 3 months, newly formed bone surrounds part of the grafted IBB particles mainly close to the native bony walls (Stevenel's blue and Van Gieson's picro fuchsin staining, original magnification x20). b. At 6 months, most of the defects are filled with newly formed bone incorporated around the grafted particles (Stevenel's blue and Van Gieson's picro fuchsin staining, original magnification x20). c. At 24 months, bovine bone mineral particles dominate the grafted site and completely incorporate with the newly formed bone to achieve complete healing site configuration (Stevenel's blue and Van Gieson's picro fuchsin staining, original magnification x20).

Under high power magnification, osteoid formation was noted even after 1 month. At 3 months, highly cellular newly formed bone was observed mainly around the grafted particles. At 6 months most of the particles were surrounded by newly-formed bone that filled the majority part of the defect.(Fig 15)

At the β-TCP sites, aggregates of β-TCP particles were still predominated at 3 months, while newly formed bone was noted primarily near the native bone (Fig16a.). At 6 months, defects showed almost complete bone fill. The grafted particles were completely embedded in the newly formed regenerated bone (Fig 16b). At 12 months, there were only remnants of the particles, particularly in the center of the defect, distal from the bony walls and at 24 months, particles were completely resorbed and the entire defect was filled with new bone in both in membrane-protected and unprotected defects. (Fig. 16c)

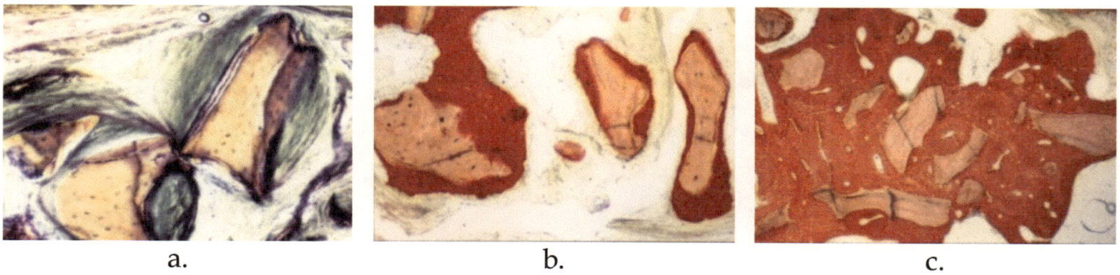

Fig. 15. Photomicrograph of a bony defect grafted with bovine bone mineral particles. a. At 1 month, bovine bone mineral is already surrounded by the greenish staining of osteoid formation (Stevenel's blue and Van Gieson's picro fuchsin staining, original magnification x200). b. Newly formed bone primarily surrounding the grafted particles (Stevenel's blue and Van Gieson's picro fuchsin staining, original magnification x100). c. At 3 months, newly formed bone is filled the space and interconnecting the grafted particles (Stevenel's blue and Van Gieson's picro fuchsin staining, original magnification x40).

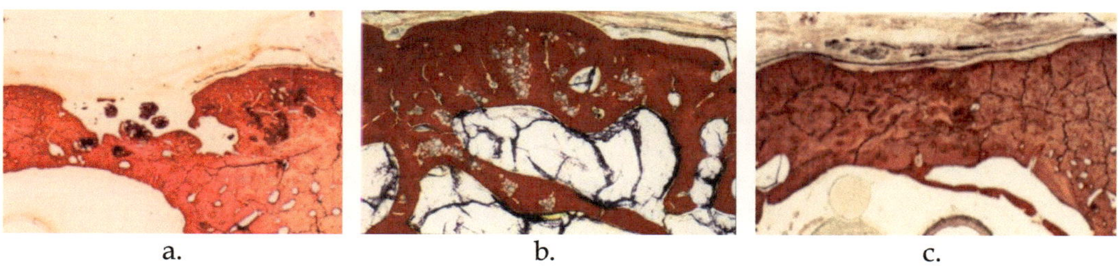

Fig. 16. Photomicrograph of a bony defect grafted with β-tri-calcium phosphate particles (β-TCP). a. At 3 months at the membrane-protected sites, newly formed bone (Stevenel's blue and Van Gieson's picro fuchsin staining, original magnification x20). b. At 6 months, there was complete newly formed bone bridging the defect (Stevenel's blue and Van Gieson's picro fuchsin staining, original magnification x20). c. At 24 months healing period, β-TCP particles were fully resorbed and the defect was completely regenerated by newly formed bone (Stevenel's blue and Van Gieson's picro fuchsin staining, original magnification x20).

High power magnification reveals that, β-TCP particles were surrounded by highly cellular newly formed bone showed grafted particles in advancing stages of resorption and/or significant degradation. Osteoclasts were observed near the resorbed particles (Fig 17).

Additional osseoconductive bone-graft substitute which is noteworthy is a biphasic hydroxyapatite/β-tricalcium phosphate (HA/TCP) produced by a single process to prevent clustering and to establish a new homogeneous molecule. Its 60:40 ratio of hydroxyapatite:β-tricalcium phosphate, gives it two phases of activity. HA/TCP offers an interconnected porosity of 90% (pores ranging from 100-500 μm in diameter) to support cellular penetration. While the HA - biphasic TCP compound have shown promising results extra-orally (Russotti *et al.* 1987; Brook *et al.* 1991; St John *et al.* 1993; Emery *et al.* 1996; Gauthier *et al.* 2001; Le Nihouannen *et al.* 2005; Schopper *et al.* 2005; Blouin *et al.* 2006; Fellah *et al.* 2006) and in animal studies also intra-orally (Hashimoto-Uoshima *et al.* 1995; Boix *et al.* 2004, 2006), it still lacks clinical validation in intra-oral applications in humans.

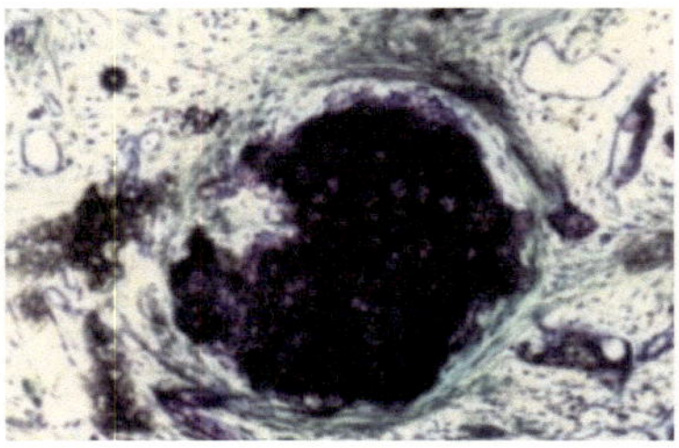

a.

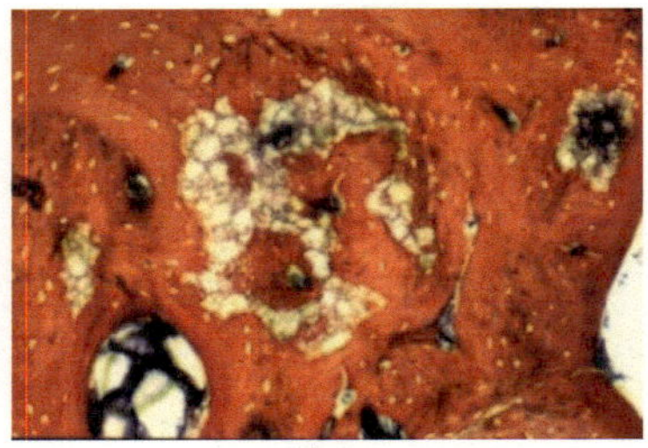

b.

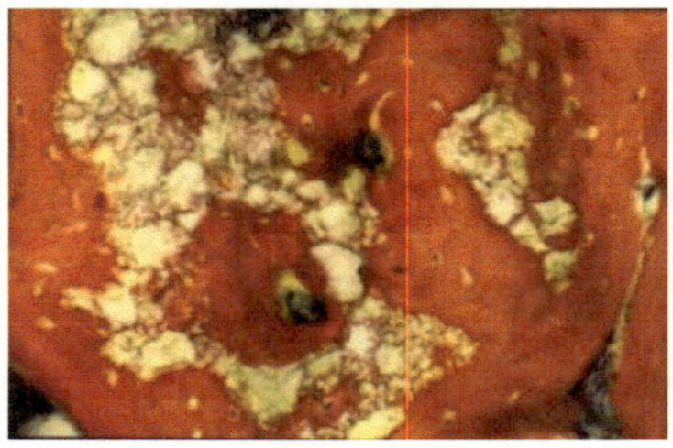

c.

Fig. 17. Photomicrograph of a bony defect grafted with β-tri-calcium phosphate particles (β-TCP). a. Greenish staining of osteoid formation was evident around β-TCP particles, which was not in proximity to the native bony walls (Stevenel's blue and Van Gieson's picro fuchsin staining, original magnification x600). b. Newly formed bone incorporated with the grafted β-TCP particles that were in an advanced stage of resorption (Stevenel's blue and Van Gieson's picro fuchsin staining, original magnification x100). c. Higher magnification of panel b. Osteoclasts observed near the resorbed β-TCP particles (Stevenel's blue and Van Gieson's picro fuchsin staining, original magnification x200).

To evaluate HA/TCP with autogenous particulate cancellous bone, this composite graft combination was examined in sinus augmentation procedures (Artzi et al 2008**).** Newly formed bone around the grafted particles was found in all samples. The encircling, highly cellular bone followed the outline of the grafted particles in direct contact (Fig. 18a-b). Both woven and lamellar types of bone were observed (Fig. 19)

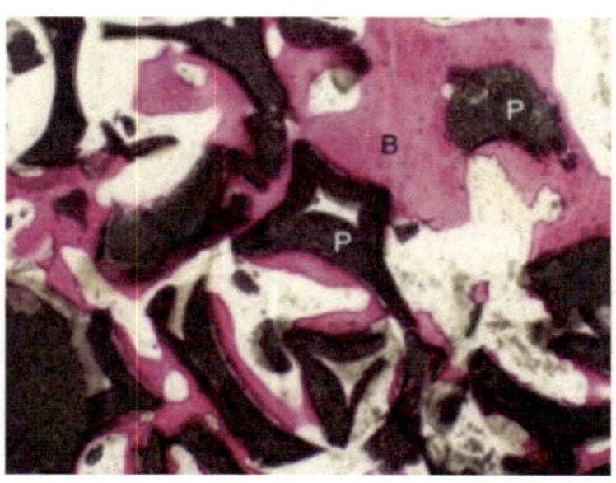

a.

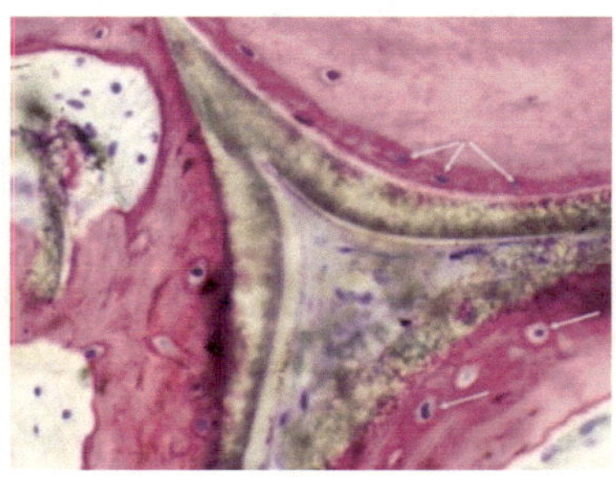

b.

Fig. 18. Photomicrograph of a bony defect grafted with a biphasic hydroxyapatite/tricalcium phosphate (HA/TCP) a. Most of the grafted particles (P) surrounded by newly formed bone (B) (Paragon staining, original magnification x150). b. Osteocytes (arrows) lining the interface osseous zone in direct contact with the grafted HA/TCP particle (Paragon staining, original magnification x600).

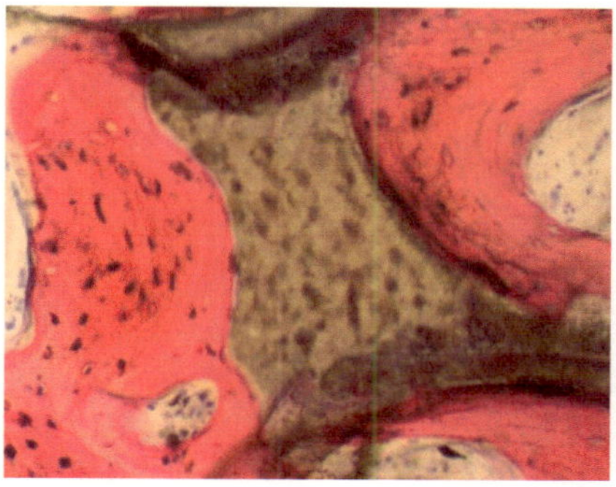

a.

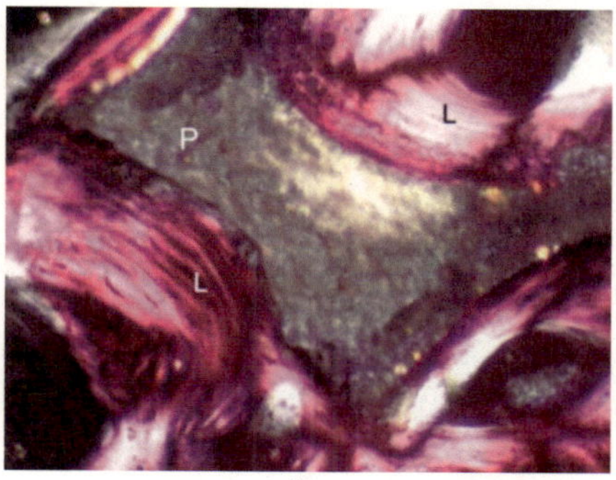

b.

Fig. 19. Photomicrograph of a bony defect grafted with a biphasic hydroxyapatite/tricalcium phosphate (HA/TCP). a. A high magnification of a 6-month grafted particle fully surrounded by newly-formed bone (Paragon staining, original magnification x600). b. Polarized light image of the specimen shown in fig. 3a.; bone lamellar structure is displayed adjacent to the grafted particle (P) (L). (Paragon staining, original magnification x600).

Morphometrically, mean bone area fraction increased from 28.6% ± 7.8 at 6 months to 41.6% ± 8.3 at 9 months. In resemblance to bovine bone mineral, this biomaterial occupied a surface average of 25% at both observation periods. This alloplast as a composite with autogenous bone chips promotes newly formed bone, which increases in its fraction along an extended healing period.

3. Alveolar ridge preservation

3.1 Alveolar ridge preservation after extractions

Since the most frequent cause for alveolar ridge augmentation is implant site development, long term studies examining different scaffold materials and GBR procedures have focused on GBR at implant sites. The survival rate of implants placed into GBR treated sites varies between 79% and 100% with over 90% survival rate after being in function for at least 1 year. These data are comparable to those reported for implants placed into native, untreated sites. For more data the reader is referred to a few systematic reviews that have focused on the subject of survival and success rates of implants placed within regenerated bone (Hammerle *et al.* 2002; Fiorellini & Nevins 2003; Chiapasco *et al.* 2006).

Augmentation and preservation of the alveolar process and ridge posses a few treatment strategies, depending on the bony defect morphology available. According to the bone morphology immediately after extraction, one out of two procedures is selected: a) the one-step (combined) approach (immediate implant placement plus GBR) is preferred if anchorage of the implant with primary stability is possible or b) the two-step (staged) approach is preferred when the defect morphology precludes primary implant stability; the two-step approach requires bone augmentation to a degree allowing implant placement in a second intervention (Figure 20).

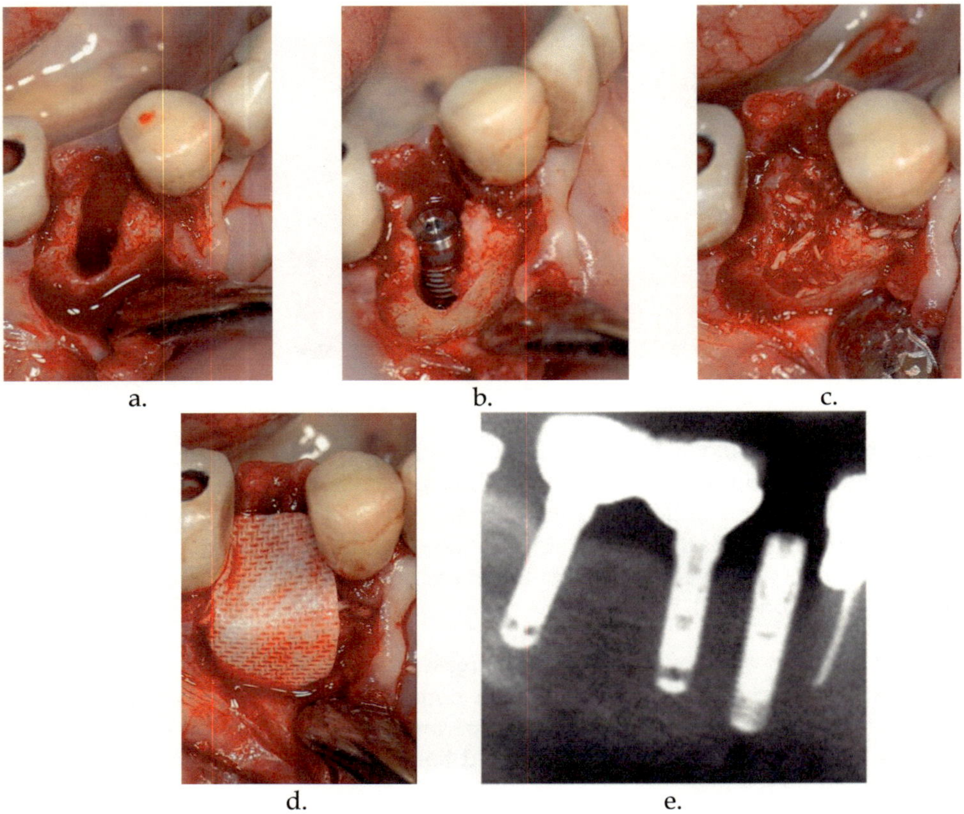

Fig. 20. One staged implant placement in a Type 2 bony defect in the mandible. a. bone defect due to extraction made 8 weeks earlier. b. endosseous implant anchored in the peripheral and apical bone. c. peri-implant gaps are filled with autogenous bone chips mixed with alloplast graft material. d. defect is covered with a cross linked collagen membrane***.

Since in most cases suffering from bone loss and/or ridge deformations there is lack of soft tissue in addition to lack of bone, it is advisable to improve the soft tissue coverage as early as possible, preferably at the time of hard tissue augmentation. Clinical attempts to maintain the ridge contours and improve the soft tissue biotype by grafting particulate autogenous bone or non-resorbable materials into the fresh extraction socket were carried out with partial success. GBR used to preserve or augment the alveolar ridge at the time of tooth extraction was employed by grafting "supporting materials" into fresh extraction sockets and covering it by non-resorbable membranes (Nemcovsky & Serfaty 1996; Lekovic *et al.* 1997; Fowler *et al.* 2000). Two main shortcomings of this procedure were a. histological findings revealed that some "supporting materials" like DFDBA presented "dead" particles with no evidence of bone formation on the surfaces of the implanted particles and no evidence of osteoclastic resorbtion of the grafted particles. (Becker et al 1994a) and b. the lack of soft tissue to completely cover the grafted site.

*** Ossix-P ™ Colbar, Ramat Hasharon, Israel

3.2 Histological and histomorphometrical studies of ridge preservation after extractions

Histologic and Histomorphometric comparison of specimens from sites treated with GBR and non treated sites revealed that more vital bone had formed in the first group. Both osteoconductivity and resorbability of the materials apparently influenced new bone formation in a positive manner (Artzi *et al.* 2000; Bolouri *et al.* 2001; Froum *et al.* 2002). Artzi et al (2000,2001a,b) have shown that when bovine bone mineral was used solely to fill the socket without applying GBR principles, the average clinical overall bone fill of the augmented socket sites was 82.3%. Newly formed bone encircled and adhered to the grafted material in most specimens. Osteoblasts were present within an osteoid layer, lining the interface zone of the bovine mineral particles and the new osseous tissue. Histomorphometric measurements showed an increase of mean bone tissue area along the histological sections from 15.9% in the coronal part to 63.9% apically with overall average of 46.3% (Fig.21 a-b). Newly formed bone was characterized by abundance of cellular woven-type bone in the coronal area, while lamellar arrangements could be identified mainly in the more apical region. Woven/lamellar bone ratio analyzed using polarized microscopy about 10 folds in favor of the woven bone at the crestal region whereas it was nearly 1:1 ratio at the apical zone of the grafted socket (Fig21 c-d).

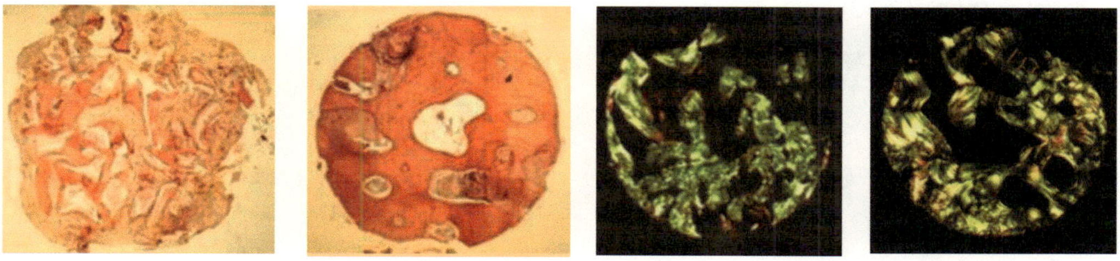

a. b. c. d.

Fig. 21. Photomicrograph of a socket site grafted with bovine bone mineral particles at 9 months. a. A crestal section cut disclosed abundant amount of connective tissue and bovine mineral particles and only occasional osseous fragments. (HE staining X 20 magnification). b. An apical section cut: most of the area fraction is occupied by osseous tissue, while bovine mineral particles are well demonstrated. Only a small amount of connective tissue could be identified. (HE staining X 20 magnification). c. Polarizing microscopy of a crestal section cut of bovine bone mineral grafted socket site. The bone area is woven type (Picrosirius red staining; original magnification X 20). d. An apical section cut under polarizing microscopy. Note the dominance of the yellowish-orange lamellar bone type (Picrosirius red staining; original magnification X 20).

3.3 Timing of augmentation and implant placement

As previously mentioned, early implant placement concurrent with guided bone regeneration technique has shown encouraging results (Donos *et al.*, 2008; Buser *et al.*, 2009). It has also been claimed that obtaining initial stability of the implant is a prerequisite for successful osseointegration no matter what technique is applied (Wikesjo & Nilveus 1990, Rasmusson et al., 1999, Lundgren et al., 1999, Becker 2005, Lioubavina-Hack et al., 2006). However, today there is no clear evidence whether simultaneous GBR procedure affects implants survival rate (Donos *et al.*, 2008).

Implant stability and soft tissue condition are the principal tools in evaluating healing and function. Most researchers and clinicians have used probing depth, bleeding on probing, and implant stability as the main parameters to assess and monitor implants success or failure. Despite the amenable healing response shown after implant placement in a simultaneously augmented bone sites, both simultaneously or in a 2-stage mode, we found a great interest in following up implants placed in either of the two techniques.

In a longitudinal study implant placement and bone augmentation as either a combined or staged procedure were monitored at 8 and 16 months post implant placement (Artzi et al 2012). Using clinical parameters such as peri-implant soft tissue conditions and implant stability. It was found that while probing depth and bleeding on probing improved along the time, implant stability was significantly higher when using the delayed mode. We concluded that both techniques may be accepted safe.

In another study in the dog, a qualitative and quantitative evaluation on the degree of osseointegration was conducted to explore the efficacy of implant placement and GBR procedure performed simultaneously or as a 2-stage procedure (Artzi et al 2010). Morphometric analysis disclosed that a similar osseointegration level over time was shown at the simultaneous (mean of 77.95% ± 11.24) and delayed (79.82% ± 7.54) techniques (Figs 22, 23).

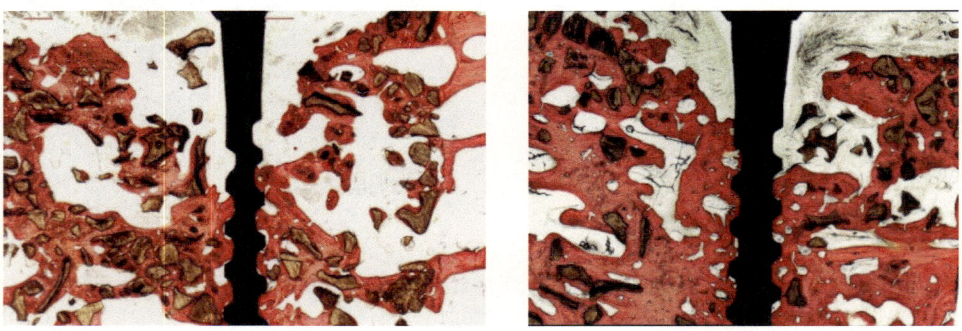

a. b.

Fig. 22. Photomicrograph of an experimental intrabony site grafted with bovine bone mineral (BBM) a. At 8 months, a coronal part of a simultaneous implant placement and bone augmentation procedure using bovine bone mineral particles. Note the crestal bone level in reference to the implant neck. (The implant core was trimmed due to lack of interest and to allow an expanded view at the periphery). (Stevenel's blue and Van Gieson's picro fuchsin staining x35 original magnification). b. A coronal part of an implant placed at a 6-month regenerated grafted BBM site at 8 months (Stevenel's blue and Van Gieson's picro fuchsin staining x35 original magnification).

In both techniques, newly-formed bone enhancement was observed proximal to the rough surface of the implant. However, the staged approach showed enhanced newly-formed bone (63.42±9.41 vs. 55.04 ±5.60; $p < 0.05$), less crestal bone resorption (0.92 ± 0.33 vs. 1.11 ± 0.26; $p < 0.05$), and smaller vertical bone defect (0.50 ± 0.37 vs. 0.88 ± 0.43; $p < 0.05$) over time compared to the combined approach. The staged approach showed also better significant results in regard to higher osteoconduction around the grafted mineral particles (71.42 ±

18.29 vs. 37.71 ± 24.31; $p < 0.05$), however, only at 8-month period. Practically, although the staged approach showed enhanced bone level and higher bone density, timing of the augmentation procedure did not influence the degree of osseointegration or the clinical outcome.

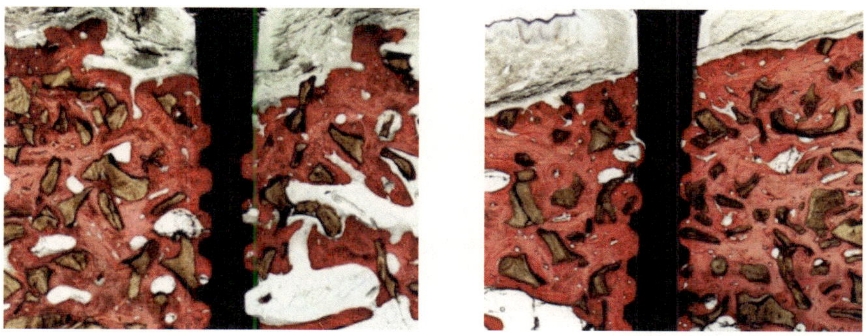

a. b.

Fig. 23. Photomicrograph of an experimental intrabony site grafted with bovine bone mineral (BBM) a. At 16 months, a coronal part of a simultaneous implant placement and bone augmentation procedure. Note the improved crestal bone level (Stevenel's blue and Van Gieson's picro fuchsin staining x35 original magnification). b. A coronal part of an implant placed at a 6-month regenerated grafted BBM site at 16 months (Stevenel's blue and Van Gieson's picro fuchsin staining x35 original magnification).

3.4 Soft tissue management in alveolar ridge preservation procedures

Applying GBR procedures resulted in reduced rate of resorpbion of the alveolar process in comparison with untreated control sites (Lekovic *et al.* 1997, 1998; Yilmaz *et al.* 1998; Camargo *et al.* 2000). Complications with soft tissue dehiscences, however, frequently occurred in GBR-treated sites (Fowler *et al.* 2000; Yang *et al.* 2000). GBR procedures applied for ridge volume preservation have two main shortcomings: a. it requires a 5-6 months healing period of time before endosseous implants can be placed; b. soft tissue coverage are technique sensitive procedures and may lead to a compromised esthetic result. Significant improvement regarding the profile of the alveolar ridge has been achieved with the introduction of various techniques aimed at improving the soft tissue conditions. Evian & Cutler (1994) described the use of autogenous soft tissue grafts to seal extraction sites at the time of implant placement. The technique was further improved by using free gingival grafts as socket sealers before (Landsberg & Bichacho 1994) or at the time of implant placement (Landsberg 1997) (Figs.24,25,26). The main problem associated with this technique was necrosis of the transplanted mucosa (Tal 1999) and poor color integration at the recipient site.

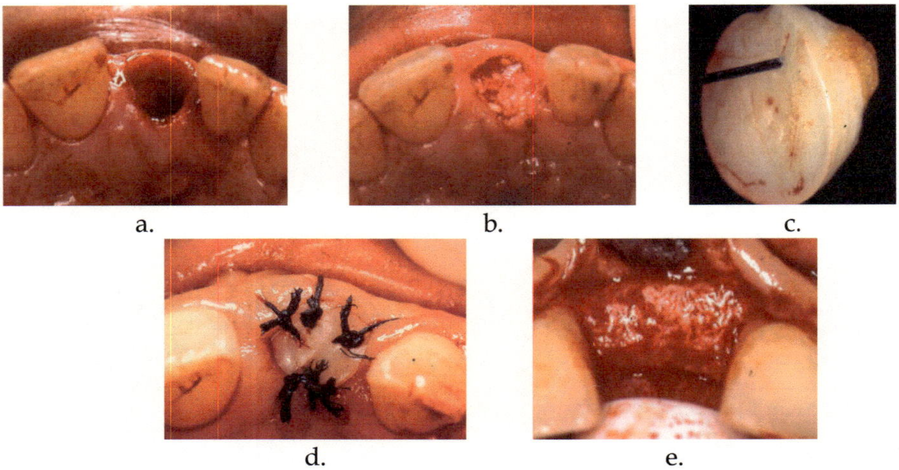

Fig. 24. Ridge preservation (implant site development) at the time of extraction using a free gingival graft as socket sealer. a. atraumatic extraction and complete circumferential curettage of the pocket epitheilium. b grafting the fresh extraction sockets with demineralized freezed dried bone allograft. c. free connective tissue graft obtained from an upper distal edentulous ridge d. socket orifice is sealed with the free gingival graft e. six months post extraction ridge preservation is demonstrated at the time of implant placement.

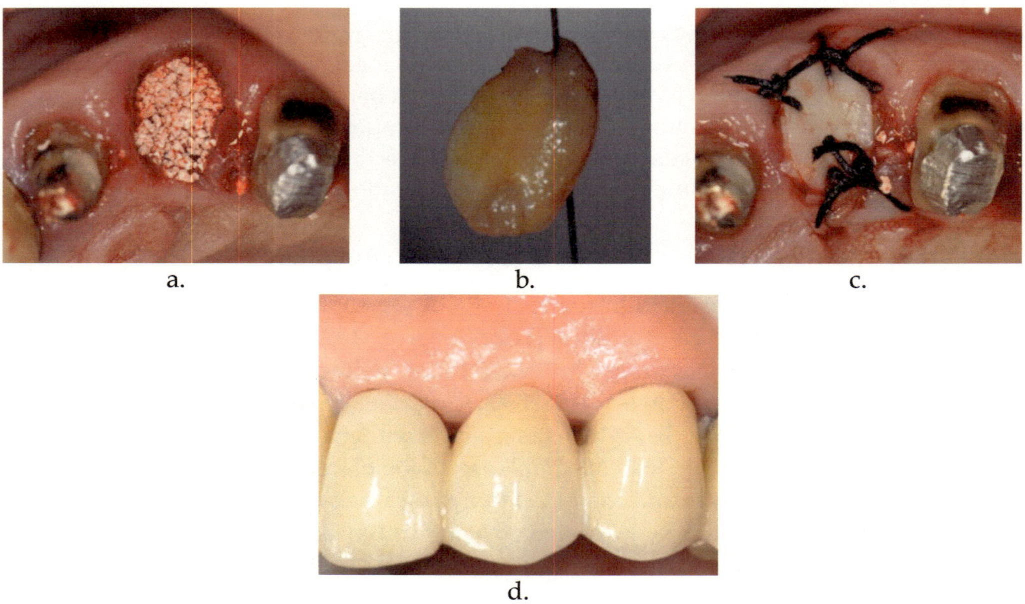

Fig. 25. Ridge preservation at the time of extraction using a free gingival graft as socket sealer. a. grafting a fresh extraction sockets with 4Bone alloplast ** material b. inner side of a free connective tissue graft obtained from the palate c. socket orifice is sealed with the free gingival graft d. Five months post extraction the alveolar ridge preserves it's natural contours, and is ready for a tooth supported 3 units bridge.

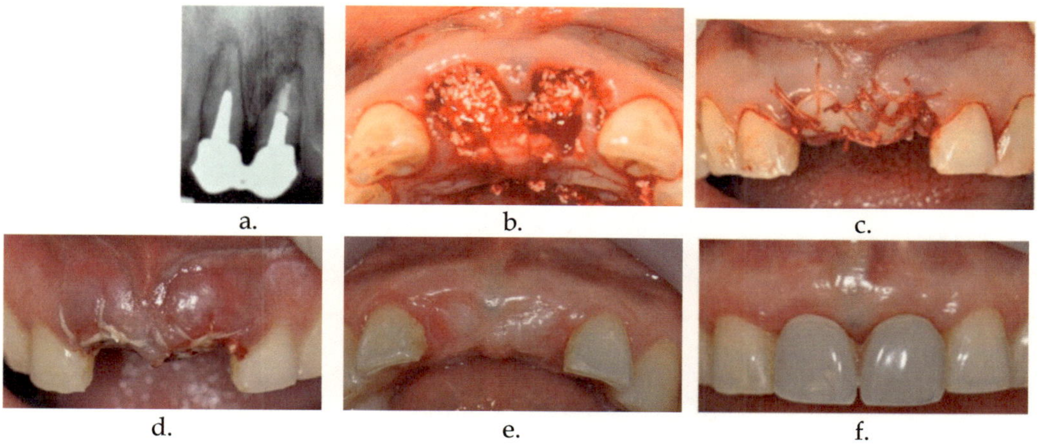

Fig. 26. Ridge preservation at the time of extraction using free gingival graft (left) and granulation reactive connective tissue (right) as socket sealers. a. radiographic view of the upper central incisors before extraction b. atraumatic extraction is followed by grafting the fresh extraction sockets with 4Bone alloplast**. c. sockets orifice is sealed with a free gingival graft (right) and socket granulation reactive connective tissue and a free gingival graft (left). e. and f. six months post extraction complete healing is demonstrated without and with a temporary restoration.

In search for a more predictable technique two additional approaches were applied. Mardinger et.al.(2010) have used intra-socket reactive soft tissue for primary closure during augmentation of infected extraction sites exhibiting severe bone loss prior to implant placement or as part of ridge preservation procedures. Porous bovine xenograft bone mineral was grafted into extraction sites demonstrating extensive bone loss. The intra-socket reactive soft tissue was sutured over the grafting material to seal the coronal portion of the socket. Biopsies of the healed mucosa and bone cores retrieved at implant placement revealed that the intrasocket reactive soft tissue demonstrated features compatible with granulation tissue and long junctional epithelium. The mucosal samples at implant placement demonstrated histopathological characteristics of keratinized mucosa with no residual elements of granulation tissue. The mean composition of the bone cores was - vital bone 40 ± 19% (13.7-74.8%); bone substitute 25.7 ± 13% (0.6-51%); connective tissue 34.3 ± 15% (13.8-71.9%). These authors concluded that intrasocket reactive soft tissue may successfully be used for primary closure of grafted fresh extraction sockets aiming to preserve the edentulous ridge. (Fig. 27)

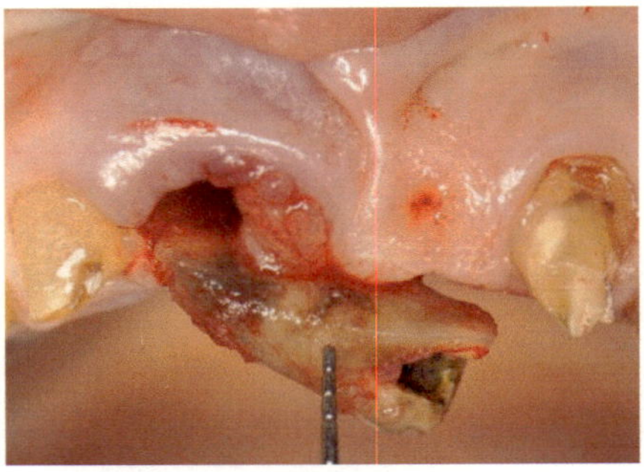

a.

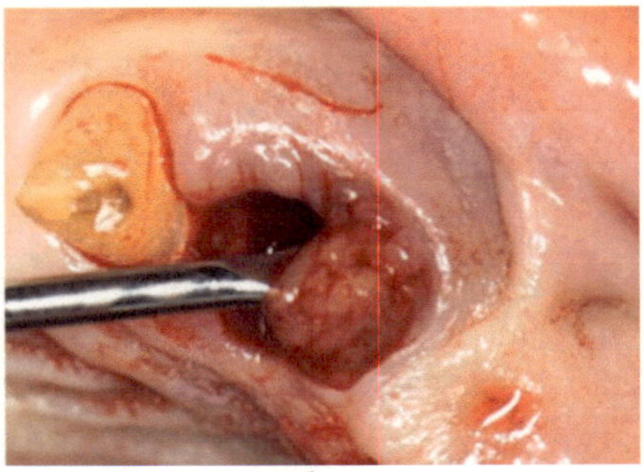

b.

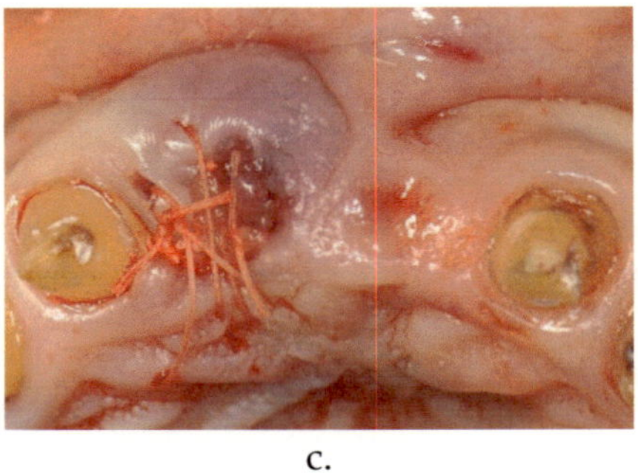

c.

Fig. 27. Intra-socket reactive soft tissue used for primary closure during augmentation of an infected extraction site exhibiting severe bone loss as part of ridge preservation procedures. a. during tooth extraction care is taken to avoid disconnecting the reactive tissue from it's blood sources. b. the intra-socket reactive soft tissue is prepared and moved aside to allow insertion of graft material. c. the reactive soft tissue sutured over the grafting material to seal the coronal portion of the socket.

A coronal and lateral sliding pedicle flaps to cover the orifice of the grafted extraction socket have been employed by Nemcovsky & Serfaty (1996). Their technique resulted in almost 100% survival rates of the connective tissue grafts (Tal et al 2004). Figures 28,29 describe the socket seal surgery using the lateral palatal pedicle flap technique employed in an immediate implant placement procedures (Fig.28) and augmentation of edentulous sites for esthetic purposes (Fig. 29).

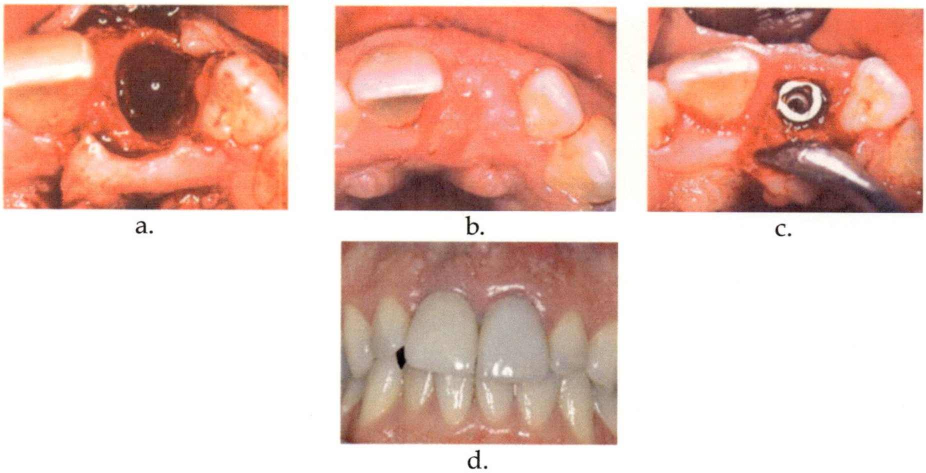

Fig. 28. Socket seal surgery using the lateral pedicle flap technique employed in an immediate implant placement procedure at the upper left central incisor. a. palatal pedicle flap is prepared for the sealing of a fresh extraction socket after placement of an endosseous implant and bone scaffold material. b. occlusal view six month following extraction c. implant exposure allows to keep the papillae and buccal soft tissue untouched d. Clinical view 8 years after restoration.

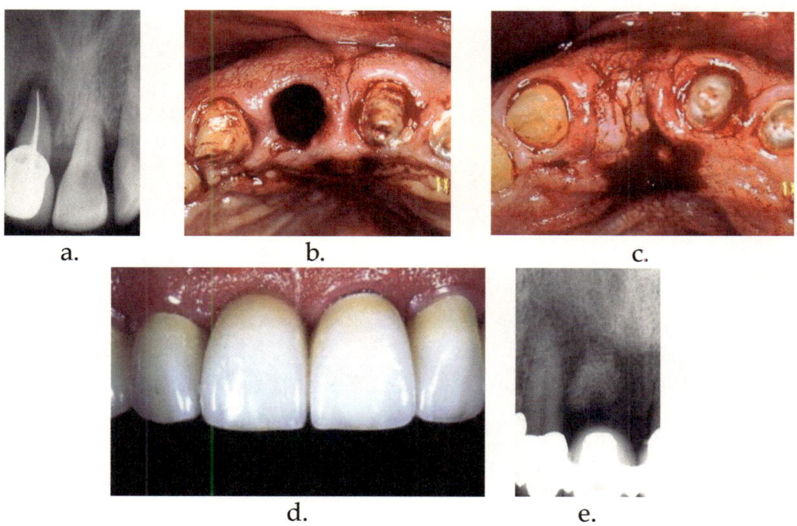

Fig. 29. Augmentation of an edentulous sites for esthetic purposes applying a modification of the palatal pedicle flap technique for covering a grafted fresh extraction socket. a. periapical lesion in an hopeless upper right central incisor. b. the fresh extraction socket is debrided and pocket epithelium removed. c. rotated palatal flap cover the orifice of the socket, after grafting with bone scaffold material. d. clinical view of a 4 unit porcelain to metal fused bridge after 6 years shows very pleasing adaptation of the pontic to the underlying soft tissue. e. radiographic view 6 years after grafting presenting a radio-opaque area at the previous extraction site.

In defects which combine a fresh extraction socket and a bony dehiscence resulted in by partial destruction of the buccal bone, a membrane may be placed within the socket against the buccal wall and the dehiscence, and the socket is filled with a membrane-supporting material which is adapted to support the membrane. If an implant is placed simultaneously, the material is placed into the space between the walls of the socket and the implant surface. In the esthetic zone, additional augmentation of the bone, beyond the labial wall of the socket is indicated; in that case, correction of the ridge contour is an additional task, added to the preservation of the volume of the socket. Such improvement can be achieved by combining the ridge preservation/augmentation procedures and the socket seal free gingival graft technique with a facially placed sub-epithelial connective tissue graft; this modification aims at improving the vertical and labial contours of the extraction site (Grunder 2011). Figure 30 presents a modified socket seal and subepithelial CT graft procedure, when both are indicated.

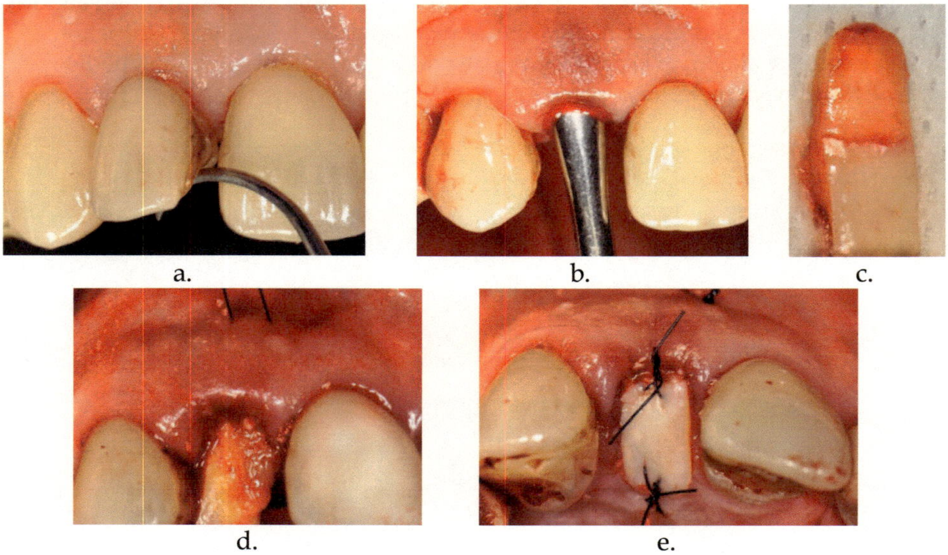

Fig. 30. Clinical view of a combined socket seal - subepithelial graft procedure. a. The upper right lateral incisor has to be removed following a traumatic injury. b. the mucosa buccal to the extraction socket is separated from the buccal plate of the bone, creating a pouch. c. a connective tissue graft is obtained from the palate; epithelium is removed from the portion which is designed to be placed into the pouch. d. the connective tissue graft is inserted into the pouch, directed to place and stabilized by the mattress suture which is connected to it and penetrates the buccal gingiva. e. once in place, the margins of the epithelialized portion of the graft are adapted to the orifice of the socket and stabilized by a few additional peripheral simple interrupted sutures.

5. Vertical bone augmentation procedures using extra-oral bone blocks

The augmentation of horizontal or vertical bone loss of the alveolar procedures are beyond the scope of this chapter. Briefly, it is believed that for the augmentation of this type of bone defects intraoral or Extraoral autogenous block transplants are preferred(Becker *et al.* 1994a; Buser *et al.* 1996; von Arx *et al.* (2005). The advantages of autogenic block grafts are mainly its handling properties, stabilization of the healing site and optimal biologic properties. The disadvantages include donor site morbidity, technical difficulties of the harvesting procedures, and the impossibility of using the graft as a carrier for growth factors. Harvesting procedures from intraoral sites have generally been preferred over extraoral sources since it may be performed under intraoral local anesthesia, it results in less morbidity, and it may provide sufficient amount of bone for the treatment of localized bone defects (Joshi & Kostakis 2004). Common intraoral donor sites are the chin and the retromolar region in the mandible. The limitations and disadvantages of intraoral bone harvesting were described by Nkenke *et al.* (2001) and von Arx *et al.* (2005). Figures 31 and 32 describe vertical bone augmentation procedures using bone blocks harvested from the iliac crest and from the skull respectively.

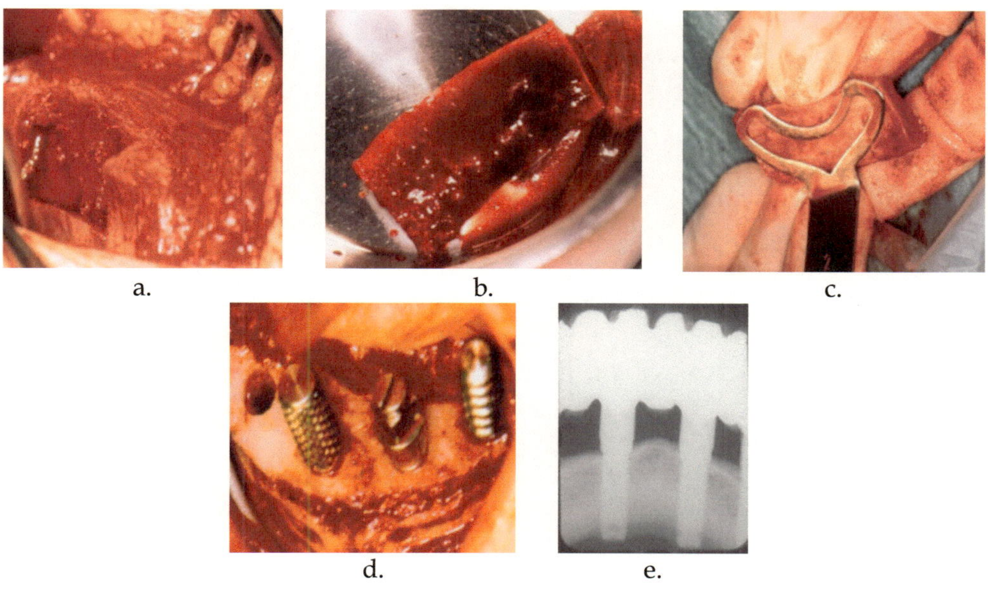

Fig. 31. Vertical bone augmentation using a bone block from the iliac crest. a. A block of bone measuring 4x1.5x2cm is removed from the iliac crest. b. clinical view of the harvested bone block c. the block is being molded using a prefabricated or custom made shablone. d. the trimmed block is placed keeping intimate contact with the exposed jaw and stabilized with 4 endosseous implants (operator P.I. Brenemark) e. radiographic view 19 years after rehabilitation. (Reprinted from Moses & Tal 2007).

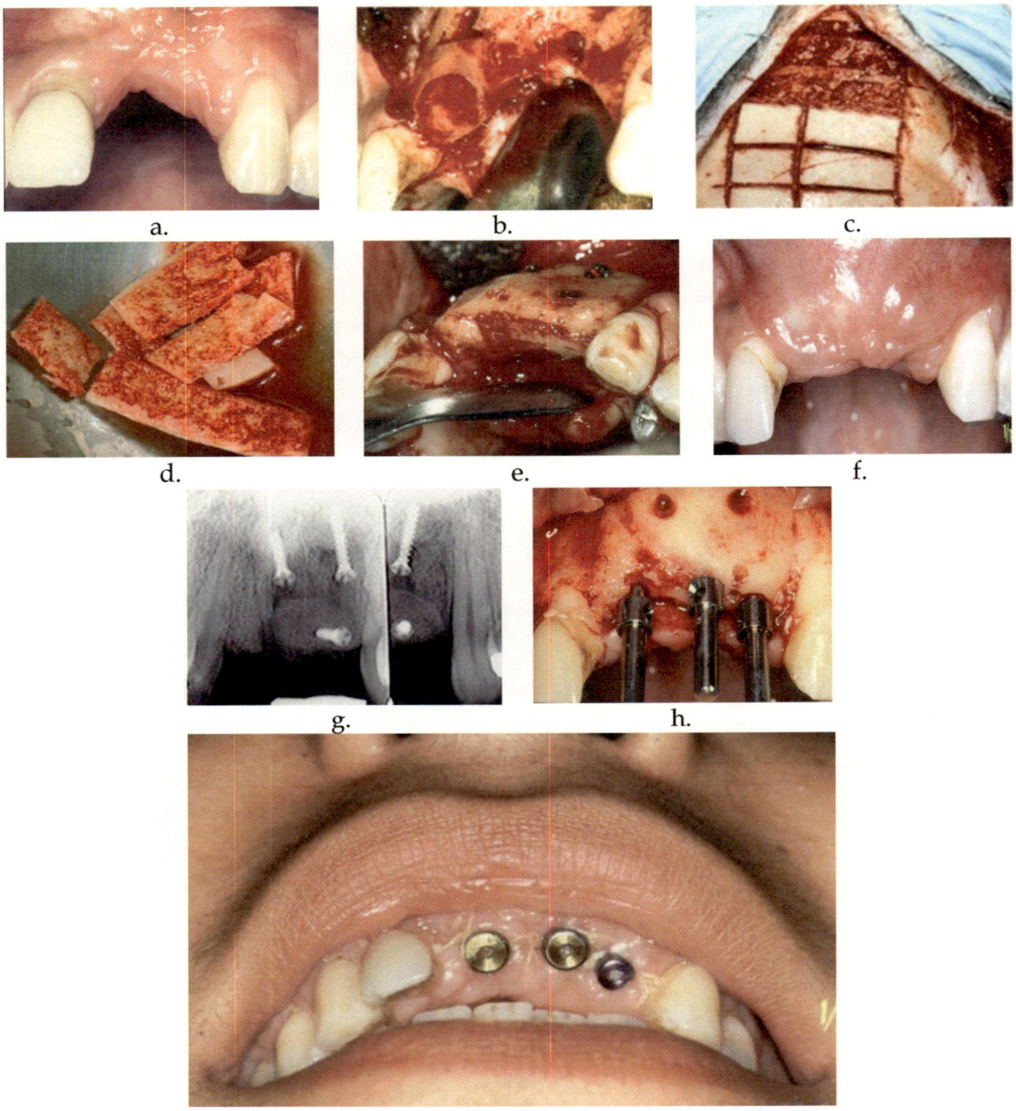

Fig. 32. Vertical bone augmentation using bone blocks from the skull. a. severe defect in the alveolar process following implant failure. b. surgical exposure of the alveolar process reveals severe bone loss requiring vertical and horizontal bone augmentation before implant placement can be considered. c. and d. bone blocks and particulated bone are removed from the skull e. bone blocks are trimmed and stabilized to the edentulous bony ridge with titanium mini screws; gaps between the blocks and jaw are filled with particulate bone. f. clinical view 6 month after the procedure g. radiographic examination 6 months after grafting.
h. implant placement procedure 8 months after grafting. h. clinical view of healing 10 days after implant placement.

6. Growth and differentiation factors for alveolar ridge augmentation

Scaffold osteoconductive materials lack osseoinductive components. Therefore, an attempt to enhance bovine bone properties was conducted by adding the synthetic peptide component P-15 (Qian and Bhatnagar 1996, Bhatnagar et al 1999). P-15, a synthetic peptide analog of collagen is a replica of the organic 15 amino-acid sequences within the sequential residues involved in bone formation in type I collagen (Bhatnagar et al 1997). A combination of this cell-binding peptide (P-15) attached to the mineral particles has been developed. It is assumed that the addition of such an organic replica component to an osteoconductive material, such as bovine bone, may enhance cell attachment by cell binding and differentiation, eventually resulting in accelerated periodontal ligament fibroblasts attachment (Lallier et al 2001) and enhanced osseous formation (Bhatnagar et al 1999). Initially produced as a particulate material, PepGen/P-15 was later replaced by flow and putty forms(Nguyen et al 2003). PepGen/P-15® (Dentsply Friadent Ceramed, Lakewood, CO, USA) has been used in several bone augmentation procedures, such as socket site preservation, (Hahn et al 2003, Tehemar et al 2003) ridge deficiency (Barboza et al 2002), maxillary sinus grafting, (Krauser et al 2000, Degidi et al 2004, Gelbart et al 2005, Philippart et al 2005) and periodontal defects (Yukna et al 1998, 2000, 2002, Walters et al 2003). However, to a large extent, the above data rely on in vitro observations (Qian and Bhatnagar 1996, Bhatnagar et al 1999, Bhatnagar et al 1997, Lallier et al 2001, Acil et al 2002, Kubler et al 2004, Trasatti et al 2004, Hole et al 2005, Turhani et al 2005).

When PepGen/P-15 was examined in surgical fenestrated membrane-protected periodontal defects in dogs (Artzi et al 2006), it proved to be biocompatible and osteoconductive material (Fig. 33). While newly-formed bone achieved similar outcome (36.1% ± 3.6 and 31.4% ± 1.9, at grafted and non-grafted sites), the non-grafted membrane-protected sites showed greater amount of new cementum (73.9% ± 2.0 vs. 59.5% ± 3.2; $p < 0.02$). It appears that PepGen/P-15 application in membrane-protected defects did not enhance regeneration.Similar findings were obtained in critical size defects (CSD) in the rat skull (Artzi et al 2008). In that study, Pepgen/P15 was applied with and without a GTR membrane while non-grafted membrane-protected and non-protected served as positive and negative controls. At 12 weeks, histomorphometric measurements showed CSD osseous build-up at mean of 60.6% ± 4.5 at the membrane-protected non- grafted sites which was greater ($p<0.05$) than at the grafted protected (50.6% ± 4.4) and grafted non-protected(44.2% ± 5.5) sites. While anorganic bovine mineral/cell-binding peptide contributes in volume, apparently, membrane application is the determinant factor to establish the gain in bone regeneration (Fig. 34).

In search for more effective techniques that predictably promote the bone natural regenerative abilityability current research is focused on the application of natural proteins and polypeptide that regulate tissue regeneration. Growth and differentiation factors are currently believed to contribute to alveolar ridge augmentation include platelet-derived growth factor (PDGF), insulin-like growth factor (IGF-I and IGF-II), transforming growth factor beta (TGF-®), fibroblast growth factor (a-FGF and b-FGF), and bone morphogenetic proteins (BMPs 1–15). Among these, bone morphogenetic protein (BMP) is the most widely considered in the dental literature. From a biologic point of view, the growth and

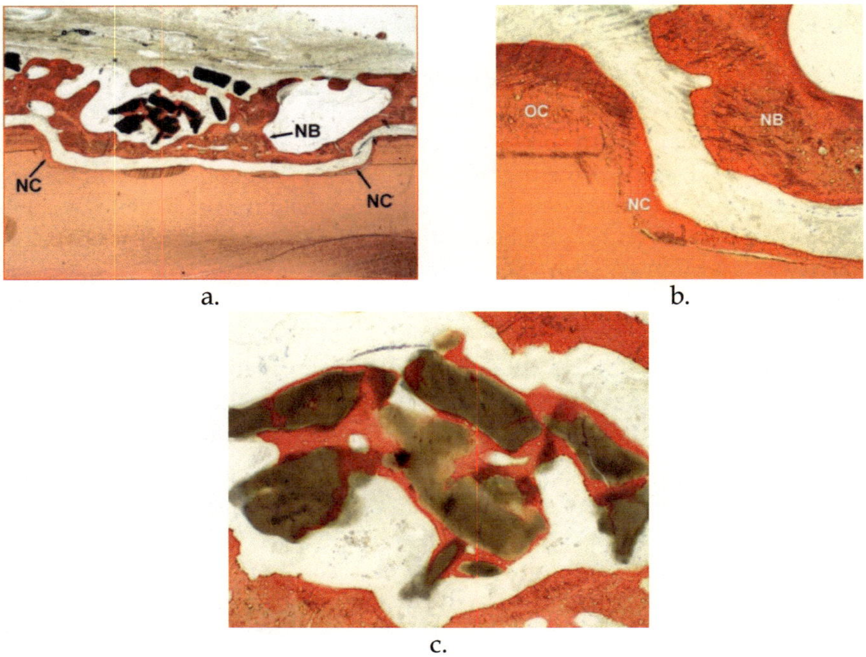

Fig. 33. Photomicrograph of an experimental fenestrated-type defect along the canine root surface grafted with PepGen/p-15 in the dog. a. Segments of new cementum (NC) are evident along the fenestrated and planed root surface. Residual PepGen/P-15 particles (black) are defined from the root surface by new bone (NB) formation. b. High-power magnification of panel a. New cellular cementum (NC) runs continuously out of the old cementum (OC) and lines the defected root surface floor concurrent with NB formation in the vicinity. However, the connective tissue arrangement in between is not yet defined. c. The grafted PepGen/P-15 particles are almost completely surrounded by NB. (Stevenel's blue and Van Gieson's picro fuchsin; original magnification: a - X20; b and c - X100.)

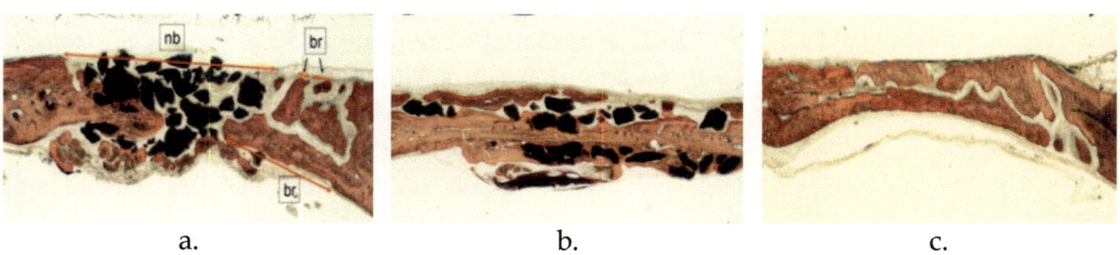

Fig. 34. Photomicrograph of an experimental critical sized defect grafted with PepGen/p-15 in the rat skull. a. Partial internal and external bone bridging (br) and a remarkable no external bridging (nb) in the Pepgen/P15 uncovered CSD site (Stevenel's blue and Van Gieson's picro fuchsin x25 original magnification). b. Non-decalcified section of the Pepgen/P15 membrane-protected CSD site. Newly formed bone surrounds Pepgen/P15 particles. (Stevenel's blue and Van Gieson's picro fuchsin x25 original magnification). c. Complete bone bridging evident at the non-grafted membrane-protected sites (Stevenel's blue and Van Gieson's picro fuchsin x25 original magnification).

differentiation factors may induce earlier bone growth into the area to be regenerated. Figure 35 presents an experimental regenerative procedure in the dog mandible using a BMP based "biologic glue". For more information on tissue engineering of bone the reader is referred to chapters 1-5.

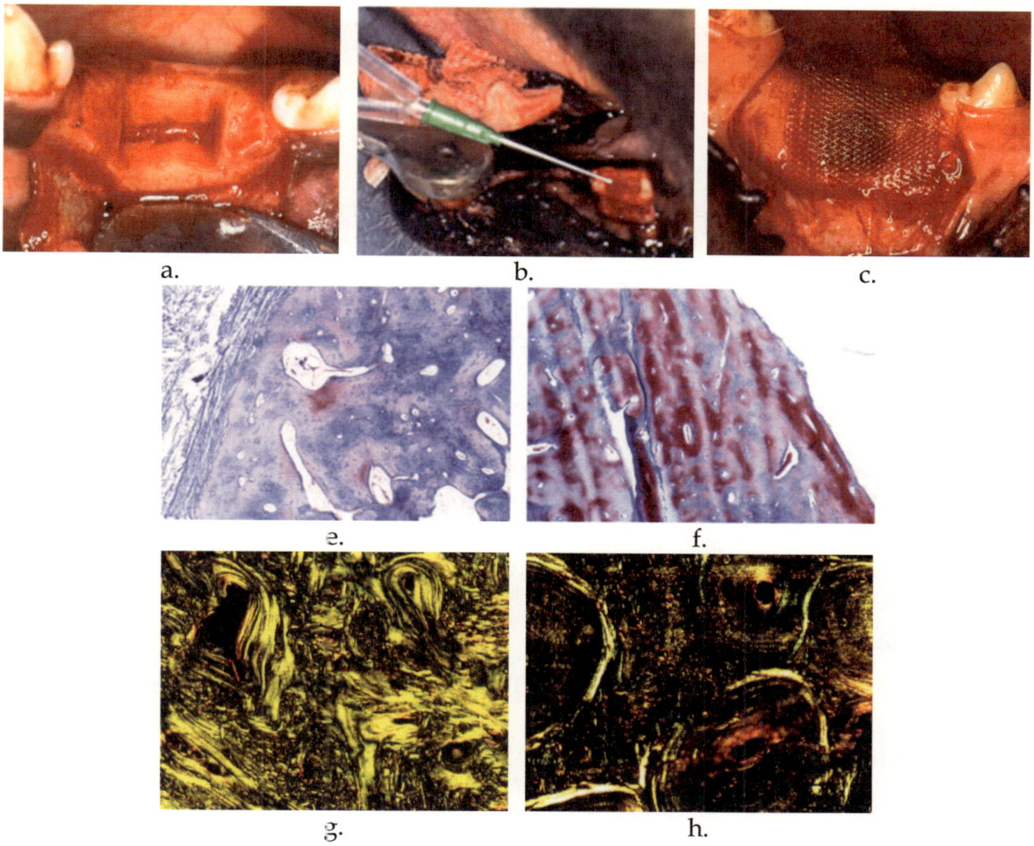

Fig. 35. Experimental regenerative procedure in the dog mandible using a. BMP based "biologic glue" a. experimental defect in an edentulous ridge in the mandible of a dog. b. experimental BMP based "biologic glue" gel is injected into the bony defect c.the gel filled defect is covered with an absorbable polyglactin membrane d.,e. bucco-lingual histological sections through the jaw show newly regenerated woven bone (left) filling the defect vs. mature pristine bone (right) from the lingual aspect of the jaw. f.,g. histological specimens treated with picrosirius red stain observed under polarized illumination show "young" collagen bundles with organizing osteons (f) vs. well established osteons and mature collagen fibers (g).

The possible relationship between susceptibility to periodontal disease and other systemic diseases and bone regeneration in the oral cavity has not been established. It has been demonstrated that implant therapy in patients who have lost their teeth due to advanced periodontitis are subject to higher rates of implant failure and complications involving the supporting tissues, compared with those who have lost their teeth due to other reasons

(Mengel *et al.* 2001; Hardt *et al.* 2002; Karoussis *et al.* 2003; Wennstrom *et al.* 2004). It is generally agreed that certain general health conditions represent a risk for successful GBR procedures. However there are no conclusive data with respect to bone augmentation procedures in patients suffering from systemic diseases which cause impaired tissue healing. Similarly, there seems to be no proof that patients who show behaviors (e.g. smoking, poor compliance) which lead to impaired tissue healing or to a higher susceptibility for disease development, (Mombelli & Cionca 2006) should be performed with these uncertainties in mind, when planning implant therapy in the presence of bone defects.

7. Conclusions

The alveolar process and alveolar ridge of bone contain the supporting attachment apparatus of teeth; therefore it's primary function is provision of anchorage to the dentition. The major development in esthetic dentistry, and more so the introduction of implant dentistry, led to significant developments aimed to regenerate or restore bony defects and bone loss in the edentulous ridge. Most clinical efforts in the developments in bone augmentation procedures are related to either simplifying clinical handling or influencing of biologic processes. These include constant improvements of the tissue barriers in use, new membrane supporting materials providing space for tissue regeneration, and finally growth and differentiation factors that induce earlier and rapid bone growth into the healing site. It is believed that the new developments would allow treatment of larger bone defects, will reduce the need for autogenous block grafts and membranes, and would reduce the technique sensitivity of the different procedures.

8. Acknowledgement

Production of this chapter was supported by The Gerald A Niznick Chair of Implantology.

9. References

Acil Y, Springer IN, Broek V, Terheyden H, Jepsen S (2002). Effects of bone morphogenetic protein-7 stimulation on osteoblasts cultured on different biomaterials. J Cell Biochem;86:90-98.

Amler, M.H. (1969). The time sequence of tissue regeneration. in human extraction wounds. *Oral Surgery, Oral Medicine and Oral Pathology* 27, 309–318.

Araújo, M.G. & Lindhe, J. (2005). Dimensional ridge alterations following tooth extraction. An experimental study in the dog. *Journal of Clinical Periodontology* 32, 212–218.

Araujo, M.G., Sukekava, F., Wennstrom, J.L. & Lindhe, J. (2006a). Tissue modeling following implant placement in fresh extraction sockets. *Clinical Oral Implants Research* 17, 615–624.

Araujo, M.G., Wennstrom, J.L. & Lindhe, J. (2006b). Modeling of the buccal and lingual bone walls of fresh extraction sites following implant installation. *Clinical Oral Implants Research* 17, 606–614.

Artzi, Z. & Nemcovsky, C.E. (1998) The application of deproteinized bovine bone mineral for ridge preservation prior to implantation. Clinical and histological observations in a case report. Journal of Periodontology 69, 1062-1067.

Artzi, Z., Tal, H. & Dayan, D. (2000) Porous bovine bone mineral in healing of human extraction sockets. Part 1. Histomorphometric evaluation at 9 months. Journal of Periodontology 71, 1015-1023.

Artzi Z, Tal H, Dayan D. (2001a) Porous bovine bone mineral in healing of human extraction sockets. Part 2. Histochemical observations at 9 months. J Periodontol Vol. 72, pp.152-159

Artzi, Z., Nemcovsky, C.E., Tal, H. & Dayan, D. (2001b) Histopathological morphometric evaluation of 2 different hydroxyapatite-bone derivatives in sinus augmentation procedures: a comparative study in humans. Journal of Periodontology 72, 911-920.

Artzi, Z., Nemcovsky, C.E. & Tal, H. (2001c) Efficacy of porous bovine bone mineral in various types of osseous deficiencies. Clinical observations and literature review. International Journal of Periodontics & Restorative Dentistry 21, 395-405.

Artzi, Z., Nemcovsky, C.E. & Dayan, D. (2002) Bovine-HA spongiosa blocks and immediate implant placement in sinus augmentation procedures. Histopathological and histomorphometric observations on different histological stainings in 10 consecutive patients. Clinical Oral Implants Research 13, 420-427

Artzi, Z., Givol, N., Rohrer, M.D., Nemcovsky, C.E., Prasad, H.S. & Tal, H. (2003a) Qualitative and quantitative expression of bovine bone mineral in experimental bone defects. Part 1: Description of a dog model and histological observations. Journal of Periodontology 74, 1143-1152.

Artzi, Z., Givol, N., Rohrer, M.D., Nemcovsky, C.E., Prasad, H.S. & Tal, H. (2003b) Qualitative and quantitative expression of bovine bone mineral in experimental bone defects. Part 2. Morphometric analysis. Journal of Periodontology 74, 1159-1160

Artzi Z., Nemcovsky C.E., Dayan D., Alpern Y. (2003) Vertical ridge augmentation using xenogenic material supported by a configured titanium mesh: Clinicohistopathologic and histochemical study. International Journal of Oral & Maxillofacial Implants 18, 440-446

Artzi, Z., Weinreb, M., Givol, N., Rohrer, M.D., Nemcovsky, C.E., Prasad, H.S. & Tal, H. (2004) Biomaterial resorbability and healing site morphology of inorganic bovine bone and beta tricalcium phosphate in the canine. A 24-month longitudinal histologic study and morphometric analysis. International Journal of Oral & Maxillofacial Implants 19:357-68.

Artzi Z, Kozlovsky A, Nemcovsky CE, Weinreb M (2005) The amount of newly formed bone in sinus grafting procedures depends on tissue depth as well as the type and residual amount of the grafted material. J of Clinical Periodontology 32,193-9.

Artzi Z, Weinreb M, Tal H, Nemcovsky CE, Rohrer MD, Prasad HS, Kozlovsky A.(2006) Experimental intrabony and periodontal defects treated with natural mineral combined with a synthetic cell-binding Peptide in the canine: morphometric evaluations J Periodontol. 77:1658-64.

Artzi Z, Kozlovsky A, Nemcovsky CE, Moses O, Tal H, Rohrer MD, Prasad HS, Weinreb M.(2008) Histomorphometric evaluation of natural mineral combined with a synthetic cell-binding peptide (P-15) in critical-size defects in the rat calvaria Int J Oral Maxillofac Implants. 23:1063-70

Artzi Z, Weinreb M, Lev Dor R, Carmeli G ,. Dard M, Nemcovsky CE (2008) Histomorphometric assessment of bone formation in sinus augmentation utilizing a combination of autogenous and hydroxyapatite/biphasic tricalcium phosphate graft materials: at 6 and 9 months in humans. Clinical Oral Implant Research 2008 Jul;19(7):686-92

Artzi Z, Nemcovsky CE, Weinreb M, Tal H, Rohrer MD, Prasad HS, Kozlovsky A (2010) Simultaneous vs. 2-stage implant placement and bone augmentation in the canine. Histomorphometry at 8 and 16 months *Journal of Clinical Periodontology* 37:1029-38

Artzi Z, Nemcovsky CE, Tal H, Kozlovsky A (2012) Timing of implant placement and augmentation with bone replacement material: Clinical assessment at 8 and 16 months *Clinical Implant Dentistry and Related Research*

Barboza EP, de Souza RO, Caula AL, Neto LG, Caula F de O, Duarte ME. (2002) Bone regeneration of localized chronic alveolar defects utilizing cell binding peptide associated with anorganic bovine-derived bone mineral: A clinical and histological study. J Periodonto 1159-73:1153;

Barzilay, I. (1993). Immediate implants: their current status. *International Journal of Prosthodontics* 6, 169–175.

Becker, W., Becker, B.E. & Caffesse, R. (1994a). A comparison of demineralized freeze-dried bone and autologous bone to induce bone formation in human extraction sockets. *Journal of Periodontology* 65, 1128–1133.

Becker W, Sennerby L, Bedrossian E, Becker BE, Lucchini JP Implant stability measurements for implants placed at the time of extraction: a cohort, prospective clinical trial. *J Periodontol* 2005; 76:391-7.

Berglundh, T. & Lindhe, J. (1997) Healing around implants placed in bone defects treated with Bio Oss. Clinical Oral Implants Research 8, 117-124.

Bhatnagar RS, Qian JJ, Wedrychowska A, Sadeghi M,Wu YM, Smith N. (1999) Design of biomimetic habitats for tissue engineering with P-15, a synthetic peptide analogue of collagen. Tissue Eng;5:53-65.

Bhatnagar RS, Qian JJ, Gough CA (1997). The role in cell binding of a beta-bend within the triple helical region in collagen alpha 1 (I) chain: Structural and biological evidence for conformational tautomerism on fiber surface. J Biomol Struct Dyn;14:547-560.

Blouin, S., Moreau, M.F., Weiss, P., Daculsi, G., Basle, M.F. & Chappard, D. (2006) Evaluation of an injectable bone substitute (β-TCP/hydroxyapatite/hydroxy-propyl-methyl-cellulose) in severely osteopenic and aged rats. *Journal of Biomedical Materials Research A* 78: 570-580.

Boix, D., Gauthier, O., Guicheux, J., Pilet, P., Weiss, P., Grimandi, G. & Daculsi, G. (2004) Alveolar bone regeneration for immediate implant placement using an injectable bone substitute: an experimental study in dogs. *Journal of Periodontology* 75: 663-671.

Boix, D., Weiss, P., Gauthier, O., Guicheux, J., Bouler, J.M., Pilet, P., Daculsi, G. & Grimandi, G. (2006) Injectable bone substitute to preserve alveolar ridge resorption after tooth extraction: A study in dog. *Journal of Material Science and Material Medicine* 17: 1145-1152.

Bolouri, A., Haghighat, N. & Frederiksen, N. (2001). Evaluation of the effect of immediate grafting of mandibular postextraction sockets with synthetic bone. *compendium of Continuing Education Dentistry* 22, 955–958.

Botticelli, D., Berglundh, T. & Lindhe, J. (2004). Hard-tissue alterations following immediate implant placement in extraction sites. *Journal of Clinical Periodontology* 31, 820–828.

Boyne, P.J. (1997) Bone grafts: Materials. In: Osseous reconstruction of the maxilla and the mandible, ed. Evensen, L., pp. 13-21. Carol Stream, IL: Quintessence Publishing Co.

Breitbart, A.S., Staffenberg, D.A., Thorne, C.H., Glat, P.M., Cunningham, N.S., Reddi, A.H., Ricci, J. & Steiner, G. (1995) Tricalcium phosphate and osteogenin: a bioactive onlay bone graft substitute. Plastic and Reconstructive Surgery 96, 699-708.

Brook, I.M., Craig, G.T. & Lamb, D.J. (1991) In vitro interaction between primary bone organ cultures, glass-ionomer cements and hydroxyapatite/tricalcium phosphate ceramics. Biomaterials 12: 179-186.

Buser, D., Dula, K., Hirt, H.P. & Schenk, R.K. (1996). Lateral ridge augmentation using autografts and barrier membranes. A clinical study in 40 partially edentulous patients. *International Journal of Oral & Maxillofacial Surgery* 54, 420–432.

Buser, D., Hoffmann, B., Bernard, J.P., Lussi, A., Mettler, D. & Schenk, R.K. (1998) Evaluation of filling materials in membrane-protected bone defects. A comparative histomorphometric study in the mandible of miniature pigs. Clinical Oral Implants Research 9, 137-150.

Buser, D., Halbritter, S., Hart, C., Bornstein, M.M., Grütter, L., Chappuis, V. & Belser, U.C. (2009) Early implant placement with simultaneous guided bone regeneration following single-tooth extraction in the esthetic zone: 12-month results of a prospective study with 20 consecutive patients. *Journal of Periodontology* 80, 152-162.

Buch, F., Albrektsson, T. & Herbst, E. (1986). The bone growth chamber for quantification of electrically induced osteogenesis. *Journal of Orthopedic Research* 4, 194–203.

Camargo, P., Lekovic, V., Weinlaender, M., Klokkevold, P., Kenney, E., Dimitrijevic, B., Nedic, M. & Jancovic, S.M.O. (2000). Infl uence of bioactive glass on changes in alveolar process dimensions after exodontia. *Oral Surgery Oral Medicine Oral Pathology Oral Radiology and Endodontics* 90, 581–586.

Cardaropoli, G., Araújo, M. & Lindhe, J. (2003). Dynamics of bone tissue formation in tooth extraction sites. An experimental study in dogs. *Journal of Clinical eriodontology* 30, 809–818.

Chen, S.T., Wilson, T.G., Jr. & Hammerle, C.H. (2004). Immediate or early placement of implants following tooth extraction: review of biologic basis, clinical procedures, and outcomes. *International Journal of Oral and Maxillofacial Implants* 19 (Suppl), 12–25.

Chiapasco, M., Romeo, E., Casentini, P. & Rimondini, L. (2004). Alveolar distraction osteogenesis vs. vertical guided bone regeneration for the correction of vertically defi cient edentulous ridges: a 1–3-year prospective study on humans. *Clinical Oral Implants Research* 15, 82–95.

Chiapasco, M., Zaniboni, M. & Boisco, M. (2006). Augmentation procedures for the rehabilitation of defi cient edentulous ridges with oral implants. *Clinical Oral Implants Research* 17 (Suppl 2), 136–159.

Covani, U., Crespi, R., Cornelini, R. & Barone, A. (2004). Immediate implants supporting single crown restoration: a 4-year prospective study. *Journal of Periodontology* 75, 982–988.

Dahlin, C., Linde, A., Gottlow, J. & Nyman, S. (1988). Healing of bone defects by guided tissue regeneration. *Plastic andReconstructive Surgery* 81, 672–677.

Dahlin, C., Alberius, P. & Linde, A. (1991a). Osteopromotion for cranioplasty. An experimental study in rats using a membrane technique. *Journal of Neurosurgery* 74 (3),487–491

Dahlin, C., Andersson, L. & Linde, A. (1991b). Bone augmentation at fenestrated implants by an osteopromotive membrane technique. A controlled clinical study. *Clinical Oral Implants Research* 2, 159–165.

Degidi M, Piattelli M, Scarano A, Iezzi G, Piattelli A (2004.) Maxillary sinus augmentation with a synthetic cellbinding peptide: Histological and histomorphometrical results in humans. J Oral Implantol;30:376-383.

Denissen, H.W., Kalk, W., Veldhuis, H.A. & van Waas, M.A. (1993). Anatomic consideration for preventive implantation. *International Journal of Oral and Maxillofacial Implants*8(2), 191–196.

Dies, F., Etienne, D., Bou Abboud, N. & Ouhayoun, J.P. (1996) Bone regeneration in extraction sites after immediate placement of an e-PTFE membrane with or without a biomaterial. Report on 12 consecutive cases. Clinical Oral Implants Research 7, 277-285.

Donos, N., Mardas, N. & Chadha, V. (2008) Clinical outcomes of implants following lateral bone augmentation: systematic assessment of available options (barrier membranes, bone grafts, split osteotomy). *Journal of Clinical Periodontology* **35** (suppl.), 173-202.

Emery, S.E., Fuller, D.A. & Stevenson, S. (1996) Ceramic anterior spinal fusion. Biologic and biomechanical comparison in a canine model. Spine 21: 2713-2719 .

Evian, C.I., Rosenberg, E.S., Cosslet, J.G. & Corn, H. (1982). The osteogenic activity of bone removed from healing extraction sockets in human. *Journal of Periodontology* 53, 81–85.

Evian, C. & Cutler, S. (1994). Autogenous gingival grafts as epithelial barriers for immediate implants: case reports. *Journal of Periodontology* 65, 201–210.

Fellah, B.H., Weiss, P., Gauthier, O., Rouillon, T., Pilet, P., Daculsi, G. & Layrolle, P. (2006) Bone repair using a new injectable self-crosslinkable bone substitute. *Journal of Orthopaedic Research* 24: 628-635.

Fiorellini, J.P. & Nevins, M.L. (2003). Localized ridge augmentation/preservation. A systematic review. *Annals of Periodontology* 8, 321–327.

Fowler, E., Breault, L. & Rebitski, G. (2000). Ridge preservation utilizing an acellular dermal allograft and demineralized freeze-dried bone allograft: Part II. Immediate endosseous implant placement. *Journal of Periodontology* 71, 1360–1364.

Froum, S., Wallace, S., Tarnow, D. & Cho, S. (2002). Effect of platelet-rich plasma on bone growth and osseointegration in human maxillary sinus grafts: three bilateral case reports. *International Journal of Periodontics and Restorative Dentistry* 22, 45–53.

Gao, T.J., Lindholm, T.S., Kommonen, B., Ragni, P., Paronzini, A. & Lindholm, T.C. (1997) Stabilization of an inserted tricalcium phosphate spacer enhances the healing of a segmental tibial defect in sheep. Archives of Orthopaedic and Trauma Surgery 116, 290-294.

Gauthier, O., Goyenvalle, E., Bouler, J.M., Guicheux, J., Pilet, P., Weiss, P. & Daculsi, D. (2001) Macroporous biphasic calcium phosphate ceramics versus injectable bone substitute: a comparative study 3 and 8 weeks after implantation in rabbit bone. *Journal of Material Science and Material Medicine* 12: 385-390.

Gelbart M, Friedman R, Burlui V, Rohrer M, Atkinson B. (2005) Maxillary sinus augmentation using a peptidemodified graft material in three mixtures: A prospective human case series of histologic and histomorphometric results. Implant Dent 2005;14:185-193

Gottlow J. (1993) Guided tissue regeneration using bioresorbable and non-resorbable devices: initial healing and long-term results. J Periodontol. 64 (Suppl):1157-65. Review.

Grunder U. (2011) Crestal ridge width changes when placing implants at the time of tooth extraction with and without soft tissue augmentation after a healing period of 6 months: report of 24 consecutive cases. Int J Periodontics Restorative Dent. 2011; 31(1):9-17.

Hahn J, Rohrer MD, Tofe AJ.(2003) Clinical, radiographic, histologic, and histomorphometric comparison of PepGen P-15 particulate and PepGen P-15 flow in extraction sockets: A same-mouth case study. Implant Dent ;12:170-174.

Hallman, M., Lundgren, S. & Sennerby, L. (2001b) Histologic analysis of clinical biopsies taken 6 months and 3 years after maxillary sinus floor augmentation with 80% bovine hydroxyapatite and 20% autogenous bone mixed with fibrin glue. Clinical Implant Dentistry & Related Research 3, 87-96.

Hallman, M., Hedin, M., Sennerby, L. & Lundgren, S. (2002a) A prospective 1-year clinical and radiographic study of implants placed after maxillary sinus floor augmentation with bovine hydroxyapatite and autogenous bone. Journal of Oral and Maxillofacial Surgery 60, 277-284.

Hämmerle, C.H.F., Schmid, J., Olah, A.J. & Lang, N.P. (1996).A novel model system for the study of experimental bone formation in humans. *Clinical Oral Implants Research* 7, 38–47.

Hämmerle, C.H.F., Olah, A.J., Schmid, J., Fluckiger, L., Gogolewski, S., Winkler, J.R. & Lang, N.P. (1997) The biological effect of deproteinized bovine bone on bone neoformation on the rabbit skull. Clinical Oral Implants Research 8, 198-207.

Hämmerle, C.H.F. (1999). Membranes and bone substitutes in guided bone regeneration. Paper presented at the 3rd European Workshop on Periodontology, Ittingen, Switzerland

Hämmerle, C.H., Jung, R.E. & Feloutzis, A. (2002). A systematic review of the survival of implants in bone sites augmented with barrier membranes (guided bone regeneration) in partially edentulous patients. *Journal of Clinical Periodontology* 29 (Suppl 3), 226–231; discussion 232–223.

Hämmerle, C.H., Chen, S.T. & Wilson, T.G., Jr. (2004). Consensus statements and recommended clinical procedures regarding the placement of implants in extraction sockets. *International Journal of Oral and Maxillofacial Implants* 19 Suppl, 26–28.

Hämmerle CH, van Steenberghe D. In: The first EAO Consensus Conference 16-19 February 2006, Pfäffikon, Switzerland. Clinical Oral Implants Research. 2006 Oct;17 Suppl 2:1.

Hämmerle, C.H.F., Araújo M & Lindhe J .Timing of Implant Placement, In: Clinical periodontology and implant dentistry: Ed: Jan Lindhe, Niklaus P. Lang & Thorkild Karring. —Fifth edition 2008(pp. 1053-76). : Blackwell Publishing Ltd

Hardt, C.R., Grondahl, K., Lekholm, U. & Wennstrom, J.L. (2002). Outcome of implant therapy in relation to experienced loss of periodontal bone support: a retrospective 5- year study. *Clinical Oral Implants Research* 13, 488–494.

Hardwick, R., Scantlebury, T.V., Sanchez, R., Whitley, N. & Ilizarov, G.A. (1989a). The tension-stress effect on the genesis and growth of tissues: Part I. The infl uence of stability of fi xation and soft tissue preservation. *Clinical Orthopaedics* 238, 249–281.

Hashimoto-Uoshima, M., Ishikawa, I., Kinoshita, A., Weng, H.T. & Oda, S. (1995) Clinical and histologic observation of replacement of biphasic calcium phosphate by bone tissue in monkeys. *International Journal of Periodontics and Restorative Dentistry* 15: 205-213.

Hole BB, Schwarz JA, Gilbert JL, Atkinson BL. (2005) A study of biologically active peptide sequences (P-15) on the surface of an ABM scaffold (PepGen P-15) using AFM and FTIR. J Biomed Mater Res A; 74:712-721.

Hossain MZ, Kyomen S, Tanne K (1996). Biologic responses of autogenous bone and beta-tricalcium phosphate ceramics transplanted into bone defects to orthodontic forces. Cleft Palate-Craniofac J;33:277-283.

Hürzeler, M.B., Quiñones, C.R., Kirsch, A., Gloker, C., Schupbach, P., Strub, J.R. & Caffesse, R.G. (1997) Maxillary sinus augmentation using different grafting materials and dental implants in monkeys. Part I. Evaluation of anorganic bovine-derived bone matrix. Clinical Oral Implants Research 8, 476-486.

Ilizarov GA. (1989a). The tension-stress effect on the genesis and growth of tissues. Part I. The influence of stability of fixation and soft-tissue preservation. Clinical Orthopedics Related Research. 238:249-81.

Ilizarov, G.A. (1989b). The tension-stress effect on the genesis and growth of tissues: Part II. The influence of the rate and frequency of distraction. *Clinical Orthopaedics Related Research.* 239, 263–285.

Jensen, S.S., Aaboe, M., Pinholt, E.M., Hjørting-Hansen, E., Melsen, F. & Ruyter, I.E. (1996) Tissue reaction and material characteristics of four bone substitutes. International Journal of Oral & Maxillofacial Implants 11, 55-66.

Joshi, A. & Kostakis, G. (2004). An investigation of postoperative morbidity following iliac crest graft harvesting. *British Dental Journal* 196, 167–171.

Karoussis, I.K., Salvi, G.E., Heitz-Mayfi eld, L.J., Bragger, U. Hammerle, C.H. & Lang, N.P. (2003). Long-term implant prognosis in patients with and without a history of chronic periodontitis: a 10-year prospective cohort study of the ITI Dental Implant System. *Clinical Oral Implants Research* 14, 329–339.

Karring, T., Nyman, S. & Lindhe, J. (1980). Healing following implantation of periodontitis affected roots into bone tissue. *Journal of Clinical Periodontology* 7, 96–105.

Kostopoulos, L. & Karring, T. (1994). Augmentation of the rat mandible using guided tissue regeneration. *Clinical Oral Implants Research* 5, 75–82.

Krauser JT, Rohrer MD, Wallace SS (2000). Human histologic and histomorphometric analysis comparing OsteoGrafN with PepGen P-15 in the maxillary sinus elevation procedure: A case report. Implant Dent;9: .302-298

Kubler A, Neugebauer J, Oh JH, Scheer M, Zoller JE.)2004(Growth and proliferation of human osteoblasts on different bone graft substitutes: An in vitro study. Implant Dent;13:171-179.

Lallier TE, Yukna R, St Marie S, Moses R (2001). The putative collagen binding peptide hastens periodontal ligament cell attachment to bone replacement graft materials. J Periodontol;72:990-997.

Landsberg, C. & Bichacho, N. (1994). A modifi ed surgical/prosthetic approach for optimal single implant supported crown. Part I – The socket seal surgery. *Practical Periodontics and Aesthetic Dentistry* 6, 11–17

Landsberg, C. (1997). Socket seal surgery combined with immediate implant placement: a novel approach for single-tooth replacement. *International Journal of Periodontics and Restorative Dentistry* 17, 140–149.

Le Nihouannen, D., Daculsi, G., Saffarzadeh, A., Gauthier, O., Delplace, S., Pilet, P. & Layrolle, P. (2005) Ectopic bone formation by microporous calcium phosphate ceramic particles in sheep muscles. *Bone* 36: 1086-1093.

Lekholm, U. & Zarb, G.A. (1985). Patient selection. In: Brånemark, P-I., Zarb, G.A. & Albreksson, T., eds. *Tissue Integrated Prostheses. Osseointegrationin Clinical Dentistry.* Chicago: Quintessence, pp. 199–209.

Lekholm, U., Adell, R. & Lindhe, J. (1986). Marginal tissue reactions at osseointegrated titanium fi xtures. (II) A crosssectional retrospective study. *International Journal of Oral & Maxillofacial Surgery* 15, 53–61.

Lekovic, V., Kenney, E.B., Weinlaender, M., Han, T., Klokkevold, P.R., Nedic, M. & Orsini, M. (1997). A bone regenerative approach to alveolar ridge maintenance following tooth extraction. Report of 10 cases. *Journal of Periodontology* 68,563–570.

Lekovic, V., Camargo, P.M., Klokkevold, P.R., Weinlaender, M., Kenney, E.B., Dimitrijevic, B. & Nedic, M. (1998). Preservation of alveolar bone in extraction sockets using bioabsorbable membranes. *Journal of Periodontology* 69, 1044–1049.

Lioubavina-Hack N, Lang NP, Karring T Significance of primary stability for osseointegration of dental implants. *Clin Oral Implants Res* 2006; 17:244-250.

Lorenzoni, M., Pertl, C., Keil, K. & Wegscheider, W.A. (1998). Treatment of peri-implant defects with guided bone regeneration: a comparative clinical study with various membranes and bone grafts. *International Journal of Oral & Maxillofacial Implants* 13, 639–646.

Lundgren, D., Lundgren, A.K., Sennerby, L. & Nyman, S. (1995). Augmentation of intramembranous bone beyond the skeletal envelope using an occlusive titanium barrier. An experimental study in the rabbit. *Clinical Oral Implants Research* 6, 67–72.

Lundgren S, Rasmusson L, Sjostrom M, Sennerby L Simultaneous or delayed placement of titanium implants in free autogenous iliac bone grafts. Histological analysis of the bone graft-titanium interface in 10 consecutive patients. *Inl J Oral Maxillofac Impl* 1999; 28:31-37

Mardinger O, Vered M, Chaushu G, Nissan J. (2010) Histomorphometrical Analysis following Augmentation of Infected Extraction Sites Exhibiting Severe Bone Loss and Primarily Closed by Intrasocket Reactive Soft Tissue. Clin Implant Dent Relat Res.

Mengel, R., Schroder, T. & Flores-de-Jacoby, L. (2001). Osseointegrated implants in patients treated for generalized chronic periodontitis and generalized aggressive periodontitis: 3- and 5-year results of a prospective long-term study. Journal of Periodontology 72, 977–989.

Merten HA, Wiltfang J, Grohmann U, Hoenig JF.(2001) Intraindividual comparative animal study of alpha- and beta-tricalcium phosphate degradation in conjunction with simultaneous insertion of dental implants. J Craniofac Surg;12:59-68.

Mombelli, A. & Cionca, N. (2006). Systemic diseases affecting osseointegration therapy. Clinical Oral Implants Research 17 (Suppl 2), 97–103.

Moses O, Nemcovsky CE, Langer Y, Tal H. (2007). Severely resorbed mandible treated with iliac crest autogenous bone graft and dental implants: 17-year follow-up. Int J Oral Maxillofac mplants.;22:1017-21

Nemcovsky, C. & Serfaty, V. (1996). Alveolar ridge preservation following extraction of maxillary anterior teeth. Report on 23 consecutive cases. *Journal of Periodontology* 67, 390–395.

Nemcovsky, C.E. & Artzi, Z. (2002). Comparative study of buccal dehiscence defects in immediate, delayed, and late maxillary implant placement with collagen embranes: clinical healing between placement and second-stage surgery. *Journal of eriodontology* 73, 754–761.

Nguyen H, Qian JJ, Bhatnagar RS, Li S.(2003) Enhanced cell attachment and osteoblastic activity by P-15 peptidecoated matrix in hydrogels. Biochem Biophys Res Commun ;311:179-186.

Nkenke, E., Schultze-Mosgau, S., Radespiel-Tröger, M., Kloss, F. & Neukam, F.W. (2001). Morbidity of harvesting of chin grafts: a prospective study. *Clinical Oral Implants Research* 12, 495–502.

Nyman, S., Karring, T., Lindhe, J. & Planten, S. (1980). Healing following implantation of periodontitis affected roots into gingival connective tissue. *Journal of Clinical Periodontology* 7, 394–401.

Nyman, S., Lindhe, J. & Karring, T. (1989). Reattachment – new attachment. In: Lindhe, J., ed. *Textbook of Clinical Periodontology*, 2nd edn. Copenhagen: Munksgaard, pp. 450–473.

Nyman, S., Lang, N.P., Buser, D. & Brägger, U. (1990). Bone regeneration adjacent to titanium dental implants using guided tissue regeneration. A report of 2 cases. *International Journal of Oral & Maxillofacial Implants* 5, 9–14.

Nyman, S.R. & Lang, N.P. (1994). Guided tissue regeneration and dental implants. *Periodontology 2000* 4, 109–118

Ohsawa K, Neo M, Matsuoka H, Akiyama H, Ito H, Kohno H, et al. (2000) The expression of bone matrix protein mRNAs around beta-TCP particles implanted into bone. J Biomed Mater Res 2000;52:460-466.

Philippart P, Daubie V, Pochet R(2005). Sinus grafting using recombinant human tissue factor, platelet-rich plasma gel, autologous bone, and anorganic bovine bone mineral xenograft: Histologic analysis and case reports. Int J Oral Maxillofac Implants;20:274-281.

Piattelli, M., Favero, G.A., Scarano, A., Orsini, G. & Piattelli, A. (1999) Bone reactions to anorganic bovine bone (Bio-Oss) used in sinus procedures: A histologic long-term report of 20 cases in humans. International Journal of Oral & Maxillofacial Implants 14, 835-840.

Pietrokovski, J. & Massler, M. (1967). Alveolar ridge resorption following tooth extraction. *Journal of Prosthetic Dentistry* 17, 21–27.

Qian JJ, Bhatnagar RS. (1996) Enhanced cell attachment to anorganic bone mineral in the presence of a synthetic peptide related to collagen. J Biomed Mater Res;31:545-554.

Rasmusson L, Meredith N, Cho IH, Sennerby L The influence of simultaneous versus delayed placement on the stability of titanium implants in onlay bone grafts. A histologic and biomechanic study in the rabbit. *Int J Oral Maxillofac Surg* 1999; 28:224-231.

Reddi, A.H. (1981). Cell biology and biochemistry of endochondral bone development. *Collagen Related Research* 1, 209–226.

Reddi, A.H., Weintroub, S. & Muthukumaran, N. (1987). Biologic principles of bone induction. *Orthopedic Clinics of North America* 18, 207–212.

Rosa AL, Brentegani LG, Grandini SA.(1995) Hydroxylapatite and tricalcium phosphate implants in the dental alveolus of rats. A histometric study. Brazilian Dent J; 6:103-109.

Russotti, G.M., Okada, Y., Fitzgerald, R.H., Jr, Chao, E.Y. & Gorski, J.P. (1987) Efficacy of using a bone graft substitute to enhance biological fixation of a porous metal femoral component. Hip 15: 120-154.

Schenk, R.K., Buser, D., Hardwick, W.R. & Dahlin, C. (1994). Healing pattern of bone regeneration in membrane protected defects: a histologic study in the canine mandible. *International Journal of Oral and Maxillofacial Implants* 9, 13–29.

Schliephake, H. & Kracht, D. (1997). Vertical ridge augmentation using polylactic membranes in conjunction with immediate implants in periodontally compromized extraction sites: an experimental study in dogs. *International Journal of Oral & Maxillofacial Implants* 12, 325–334.

Schmid, J., Hämmerle, C.H.F., Flückiger, L., Gogolewski, S., Winkler, J.R., Rahn, B. & Lang, N.P. (1997). Blood filled spaces with and without fi ller materials in guided bone regeneration. A comparative experimental study in the rabbit using bioresorbable membranes. *Clinical Oral Implants Research* 8, 75–81.

Schopper, C., Ziya-Ghazvini, F., Goriwoda, W., Moser, D., Wanschitz, F., Spassova, E., Lagogiannis, G., Auterith, A. & Ewers, R. (2005) HA/TCP compounding of a porous CaP biomaterial improves bone formation and scaffold degradation--a long-term histological study. *Journal of Biomedical Materials Research B Applied Biomaterials* 74: 458-467.

Schropp, L., Wenzel, A., Kostopoulos, L. & Karring, T. (2003). Bone healing and soft tissue contour changes following single-tooth extraction: a clinical and radiograhic 12-month prospective study. *International Journal of Periodontics & Restorative Dentistry* 23, 313–323.

Schwartz-Arad, D. & Chaushu, G. (1997a). The ways and wherefores of immediate placement of implants into fresh extraction sites: A literature review. *Journal of Periodontology* 68, 915–923.

Skoglund, A., Hising, P. & Young, C. (1997) A clinical and histologic examination in humans of the osseous response to implanted natural bone mineral. International Journal of Oral & Maxillofacial Implants 12, 194-199.

Seibert, J. & Nyman, S. (1990). Localized ridge augmentation in dogs: a pilot study using membranes and hydroxyapatite. *Journal of Periodontology* 3, 157–165.

Simion, M., Baldoni, M., Rossi, P. & Zaffe, D. (1994a). A comparative study of the effectiveness of e-PTFE membranes with and without early exposure during the healing period. *International Journal of Periodontics and Restorative Dentistry* 14, 167–180.

Simion, M., Trisi, P. & Piattelli, A. (1994b). Vertical ridge augmentation using a membrane technique associated with osseointegrated implants. *International Journal of Periodontics and Restorative Dentistry* 14, 497–511.

Smiler, D.G., Johnson, P.W., Lozada, J.L., Misch, C., Rosenlicht, J.L., Tatum, O.H., Jr. & Wagner, J.R. (1992) Sinus lift grafts and endosseous implants: Treatment of the atrophic posterior maxilla. Dental Clinics of North America 36, 151-186.

Smukler, H., Landi, L. & Setayesh, R. (1999). Histomorphometric evaluation of extraction sockets and defi cient alveolar ridges treated with allograft and barrier membrane: a pilot study. *International Journal of Oral & Maxillofacial Implants* 14, 407–416.

Spector, M. (1994) Anorganic bovine bone and ceramic analogs of bone mineral as implants to facilitate bone regeneration. Clinics in Plastic Surgery 21, 437-444.

St John, K.R., Zardiackas, L.D., Black, R.J. & Armstrong, R. (1993) Response of canine bone to a synthetic bone graft material. Clinical Materials 12: 49-55.

Szabo G, Suba Z, Hrabak K, Barabas J, Nemeth Z. (2001) Autogenous bone versus beta-tricalcium phosphate graft alone for bilateral sinus elevations (2-and 3-dimensional computed tomographic, histologic, and histomorphometric observations): preliminary results. Int J Oral Maxillofac Implants;16:681-692.

Tal, H. (1999). Autogenous masticatory mucosal grafts in extraction socket seal procedures: a comparison between sockets grafted with demineralized freeze-dried bone and deproteinized bovine bone mineral. *Clinical Oral Implants Research* 10, 289–296.

Tal, H, Bichacho N, Imber S, Kornowski Y, Nemcovsky CE. (2004) Rotated palatal flaps: A functional and aesthetic solution in endentulous sites. Pract Proced Aesthet Dent. 16(8):599-606.

Tehemar S, Hanes P, Sharawy M.(2003) Enhancement of osseointegration of implants placed into extraction sockets of healthy and periodontally diseased teeth by using graft material, an ePTFE membrane, or a combination. Clin Implant Dent Relat Res 2003;5:193-211.

Tinti, C., Parma-Benfenati, S. & Polizzi, G. (1996). Vertical ridge augmentation: what is the limit? *International Journal of Periodontics and Restorative Dentistry* 16, 220–229.

Tinti, C. & Parma-Benfenati, S. (1998). Vertical ridge augmentation: surgical protocol and retrospective evaluation of 48 consecutively inserted implants. *International Journal of Periodontics and Restorative Dentistry* 18, 445–443.

Trasatti C, Spears R, Gutmann JL, Opperman LA.(2004) Increased TGF-b1 production by rat osteoblasts in the presence of PepGen P-15 in vitro. J Endod ;30: .217-213

Turhani D, Weissenbock M, Watzinger E, et al.(2005) In vitro study of adherent mandibular osteoblast-like cells on carrier materials. Int J Oral Maxillofac Surg ;34: .550-543

Urist, M.R. (1965). Bone: formation by autoinduction. *Science* 150, 893–899.

von Arx, T., Cochran, D.L., Hermann, J.S., Schenk, R.K., Higginbottom, F.L. & Buser, D. (2001). Lateral ridge augmentation and implant placement: An experimental study evaluating implant osseointegration in different augmentation materials in the canine mandible. *International Journal of Oral & Maxillofacial Implants* 16, 343–354.

Valentini, P. & Abensur, D. (1997) Maxillary sinus floor elevation for implant placement with demineralized freeze-dried bone and bovine bone (Bio-Oss): A clinical study of 20 patients. International Journal of Periodontics & Restorative Dentistry 17, 233-241.

von Arx, T., Hafl iger, J. & Chappuis, V. (2005). Neurosensory disturbances following bone harvesting in the symphysis: a prospective clinical study. *Clinical Oral Implants Research* 16, 432–439.

Walters SP, Greenwell H, Hill M, Drisko C, Pickman K, Scheetz J.P.(2003) Comparison of porous and non-porous teflon membranes plus a xenograft in the treatment of vertical osseous defects: A clinical reentry study. J Periodontol;74:1161-1168.

Watzek, G., Haider, R., Mensdorff-Pouilly, N. & Haas, R. (1995). Immediate and delayed implantation for complete restoration of the jaw following extraction of all residual teeth: A retrospective study comparing different types of serial immediate implantation. *International Journal of Periodontics and Restorative Dentistry* 12, 206–217.

Wennstrom, J.L., Ekestubbe, A., Grondahl, K., Karlsson, S. & Lindhe, J. (2004). Oral rehabilitation with implant supported fi xed partial dentures in Periodontitis susceptible subjects. A 5-year prospective study. *Journal of Clinical Periodontology* 31, 713–724.

Werbitt, M.J. & Goldberg, P.V. (1992). The immediate implant: bone preservation and bone regeneration. *International Journal of Periodontics and Restorative Dentistry* 12, 206–217.

Wetzel, A.C., Stich, A. & Caffesse, R.G. (1995) Bone apposition onto oral implants in the sinus area filled with different grafting materials. A histological study in beagle dogs. Clinical Oral Implants Research 6, 155-163.

Wikesjo UM, Nilveus R Periodontal repair in dogs: Effect of wound stabilization on healing. *J Periodontol* 1990; 61: 719-724.

Wiltfang J, Merten HA, Schlegel KA, Schultze-Mosgau S, Kloss FR, Rupprecht S, et al. Degradation characteristics of alpha and beta tri-calcium-phosphate (TCP) in minipigs. J Biomed Mater Res 2002;63:115-121.

Yamada S, Heymann D, Bouler JM, Daculsi G. Osteoclastic resorption of calcium phosphate ceramics with different hydroxyapatite/beta-tricalcium phosphate ratios. Biomater 1997;18:1037-1041.

Yang, J., Lee, H. & Vernino, A. (2000). Ridge preservation of dentition with severe periodontitis. *Compendium of Continuing Education Dentistry* 21, 579–583.

Yilmaz, S., Efeoglu, E. & Kilic, A. (1998). Alveolar ridge reconstruction and/or preservation using root form bioglass cones. *Journal of Clinical Periodontology* 25, 832–839.

Yukna RA, Callan DP, Krauser JT, et al(1998). Multi-center clinical evaluation of combination anorganic bovinederived hydroxyapatite matrix (ABM)/cell binding peptide (P-15) as a bone replacement graft material in human periodontal osseous defects. 6-month results. J Periodontol ;69:655-663.

Yukna RA, Krauser JT, Callan DP, Evans GH, Cruz R, Martin M.(2000) Multi-center clinical comparison of combination anorganic bovine-derived hydroxyapatite matrix (ABM)/cell binding peptide (P-15) and ABM in human periodontal osseous defects. 6-month results. J Periodontol 2000;71:1671-1679.

Yukna RA, Krauser JT, Callan DP, Evans GH, Cruz R Martin M (2002). Thirty-six month follow-up of 25 patients treated with combination anorganic bovine-derived hydroxyapatite matrix (ABM)/cell-binding peptide (P-15(bone replacement grafts in human infrabony defects. I. Clinical findings. J Periodontol;73:123-128.

Yukna R, Salinas TJ, Carr RF.(2002) Periodontal regeneration following use of ABM/P-15: A case report. Int J Periodontics Restorative Dent ;22:146-155.

Zitzmann, N.U., Naef, R. & Schärer, P. (1997). Resorbable versus nonresorbable membranes in combination with Bio-Oss for guided bone regeneration. *International Journal of Oral & Maxillofacial Implants* 12, 844–852.

Zitzmann, N.U., Schärer, P. & Marinello, C.P. (1999). Factors influencing the success of GBR. *Journal of Clinical Periodontology* 26, 673–682.

8

Distraction Osteogenesis and Its Challenges in Bone Regeneration

Reggie C. Hamdy, Juan S. Rendon and Maryam Tabrizian
McGill University/Department of Orthopaedic Surgery and Biomechanical Engineering
Canada

1. Introduction

1.1 Distraction osteogenesis

Bone is amongst the very few tissues in the human body that possess intrinsic capacity to heal spontaneously following injury. However, beyond a certain critical size defect, bone cannot heal by itself and outside intervention is required. Numerous techniques are available for the management of these defects, including the gold standard autogenous bone grafts, allografts, bone graft substitutes, vascularized fibular bone grafts and systemic administration of anabolic agents. All these techniques, however, do have limitations (Dimitriou et al., 2011; Nauth et al., 2011). Such instances of severe bone loss, whether due to congenital bony deficiencies or acquired causes, pose an immense challenge to the treating physicians, and it is in these cases that distraction osteogenesis could offer a viable and successful alternative to these techniques. Distraction osteogenesis (DO) is a surgical technique in which the intrinsic capacity of bone to regenerate is being harnessed to lengthen bones or to replace large segments of bone. It consists of the application of an external fixator to the affected bone (Figure 1), followed by an osteotomy of the bone and then gradual and controlled distraction is applied to the two bone segments. This controlled distraction, usually by an external fixator, generates new bone within the distracted gap. When the desired lengthening is obtained, distraction is stopped and the external fixator is kept on until the newly formed in the distracted gap is mechanically strong enough to allow removal of the fixator. DO is considered a type of in vivo bone tissue engineering and is superior to other methods of bone regeneration in the management of cases of bone loss, because this technique allows the spontaneous formation of de novo native bone without the need for bone grafts. DO also has the unique ability to regenerate bone and soft tissues simultaneously.

2. Historical aspect of distraction osteogenesis

Codivilla from Italy is credited for having performed the first lengthening procedure by applying skeletal traction through a calcaneal pin, following osteotomy of the femur (Codivilla, 1905). However, it was a Russian surgeon, Gavriil Ilizarov, who pioneered the biological principles of bone and soft tissue regeneration and popularized the technique of distraction osteogenesis, when he discovered that under slow and gradual distraction, new bone will regenerate in the distracted gap (Ilizarov, 1989a; Ilizarov, 1989b). Starting in the early 1950s, he worked in a village in Siberia, Kurgan, unknown to the rest of the world. Then, in 1982, he successfully treated a famous Italian explorer "Carlo Mauri" for a resistant non-union of his tibia and it was only then when his principles were made known to the Western World.

3. Ilizarov principles or the law of tension stress (Green, 2011)

Ilizarov developed the *law of tension-stress,* which describes the process of new bone and soft tissue regeneration under the effect of tension-stress caused by slow and gradual distraction. His biological principles can be summarized as follows:

3.1 Minimal disturbance of bone and soft tissues

Ilizarov showed that formation of new bone at the osteotomy site is definitely influenced by the amount of damage to the bone, medullary cavity, and periosteum. He described the new concept of corticotomy, where only the cortex of the bone is cut, preserving the periosteum and medullary cavity. The value of corticotomy has recently been questioned, because the medullary blood supply regenerates in 7 to 10 days following a complete osteotomy. The integrity of the periosteum is the only important factor for new bone formation at the site of the osteotomy.

3.2 Delay before distraction

(Latency phase) Duration of delay varies from 5 days in a child to about 10 days in a skeletally-mature patient. This allows the formation and organization of a hematoma.

3.3 Rate and rhythm of distraction

The optimum rate was found by Ilizarov to be 1mm/day and the optimum rhythm of distraction was 0.25mm every 6 hours. Elongation of more than 2mm/day may lead to slowing of osteogenesis, while elongation of 0.5mm/day or less may lead to premature consolidation. An autodistractor causing a continuous gradual and slow distraction of 1.0 mm/day was found to be superior to a rhythm of four times a day.

3.4 Site of lengthening

Metaphyseal lengthening leads to better osteogenesis than diaphyseal lengthening. The metaphyseal region contains much more cancellous bone than the diaphyseal region and this type of bone has a much higher potential for osteogenesis.

3.5 Stable fixation of the external fixator

Similar to fracture healing, this has been shown to be of paramount importance. Some axial micromotion is, however, beneficial to the consolidation of the regenerate bone.

3.6 Functional use of the limb and intense physiotherapy

During the whole lengthening procedure, this is of foremost importance in order to obtain a satisfactory outcome.

4. Phases of distraction osteogenesis

As shown in Figure 1, DO consists of the following phases:

- Latency phase: This is the phase immediately following the osteotomy, where there is no distraction. It lasts 5 to 10 days
- Distraction phase: the two bone segments are gradually distracted at a rate of 1.0 mm day in several increments, until the desired amount of lengthening is obtained.
- Consolidation phase: distraction is ceased and the two bone segments are held in place until the newly formed bone in the distracted gap consolidates (about one month per cm lengthened)
- Removal of the fixator.

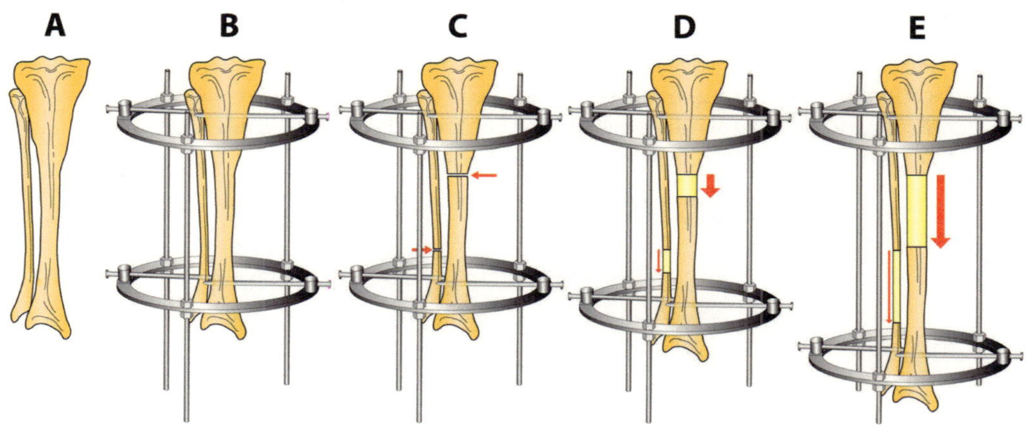

Fig. 1. Steps of the technique of distraction osteogenesis used for limb lengthening; A. Shows the bone to be lengthened; B. Application of the external fixator; C. Osteotomy of the proximal tibia; D. Start of distraction and E. End of distraction when desired amount of lengthening is obtained.

5. Clinical significance of distraction osteogenesis

DO is a very successful technique of bone regeneration, widely used for lengthening bones and in the management of bone loss secondary to congenital or acquired causes, both in long and tubular bones of the axial skeleton as well as flat bones of the craniofacial skeleton. It has many indications, including:

5.1 Distraction osteogenesis as a bone lengthening technique

DO could be used in restoring the length of bones in numerous conditions including congenital causes of bone defects specifically congenital limb deficiencies or acquired causes due to growth plate injuries leading to growth arrest and subsequent limb length discrepancy (Birch and Samchukov, 2004; Murray and Fitch, 1996), (Figure 2).

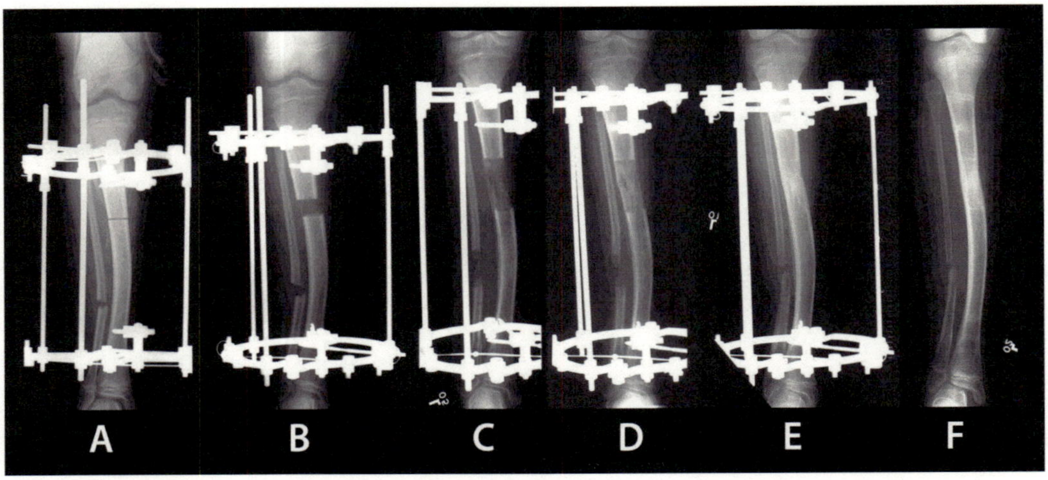

Fig. 2. Lengthening of short tibia showing various phases of the distraction process. A. Application of the fixator and osteotomy of the tibia; B. Start of distraction; C. End of distraction; D. and E. Consolidation phase, without any distraction, until bone in the distracted gap consolidates; F. Removal of the fixator.

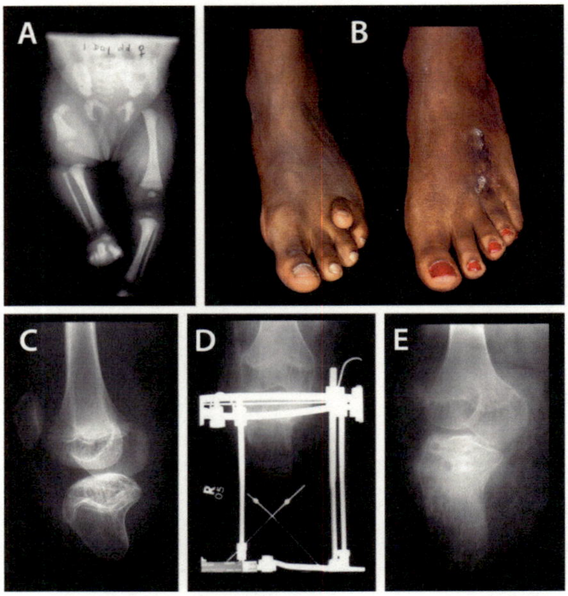

Fig. 3. Examples of short bones lengthened by distraction osteogenesis. A. A very short femur; B. Short 4th metatarsal bone pre and post lengthening; C. Amputation stump pre-lengthening, D. During distraction phase, and E. Post-lengthening.

5.2 Distraction osteogenesis as a bone transport technique

DO is also widely used for the reconstruction of large segmental skeletal defects by a special technique called bone transport (Figure 4). The magnitude of this problem is enormous.

Approximately 150,000 segmental skeletal defects are sustained in the United States each year as a result of trauma (Cierny and Zorn, 1994). To this must be added a significant number of bone defects following debridement after severe cases of osteomyelitis (Cierny and DiPasquale, 2011) and resections for malignant bone tumours (Tsuchiya, 2011).

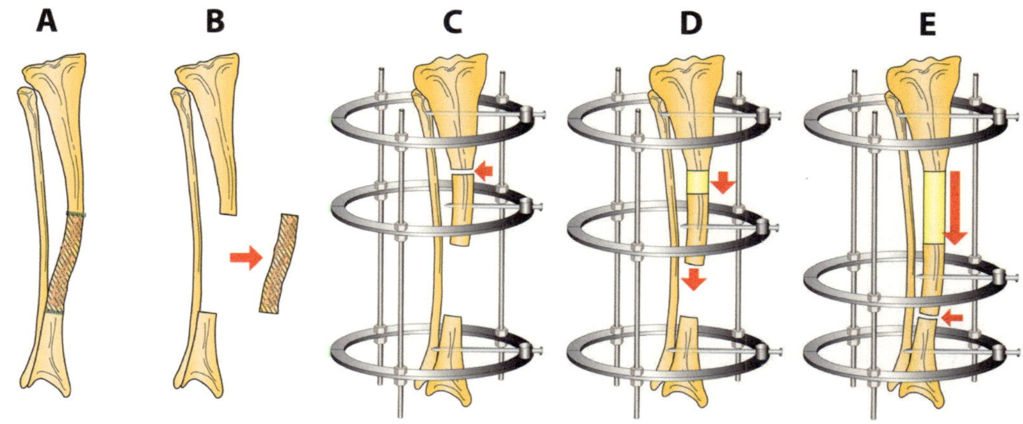

Fig. 4. A and B. Showing the segmental bone defect; C. Application of the external fixator, the bone defect and osteotomy of the proximal tibia; D. Start of distraction and transport of the healthy bone segment distally to fill the bone defect, while at the same time, new regenerate bone is formed in the distracted gap proximally; E. Completion of the bone transport.

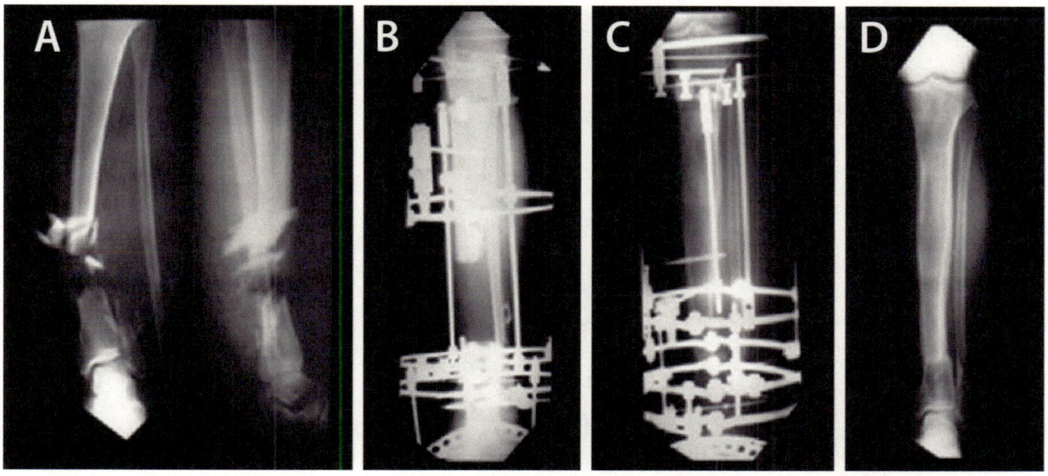

Fig. 5. A. Showing large segmental bone loss of the tibia due to severe trauma; B. Application of the external fixator; C. Completion of the bone transport and successful and successful bone formation in the distracted zone; D. Complete bone regeneration of the affected bone.

5.3 Distraction osteogenesis in the craniofacial skeleton

The clinical use of DO is not only limited to orthopaedic problems, but extends to the field of cranio-facial surgery. Mandibular distraction was first performed in 1973 in a canine model (Snyder et al., 1973). However, it was only in 1992 when it was first reported in humans (Mccarthy et al., 1992). Currently, both extraoral and intraoral devices may be used depending on the condition to be treated (Goldwaser et al., 2011). Commonly treated conditions include craniofacial deficiencies, syndromic craniosynostosis, Pierre Robin Sequence, posttraumatic deformities, and sleep-related breathing disorders (Vander Kolk et al., 2001).

6. Types of external fixators used in distraction osteogenesis for long bones

There are two types of external fixators used in DO: circular and monolateral. Circular fixators, also called ring fixators, include the standard circular frames (Figure 6A) described by Ilizarov and the more recent Taylor Spatial Frames (Feldman and Chaudhry, 2011). Monolateral fixators consist of a single bar transfixing bone with screws and pins (Figure 6B). Hybrid fixators also have been developed, consisting of a combination of circular and monolateral constructs.

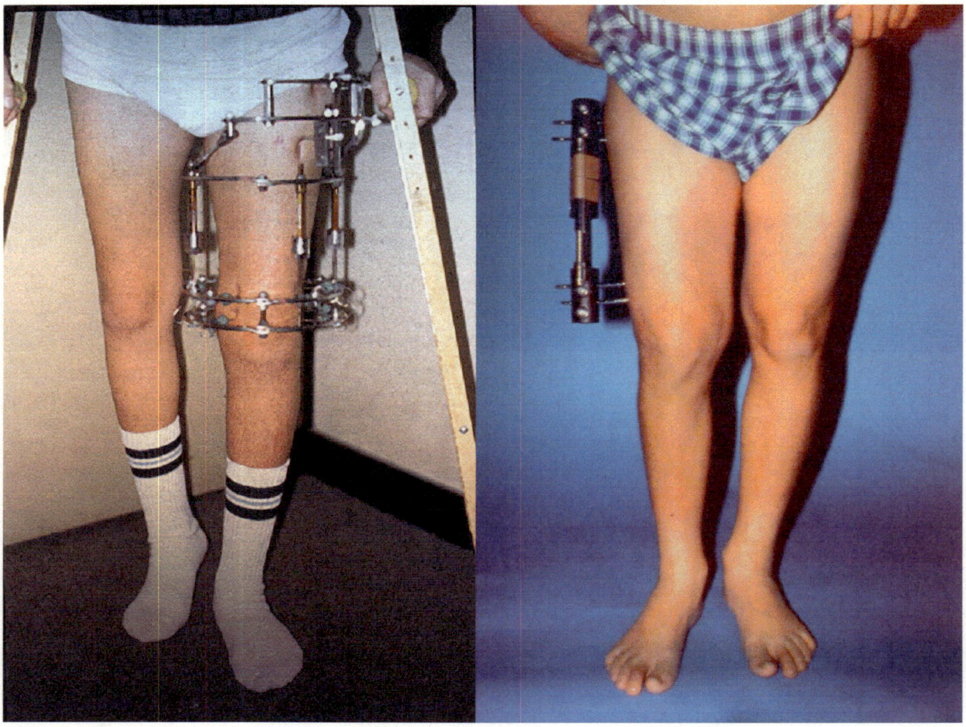

Fig. 6. A. (Circular); B. (Monolateral) External fixators.

7. Cellular events in distraction osteogenesis

The histological features of DO closely resemble those of fracture healing, as shown in Figure 7. Immediately after the osteotomy, a hematoma is formed. As distraction progresses, this hematoma is organized into fibrous and fibro-cartilaginous tissue in a longitudinal pattern along the direction of distraction and gives a striated appearance. New bone starts to be formed as early as two weeks after distraction. This new bone is formed from the periosteum, from the cortex at the site of the osteotomy and from the spongiosa and proceeds from the osteotomy cuts towards the center, always forming a fibrous, radiolucent interzone between the two advancing edges of the mineralization front (Aronson et al., 1990; Aronson et al., 1989; Aronson et al., 1997; Hamdy et al., 1997). Besides the formation of new bone in the distracted gap, all the surrounding soft tissues are also stimulated and lengthened, including skin, muscles, connective tissue, nerves and vessels (Makarov et al., 2009).

8. Molecular events and mechanism of bone formation in distraction osteogenesis

New bone formation in DO occurs through mechanotransduction, the process by which the mechanical tension-stress forces induced by distracting the bony segments at the site of the osteotomy, are converted into a cascade of molecular signals, which in turn activate numerous cellular events (differentiation, proliferation and secretory functions), that ultimately lead to new bone formation (Huang and Ogawa, 2010). We and others (Ai-Aql et al., 2008; Haque et al., 2007; Haque et al., 2008; Haque et al., 2006; Rauch et al., 2000) have shown that the mechanical forces applied during DO lead to the temporal and spatial expression of numerous cytokines, growth factors including BMP (Bone Morphogenetic Proteins), FGF (Fibroblast Growth Factor), IGF (Insulin Growth Factor), TGFB (Transforming Growth Factor B), PDGF (Platelet Derived Growth Factor), vascular factors VEGF (Vascular Endothelial Growth Factor), HIF (Hypoxia Induced Factor), as well as extracellular matrix proteins (Sato et al., 1998) and matrix metalloproteinases (Marucci et al., 2002). During the distraction phase, while the distraction forces are still being applied, many of these molecules are up-regulated and then, once the distraction forces cease at the end of the distraction phase, they become down-regulated.

9. The role of angiogenesis in distraction osteogenesis

DO is a vascular dependent process and new bone formation in DO is associated with robust neo-angiogenesis and neo-vascularity (Choi et al., 2000; Choi et al., 2002). There is an increased expression of numerous vascular growth factors in the distracted zone, including VEGF, HIF, basic FGF and Angiopoietin (Pacicca et al., 2003). Mobilization of endothelial progenitor cells have been described to play a major role in new bone formation in DO (Lee et al., 2008).

1. Haematoma formation after osteotomy and start of distraction
2. Mid-distraction showing start of bone formation
3. End of distraction (2.0 cms) showing fibrous interzone in the middle of the distracted zone. New bone formation is laid down longitudinally along the direction of distraction stress

4-5. Fibrous interzone decreases as new bone formation advances from the osteotomy ends towards the centre of the distracted zone
6. New bone completely bridges the distraction gap.

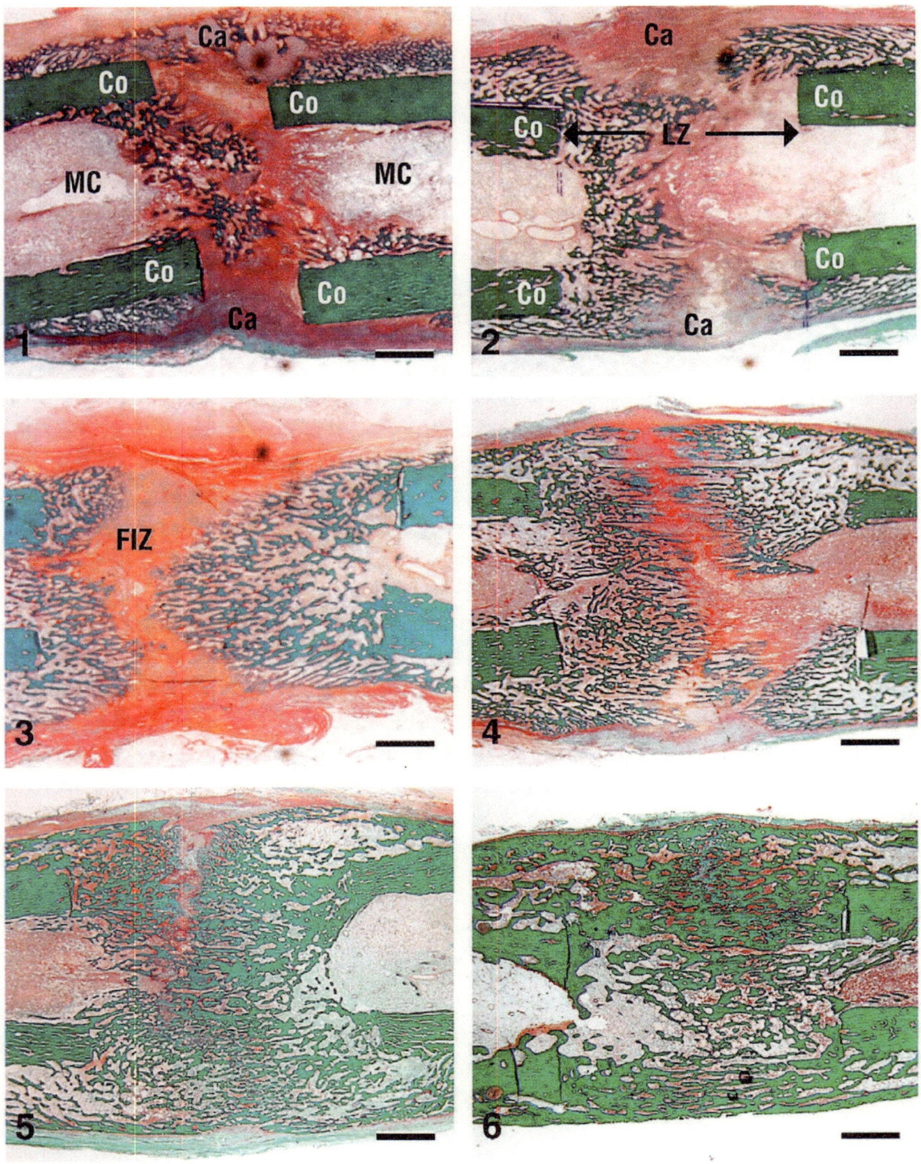

Co: Cortex, Mc: Medullary cavity, Ca: callus, FIZ: fibrous interzone.

Fig. 7. Cellular changes during distraction osteogenesis of the tibia of 2.0 cms in a rabbit model of DO using modified uniplanar fixator (trichrome staining), *Reprint with permission from Bone. 2000 Jun;26(6):611-7:*

10. Types of bone formation in distraction osteogenesis

While many aspects of DO, both at the cellular and molecular level have been elucidated, the exact mechanism and type of bone formation in DO is still being debated. Numerous authors have reported that regenerate bone formation in DO is mostly intramembranous (Aronson et al., 1990; Fink et al., 2003). However, others reported the presence of predominantly endochondral bone (Delloye et al., 1990; Kojimoto et al., 1988). Ilizarov believed that new bone formation in a canine model of DO, was mostly intramembranous, although he also described the presence of islands of cartilage-like cells (Ilizarov, 1989a; Ilizarov, 1989b). Our own results in several animal models of DO (dogs, rabbits and mice), revealed a combination of both types of ossification (Hamdy et al., 2003; Hamdy et al., 1997). Yasui et al (Yasui et al., 1997) in a rat model of DO, described a third type of ossification called transchondroid. It is also possible that the type of intra-membranous formation in DO is different from that of bone development, as reported by Isefuku et al (Isefuku et al., 2004), who showed that DO in Cbfa1 heterozygous knockout mice, new bone formation was the same as that of the controls. While the debate regarding the type of bone formation in DO will likely continue, many factors have been identified as playing a major role in determining which type of bone formation will predominate. These include stability of the fixator, vascularity of the surrounding tissues, rate and rhythm of distraction and the animal species in which DO was tested.

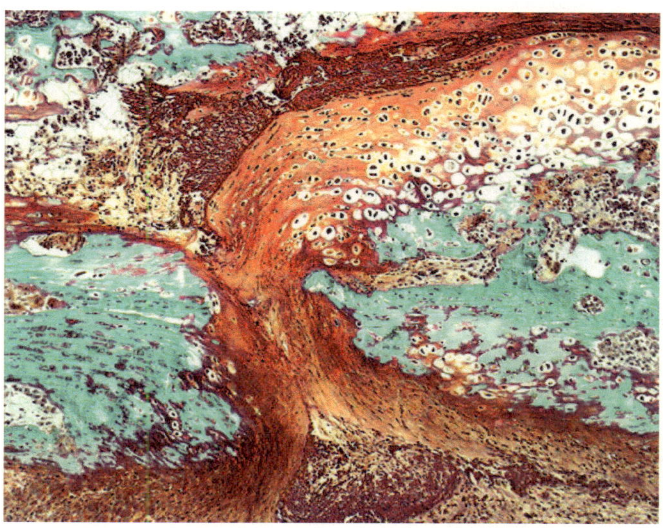

Fig. 8. Histological sections at the middle of distracted zone of rabbits sacrificed at 5 weeks post surgery. Trichrome Goldner staining showing mineralized tissue (bone and cartilage) in green, unmineralized cartilage in orange red. Chondrocyte-like cartilage is seen in large amounts.

11. Problems and issues associated with the technique of distraction osteogenesis and current research aimed at addressing these issues

Although very popular and very successful worldwide as a method of bone regeneration, this technique has several problems, specifically the long period of time that the external fixator needs to be kept on until the newly formed bone in the distracted zone consolidates. It has been shown by Ilizarov that the distraction phase cannot be increased, as this may lead to poor regenerate bone formation and soft tissue problems, such as contractures and neurovascular problems. The consolidation phase is long, requiring the fixator to be kept on about one month for every cm lengthened. For example, a lengthening of 6.0 cms, would require the fixator to be kept for about 6 months. This, in turn, can lead to or exacerbate many medical, psychological, social and financial problems for the patient and his family (Paley, 1990). These include pin site infections; pain that may require narcotic medications; edema of the lengthened limb causing discomfort and pain; psychological complications; missing school days for children or working days for adults; long and continuous follow-up in the form of regular outpatient visits, frequent radiological examinations and re-adjustment of the frame. The question then arises: how to accelerate the consolidation of the regenerate bone, so that the external fixator could be removed at an earlier time?

12. Attempts at accelerating distraction osteogenesis

Numerous attempts at enhancing newly formed bone have been described and include the application of external biophysical stimuli (i.e. mechanical loading), and administration of biological agents, systemically or locally. Most of these studies have been performed in animal models of DO (Sabharwal, 2011). These include;

12.1 Mechanical loading, axial compression and dynamization

The effect of mechanical loading on the maintenance of bone mass has long been recognized. The mechanical stresses induced in the soft tissues during distraction are converted into a cascade of signals (mechanotransduction) that lead to the activation of numerous osteogenic pathways. At the cellular level, there is cellular differentiation, angiogenesis formation, mineralization of bone matrix as well of bone remodeling. There are no strict guidelines as to the amount of compression or shortening necessary to enhance bone regeneration, however many authors recommend compression of three to five days after completing the lengthening. Some authors also recommended over lengthening by about 5mm followed by compression of 5mm (Hwang et al., 2009; Mori et al., 2006).

12.2 Accordion maneuver

This technique consists of alternating cycles of compression and distraction. Numerous variations of this technique have been described: alternated cycles of compression and distraction the same day, and compression for a period of 1 or more weeks followed by distraction for the same duration (Claes et al., 2008).

12.3 Vibrations

Numerous studies have shown that application of low magnitude and high frequency mechanical stimuli may have a beneficial effect on bone formation. Vibration plates have been and are being used to enhance bone formation. However, there had been no studies in cases of distraction osteogenesis in humans (Hou et al., 2011).

12.4 Ultrasound

Numerous studies have shown the efficiency of LIPUS (Low Intensity Pulsed Ultrasound). The device is applied to the skin previously covered by gel corresponding to the point of fracture or distraction osteogenesis for 20 minutes a day. The treatment is usually self administered by the patient at home and the period of treatment ranges from 2 to 6 months or more, until healing is completed. Ultrasound may also enhance angiogenesis and increase the blood flow around the site of new bone formation (Busse et al., 2009; Claes and Willie, 2007; Romano et al., 2009; Watanabe et al., 2010).

12.5 Extra corporal Shock Wave (ESW)

Positive effects of ESW on regenerate bone formation have been reported in rat mandibular and rabbit tibia models of DO (Lai et al., 2010; Narasaki et al., 2003).

12.6 Electrical stimulation

Electrical stimulation could be applied using capacitively coupled electric field (CCEF) method, direct current (DC) stimulation, electromagnetic stimulation or alternating current (AC) stimulation and has been shown to give positive results in animal studies (Hagiwara and Bell, 2000; Kawamoto et al., 2005; Pepper et al., 1996).

12.7 Bisphosphonates

As both, bone formation and bone resorption occur in distraction osteogenesis, it is reasonable to assume that blocking bone resorption by anti-resorptive agents, such as bisphosphonates, may lead to increased bone formation. Numerous animal studies have documented the positive effect of bisphosphonates in distraction process (Abbaspour et al., 2009). In a case series of 7 patients, bisphosphonates were used for the treatment of poor regenerate bone (Kiely et al., 2007); six patients responded well and completely healed. However, despite the reported good results, there has been no other published series on the positive effect of bisphosphonates on enhancement of poor bone regeneration.

12.8 Systemic drugs

Other systemically administered drugs that have been investigated including, *Calcitonin* (Sen et al., 2006), *Prostaglandin E2* (Yamane et al., 1999).

12.9 Locally applied agents

These include *alpha-tocopherol* (Kurklu et al., 2011); *Adiponectin* (Jiang et al., 2011); *Inhibin A* (Perrien et al., 2011); *Nerve growth factor via a hydrogel* (Cao et al., 2011); *Thrombin peptide 508* (Cakarer et al., 2010); *Bone marrow progenitor cell mobilizing agent* (Davidson et al., 2011); *Calcium sulphate injection* (Song et al., 2004); *Osteoblast-like cells* (Shao et al., 2007); *Stromal cell derived factor-1* (Fujio et al., 2011); *NEL-like molecule-1* (Xue et al., 2011).

12.10 Cell therapy and platelet-rich plasma

The use of bone marrow cells to enhance bone healing has been in use from many years. The problem with direct bone marrow injections is that the number of active osteogenic cells is very low, and therefore special techniques have been developed to aspirate bone marrow, culture the cells in vitro so as to increase and expand their number and then inject them in the desired area of poor bone formation (fracture site, non-union, distracted zone). The use of platelet rich plasma alone was reported to give positive results in a human study (Latalski et al., 2011). The positive osteogenic effect of culture expanded bone marrow cells has also been reported when combined with platelet-rich plasma in humans and with bFGF in rabbits (Jiang et al., 2010; Kitoh et al., 2007).

12.11 Combination of methods

These include rhBMP-2 combined with HA-TCP biomaterial (Ni et al., 2011); BMPs and NEL-1 (Zhu et al., 2011); Autografts and demineralized bone matrix (Canter et al., 2007); autologous bone marrow demineralized bone matrix (Hatzokos et al., 2011). Another experimental study reported also on the positive effects of an internal drug releasing distractor where small doses of BMP-2 are released from chitosan gel with every distraction (Konas et al., 2009).

12.12 Peptide growth factors

These include TGF-β (Transforming Growth Factor Beta), IGF (Insulin Growth Factors) (Bernstein et al., 2010), FGFs (Fibroblast Growth Factors) (Okazaki et al., 1999), PDGF (Platelet Derived Growth Factor) (Moore et al., 2009), Hypoxia Induced Factor (Wan et al., 2008).

13. Bone Morphogenetic Proteins (BMPs)

Of all the osteogenic growth factors, BMPs seem to be the most promising in stimulating bone formation in the context of DO. BMPs are members of the TGF-β superfamily acting on many systems including the kidney, heart and bone. These molecules are amongst the most powerful osteogenic growth factors and the only osteo-inductive ones that act on undifferentiated mesenchymal cells very early in the differentiation process (Gazzerro and Canalis, 2006; Miyazono et al., 2010; Rosen, 2006) (Figure 9).

BMPs are produced by many cells, including osteoblasts, chondrocytes and platelets. BMP signaling is shown in Figure 14 and involves binding to specific membrane receptors, phosphorylation of Smads 1, 5 and 8 and binding with Co-Smad 4, translocation into the nucleus and activation of transcription factors. In bone, BMPs trigger a cascade of events leading to osteogenesis, chondrogenesis, angiogenesis and up-regulation of numerous growth factors and cytokines. Cross-talk between BMP, FGF and Wnt pathways has been reported. Numerous studies have shown that recombinant BMP2 and BMP7 stimulate new bone formation in critical size defects, long bone non-unions, fracture healing, spine fusion, and augmentation of autografts and allografts (Einhorn, 2003).

13.1 Importance of BMPs in distraction osteogenesis

There is enough evidence, today, to suggest that BMPs, specifically BMP2 and BMP7, do play a major role in regenerate bone formation in DO. First, the expression of various members of the BMP signaling pathway was extensively analyzed, at the protein level using immunohistochemistry and at the mRNA level using RT-PCR, and it was shown by us (Haque et al., 2007; Haque et al., 2008; Haque et al., 2006; Rauch et al., 2000) and others (Ai-Aql et al., 2008) that BMP ligands, receptors, transcription factors and downstream targets are up-regulated during the distraction phase and then down-regulated once the mechanical forces of distraction cease.

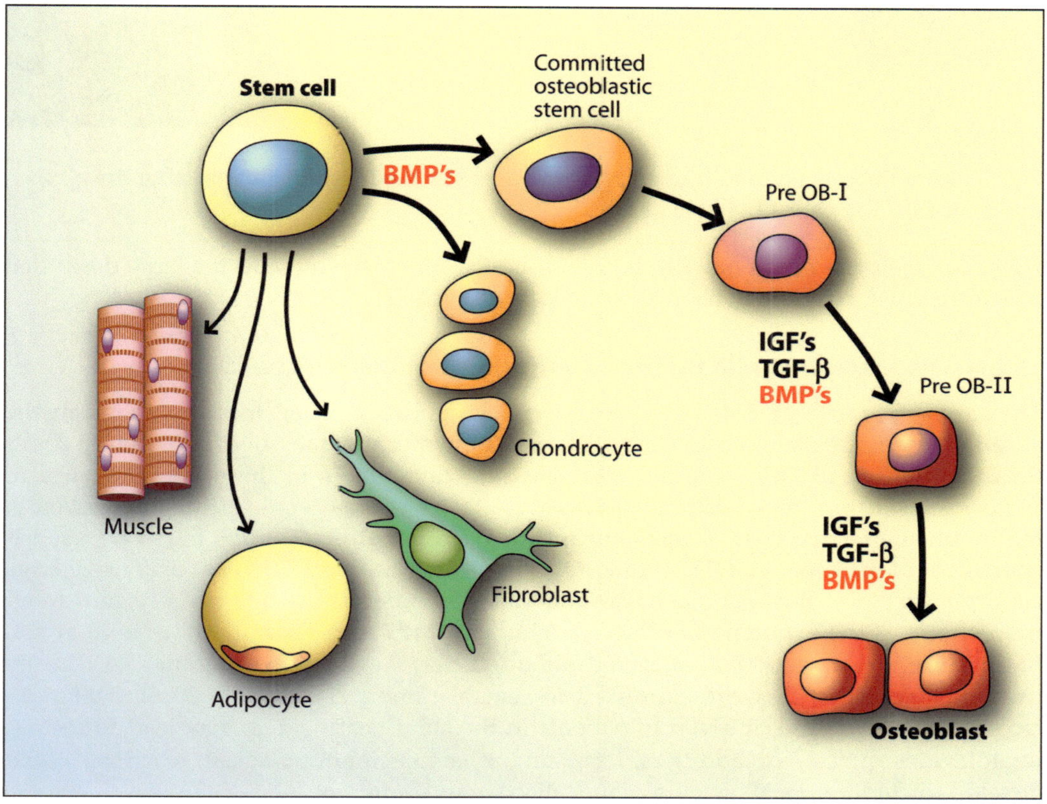

Fig. 9. BMPs acting on undifferentiated cells.

Second, to confirm the important role of BMPs in DO in a mechanistic way, we studied the process of DO in conditional BMP2 knockout mice (supplied by Dr. V. Rosen, Boston) and found that there was a delay in bone formation in the distracted gap of the heterozygous mice (Figure 10), when compared to the control littermates, thus showing that BMP2 is essential for bone formation in DO (Haque et al., 2008) and also supporting the results of (Tsuji et al., 2006) who reported that BMP2 is required for the initiation of fracture healing.

Third, we (Mandu-Hrit et al., 2006) and others (Lesaichot et al., 2011; Li et al., 2002; Mizumoto et al., 2003) have also reported that application of rhBMP7 and rhBMP2 in various animal models of DO, had a significant effect on the acceleration of bone formation in DO.

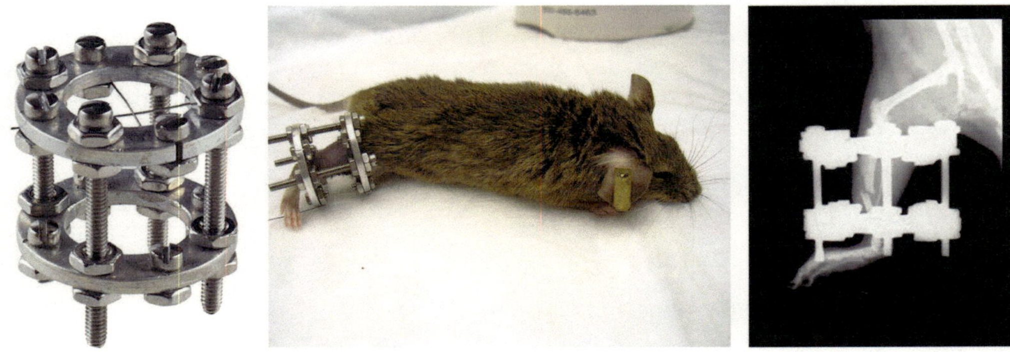

Fig. 10. Mini Ilizarov circular frame applied to a mouse model of DO in our laboratory , and developed by Tay et al. (Tay et al., 1998).

However, the main problem with the use of BMPs in humans remains the large doses that have to be used in order to obtain clinical significant results.

13.2 Issues associated with the use of large doses of BMPs in the clinic

The rapid clearance of BMPs from the site of application when locally applied in the absence of an adequate system for sustained delivery and the short half life of BMPs mandate the use of large, supraphysiological doses of BMPs in the milligram range to achieve satisfactory bone healing (Haidar et al., 2009b; Haidar et al., 2010b). Importantly, the minimal effective dose of recombinant BMPs (rhBMP2 or rhBMP7) currently used is equivalent to the sum of all BMPs present in 1000 human skeletons! These expensive large supraphysiological doses could have numerous side effects and unknown serious long-term effects as the expression of BMPs, including BMP2 and 7 is not exclusive to bones. BMPs have a neoplastic potential and although they have not been shown to directly cause malignancies, they are expressed in several tumors (Hsu et al., 2005). BMPs can cross the placenta, and, if given to patients in the child bearing age, may lead to serious teratologic sequelae. In addition, local and systemic immunological reactions could develop including local swelling and discharge (Dohin et al., 2009). Also, ectopic ossification could develop. In addition, the effects of BMPs on the growth plate remain unknown. Due to all these concerns, the use of BMPs is contraindicated in children and skeletally immature patients, thus precluding their use in a large population of patients that may benefit from BMPs.

Two compelling questions then arise: first, how can we decrease these large doses of exogenous BMPs without altering their efficiency? Second, as an alternative, would manipulating the endogenous BMP pathway in DO by decreasing BMP antagonist expression increase the bioavailability of endogenous BMPs and enhance osteogenesis (without having to apply exogenous BMPs). We believe that a local sustained and prolonged release of low doses of rhBMPs would address the first question. To provide a sustained and prolonged release of BMPs, several options are currently being investigated: gene therapy, cell therapy, multiple repeated injections of rhBMPs or cytokine therapy. Gene therapy is still in the experimental stages of research (Long et al., 2011). Multiple repeated injections of rhBMPs are not an option from a practical point of view. Cell therapy is a viable option, as was previously mentioned. However, in order to be effective, bone marrow mesenchymal stem cells have to be cultured before local application. Cytokine therapy is feasible provided, there is a delivery system that would ensure a sustained and prolonged delivery of adequate protein concentrations to the desired site (Haidar et al., 2009b; Haidar et al., 2010b; Schmidmaier et al., 2008).

14. Delivery systems for BMPs and other growth factors

There is no question that the efficacy of BMPs in enhancing bone formation is dependent on its mode of delivery (Boerckel et al., 2011; La et al., 2010). In the absence of a suitable delivery system, huge doses of BMPs have to be used in order to overcome the rapid clearance and the very short life of BMPs. In order to be able to use low doses of BMPs that would be equally effective as a single large dose, an adequate delivery system that would allow the slow and controlled release of adequate concentrations (in low doses) of BMPs over the desired period of time becomes necessary. Probably, there is no single delivery system that would be suitable for all conditions. Rather, different delivery systems will be required for different pathologies (for example an injectable system may be required for some cases, while in others a locally applied one at the time of surgery would be indicated).

Furthermore, as BMP2 and BMP7 have different temporal and spatial expression as well as different modes of action and different cellular targets, the use of a delivery system that would allow the sequential delivery of these two factors may present a huge step towards improving bone formation and a huge advantage over the use of a single factor.

The literature is abundant on studies describing various delivery systems for BMPs and other growth factors, including various biomaterials, scaffolds, gene delivery and numerous tissue engineering techniques. Recently, the use of nanoparticles as a delivery system has gained popularity (Facca et al., 2011; Yilgor et al., 2010).

14.1 Nanoparticles as delivery system

Besides the known advantages including the size, long shelf life and ability to entrap more drugs (Gref et al., 1994), nano-sized systems reside longer in circulation, therefore greatly extend the biological activity when compared to microparticles (Desai et al., 1996). The literature provides evidence that particles (<500 nm) cross membranes of epithelial cells through endocytosis while larger particles (>5 mm) are taken up via the lymphatic system (Dong and Feng, 2004; Zhang and Feng, 2006).

15. Ongoing research in our laboratory

Our current research is focused on the use of BMPs in DO, specifically in two areas: first, the development of a delivery system for BMPs that would allow sustained and prolonged delivery of low doses of exogenous BMPs, with the same osteogenic effect of a single large dose, and second, to investigate if suppression of BMP antagonists could up-regulate the biological activity of endogenous BMPs so as to decrease or avoid the use the use of exogenous BMPs.

15.1 Development of a delivery system for BMPs

Over the last 5 years, we have been working on the development of a unique hybrid core-shell nanoparticle layer-by-layer (chitosan-alginate) delivery system. The core is composed of charged large unilamellar liposomes (LUVs) and the shell is constructed through the layer-by-layer (L-b-L) self-assembly of alternating layers of sodium alginate and chitosan. Alginates are water-soluble linear un-branched anionic polymers (marine sources, algae). Chitosan is a linear cationic polysaccharide (derived by the N-deacetylation of chitin, a product found in the shells of crustaceans). It is a biocompatible, non-immunogenic, and biodegradable polymer with bioadhesive, wound healing, antimicrobial and even osteogenic properties; making it a favorable option for biomedical applications. Both, alginate and chitosan have been extensively studied for drug delivery in different forms, such as microcapsules, beads or even wound dressing membranes (Haidar et al., 2010b).

In our laboratory, we have successfully completed all the *in vitro* characterization, formulation (Figure 11) and release kinetics studies (Haidar et al., 2008; Haidar et al., 2010a) of our NP system. Then, we showed the capability of the novel core-shell NPs to efficiently encapsulate and release a range of concentrations of rhBMP7 up to 45 days (Haidar et al., 2009a) (Figure 12). In order to better understand the fate of the NPs and their targeted delivery, we have developed an imaging tool using quantum dots (QDs) and reported on the development of biocompatible chitosan-QD nanoparticles (Sandros et al., 2007). We then completed a toxicology study in rats, where we determined that our delivery system was safe and biocompatible (Haidar et al., 2010a). We then showed that a single injection of NPs loaded with low doses of rhBMP7 administered early in the distraction phase accelerated osteogenesis (Haidar et al., 2010b) and that some nanoparticles were detected in the distracted gap (Figure 13).

Ongoing research in our laboratory is focused now on the optimization of the timing and dosage of the nanoparticle–BMP injections.

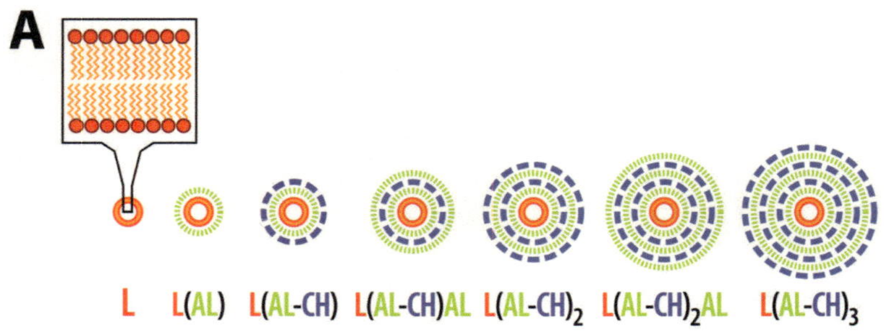

Fig. 11. Layer-by-Layer self-assembly of Alginate (AL) and Chitosan (CH) on Liposomes (L) up to six layers are shown. Reprint with kind permission from Springer Science+Business Media B.V; *Biotechnol Lett.* 2009 Dec;31(12):1817-24. *Delivery of recombinant bone morphogenetic proteins for bone regeneration and repair. Part A: Current challenges in BMP delivery. Haidar ZS, Hamdy RC, Tabrizian M.*

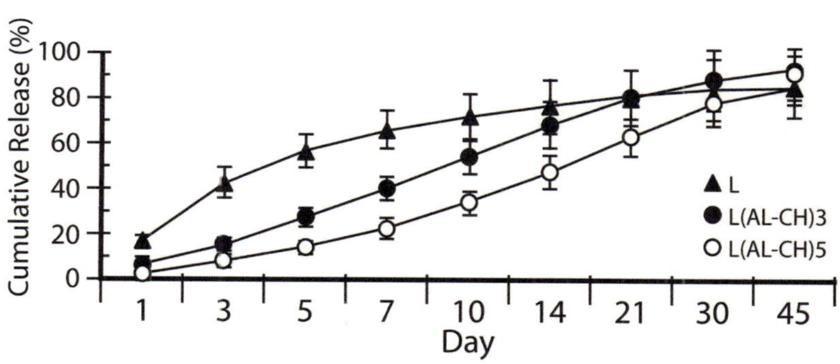

Fig. 12. Cumulative rhBMP7 release kinetic profile for uncoated liposomes (L), 3 bi-layered [L(AL-CH)3 or NPs3] and 5 bi-layered [L9AL-CH)5 or NPs5] core-shell NPs over an extended period of 45 days, *in vitro*. Controlled initial burst and rhBMP7 release from NPs is evident. *Reprint with permission from J Biomed Mater Res A. 2009 Dec;91(3):919-28.*

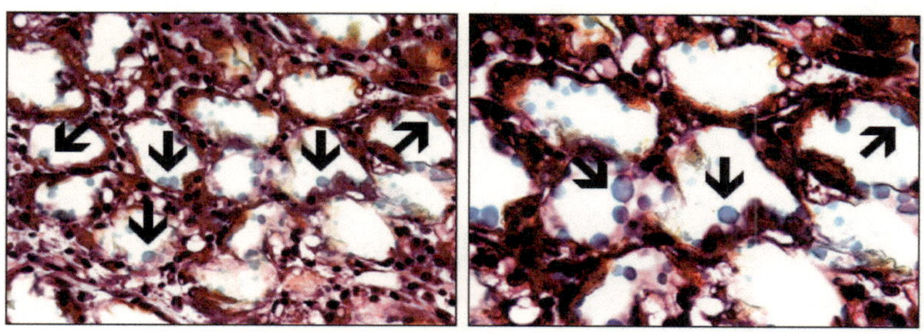

Fig. 13. Histological images (Goldner-Trichrome stain) of the distracted gap of a rabbit model of DO that received a single injection of NPs with 0.5 µg of rhBMP7 two weeks post-surgery, showing the persistence of NPs as indicated by the arrows. Animals were sacrificed five weeks after surgery (Data not published).

16. Suppression of BMP antagonists

As stated previously, the use of exogenous BMPs to enhance bone repair suffers from several drawbacks. Instead of administering exogenous BMPs to speed to up the regeneration process, an alternative strategy envisioned would be to inhibit BMP antagonists. This would allow increased levels of biologically active endogenous BMPs, which would in turn speed up the osteogenesis repair process. Attenuating BMP antagonist expression might also help to reduce the effective dose of exogenously applied BMP. This raises the question: do BMP antagonists play a role in fracture healing and new bone formation in DO?

BMP antagonists have been identified as a broad class of molecules, (Figure 14), which control BMP activity through a negative feedback mechanism. Some BMP antagonists, such as Noggin and Chordin, interact directly with BMP ligands (mostly BMP2, 4, 6 and 7) to restrict their biological activities extracellularly. Once bound to BMP antagonists, BMPs are prevented from interacting with their cognate membrane receptors and inducing intracellular signaling. BMP antagonists may also act at the membrane level (such as Bambi) or intracellularly (Smad 6 and 7) (Cao and Chen, 2005; Gazzerro and Canalis, 2006; Rosen, 2006).

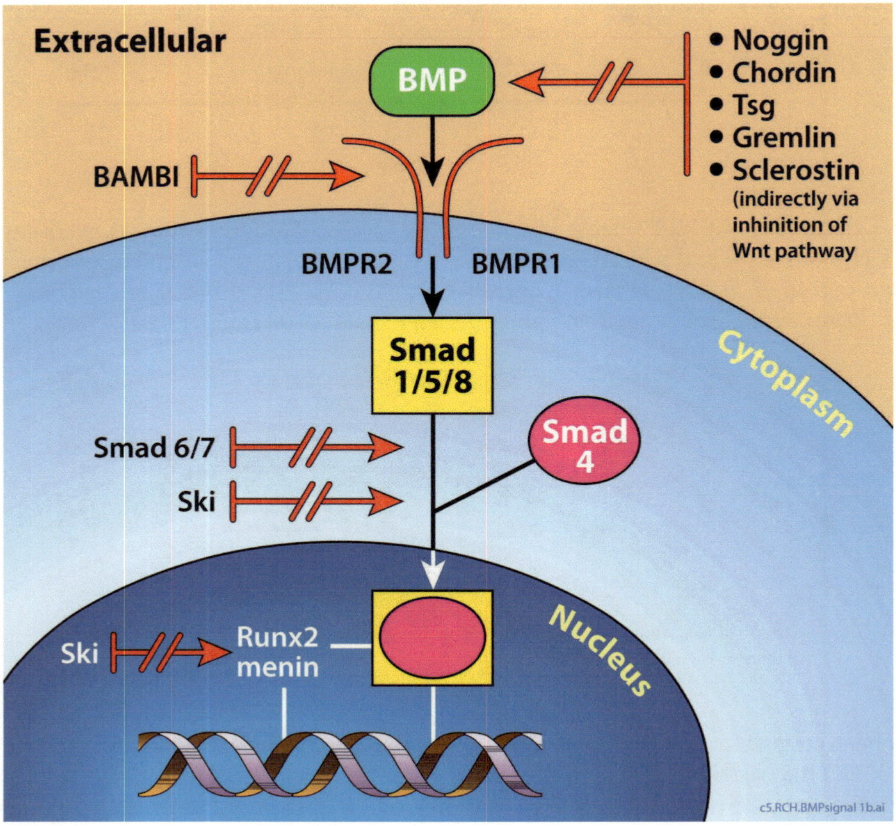

Fig. 14. BMP Signaling and BMP Antagonists.

Many *in vitro* studies have been conducted to investigate the role of BMP antagonists on osteoblast function and various aspects of osteogenesis and chondrogenesis. Noggin and Chordin have been particularly well studied in this context. Exogenous addition of Noggin or Chordin to osteoblast cell culture models results in the inhibition of expression for bone-specific genes, alkaline phosphatase, osteocalcin and bone sialoprotein, with a concomitant decrease in mineralization potential (Tsialogiannis et al., 2009).

Using transgenic and knockout animal models, many studies have revealed the crucial roles played by BMP antagonists in bone formation and repair. Over-expression of BMP antagonists has been shown to have detrimental effects on various bone parameters (Wu et al., 2003). In contrast, suppression of BMP antagonists using RNA interference caused increased osteogenic differentiation of cultured pre-osteoblastic cells, stromal cells and myoblastic cells (Kwong et al., 2008; Takayama et al., 2009). These findings strongly suggest that Noggin suppression via RNA interference stimulates bone formation in a mouse model and may potentiate the effects of endogenous BMPs. Furthermore, as Noggin binds to several BMPs (BMP2, 4, 6 and 7), reducing Noggin expression may lead to increased availability of several, and not one BMP, thus reconstituting the normal biological environment, as compared to the exogenous application of BMPs, where only a single BMP, - either BMP2 or BMP7 can be locally applied.

Numerous reports have emphasized the role BMP antagonists may play in fracture healing and the potential acceleration of new bone formation by inhibiting the inhibitors (Tsialogiannis et al., 2009). The expression of various BMP antagonists has been reported in human specimen of fracture healing and cases of non-unions (Kloen et al., 2002). However, there have been very few studies in humans, analysing the effects of blocking BMP antagonists on fracture healing (Lissenberg-Thunnissen et al., 2011).

In the context of DO, we have previously shown in a mouse model of DO, that BMP antagonists, including BMP3, Noggin, Chordin and Inhibin showed significant increases in expression during the distraction process (Haque et al., 2008).

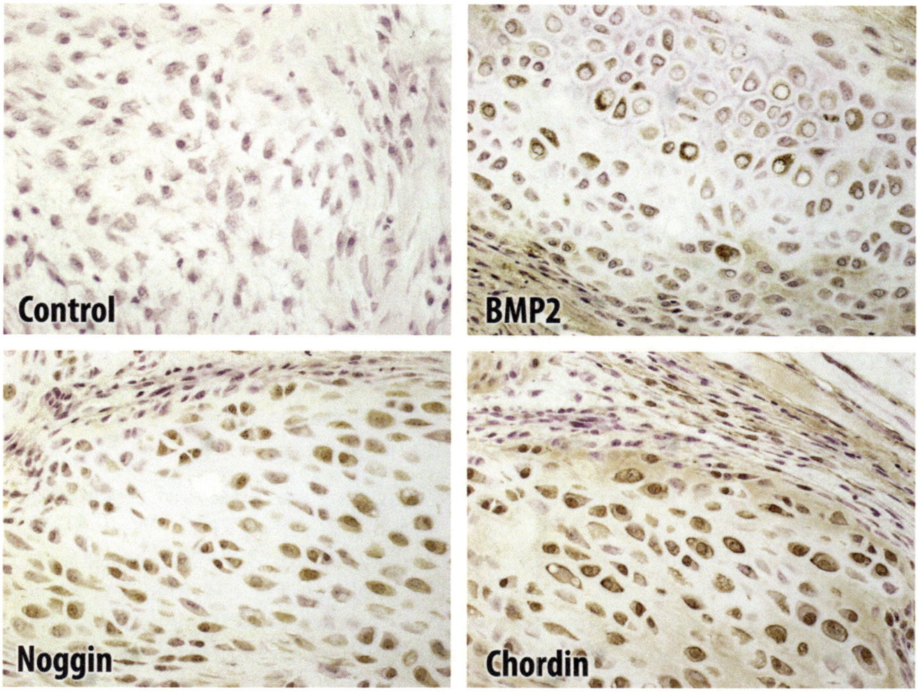

Fig. 15. Immunohistochemical staining of regenerate bone in the distracted gap in a rabbit model of DO, five weeks after osteotomy of the tibia and distraction of 2.0 cms, showing intense staining for BMP antagonists Noggin and Chordin in cartilage-like cells.

Based on these results, and a review of other reports in the literature, it would seem logical to hypothesize that these antagonists would be ideal candidates for therapeutic manipulation; blocking or inhibiting these antagonists, would in turn lead to up-regulation of BMPs and hence increased osteogenesis. The question then arises: which BMP antagonists to block and how to block them?

16.1 Methods to block BMP antagonists

These include naturally occurring substances, such as Heparan sulphate, monoclonal antibodies, and small interfering RNAs. In our laboratory, we first investigated if locally applied heparin sulphate in a mouse model of DO would enhance bone formation. Our results did not show any benefit with the use of heparin sulphate (results submitted for publication). We then decided to attempt inhibit BMP antagonists with the use of RNA technology.

16.1.1 RNA interference

As a method to block BMP antagonists, the use of RNA interference techniques is of particular interest. RNA interference is a powerful new strategy that has been broadly utilized in recent years to elucidate gene function (Pushparaj et al., 2008). It consists of delivering gene-specific double stranded RNA into cells that eventually causes silencing (knockdown) of the expression of a target gene. The mechanism leading to mRNA degradation involves a complex series of steps mediated by the host cell post-transcriptional machinery (Rana, 2007). Although this technology is easily applicable *in vitro*, directly by transfecting cells with synthetic double stranded small interfering RNA (siRNA), it requires a different delivery system for it to be used *in vivo*. Viral-mediated gene delivery of small hairpin RNA (shRNA) is one of the methods of choice to achieve gene silencing *in vivo*.

Lentiviruses represent a very flexible tool to modulate sustained gene expression both *in vitro* and *in vivo*. Lentiviral vectors are a method of choice because they: 1- infect both dividing and nondividing cells, 2- are less immunogenic because they are not encapsulated, 3- have a wide tropism, 4- can be concentrated to high titers, and 5-incorporate into the host genome (Kootstra and Verma, 2003). Lentiviruses have been used and showed that they can infect osteoblasts with high efficiency *in vitro*. When directly applied *in vivo* into surgically created lesions of the mandible and tibia, lentiviruses were found to infect most cell types present in bone, especially in the regenerating chondrocytic callus of the tibia (Wazen et al., 2006). A recent study showed success with lentiviral vectors as delivery system for shRNA-mediated RNA interference in MC3T3 cells (Moffatt et al., 2008).

Ongoing research in our laboratory aims at investigating the effects of blocking the BMP antagonists Noggin and Chordin using siRNA (small interfering RNA). We have chosen to focus our attention on Noggin and Chordin mainly because shRNA sequences against Noggin and Chordin have been identified previously to repress their expression with efficiencies greater than 80%. Furthermore, previous studies using knockout and transgenic mice revealed that Noggin suppression via RNA interference stimulates bone formation in a mouse model and may potentiate the effects of endogenous BMPs.As a first step and as proof of concept, we are using lentivirus to deliver siRNA to our mouse model of DO, as this delivery vehicle has been shown to have high transfection rate. Future research will aim at using our nanoparticle system for the delivery of siRNA in the context of DO.

17. Conclusion and future perspectives in distraction osteogenesis research

DO is a fascinating technique, that has become a standard method for bone regeneration used worldwide for the treatment of numerous orthopaedic conditions associated with bone loss, deficiencies and poor bone formation. The clinical importance of DO also extends to the fields of craniofacial surgery and dentistry, where it has revolutionized the treatment of numerous previously untreatable pathologies.

DO is not only of interest to clinicians, but is equally attractive to many scientists and researchers in various specialties including developmental and molecular biologists, chemists, protein scientists as well biomedical and tissue engineers.

Opportunities for future research in DO are immense and include not only efforts aiming at enhancing bone formation but also the development of new devices for distracting bones, both internal and external, and novel methods to assess the quality and quantity of newly formed bone.

In this chapter, various methods used to accelerate bone formation in DO were reviewed and the use of BMPs was specifically addressed. BMPs are probably the most potent osteogenic factors known to date, however, the huge doses that need to be used in humans may pose serious problems and adverse effects that limit their widespread use. How to improve the efficacy of BMPs in order to optimize their clinical use is the challenge that we, scientists and clinicians, are facing. Can we meet this challenge? Much more work needs to be done in that respect. Only two avenues, as pertained to our research program were discussed here: the development of an adequate delivery system and methods to inhibit the BMP antagonists. However, there are many more methods to increase the efficacy of BMPs that have been recently reported. BMP2 and BMP7 are the two most extensively studied BMPs, however, less known BMPs such as BMP6 and BMP9 may possess similar or even greater osteogenic properties than BMP2 and 7 and more experiments are needed in that area. Wnt signaling is emerging as a new and powerful osteogenic pathway that needs to be investigated, specifically in the context of DO. More experiments need to be performed in order to assess if the combined local application of growth factors BMP2 and BMP7 or BMP and TGFB may be more effective than the use of a single growth factor. The sequential application of BMP2 and BMP7 via a suitable delivery system is a very attractive concept that needs to be further analyzed in order to determine its clinical efficacy.

There is no doubt that improving the technique of DO will have a huge impact medically, on the quality of life of patients and financially, on patients, their families and the health care system. Not only patients undergoing DO will benefit from advances in that field, but also patients with delayed fracture healing, non-unions and possibly also in patients with poor bony conditions, such as osteoporosis.

More than half a century after the Magician of Kurgan, Ilizarov, unraveled the biological principles of DO, many questions still remain to be answered.

18. Acknowledgment

The authors thank Guylaine Bedar and Mark Lepik for their collaboration with the illustrations on this chapter.

19. References

Abbaspour, A., M. Takahashi, K. Sairyo, S. Takata, K. Yukata, A. Inui, and N. Yasui. 2009. Optimal increase in bone mass by continuous local infusion of alendronate during distraction osteogenesis in rabbits. *Bone*. 44:917-923.

Ai-Aql, Z.S., A.S. Alagl, D.T. Graves, L.C. Gerstenfeld, and T.A. Einhorn. 2008. Molecular mechanisms controlling bone formation during fracture healing and distraction osteogenesis. *J Dent Res*. 87:107-118.

Aronson, J., B. Good, C. Stewart, B. Harrison, and J. Harp. 1990. Preliminary studies of mineralization during distraction osteogenesis. *Clin Orthop Relat Res*:43-49.

Aronson, J., B.H. Harrison, C.L. Stewart, and J.H. Harp, Jr. 1989. The histology of distraction osteogenesis using different external fixators. *Clin Orthop Relat Res*:106-116.

Aronson, J., X.C. Shen, R.A. Skinner, W.R. Hogue, T.M. Badger, and C.K. Lumpkin, Jr. 1997. Rat model of distraction osteogenesis. *J Orthop Res*. 15:221-226.

Bernstein, A., H.O. Mayr, and R. Hube. 2010. Can bone healing in distraction osteogenesis be accelerated by local application of IGF-1 and TGF-beta1? *J Biomed Mater Res B Appl Biomater*. 92:215-225.

Birch, J.G., and M.L. Samchukov. 2004. Use of the Ilizarov method to correct lower limb deformities in children and adolescents. *J Am Acad Orthop Surg*. 12:144-154.

Boerckel, J.D., Y.M. Kolambkar, K.M. Dupont, B.A. Uhrig, E.A. Phelps, H.Y. Stevens, A.J. Garcia, and R.E. Guldberg. 2011. Effects of protein dose and delivery system on BMP-mediated bone regeneration. *Biomaterials*. 32:5241-5251.

Busse, J.W., J. Kaur, B. Mollon, M. Bhandari, P. Tornetta, 3rd, H.J. Schunemann, and G.H. Guyatt. 2009. Low intensity pulsed ultrasonography for fractures: systematic review of randomised controlled trials. *BMJ*. 338:b351.

Cakarer, S., V. Olgac, N. Aksakalli, A. Tang, and C. Keskin. 2010. Acceleration of consolidation period by thrombin peptide 508 in tibial distraction osteogenesis in rats. *Br J Oral Maxillofac Surg*. 48:633-636.

Canter, H.I., I. Vargel, and M.E. Mavili. 2007. Reconstruction of mandibular defects using autografts combined with demineralized bone matrix and cancellous allograft. *J Craniofac Surg*. 18:95-100; discussion 101-103.

Cao, J., L. Wang, D.L. Lei, Y.P. Liu, Z.J. Du, and F.Z. Cui. 2011. Local injection of nerve growth factor via a hydrogel enhances bone formation during mandibular distraction osteogenesis. *Oral Surg Oral Med Oral Pathol Oral Radiol Endod*:[Epub ahead of print].

Cao, X., and D. Chen. 2005. The BMP signaling and in vivo bone formation. *Gene*. 357:1-8.

Choi, I.H., J.H. Ahn, C.Y. Chung, and T.J. Cho. 2000. Vascular proliferation and blood supply during distraction osteogenesis: a scanning electron microscopic observation. *J Orthop Res*. 18:698-705.

Choi, I.H., C.Y. Chung, T.J. Cho, and W.J. Yoo. 2002. Angiogenesis and mineralization during distraction osteogenesis. *J Korean Med Sci*. 17:435-447.

Cierny, G., 3rd, and K.E. Zorn. 1994. Segmental tibial defects. Comparing conventional and Ilizarov methodologies. *Clin Orthop Relat Res*:118-123.

Cierny, G., and D. DiPasquale. 2011. Treatment of segmental bone loss due to infection. *In* Management of limb-length discrepancies. R. Hamdy and J.J. McCarthy, editors. American Academy of Orthopaedic Surgeons, Rosemont, IL. 159-172.

Claes, L., P. Augat, S. Schorlemmer, C. Konrads, A. Ignatius, and C. Ehrnthaller. 2008. Temporary distraction and compression of a diaphyseal osteotomy accelerates bone healing. *J Orthop Res*. 26:772-777.

Claes, L., and B. Willie. 2007. The enhancement of bone regeneration by ultrasound. *Prog Biophys Mol Biol*. 93:384-398.

Codivilla, A. 1905 On the means of lengthening, in the lower limbs, the muscles and tissues, which are shortened through deformity. *J Bone Joint Surg Am..* s2-2:353-369.

Davidson, E.H., S.M. Sultan, P. Butala, J.P. Tutela, O. Canizares, I.J. Wagner, D. Knobel, P.B. Saadeh, and S.M. Warren. 2011. Augmenting neovascularization accelerates distraction osteogenesis. *Plast Reconstr Surg*. 128:406-414.

Delloye, C., G. Delefortrie, L. Coutelier, and A. Vincent. 1990. Bone regenerate formation in cortical bone during distraction lengthening. An experimental study. *Clin Orthop Relat Res*:34-42.

Desai, M.P., V. Labhasetwar, G.L. Amidon, and R.J. Levy. 1996. Gastrointestinal uptake of biodegradable microparticles: effect of particle size. *Pharm Res*. 13:1838-1845.

Dimitriou, R., E. Jones, D. McGonagle, and P.V. Giannoudis. 2011. Bone regeneration: current concepts and future directions. *BMC Med*. 9:66.

Dohin, B., N. Dahan-Oliel, F. Fassier, and R. Hamdy. 2009. Enhancement of difficult nonunion in children with osteogenic protein-1 (OP-1): early experience. *Clin Orthop Relat Res*. 467:3230-3238.

Dong, Y., and S.S. Feng. 2004. Methoxy poly(ethylene glycol)-poly(lactide) (MPEG-PLA) nanoparticles for controlled delivery of anticancer drugs. *Biomaterials*. 25:2843-2849.

Einhorn, T.A. 2003. Clinical applications of recombinant human BMPs: early experience and future development. *J Bone Joint Surg Am*. 85-A Suppl 3:82-88.

Facca, S., A. Ferrand, C. Mendoza-Palomares, F. Perrin-Schmitt, P. Netter, D. Mainard, P. Liverneaux, and N. Benkirane-Jessel. 2011. Bone formation induced by growth factors embedded into the nanostructured particles. *J Biomed Nanotechnol*. 7:482-485.

Feldman, D.S., and S. Chaudhry. 2011. Distraction osteogenesis: lengthening with external fixators. *In* Management of limb-length discrepancies. R. Hamdy and J.J. McCarthy, editors. American Academy of Orthopaedic Surgeons, Rosemont, Il. 45-56.

Fink, B., C. Pollnau, M. Vogel, R. Skripitz, and A. Enderle. 2003. Histomorphometry of distraction osteogenesis during experimental tibial lengthening. *J Orthop Trauma*. 17:113-118.

Fujio, M., A. Yamamoto, Y. Ando, R. Shohara, K. Kinoshita, T. Kaneko, H. Hibi, and M. Ueda. 2011. Stromal cell-derived factor-1 enhances distraction osteogenesis-mediated skeletal tissue regeneration through the recruitment of endothelial precursors. *Bone*:[Epub ahead of print].

Gazzerro, E., and E. Canalis. 2006. Bone morphogenetic proteins and their antagonists. *Rev Endocr Metab Disord*. 7:51-65.

Goldwaser, B.R., M.E. Papadaki, L.B. Kaban, and M.J. Troulis. 2011. Automated Continuous Mandibular Distraction Osteogenesis: Review of the Literature. *J Oral Maxillofac Surg*:[Epub ahead of print].

Green, A.A. 2011. The Ilizarov method of distraction osteogenesis. *In* Management of limb-length discrepancies. R. Hamdy and J.J. McCarthy, editors. American Academy of Orthopaedic Surgeons, Rosemont, IL. 39-44.

Gref, R., Y. Minamitake, M.T. Peracchia, V. Trubetskoy, V. Torchilin, and R. Langer. 1994. Biodegradable long-circulating polymeric nanospheres. *Science*. 263:1600-1603.

Hagiwara, T., and W.H. Bell. 2000. Effect of electrical stimulation on mandibular distraction osteogenesis. *J Craniomaxillofac Surg*. 28:12-19.

Haidar, Z.S., F. Azari, R.C. Hamdy, and M. Tabrizian. 2009a. Modulated release of OP-1 and enhanced preosteoblast differentiation using a core-shell nanoparticulate system. *J Biomed Mater Res A*. 91:919-928.

Haidar, Z.S., R.C. Hamdy, and M. Tabrizian. 2008. Protein release kinetics for core-shell hybrid nanoparticles based on the layer-by-layer assembly of alginate and chitosan on liposomes. *Biomaterials*. 29:1207-1215.

Haidar, Z.S., R.C. Hamdy, and M. Tabrizian. 2009b. Delivery of recombinant bone morphogenetic proteins for bone regeneration and repair. Part B: Delivery systems for BMPs in orthopaedic and craniofacial tissue engineering. *Biotechnol Lett*. 31:1825-1835.

Haidar, Z.S., R.C. Hamdy, and M. Tabrizian. 2010a. Biocompatibility and safety of a hybrid core-shell nanoparticulate OP-1 delivery system intramuscularly administered in rats. *Biomaterials*. 31:2746-2754.

Haidar, Z.S., M. Tabrizian, and R.C. Hamdy. 2010b. A hybrid rhOP-1 delivery system enhances new bone regeneration and consolidation in a rabbit model of distraction osteogenesis. *Growth Factors*. 28:44-55.

Hamdy, R.C., M. Amako, L. Beckman, M. Kawaguchi, F. Rauch, D. Lauzier, and T. Steffen. 2003. Effects of osteogenic protein-1 on distraction osteogenesis in rabbits. *Bone*. 33:248-255.

Hamdy, R.C., A. Silvestri, C.H. Rivard, and M. Ehrlich. 1997. Histologic evaluation of bone regeneration in cases of limb lengthening by Ilizarov's technique. An experimental study in the dog. *Ann Chir*. 51:875-883.

Haque, T., M. Amako, S. Nakada, D. Lauzier, and R.C. Hamdy. 2007. An immunohistochemical analysis of the temporal and spatial expression of growth factors FGF 1, 2 and 18, IGF 1 and 2, and TGFbeta1 during distraction osteogenesis. *Histol Histopathol*. 22:119-128.

Haque, T., F. Hamade, N. Alam, M. Kotsiopriftis, D. Lauzier, R. St-Arnaud, and R.C. Hamdy. 2008. Characterizing the BMP pathway in a wild type mouse model of distraction osteogenesis. *Bone*. 42:1144-1153.

Haque, T., M. Mandu-Hrit, F. Rauch, D. Lauzier, M. Tabrizian, and R.C. Hamdy. 2006. Immunohistochemical localization of bone morphogenetic protein-signaling Smads during long-bone distraction osteogenesis. *J Histochem Cytochem*. 54:407-415.

Hatzokos, I., S.I. Stavridis, E. Iosifidou, D. Karataglis, and A. Christodoulou. 2011. Autologous bone marrow grafting combined with demineralized bone matrix improves consolidation of docking site after distraction osteogenesis. *J Bone Joint Surg Am*. 93:671-678.

Hou, W.W., Z.L. Zhu, Y. Zhou, C.X. Zhang, and H.Y. Yu. 2011. Involvement of Wnt activation in the micromechanical vibration-enhanced osteogenic response of osteoblasts. *J Orthop Sci*: [Epub ahead of print].

Hsu, M.Y., S. Rovinsky, S. Penmatcha, M. Herlyn, and D. Muirhead. 2005. Bone morphogenetic proteins in melanoma: angel or devil? *Cancer Metastasis Rev*. 24:251-263.

Huang, C., and R. Ogawa. 2010. Mechanotransduction in bone repair and regeneration. *FASEB J*. 24:3625-3632.

Hwang, S.J., S. Lublinsky, Y.K. Seo, I.S. Kim, and S. Judex. 2009. Extremely small-magnitude accelerations enhance bone regeneration: a preliminary study. *Clin Orthop Relat Res*. 467:1083-1091.

Ilizarov, G.A. 1989a. The tension-stress effect on the genesis and growth of tissues. Part I. The influence of stability of fixation and soft-tissue preservation. *Clin Orthop Relat Res*:249-281.

Ilizarov, G.A. 1989b. The tension-stress effect on the genesis and growth of tissues: Part II. The influence of the rate and frequency of distraction. *Clin Orthop Relat Res*:263-285.

Isefuku, S., C.J. Joyner, A.A. Reed, and A.H. Simpson. 2004. Distraction osteogenesis in the Cbfa-1+/- mouse. *J Orthop Res*. 22:1276-1282.

Jiang, X., D. Song, B. Ye, X. Wang, G. Song, S. Yang, and J. Hu. 2011. Effect of intermittent administration of adiponectin on bone regeneration following mandibular osteodistraction in rabbits. *J Orthop Res*. 29:1081-1085.

Jiang, X., S. Zou, B. Ye, S. Zhu, Y. Liu, and J. Hu. 2010. bFGF-Modified BMMSCs enhance bone regeneration following distraction osteogenesis in rabbits. *Bone*. 46:1156-1161.

Kawamoto, K., W.C. Kim, Y. Tsuchida, Y. Tsuji, M. Fujioka, M. Horii, Y. Mikami, D. Tokunaga, and T. Kubo. 2005. Effects of alternating current electrical stimulation on lengthening callus. *J Pediatr Orthop B*. 14:299-302.

Kiely, P., K. Ward, C.M. Bellemore, J. Briody, C.T. Cowell, and D.G. Little. 2007. Bisphosphonate rescue in distraction osteogenesis: a case series. *J Pediatr Orthop*. 27:467-471.

Kitoh, H., T. Kitakoji, H. Tsuchiya, M. Katoh, and N. Ishiguro. 2007. Transplantation of culture expanded bone marrow cells and platelet rich plasma in distraction osteogenesis of the long bones. *Bone*. 40:522-528.

Kloen, P., S.B. Doty, E. Gordon, I.F. Rubel, M.J. Goumans, and D.L. Helfet. 2002. Expression and activation of the BMP-signaling components in human fracture nonunions. *J Bone Joint Surg Am*. 84-A:1909-1918.

Kojimoto, H., N. Yasui, T. Goto, S. Matsuda, and Y. Shimomura. 1988. Bone lengthening in rabbits by callus distraction. The role of periosteum and endosteum. *J Bone Joint Surg Br*. 70:543-549.

Konas, E., M. Emin Mavili, P. Korkusuz, D. Demir, F. Oner, and H.I. Canter. 2009. Acceleration of distraction osteogenesis with drug-releasing distractor. *J Craniofac Surg*. 20:2041-2048.

Kootstra, N.A., and I.M. Verma. 2003. Gene therapy with viral vectors. *Annu Rev Pharmacol Toxicol*. 43:413-439.

Kurklu, M., C. Yildiz, O. Kose, Y. Yurttas, O. Karacalioglu, M. Serdar, and S. Deveci. 2011. Effect of alpha-tocopherol on bone formation during distraction osteogenesis: a rabbit model. *J Orthop Traumatol*:[Epub ahead of print].

Kwong, F.N., S.M. Richardson, and C.H. Evans. 2008. Chordin knockdown enhances the osteogenic differentiation of human mesenchymal stem cells. *Arthritis Res Ther*. 10:R65.

La, W.G., S.W. Kang, H.S. Yang, S.H. Bhang, S.H. Lee, J.H. Park, and B.S. Kim. 2010. The efficacy of bone morphogenetic protein-2 depends on its mode of delivery. *Artif Organs*. 34:1150-1153.

Lai, J.P., F.S. Wang, C.M. Hung, C.J. Wang, C.J. Huang, and Y.R. Kuo. 2010. Extracorporeal shock wave accelerates consolidation in distraction osteogenesis of the rat mandible. *J Trauma*. 69:1252-1258.

Latalski, M., Y.A. Elbatrawy, A.M. Thabet, A. Gregosiewicz, T. Raganowicz, and M. Fatyga. 2011. Enhancing bone healing during distraction osteogenesis with platelet-rich plasma. *Injury*. 42:821-824.

Lee, D.Y., T.J. Cho, J.A. Kim, H.R. Lee, W.J. Yoo, C.Y. Chung, and I.H. Choi. 2008. Mobilization of endothelial progenitor cells in fracture healing and distraction osteogenesis. *Bone*. 42:932-941.

Lesaichot, V., D. Leperlier, V. Viateau, D. Richarme, H. Petite, and F. Sailhan. 2011. The influence of Bone Morphogenic Protein-2 on the consolidation phase in a distraction osteogenesis model. *Injury*:[Epub ahead of print].

Li, G., M.L. Bouxsein, C. Luppen, X.J. Li, M. Wood, H.J. Seeherman, J.M. Wozney, and H. Simpson. 2002. Bone consolidation is enhanced by rhBMP-2 in a rabbit model of distraction osteogenesis. *J Orthop Res*. 20:779-788.

Lissenberg-Thunnissen, S.N., D.J. de Gorter, C.F. Sier, and I.B. Schipper. 2011. Use and efficacy of bone morphogenetic proteins in fracture healing. *Int Orthop*. 35:1271-1280.

Long, J., P. Li, H.M. Du, L. Liu, X.H. Zheng, Y.F. Lin, H. Wang, W. Jing, W. Tang, W.H. Chen, and W.D. Tian. 2011. Effects of bone morphogenetic protein 2 gene therapy on new bone formation during mandibular distraction osteogenesis at rapid rate in rabbits. *Oral Surg Oral Med Oral Pathol Oral Radiol Endod*. 112:50-57.

Makarov, M., J. Birch, and M. Samchukov. 2009. The role of variable muscle adaptation to limb lengthening in the development of joint contractures: an experimental study in the goat. *J Pediatr Orthop*. 29:175-181.

Mandu-Hrit, M., T. Haque, D. Lauzier, M. Kotsiopriftis, F. Rauch, M. Tabrizian, J.E. Henderson, and R.C. Hamdy. 2006. Early injection of OP-1 during distraction osteogenesis accelerates new bone formation in rabbits. *Growth Factors*. 24:172-183.

Marucci, D.D., Y. Yu, J. McTavish, S. Fiona Bonar, M.D. Poole, and W.R. Walsh. 2002. Matrix metalloproteinases and their inhibitors in bone remodelling following distraction osteogenesis of the sheep mandible. *J Craniomaxillofac Surg*. 30:208-212.

Mccarthy, J.G., J. Schreiber, N. Karp, C.H. Thorne, and B.H. Grayson. 1992. Lengthening the Human Mandible by Gradual Distraction. *Plast Reconstr Surg*. 89:1-8.

Miyazono, K., Y. Kamiya, and M. Morikawa. 2010. Bone morphogenetic protein receptors and signal transduction. *J Biochem*. 147:35-51.

Mizumoto, Y., T. Moseley, M. Drews, V.N. Cooper, 3rd, and A.H. Reddi. 2003. Acceleration of regenerate ossification during distraction osteogenesis with recombinant human bone morphogenetic protein-7. *J Bone Joint Surg Am*. 85-A Suppl 3:124-130.

Moffatt, P., M.H. Gaumond, P. Salois, K. Sellin, M.C. Bessette, E. Godin, P.T. de Oliveira, G.J. Atkins, A. Nanci, and G. Thomas. 2008. Bril: a novel bone-specific modulator of mineralization. *J Bone Miner Res*. 23:1497-1508.

Moore, D.C., M.G. Ehrlich, S.C. McAllister, J.T. Machan, C.E. Hart, C. Voigt, A.M. Lesieur-Brooks, and E.W. Weber. 2009. Recombinant human platelet-derived growth factor-BB augmentation of new-bone formation in a rat model of distraction osteogenesis. *J Bone Joint Surg Am*. 91:1973-1984.

Mori, S., M. Akagi, A. Kikuyama, Y. Yasuda, and C. Hamanishi. 2006. Axial shortening during distraction osteogenesis leads to enhanced bone formation in a rabbit model through the HIF-1alpha/vascular endothelial growth factor system. *J Orthop Res*. 24:653-663.

Murray, J.H., and R.D. Fitch. 1996. Distraction Histiogenesis: Principles and Indications. *J Am Acad Orthop Surg*. 4:317-327.

Narasaki, K., H. Shimizu, M. Beppu, H. Aoki, M. Takagi, and M. Takashi. 2003. Effect of extracorporeal shock waves on callus formation during bone lengthening. *J Orthop Sci*. 8:474-481.

Nauth, A., M.D. McKee, T.A. Einhorn, J.T. Watson, R. Li, and E.H. Schemitsch. 2011. Managing bone defects. *J Orthop Trauma*. 25:462-466.

Ni, M., G. Li, P.F. Tang, K.M. Chan, and Y. Wang. 2011. rhBMP-2 not alendronate combined with HA-TCP biomaterial and distraction osteogenesis enhance bone formation. *Arch Orthop Trauma Surg*:[Epub ahead of print].

Okazaki, H., T. Kurokawa, K. Nakamura, T. Matsushita, K. Mamada, and H. Kawaguchi. 1999. Stimulation of bone formation by recombinant fibroblast growth factor-2 in callotasis bone lengthening of rabbits. *Calcif Tissue Int*. 64:542-546.

Pacicca, D.M., N. Patel, C. Lee, K. Salisbury, W. Lehmann, R. Carvalho, L.C. Gerstenfeld, and T.A. Einhorn. 2003. Expression of angiogenic factors during distraction osteogenesis. *Bone*. 33:889-898.

Paley, D. 1990. Problems, obstacles, and complications of limb lengthening by the Ilizarov technique. *Clin Orthop Relat Res*:81-104.

Pepper, J.R., M.A. Herbert, J.R. Anderson, and W.P. Bobechko. 1996. Effect of capacitive coupled electrical stimulation on regenerate bone. *J Orthop Res*. 14:296-302.

Perrien, D.S., K.M. Nicks, L. Liu, N.S. Akel, A.W. Bacon, R.A. Skinner, F.L. Swain, J. Aronson, L.J. Suva, and D. Gaddy. 2011. Inhibin A enhances bone formation during distraction osteogenesis. *J Orthop Res*:[Epub ahead of print].

Pushparaj, P.N., J.J. Aarthi, J. Manikandan, and S.D. Kumar. 2008. siRNA, miRNA, and shRNA: in vivo applications. *J Dent Res*. 87:992-1003.

Rana, T.M. 2007. Illuminating the silence: understanding the structure and function of small RNAs. *Nat Rev Mol Cell Biol*. 8:23-36.

Rauch, F., D. Lauzier, S. Croteau, R. Travers, F.H. Glorieux, and R. Hamdy. 2000. Temporal and spatial expression of bone morphogenetic protein-2, -4, and -7 during distraction osteogenesis in rabbits. *Bone*. 27:453-459.

Romano, C.L., D. Romano, and N. Logoluso. 2009. Low-intensity pulsed ultrasound for the treatment of bone delayed union or nonunion: a review. *Ultrasound Med Biol*. 35:529-536.

Rosen, V. 2006. BMP and BMP inhibitors in bone. *Ann N Y Acad Sci*. 1068:19-25.

Sabharwal, S. 2011. Enhancement of bone formation during distraction osteogenesis: pediatric applications. *J Am Acad Orthop Surg*. 19:101-111.

Sandros, M.G., M. Behrendt, D. Maysinger, and M. Tabrizian. 2007. InGaP@ZnS-Enriched Chitosan Nanoparticles: A Versatile Fluorescent Probe for Deep-Tissue Imaging. *Advanced Functional Materials*. 17:3724-3730.

Sato, M., N. Yasui, T. Nakase, H. Kawahata, M. Sugimoto, S. Hirota, Y. Kitamura, S. Nomura, and T. Ochi. 1998. Expression of bone matrix proteins mRNA during distraction osteogenesis. *J Bone Miner Res*. 13:1221-1231.

Schmidmaier, G., P. Schwabe, C. Strobel, and B. Wildemann. 2008. Carrier systems and application of growth factors in orthopaedics. *Injury*. 39 Suppl 2:S37-43.

Sen, C., T. Gunes, M. Erdem, R.D. Koseoglu, and N.O. Filiz. 2006. Effects of calcitonin and alendronate on distraction osteogenesis. *Int Orthop*. 30:272-277.

Shao, Z., B. Liu, Q. Peng, W. Liu, Y. Liu, R. Liu, Y. Xu, and L. Liu. 2007. Transplantation of osteoblast-like cells to the distracted callus in the rabbit mandible. *Plast Reconstr Surg*. 119:500-507.

Snyder, C.C., G.A. Levine, H.M. Swanson, and E.Z. Browne. 1973. Mandibular Lengthening by Gradual Distraction - Preliminary Report. *Plast Reconstr Surg*. 51:506-508.

Song, H.R., C.W. Oh, H.S. Kyung, I.H. Park, P.T. Kim, S.H. Baek, S.J. Kim, and S.T. Lee. 2004. Injected calcium sulfate for consolidation of distraction osteogenesis in rabbit tibia. *J Pediatr Orthop B*. 13:170-175.

Takayama, K., A. Suzuki, T. Manaka, S. Taguchi, Y. Hashimoto, Y. Imai, S. Wakitani, and K. Takaoka. 2009. RNA interference for noggin enhances the biological activity of bone morphogenetic proteins in vivo and in vitro. *J Bone Miner Metab*. 27:402-411.

Tay, B.K., A.X. Le, S.E. Gould, and J.A. Helms. 1998. Histochemical and molecular analyses of distraction osteogenesis in a mouse model. *J Orthop Res*. 16:636-642.

Tsialogiannis, E., I. Polyzois, Q. Oak Tang, G. Pavlou, E. Tsiridis, and M. Heliotis. 2009. Targeting bone morphogenetic protein antagonists: in vitro and in vivo evidence of their role in bone metabolism. *Expert Opin Ther Targets*. 13:123-137.

Tsuchiya, H. 2011. Treatment of segmental bone loss due to tumors. *In* Management of limb-length discrepancies. R. Hamdy and J.J. McCarthy, editors. American Academy of Orthopaedic Surgeons, Rosemont, IL. 173-178.

Tsuji, K., A. Bandyopadhyay, B.D. Harfe, K. Cox, S. Kakar, L. Gerstenfeld, T. Einhorn, C.J. Tabin, and V. Rosen. 2006. BMP2 activity, although dispensable for bone formation, is required for the initiation of fracture healing. *Nat Genet*. 38:1424-1429.

Vander Kolk, C.A., M.M. Mofid, P.N. Manson, B.C. Robertson, A.P. Tufaro, and J.J. Elias. 2001. Craniofacial distraction osteogenesis: A review of 3278 cases. *Plast Reconstr Surg*. 108:1103-1114.

Wan, C., S.R. Gilbert, Y. Wang, X. Cao, X. Shen, G. Ramaswamy, K.A. Jacobsen, Z.S. Alaql, A.W. Eberhardt, L.C. Gerstenfeld, T.A. Einhorn, L. Deng, and T.L. Clemens. 2008. Activation of the hypoxia-inducible factor-1alpha pathway accelerates bone regeneration. *Proc Natl Acad Sci U S A*. 105:686-691.

Watanabe, Y., T. Matsushita, M. Bhandari, R. Zdero, and E.H. Schemitsch. 2010. Ultrasound for fracture healing: current evidence. *J Orthop Trauma*. 24 Suppl 1:S56-61.

Wazen, R.M., P. Moffatt, S.F. Zalzal, N.G. Daniel, K.A. Westerman, and A. Nanci. 2006. Local gene transfer to calcified tissue cells using prolonged infusion of a lentiviral vector. *Gene Ther*. 13:1595-1602.

Wu, X.B., Y. Li, A. Schneider, W. Yu, G. Rajendren, J. Iqbal, M. Yamamoto, M. Alam, L.J. Brunet, H.C. Blair, M. Zaidi, and E. Abe. 2003. Impaired osteoblastic differentiation, reduced bone formation, and severe osteoporosis in noggin-overexpressing mice. *J Clin Invest*. 112:924-934.

Xue, J., J. Peng, M. Yuan, A. Wang, L. Zhang, S. Liu, M. Fan, Y. Wang, W. Xu, K. Ting, X. Zhang, and S. Lu. 2011. NELL1 promotes high-quality bone regeneration in rat femoral distraction osteogenesis model. *Bone*. 48:485-495.

Yamane, K., T. Okano, H. Kishimoto, and H. Hagino. 1999. Effect of ED-71 on modeling of bone in distraction osteogenesis. *Bone*. 24:187-193.

Yasui, N., M. Sato, T. Ochi, T. Kimura, H. Kawahata, Y. Kitamura, and S. Nomura. 1997. Three modes of ossification during distraction osteogenesis in the rat. *J Bone Joint Surg Br*. 79:824-830.

Yilgor, P., N. Hasirci, and V. Hasirci. 2010. Sequential BMP-2/BMP-7 delivery from polyester nanocapsules. *J Biomed Mater Res A*. 93:528-536.

Zhang, Z., and S.S. Feng. 2006. The drug encapsulation efficiency, in vitro drug release, cellular uptake and cytotoxicity of paclitaxel-loaded poly(lactide)-tocopheryl polyethylene glycol succinate nanoparticles. *Biomaterials*. 27:4025-4033.

Zhu, S., D. Song, X. Jiang, H. Zhou, and J. Hu. 2011. Combined effects of recombinant human BMP-2 and Nell-1 on bone regeneration in rapid distraction osteogenesis of rabbit tibia. *Injury*:[Epub ahead of print].

9

Skull Expansion by Spring-Mediated Bone Regeneration

Rodrigo Dornelles, Vera Cardim
and Nivaldo Alonso
University of São Paulo
Brazil

1. Introduction

Paleontologists discovered the skull of a Homo heildelbergensis individual, ancestor of the 530-thousand-year-old Neanderthal, that belonged to a craniosynostotic child aged between five and 12 years, in the Spanish province of Burgos, and dated the oldest case of this condition. There was a time when cranial malformations were considered a result of divine or supernatural reasons and there are pictures of Chinese gods that display tall heads or great projections in their frontal region. (Guimarães-Ferreira, Miguéns et al. 2004) To such an extent that this intentional cranial deformation became part of the culture of ancient peoples. In the Americas, cases dated 8000 BC and this tradition only ceased in 1752 under the Spanish colonial domain. As a result of this intentional deformation, by distorting the normal directions of cranial growth, there is a change in the normal process of cranial suture closure. Many of these skulls presented premature suture ossification whereas others presented sutures that remained with no closure until more advanced age(Abreu 2002; Tubino and Alves 2009).

The famous work "De Vulneribus Capitis Líber", by Hippocrates (460-367 BC) apud Guimarães-Ferreira(Guimarães-Ferreira, Miguéns et al. 2004), begins with the sentences: "Human heads are not similar among themselves. Neither are sutures the same in terms of number and location", as an evidence that the variability of cranial sutures and their closures were already observed.

Craniofacial sutures are important regions for facial and cranial bone growth with major development during embryogenesis. Their important functions are twofold:

1. Maintaining skull malleability during the passage through the birth canal.
2. Enabling the separation between cranial bones during intrauterine and perinatal life.

Craniosynostosis is the premature fusion of one or more cranial sutures with resultant cranial and/or facial deformity and affects approximately one in 1,700 to 4,000 live births (Renier, Le Merrer et al. 2006). In addition to shape, function may also be compromised because limitation in development can cause increased intracranial pressure, visual disorders, mental retardation, among other anomalies.(Guimarães-Ferreira, Miguéns et al. 2004; Renier, Le Merrer et al. 2006).

Lycosthene (1557) described a child with deformities in skull and limbs, acrocephalosyndactyly, which was later described by Apert (1906)* apud McCarthy (McCarthy, Epstein et al. 1990).

The term craniosynostosis was first used by Otto in 1830 (McCarthy, Epstein et al. 1990; Slater, Lenton et al. 2008), however, Rudolf Ludwig Carl Virchow*, in 1851, published an article still important to date describing a series of 29 skulls with malformations in an attempt to elucidate cases of cretinism. Virchow quotes Stahl, Gibson and Sommering, and Hyrtl as pioneers of the idea that the premature closure of a cranial suture leads to cranial deformation, but indicates compensations during growth and attempts to establish predictive general patterns, apud Persing(Persing, Jane et al. 1989).

- Apert, E. De l'acricephalosyndactylie. Bull. Soc. Med. Hop. Paris, 23:1310,1906
- Virchow, R. Üeber den Cretinismus, namentlich in Franken, und üeber pathologische Schädelformen. Verh. Physikalisch. Med. Ges. Wurxburg 2:230,185

1.1 Aim

The aim of this study is:

To evaluate bone regeneration and cephalometric changes with the implantation of expansible springs in the interparietal and parasagittal region in post-craniectomy rabbits through radiologic and histologic analyses.

2. History of treatments

2.1 Craniectomies

Lannelongue (1890)* apud Guimarães(Guimarães-Ferreira, Miguéns et al. 2004) reported the first surgical treatment with strip craniectomy and for many years, despite the high morbidity of this procedure, surgeries utilizing craniectomy of involved sutures were performed. In the mid-twentieth century, many surgeons produced techniques for cranial reshaping with complex osteotomies, surgeries for removing involved sutures and, in some cases, of those not involved, resulting in cases of extensive cranial bone resections with post-operative protection with helmets, taking into consideration that reossification is unpredictable. The interposition of materials between sutures and caustic substances to inhibit suture closure and other procedures also presented high morbidity with elevated rates of mortality additionally had their indications(Guimarães-Ferreira, Miguéns et al. 2004; Greensmith, Holmes et al. 2008).

The first correction techniques were developed in order to prevent the evolution towards an increase in intracranial pressure (ICP) and/or mental retardation. The modern techniques, introduced by Tessier (Tessier 1971), also tried to correct dysmorphia preserving the psychological balance of children (Arnaud, Marchac et al. 2006). In the 14th International Plastic Surgery Congress (1967), held in Rome, Paul Tessier baffled the scientific community by exhibiting his results on the treatment of craniofacial stenoses with endocranial approach. Using techniques developed from his experience in the treatment of complex craniofacial and Le Fort I, II and III fractures, he demonstrated a multidisciplinary approach with results that revolutionalized the history and changed the treatment of craniosynostoses.

Afterwards, with the introduction of rigid internal fixation materials, i.e. microfixation plates and screws, at first made of titanium and recently of absorbable material, there has been an evolution in approach, as described by Marchac(Marchac and Renier 1979), Ortiz-Monastério(Ortiz-Monasterio, Fuente del Campo et al. 1978), Van Der Meulen* apud McCarthy (McCarthy, Epstein et al. 1990), and others.

If craniosynostosis is not corrected, the deformity progresses involving the facial skeleton and may cause changes in facial symmetry and malocclusion. We may state that, therefore, in order to reduce craniofacial changes resulting from synostosis, the surgery must be performed early because(Panchal and Uttchin 2003) :

- With brain growth, altered growth vectors will cause greater deformity as individual ages.
- There is an increase in intracranial pressure in nonsyndromic single-suture craniosynostoses (Gault, Renier et al. 1992; Heller, Heller et al. 2008). In experimental models, compensatory changes in the neurocranium do not allow the adequate expansion of the neurocapsular matrix (Singhal, Mooney et al. 1997).
- Before one year of age, bone defects ossify in an efficacious manner.
- Delay in performing corrections in the first nine to 12 months causes deformities in the cranial base with abnormal facial growth with maxillary and mandibular asymmetry.
- From three to nine months cranial bones are malleable.

Current techniques present some problems related to bone detachment, both from the subperiosteum and dura mater (Arnaud, Marchac et al. 2006). Namely regarding infants, considering complementary surgeries will have a greater degree of difficulty. Whenever increase in ICP occurs, or in cases of pansynostoses, marking of cerebral convolutions in the inner table, indicated by the so-called copper-beaten appearance in radiographic evidence, greater difficulty of detachment with increased blood loss and risks of approach on the venous sinuses is added. Furthermore, the detachment of functioning sutures is equally difficult, due to adherences.

In cases of extensive cranioplasty, with great dura mater detachment, there is a risk of extradural dead space with hematoma formation. Although this is a rare event, it is especially undesirable when airways are manipulated, as in those cases for correcting craniofacial stenoses.

- Lannelongue, J. De la craniectomie dans la microcephalie. C. R. Acad. Sci., 110:1382, 1890
- Van der Meulen, J. C. Medial faciotomy. Br. J. Plast. Surg., 32:339,1979

2.2 Osteogenic distraction

The fundamentals of the so-called osteogenic distraction date back the early work conducted by Codivilla* (1905) apud McCarthy(McCarthy, Stelnicki et al. 2001), with the denominated "continuous lengthening" and further applied to long bones by Ilizarov et al. (Ilizarov 1989; McCarthy, Schreiber et al. 1992). These authors established the principles of bone elongation for the treatment of bone defects and non-union for the upper and lower extremities. Wasmund (1926)* and Rosenthal (1927)* apud Hönig(Hönig, Grohmann et al. 2001) described, respectively, bone distraction for mandibular elongation using intraoral devices with progressive mechanical activation and advancement of the maxilla mediated by elastic forces. For nearly 50 years, this possibility of treatment was disregarded by the scientific community, but interest arouse with the study conducted by Snyder et al.(Snyder, Levine et al. 1973) that published an article reporting mandibular elongation in dogs for correcting crossbite in 1973.

This technique consists in the use of an external device attached to two pins with threaded adjustment that enables lengthening of the distance between them. The pins are placed in the bone to be elongated and a corticotomy is performed between them. Afterwards, distraction is progressively and constantly carried out until reaching the desired elongation. In the gap between the separated stumps, a bone callus is formed. After distraction is performed, a period of time is allowed for consolidation and the device is removed.

McCarthy et al.(McCarthy, Schreiber et al. 1992) published their first successful cases of mandibular osteogenic distraction in humans (Figure 25) and the consolidation of this technique for the treatment of mandibular osteogenic distraction happened with the series of cases published by Ortiz-Monasterio and Molina (Ortiz-Monasterio and Molina 1994; Molina and Ortiz-Monasterio 1995) .

Osteogenic distraction progressed as treatment for changes in the craniofacial skeleton and was applied for maxillary distraction with Le Fort I and III osteotomies and frontofacial advancements. One of its last indications was cranial distraction. In 1997, Tung et al. (Tung, Robertson et al. 1999) published an experimental study of osteogenic distraction in membranous bones of the craniofacial skeleton in rabbits affecting cranial volume and skull shape. In this study an external distractor was used on the cranial suture with no osteotomy.

The advantages of this method include (Guimarães-Ferreira, Miguéns et al. 2004):

1. Bone neoformation preventing bone grafts and intermaxillary fixation.
2. Preventing dura mater dissection by creating dead space, limiting infection risks, need for transfusion and bone reabsorption.
3. Concurrent expansion of adjacent soft tissues.

Osteogenic distraction was found to be a promising form of treatment in cranial vault bones(Imai, Komune et al. 2002), so that the principles of less detachment and procedure morbidity reduction could be applied in the treatment of craniosynostosis.

- Codivilla, A. On the means of lengthening in the lower limbs, the muscles and tissues which are shortened through deformity. Am. J. Orthop. Surg. 2:353, 1905
- Wassmund, M. Frankturen und Luxationen des Gesichtschädels. Berlin : Meuser, 1926. P.360
- Rosenthal, W. In E. Sonntag and W. Rosenthal (Eds.), Lehrbuch der Mund und Kieferchirurgie. Leipzig; Georg. Thieme, 1930. Pp. 173-175

2.3 Expansible springs

In classic physics, a spring may be regarded as a device that stores potential energy by stretching the ligations among atoms from the same elastic material.

Hooke's law(wikipedia 2009) is the law of physics related to the elasticity of materials used to calculate the deformation caused by the force exerted on a body that equals to the displacement of the mass from its equilibrium position times the constant feature of the spring or body that will endure deformation:

$$F = k.\Delta l$$

In the International System of Units (SI), F is expressed in newtons, k in newton/meter and Δl in meters.

The force produced by a spring is directly proportional to its displacement from its initial position (equilibrium). Spring equilibrium occurs when it is in its natural state, i.e. without being compressed or stretched. After compressing or stretching a spring, it pulls back and its force is calculated by the equation above.

The principles were already applied in remodeling facial skeleton, for instance in palatal expansion(Haas 1970). In 1986, Persing et al. (Persing, Babler et al. 1986) described a cranial expansion using a device named spring. In this study the authors were able to reverse abnormalities produced by brachycephaly caused by the premature restriction in sutural development. Results from the study carried out in rabbits were shown to be better than those obtained with linear craniectomy of suture alone. Subsequently, an expansion of the skull base and the suture contralateral to that affected with the use of springs was observed(Persing, Morgan et al. 1991).

The first clinical description of a cranioplasty using expansible springs dates back to 1998(Lauritzen, Sugawara et al. 1998), in which the principle of the dynamic expansion of internal implantable springs was used(Lauritzen, Sugawara et al. 1998; Lauritzen and Tarnow 2003). This study consisted in linear craniectomies with spring interposition in order to stimulate bone edges in an expansive or compressive manner (or both). Unlike other external distractors, it was totally implantable and reduced the risks of using pins and screws for activation. The action of springs in infant skulls with malleable bones of membranous nature is not exclusively restricted to bone edges, but extends to adjacent cranial vault. In the first 100 cases, Lauritzen et al. (Lauritzen, Davis et al. 2008) indicated springs for treating metopic, bicoronal, multisutural synostosis, advancements of the middle third of the face, and others. No serious complications were observed.

In 2001, an experimental study was published demonstrating a change in the form of cranial growth in rabbits using springs(Gewalli, Guimaraes-Ferreira et al. 2001).

In 2008, Cardim et al., based on their experience with elastic distraction of the middle third of the face(Cardim, Dornelles et al. 2002), published their experiment using expansible springs conducted since 2002 for treating craniosynostosis and craniofacial stenoses(Dornelles, Cardim et al. 2007). In this series, scaphocephaly patients underwent spring placement through bilateral parasagittal linear craniotomy, which yielded satisfactory cranial expansion and reshaping results, even when the synostotic suture was maintained.

Mackenzie et al.(Mackenzie, Davis et al. 2009) published their first comparative study of craniosynotosis treatments with craniectomy, cranioplasty with absorbable plates and use of expansible springs. Blood loss, operative time and hospital stay were evaluated. The study concluded that the use of springs provide adequate skull shape and present minimal morbidity.

Skull expansion with totally implantable springs differs from the principles of osteogenic distraction because there is no latency period and the effect of the spring starts at the very moment it is placed by exerting a continuous force with no external control. No control of ossification at the osteotomy site is verified and spring expansion ceases prior to its total opening capacity. The indication, number and force of springs, as well as their implantation site are determined.(Pyle, Glazier et al. 2009)

3. Material and method

3.1 Material

Thirteen female New Zealand white rabbits (Oryctolagus cuniculus) aged four weeks with mean weight of 2,400 g, from the Central Animal Facility of the São Paulo University Medical School (FMUSP), were used for this study. The study was conducted in the Microsurgery Experimental Laboratory (LIM4) at FMUSP. This research complied with the ethical principles for animal experiments established by the Brazilian College of Animal Experimentation (COBEA) and the rules in the Guide for Care and Use of Laboratory Animals (Institute of Laboratory Animal Resources, Commission on Life Science, National Research Council, Washington, DC, 1996). All animals remained in the animal facility of this laboratory during the experimental period. All animals were separately housed in individual cages, were fed pellets and given water ad libitum on a 12-hour light-dark cycle at room temperature of 21°C.

In order to analyze the changes in rabbit skulls under the action of expansible springs, the animals were divided into four groups as follows:

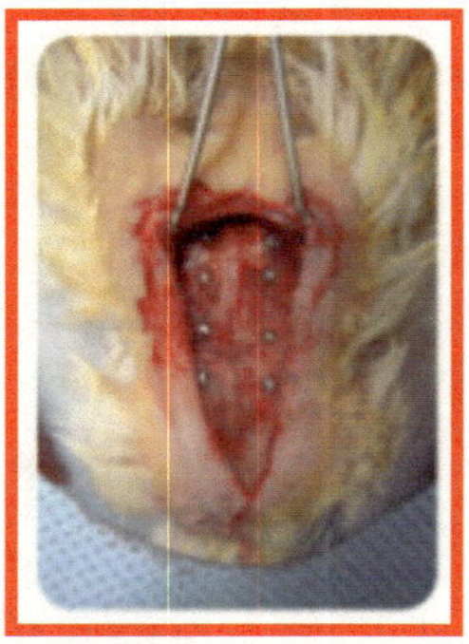	• GROUP I • Parasagittal amalgam markers were placed on the cranial vault • ONE ANIMAL
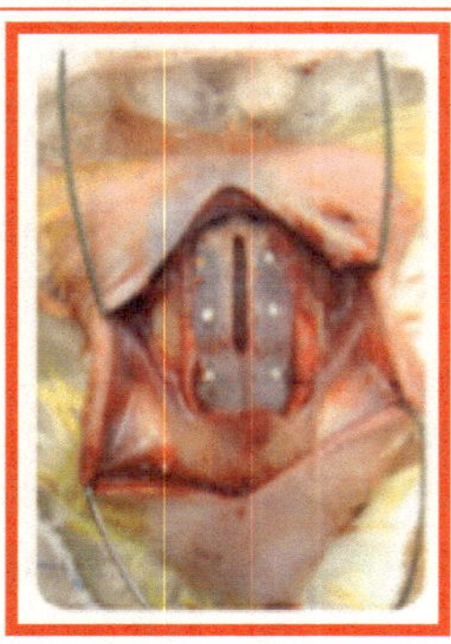	• GROUP II • Parasagittal amalgam markers were placed and a craniectomy was performed along the sagittal suture • FOUR ANIMALS

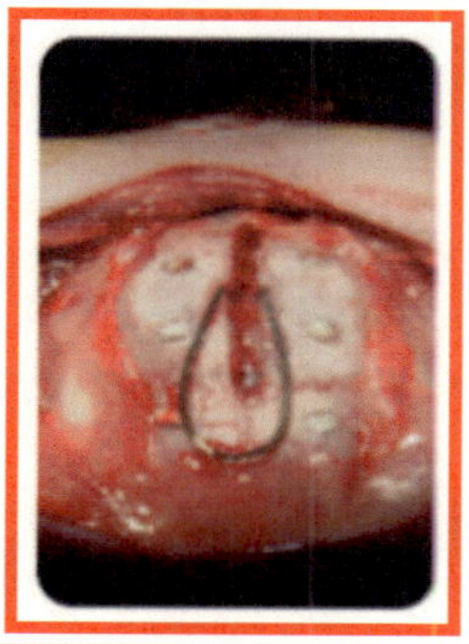	• GROUP III • Parasagittal amalgam markers were placed and springs were placed in the suturectomy site • FOUR ANIMALS
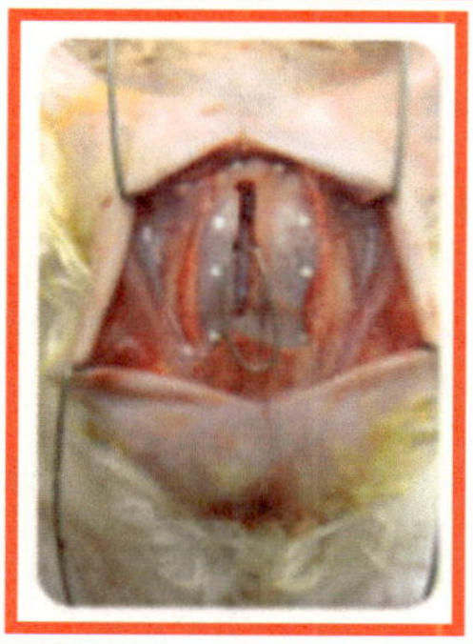	• GROUP IV • Parasagittal amalgam markers were placed and springs were implanted parallel to the sagittal suture with no suturectomy • FOUR ANIMALS

Table 1. Groups

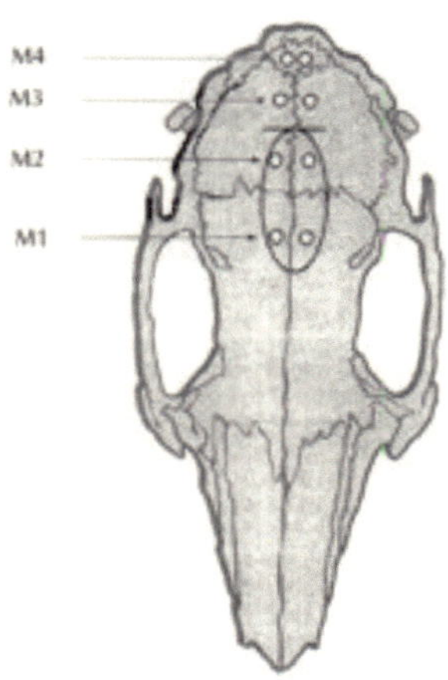

- M1 - Marker 1: 5 mm (either side) of the midline and 5 mm anterior to the coronal suture;
- M2 - Marker 2: 5 mm (either side) of the midline and 5 mm posterior to the coronal suture;
- M3 - Marker 3: 5 mm (either side) of the midline and 5 mm anterior to the lambdoid suture;
- M4 - Marker 4: 3 mm (either side) of the midline and 5 mm posterior to the lambdoid suture.
- spring placement between M2 e M3

Table 2. Schematic representation of marker and spring placement

3.2 Springs

Springs were manually manufactured by the first author from nickel-chromium stiff elastic wire for orthodontics. A 0.020″-bend diameter wire with 18-mm arm length and 25-mm deflection range at rest was used. For the purpose of obtaining prior standardization, all springs had their expansion force checked by a WAFIOS WG 3/2® dynamometer. Mean force obtained from measurements was 335 grams force (Schnaider and De Souza 2003) with 2-mm opening.

3.3 Procedure

All rabbits were anesthetized with IM injections of ketamine hydrochloride (35 mg/kg) associated with xylazine hydrochloride (5 mg/kg). Trichotomy in the cranial region from the nasion to the occipital region was performed.

After anesthesia, a local infiltration of 0.2%-lidocaine hydrochloride with 1:200,000 adrenaline (final concentration) was administered, respecting the maximum dose of 7mg/kg for lidocaine.

After trichotomy was carried out in rabbit skulls, the animals were placed in ventral decubitus position. Antisepsis with chlorhexidine gluconate was applied in all the animal heads followed by the placement of sterile drapes.

Surgical approach was performed through a 20-mm longitudinal incision in the cranial region exposing the sagittal suture with minimal periosteal detachment.

Amalgam markers were placed in all rabbit skulls through craniotomy using a low-rotation micro motor with round bur of 1.8 mm in diameter.

In order to obtain spring maintenance at the placement site, an incisure was made in the bone edges between M2 and M3.

All animals received a single dose of IM benzathine benzilpenicillin 100.000 IU/kg.

In group I amalgam markers were placed and no craniotomy was performed. The incision was sutured closed through the approximation and fixation of tissues with continuous stitches using 5-0 monofilament nylon thread.

In group II, a 1.8-mm wide strip craniotomy was done including the sagittal suture reaching one millimeter beyond the coronal and lambdoid sutures. Sagittal suturectomy was performed using a low-rotation round bur with 1.8 mm in diameter. The incision was sutured closed through the approximation and fixation of tissues with continuous stitches using 5-0 monofilament nylon thread.

In group III, a 1.8-mm wide strip craniotomy was done including the sagittal suture reaching one millimeter beyond the coronal and lambdoid sutures. Osteotomy was performed using a low-rotation round bur with 1.8 mm in diameter. The spring was positioned with minimal dura mater detachment in the stipulated position between M2 and M3. Fixation was maintained by the spring force in the bone incisures made on the edge of the craniotomy. The incision was sutured closed through the approximation and fixation of tissues with continuous stitches using 5-0 monofilament nylon thread.

In group IV, a 1.8-mm wide strip craniectomy was performed in the right parasagittal region and a spring was placed in the stipulated position. The incision was sutured closed through the approximation and fixation of tissues with continuous stitches using 5-0 monofilament nylon thread.

3.4 Evaluation

The evaluation of new bone formation of the effects from springs in the sagittal and parasagittal region was carried out for all groups at weeks two, four, eight, and twelve after the surgery for spring placement. Longitudinal body weight recordings were obtained from rabbits at each time point, radiographs were taken and all animals were euthanized with the anesthetic solution previously described through subsequent asphyxia in an individual carbon dioxide (CO_2) chamber and operated on for havesting tissue sample from the cranial vault for histopathologic analysis.

In order to obtain histopathologic data for assessing new bone formation, samples for post postmortem analysis were collected from a segment containing involved suture with anterior and posterior bone edges. A quadrangular shaped osteotomy was performed using low speed bur with 1.8 mm in diameter and the bone fragment was separated from the cranial vault preserving its continuity with the dura mater.

Fragments preserved in 10%-formaldehyde solution were sent to the Buccal Pathology Department of the Odontology School of São Paulo University where they were fixed in 10%-buffered formaldehyde solution and decalcified in 5% formic acid. The material was dehydrated in increasing ethanol concentrations, deparaffinized in xylene and embedded in paraffin. A series of tissue sections was obtained from cuts perpendicular to the osteotomy line with 5μm thickness. Sections comprised the cranial and the caudal transverse perspective of the intermediate distance from M2 and M3, consistent with spring support. Histologic analysis was carried out with material stained with hematoxylin and eosin (Alberius, Malmberg et al.).

Histologic evaluation was carried out with a Zeiss® microscope, Axio Imager A1 model, magnification 100x, by a pathologist for the areas of interest. The following parameters were chosen for assessment:

- Granulation tissue
- Osteoblasts
- Osteoclasts
- Bone trabeculae

Parameters were evaluated with the following classification:

(-) absence
(+) little/scarce
(++) moderate
(+++) abundant

For granulation tissue and bone trabeculae, the evaluation took into consideration:

(+) up to 25% of defect
(++) up to 50% of defect
(+++) more than 50% of defect

To obtain craniometric data e evaluate the variation of the distance between amalgam markers, radiographs of rabbit skulls were taken in the facilities of the Radiology Department of the Veterinary School of São Paulo University. All animals were anesthetized as previously described to allow their adequate positioning for radiographic projections with distance standardized at one meter. They were subsequently placed in ventral decubitus position to provide ventrodorsal projection and in lateral decubitus position to lateral projection.

A centimeter-scale lead marker was placed beside the skull, above the neck, at the same vertical distance as the device to provide gauging at the moment the distances between amalgam markers were measured and to standardize findings.

A 500-mA, 125-k V diagnostic x-ray unit, RT 500/125 model (RAY-TEC®, Brazil) with a microprocessor controlled system, full-wave, rectified silicon generator, radiographic table with anti-scatter grid, recipromatic Bucky, and rotating anode x-ray tube was used in radiographic examinations. TMS-1® and MXG/PLUS®, 24x30-cm radiographic films were used and placed on a metal chassis with CRONEX HI plus® intensifier screens. Films were developed and fixed in an Automatic RPX-OMAT Processor®, after appropriate luminescent identification.

All measurements were made with image digitalization and calculated by the first author using MIRROR® (Canfield Imaging Systems, Fairfield, NJ) software. Metric gauging was done using a lead marker, as well as linear measurement. In ventrodorsal projections, the distance between amalgam markers was measured using the center of the positions M1, M2, M3, and M4 as reference. In lateral projection, for the determination of points to be used as reference for cranial base measurement, the following diagram was used (Figure 1):

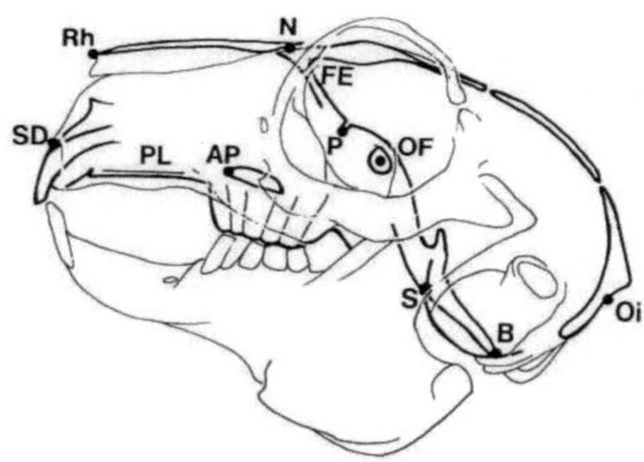

Fig. 1. Rabbit skull schematic diagram with craniometric points used for measuring cranial base length.

Where:

FE-P - Cribform plate length (anterior cranial base)
P-S - Pre-sphenoid length
S-B - Posterior cranial base length

For measuring lateral projection, the following craniometric points were used: (Figure 2):

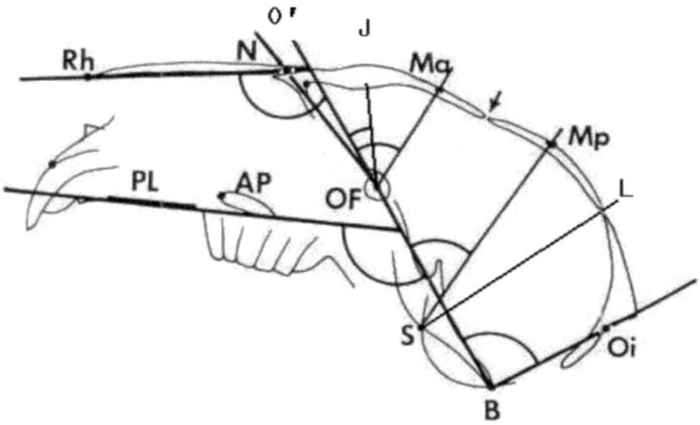

Fig. 2. Rabbit skull schematic diagram, lateral view with cephalometric points and angles.

Where:

O – a point in the center of the optic foramen
Oi – the deepest point in the outer contour of the occipital bone between the foramen magnum and the external occipital protuberance
Mp – the most posterior point on the anterior marker in the left parietal bone
Ma – the most anterior point on the posterior marker in the left frontal bone
J – the tip of the endocranial ridge in which the frontal bone separates the anterior and the middle cranial fossae
S – the deepest point on the cartilaginous spheno-occipital synchondrosis
PL – a line tangential to the lower rim of the hard palate
BO′ – the extension of the line through the points B and O

The following angles were obtained:

BO′/PL – angle formed between and below the lines BO′ and PL (maxillary-basilar angle)
BO′/RhN – angle formed between and above the lines BO′ and RhN (rhino-basilar angle)
BO′/NO – angle formed between and above the lines BO′ and NO (naso-basilar angle)
BO′/JO – angle formed between and above the lines BO′ and JO (basilar-olfactory angle)
BO′/MaO – angle formed between and above the lines BO′ and MaO (anterior mark-basilar angle)
BO′/MpS – angle formed between and above the lines BO′ and MpS (posterior mark-basilar angle)
BO′/LS – angle formed between and above the lines BO′ and LS (lambdoid-basilar angle)
BO′/BOi – angle formed between and above the lines BO′ and BOi (foraminal-basilar angle)

3.5 Method for data evaluation

This study was observational and descriptive, in which artificial intelligence techniques were used through the applications of annotated paraconsistent logic in order to parametrize and contextualize craniometric variables.

Whenever a rabbit variable was entered into one of the fields, it was compared with a normal variable from the data bank containing means of normal variables that, for the purpose of this study, were considered as those from the rabbit in group I.

Parametrization was carried out within the working interval for paraconsistent logic, i.e. values between “0” and “1” were attributed. All normal variables received value “1” and rabbit variables, proportional values according to the following criterion: if a rabbit variable equaled the normal variable, a degree of abnormality evidence = 0 was attributed to that variable. The degree of abnormality evidence, therefore, was an indicator of the discrepancy between the measurement of the animal variable and a normal variable.

The degree of abnormality evidence of a rabbit variable was close to 1 when its measurment was approximately threefold the standard deviation value established for that variable in the data bank used in this study.

All angular measurements were contextualized with cranial base measurement, so that firstly normalization of measurements was carried out using the following equation (1):

$$x = 1 / (1+(\text{ reference } / \text{ measument}))^{|\text{measurement-reference}|} \quad (1)$$

Where:

x - normalization
Reference - normal variable
Measurement - measurement variable

Next, the basic structural equation (BSE) of the paraconsistent logic was applied in order to contextualize the measurement through the equation (2):

$$\mu r = ((\mu - \lambda) + 1) / 2 \quad (2)$$

Where:

μr - resultant degree of evidence
μ - favorable evidence (μ = x, which will be the measurement of the cranial base)
λ - contrary evidence (in this case λ = y, where y is the normalized measurement)

The curve of evidence degree variation for each variable was molded with standard deviation ranging in accordance with variable values. Such effect caused the curves to have features of different inclinations according to different variable values, as well as symmetrical shape, what made them to be closer to real cases.

4. Results

4.1 Radiologic aspects

Considering the parameter of implanted amalgam markers, cranial expansion occurred so that the distance between markers at the M3 position was greater in the groups with springs than in those without springs at all time points studied. Measured values are presented in the graph. (Graph 1).

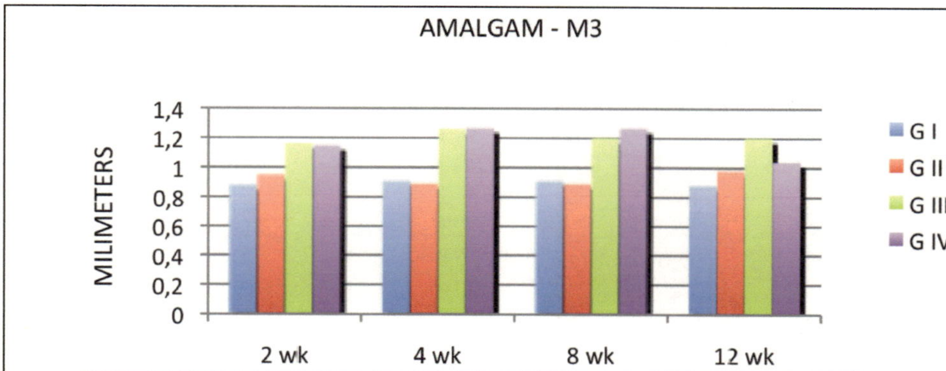

	2 wk	4 wk	8 wk	12 wk
G I	0,88	0,91	0,91	0,88
G II	0,95	0,89	0,89	0,98
G III	1,16	1,26	1,2	1,2
G IV	1,15	1,27	1,27	1,04

Graph 1. Results measured in millimeters between amalgam markers at M3.

Angle measurements were tabulated and are presented in Tables 3, 4, 5, and 6:

Group I

	2 wk		4 wk		8 wk		12 wk	
	Pré	Pós	Pré	Pós	Pré	Pós	Pré	Pós
BO'NO	9,3	13,8	9,3	12,1	9,3	15	9,3	13,6
BO'/RhN	116,8	119,9	116,8	122,8	116,8	123,2	116,8	120,9
BO'/MaO	59,1	55,3	59,1	54	59,1	55,3	59,1	61,9
BO'/PL	120,9	119,5	12,9	123,4	120,9	124,9	120,9	120,7
BO'/MpS	68,4	69	68,4	71,3	68,4	74	68,4	70,1
BO'/Boi	83	77,2	83	75,3	83	75,2	83	71
BO'/LS	68,4	79,1	68,4	82,3	68,4	85	68,4	80,5
BO'/JO	22,4	29,4	22,4	31,3	22,4	33,6	22,4	30,7

Table 3. Results of angles obtained in Group I

Group II

	2 wks		4 wks		8 wks		12 wks	
	Pré	Pós	Pré	Pós	Pré	Pós	Pré	Pós
BO'NO	12,9	10,2	11,5	11,4	13,5	10,7	13	12,1
BO'/RhN	122,4	120,2	123,5	120,5	119,7	122,9	119	122,7
BO'/MaO	64	63,5	54,2	53,5	69,6	50	68,2	62,3
BO'/PL	119,8	119,7	122,2	120,5	116,9	121,4	119,9	120
BO'/MpS	68,6	71	79,5	71,3	72,5	76,2	67,4	71,1
BO'/Boi	79,7	76,5	97,9	85,4	81,2	81,3	75,5	75,6
BO'/LS	81,3	86,6	94	87,9	84,3	92,2	78,5	86,9
BO'/JO	24,5	24,9	29,1	24	28,7	26	28,7	28,9

Table 4. Results of angles obtained in Group II

Group III

	2 wks		4 wks		8 wks		12 wks	
	Pré	Pós	Pré	Pós	Pré	Pós	Pré	Pós
BO'NO	10,5	11,5	10,4	10,1	8,6	15,1	7,5	13,2
BO'/RhN	122,1	120,7	121,1	121,8	124	120,2	128,1	121,9
BO'/MaO	69	48,5	55,6	55,1	67,2	50	51,4	58,9
BO'/PL	121,1	118,7	123,8	123	121,3	119,9	124,2	116,5
BO'/MpS	74,8	68,6	71,5	70,5	73,6	70,5	72,9	69,3
BO'/Boi	77,2	78	81,8	80,6	71,5	75,2	78,8	74,7
BO'/LS	88,7	88,1	90	90	91,9	87,7	88,2	86,1
BO'/JO	25,6	23,2	20,4	27,4	24,4	22,5	26,7	30,8

Table 5. Results of angles obtained in Group III

Group IV

	2 wks		4 wks		8 wks		12 wks	
	Pré	Pós	Pré	Pós	Pré	Pós	Pré	Pós
BO'NO	10,1	9,8	10,5	12,1	9,5	13,1	5,9	12,3
BO'/RhN	121,8	122,1	122,9	120,2	125,5	121,6	123,1	118,5
BO'/MaO	45,7	54,7	61,7	52,6	51,8	51,4	56,5	55,4
BO'/PL	122,6	121,7	122,5	119,1	123,3	121,8	124	119,1
BO'/MpS	70,3	65,7	79,8	70,3	74,9	72,2	76,3	71,7
BO'/BOi	79,5	71	81,2	78	78,5	76,6	78,3	77
BO'/LS	89,4	92,8	92,7	89,5	90,2	86,9	92,2	88,5
BO'/JO	27,1	31,3	28,9	28,8	26,4	26	28	28,5

Table 6. Results of angles obtained in Group IV

After all measurements were normalized and contextualized, the results were transferred to graphs, so that changes are presented in Graphs 2, 3 and 4.

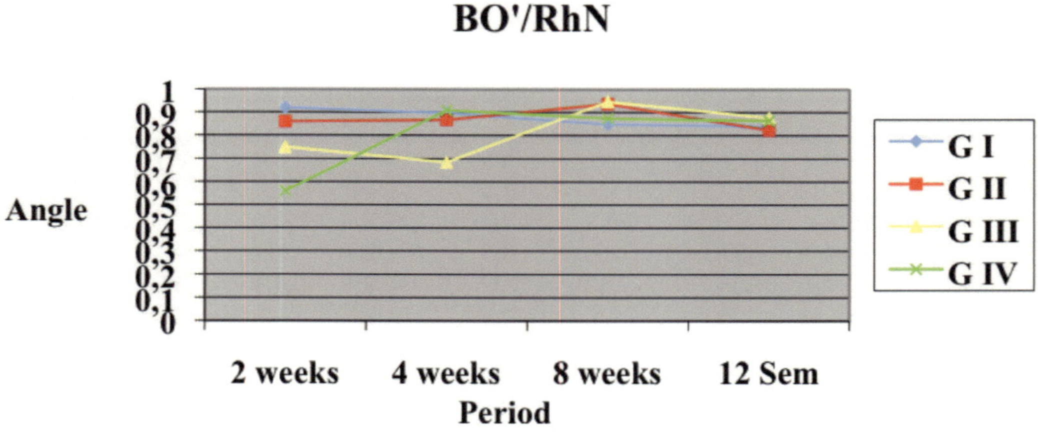

Graph 2. Graphic representation of the contextualized measurements of BO'/RhN angle.

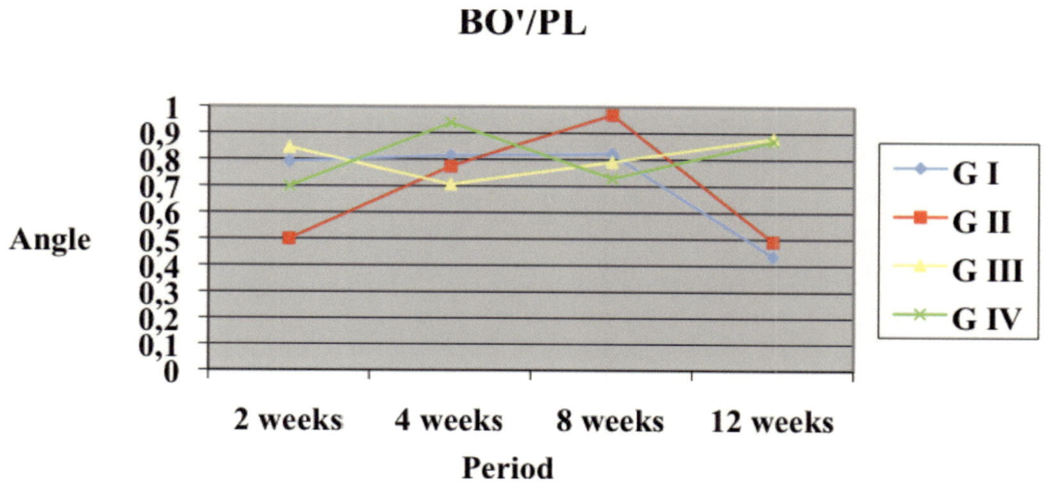

Graph 3. Graphic representation of the contextualized measurements of BO'/PL angle.

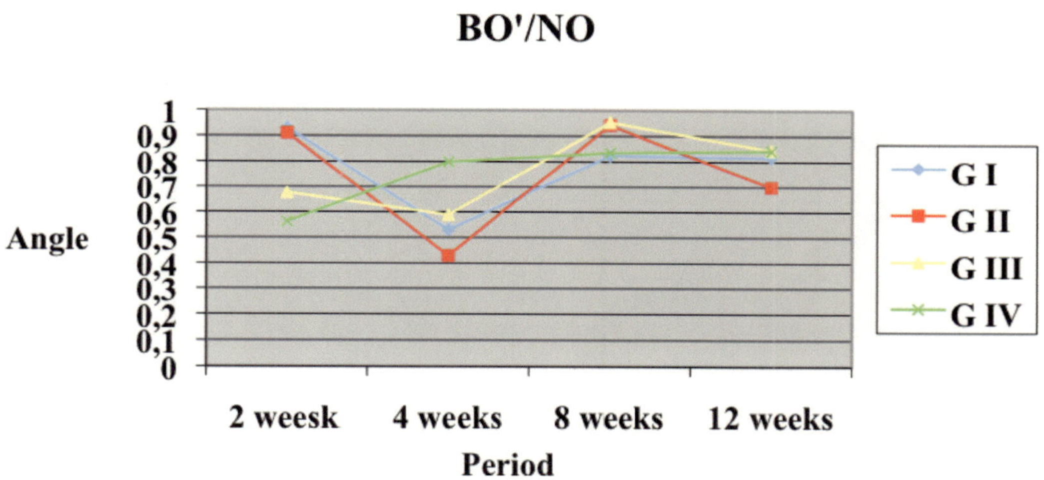

Graph 4. Graphic representation of the contextualized measurements of BO'/NO angle.

4.2 Histologic aspects

Histologic aspects:

* 2 weeks (Figure 3)

Group II: Histologic sections revealed dense, highly cellular connective tissue with new fibroblasts and newly formed vessels, in addition to moderate inflammatory mononuclear cell infiltration characterizing granulation tissue, which fills the area of bone defect. Along the edges of the defect there is intense peripheral activity, as well as initial bone deposition. Rare, small fragments of mineralized sphere-shaped tissue are observed permeating the granulation tissue.

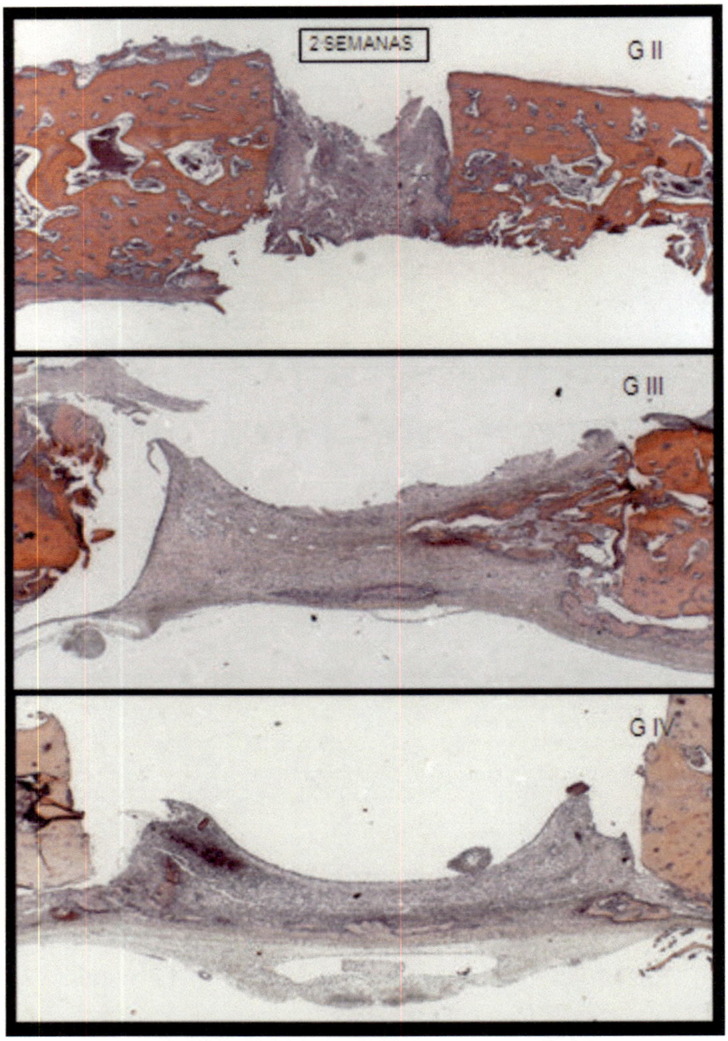

Fig. 3. Histologic sections of the skull at M3 in Groups II, III and IV after 2 weeks

Group III: The defect area was found to be totally filled with new, richly cellular connective tissue permeated by inflammatory mononuclear infiltration. Initial bone deposition may be observed in both the area of remaining bone stump and connective tissue permeating the area.

Group IV: Histologic sections revealed granulation tissue permeated by diffuse, slightly inflammatory mononuclear infiltration with many newly formed vessels showing important newly formed bone areas concentrated around one of the stumps and also in the internal area.

* 4 weeks (Figure 4)

Group II: The area of bone defect is almost totally filled with newly formed bone tissue still under intense deposition activity. The internal bone surface is found to have less new bone formation compared with the external surface.

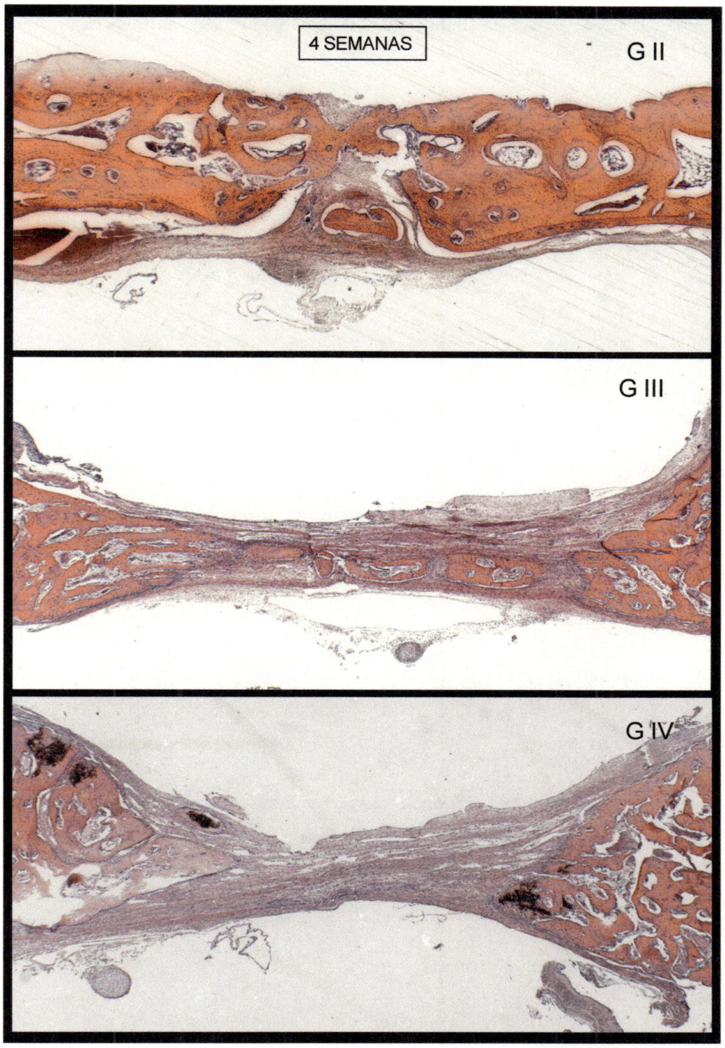

Fig. 4. Histologic sections of the skull at M3 in Groups II, III and IV after 4 weeks

Group III: The area of bone defect presents intense new bone formation in the shape of new trabeculae and spherical depositions scattered throughout the remaining granulation tissue.

Group IV: The area of bone defect is almost exclusively filled with new connective tissue. There is deposition of immature bone tissue, in the shape of trabeculae bound to the lateral stumps.

* 8 weeks (Figure 5)

Group II: Status is observed to be the same for this group at week 2.

Group III: The external area of the cortinal bone presents initial lamellar and gap formation filled with osteocytes regularly deposited. Along this cortical formation, immature trabeculae fill the total width of the defect produced. In the internal area, bone deposition in less advanced stage and new connective tissue may be observed, however with no inflammatory cell infiltration.

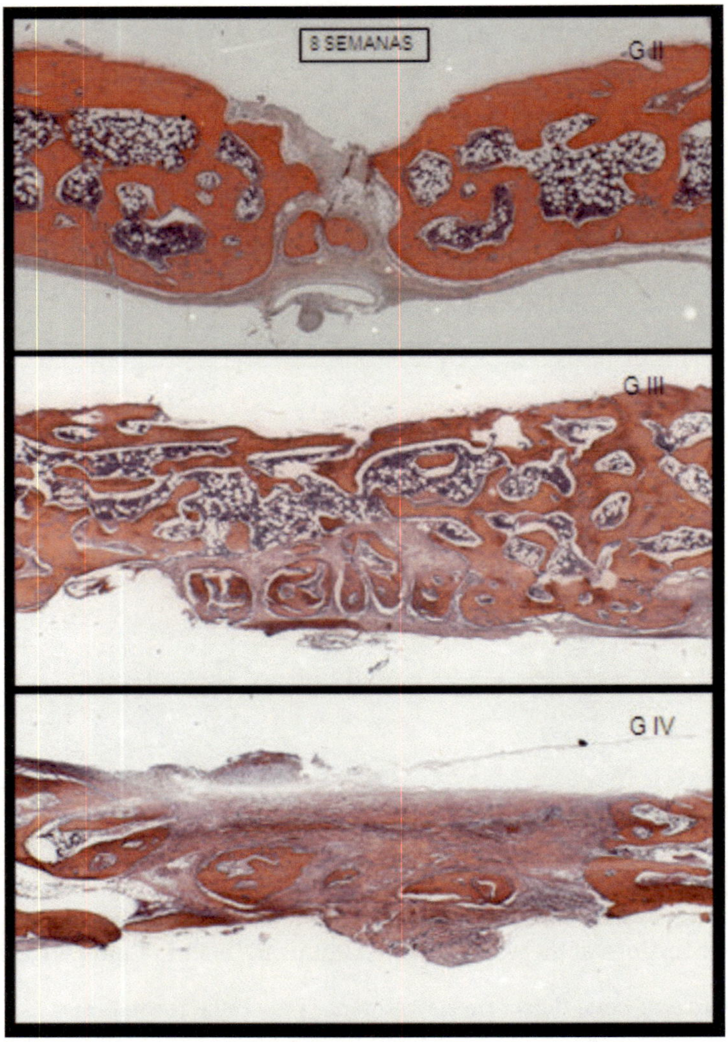

Fig. 5. Histologic sections of the skull at M3 in Groups II, III and IV after 8 weeks

Group IV: Although significant new bone formation may be observed, there is no continuous cortical presence on the external surface so far, despite the fact that the areas filled with new connective tissue are more abundant than those in group III.

* 12 weeks

Group I: The histologic analysis of the cranial vault of the animal in this group revealed normal aspect of the cortical and trabebular bone in the area.

Group II: The sections present initial bone tissue organization with sutural aspect. Bone tissue is almost totally lamellar. Hematopoietic bone marrow with normal aspect fills the gaps between trabeculae.

Group III: Status is observed to be the same for this group at week 8.

Group IV: The area of defect is almost totally filled with trabeculae of lamellar bone tissue and hematopoietic bone marrow with normal aspect. The thickness of the deposited bone is smaller than that of the remaining areas.

5. Discussion

Rabbits have been widely used as experimental model in studies of behavior of cranial sutures, intracranial pressure, craniometric variations, evolution of fixation materials, and distraction of craniofacial structures(Mabbutt and Kokich 1979; Persson, Roy et al. 1979; Alberius and Selvik 1983a; Alberius, Selvik et al. 1986; Persing, Babler et al. 1986; Alberius, Malmberg et al. 1990; Smith, Mooney et al. 1996; Singhal, Mooney et al. 1997; Losken, Mooney et al. 1998; Mooney, Siegel et al. 1998; Mooney, Siegel et al. 1998; Mooney, Siegel et al. 1999; Wendy, Fellows-Mayle et al. 2000; Gewalli, Guimaraes-Ferreira et al. 2001; Putz, Smith et al. 2001; Abreu 2002; David, Gewalli et al. 2002; Gosain, Santoro et al. 2002; David, Proffer et al. 2004; Guimarães-Ferreira, Miguéns et al. 2004; Cooper, Singhal et al. 2006; Cardoso, Cançado et al. 2007; Davis, Windh et al. 2008; Davis, Windh et al. 2009). Cerebral growth curve of rabbit specimens is similar to those of newborn human children: craniofacial growth in rabbits is accelareted immediately after birth, progressively decreasing from two to thirty-four weeks of age(Alberius, Selvik et al. 1986; Abreu 2002). These animals have perinatal brain growth, making young rabbits adequate for the study of cranial suture immobilization. Additionally, they have characteristics similiar to humans, namely the internally curved cranial base(Babler, J.A. et al. 1982; Abreu 2002). Some authors, as Wong(Wong and al. 1991), reported that a 7-week-old rabbit corresponds to a one-year-old infant. Nonetheless, the fact that rabbits were breast-fed up to week four(Cardoso, Cançado et al. 2007) limited studies from that age upwards due to difficulties in management and restrictions for releasing the rabbits from the central animal facility before they were weaned. During surgeries performed at weeks two, four, eight and twelve, a progressive thickening of the cranial vault thickness was observed as the animal aged and a higher resilience to osteotomies, considering that along the observed period of time, up to 16 weeks of life, a continuous maturation of the cranial development occurred.

The duration of anesthesia effect was completely satisfying for performing the procedures with no need for dose reinforcement in any of the cases. Surgical dissection was facilitated by vasoconstrictive action of the adrenaline used in infiltration and no hemostasis was necessary. After exposing the cranial vault, sutures were easily exposed and identified. Tissue management was facilitated with the use of self-retaining retractors developed for the procedures.

Springs were made of orthodontic wire with standard opening tension and demonstrated capacity for exerting expansion force able to promote spacing between osteotomized bone edges. Prior to the start of the experimental study, pilot surgeries were performed to test springs with thickness of various material and forms and to develop inclusion standardization using a support to immobilize springs. For this reason, no displacement or material extrusion was observed, so that in all cases springs were externally palpable during lodging period, however no pressure was caused to the soft parts, which could lead to extrusion, infection and result in the withdrawal of an individual from the experiment.

Ten Cate et al.(Ten Cate, Freeman et al. 1977), in 1977, conducted a study in rats using a spring device similar to the one used in palatal expansion as described by Haas(Haas 1970). In this study, the force applied through springs on normal sutures and expansion of bone remodeling were observed. From this study regeneration is defined as "the complete reconstitution of an injured organ or tissue with resulting complete restoration of its original archtecture" and repair as "reconstitution with new structures (scar), with variable degree of archtectural distortion" as one of the inevitable consequences. These authors additionally suggest that sutural expansion involves an injury followed by repair phenomenon, which leads to scar tissue in other tissues, nonetheless it promotes regeneration in the suture area.

The behavior of suture is compared to that of gomphosis (bone, ligament, tooth), which as fibrotic joints have a similar response to expansion techniques, i.e. regeneration after orthopedic treatment instead of scar formation or tissue repair.

These findings are compatible with the observation of cranial behavior when springs are placed in the interparietal position. However, in group IV, in which springs were positioned in a non-suture area, a significant difference in ossification would be expected, but it did not occur. The evolution was very similar in skulls in groups III and IV, that is, in suture and non-suture areas. These observations suggest that regeneration does not depend on the suture area.

The use of rabbits that did not present synostosis was not a limitation for cranial expansion under the expanding action of springs, corroborating reports from the literature(Ten Cate, Freeman et al. 1977; Persing, Morgan et al. 1991; Gewalli, Guimaraes-Ferreira et al. 2001; David, Gewalli et al. 2002; Davis, Windh et al. 2008; Davis, Windh et al. 2009). Tanaka et al. (Tanaka, Miyawaki et al. 2000; Tanaka, Miyawaki et al. 2000) published a study analyzing the local response to spring placement at the interparietal suture in healthy rats. The presented biomechanical properties validate suture response with an increase in production of type III collagen along the expansion period. Radiologic changes indicate that a cranial expansion occurred, both when changes in spacing between amalgam markers and angle changes at the cranial base are observed. Therefore, the expansible effect of springs did not appear to depend on skull physiologic changes present in craniosynostosis, as for instance in a change in intracranial pressure or responsivity of sutures adjacent to the expansible action.

During the harvesting of fragments for histopathologic analysis, springs were observed to maintain their expansible force at the moment one of their extremities was removed. Their opening was longer than the final distance obtained during expansion after ossification and returned to the distance of their original rest position. This behavior, also observed in clinical cases, suggests that the stabilization of expansion process may be determined by both the dynamic equilibrium between the pressure exerted by springs with the resistance of bone edges and tissue osteogenic action. The latter is associated with a local inflammatory process stimulated by expansion and would promote ossification between osteotomy edges even before the expansible action of springs was completed.

Hooke's law(wikipedia 2009) may be applied provided that the elastic limit of the material is not exceeded. The elastic behavior of materials comply with the elastic regime from Hooke's law up to a determined force value; from this value upwards, the relationship of proportionality cannot be defined anymore (although the body returns to its initial length after removal of the respective force). If this force continues to increase, the body loses its elasticity and deformation becomes permanent (inelasticity), leading to the rupture of the material.

The knowledge on the priniciples of spring behavior may confer an inherent property that expansion stabilization is a feature of the equilibrium reached between spring action force and bone resistance to the maintenance of a "residual" force of opening. This does not mean that expansion has necessarily ceased by ossification and consequent increase of tissue resistence under spring action, but the probability that equilibrium was established between these forces and ossification followed its natural course in the area, due to the osteogenic action of the dura mater and scalp.

In the study conducted by Persson et al.(Persson, Roy et al. 1979) a premature fusion of coronal sutures in rabbits through suture immobilization bilaterally was induced with futher removal. Changes occurred in cranial angular dimensions with anteroposterior shortening and, after a craniectomy was performed to release immobilized sutures, a return to normal patterns of measurements was observed. Also, in the control group, suture removal was shown to cause changes in measurements, however fewer than those observed in the groups with immobilized sutures. Suture removal was not followed by significant changes in angular dimensions of the skull base in nonsynostotic animals.

Accordingly, cranial expansion behavior was considered to possibly be validated in the groups of rabbits with springs. The mere fact of removing a suture does not appear to permanently change craniometric behavior. The removal of sagittal suture in groups II, III and IV was observed to alter until week 4 the total length of the cranial base with less increase relative to group I, however in remaining time points an increase regarding the latter occurred.

Mooney et al.(Mooney, Siegel et al. 1998) and Putz et al.(Putz, Smith et al. 2001) demonstrated that changes in the cranial base in intracranial pressure occur when synostotic rabbits underwent suturectomy. This is a key observation to understand skull behavior subjected to expansion because in addition to the expansile force exerted by springs, their effect is strengthened through altered intracranial pressure, which when stabilized probably accelerates the ossification process through ceasing the tensile action on bone edges.

In 1986, Persing et al.published a study with rabbits in which immobilization of the unilateral coronal suture was performed at 9 days of life, leading to cranosynostosis(Persing, Babler et al. 1986). At 60 days of age, suture osteotomy was performed in one group and the same osteotomy with expansible spring placement in another. As a result, greater expansion was observed in the group of rabbits with springs. A shortening and a distortion in cephalometric data in this group was also observed. Nonetheless, no histologic behavior was reported. Even without using a mechanism of suture contention, in the present study it was possible to observe that the expansion was followed by some singular histologic features at each time point, mainly in group III, which remained with the presence of granulation tissue and osteoblasts until week 12 with suture tissue appearance. Therefore, suture feature were maintained even after the surgical removal of bone segment corresponding to the local distracton of the bone edges.

Gewalli et al.(Gewalli, Guimaraes-Ferreira et al. 2001) report that despite the quick opening of springs regarding established bone distraction procedures(Ilizarov 1989; McCarthy, Schreiber et al. 1992), ossification occurs in the gap between bone edges separated by springs and no progressive, gradual bone distraction process is necessary. The hypothesis that a minor detachment of the dura mater may be associated with this finding is established by Mabbutt and Kokich(Mabbutt and Kokich 1979) and is clearly evidenced by Gosain et al. (Gosain, Santoro et al. 2002) who also observed that the defect created in adult animals was maximized by maintaining the contact with the pericranium (bone growth from peripheral areas) and dural contact (bone growth from depth). Similar finding is observed in Groups III and IV (Figure 4).

Although the pericranium was not approximated, bone deposition occurred from both bone edges and isolated bone islands in the lower part next to the dura mater, demonstrating a multicenter process of ossification. At first, at week two, all groups presented characteristics of inflammatory tissue and immature bone pattern. During the first four postoperative weeks an increased deposition activity occurred with further stabilization and also synchronism in the maturation of newly formed tissue, in which group III presented earlier new bone formation compared with group II and IV. After 12 weeks uniform bone regeneration occurred in the expanded area in groups III and IV, but the total thickness of the vault was smaller than that in group II.

Morphometric measurement methods are not comparable among themselves(Alberius P 1986; Alberius, Malmberg et al. 1990; Abreu 2002) to the extent that comparisons of various studies are not possible, however conclusions drawn from a specific project might be valid(Alberius P 1986). Cranial expansion evolution is indirectly affected by behavior, from the influence of minimum trauma, such as amalgam placement in cranial expansion(Alberius and Selvik 1983a), variation in body weight(Alberius P 1986), changes in growth vector of adjacent sutures and cranial base(Losken, Mooney et al. 1998; Davis, Windh et al. 2008), among others. The difference of expansion among the groups of rabbits with springs was minimal, but expansion was observed to be greater in both groups compared to that without springs, suggesting an action of cranial expansion caused by this device and indicating a similarity between the group with springs in the interparietal and parasagittal areas, even though suture was maintained in the latter.

In order to evaluate cranial expansion from the distance between amalgam markers, data collected from the measurement of the spacing between markers at the M3 position were considered to be the most representative, since they were positioned next to the spring in a site of great resistance and due to the section of histopathologic studies.

A conventional cephalometric analysis enabled the comparison of measurements from an individual with a pattern. Such information at best suggests the degree of deviation from normality for that particular variable.

A better scenario was to know how much the value of a given variable for a specific animal was deviated regarding its similar norm (Persing, Babler et al. 1986), however in a contextualized and non-individualized manner. In other words, measuring how much a suggested deviation occurred in practical terms, a significant deviation for that particular situation. In reality, the difference between these two situations only reflects a limitation in the practicality of conventional mathematical tools applied to cephalometry(Mario 2006).

With the knowledge-based system, the degree of abnormality evidence was evaluated for each value measured considering the general context in which this value was entered into. By measuring the degrees of contextualized evidence, expansion behavior may be identified.

Future perspectives:

The descriptive studies may be populational or individual, as reports of cases and case series (Dos Reis, Ciconelli et al. 2002). The data obtained through descriptive studies serve as basis for formulating hypotheses for future studies. When hypotheses are formulated, analytical studies are necessary for their confirmation or rejection. In this observational study some hypotheses could be formulated, such as:

- The ossification of the expanded cranial area occurred through local osteogenic action and not through gradual distraction of bone edges.
- The cranial expansion with springs is directly related to the interaction among tissues at the site and not to a craniosynostosis-dependent phenomenon.
- The expansible spring action is directly related to its force.

The findings corroborated the clinical results obtained so far, however the applicability of springs of non-suture areas must be better investigated regarding spring action on cranial growth vectors.

Bone distraction is a well-established phenomenon in medicine. Various designs of bone distraction devices have been presented, not only for treating craniofacial abnormalities but also for other areas. In orthopedics, for instance, there are various types of devices for enlongating long bones, correcting vertebral scoliosis, changing rib cages, and others. The use of springs is widely diffused. There is research for determining both the amount of material and its best shape. There are possibilities, including springs made of absorbable material, what would reduce even more procedure morbidity because there would be no second surgery for material removal.

6. Conclusions

Spring-mediated cranial expansion in rabbits, through the observation of the results of cephalometric and histopathologic behavior present the following features:

- Cranial expansion presents similar evolution patterns both when the device is used directly in the region of the removed suture and also when springs are used in the parasagittal region for maintaining sagittal suture. In both cases, changes are greater than those occurred with the simple removal of the sagittal suture.
- Bone regeneration fills expansion areas due to regenerative stimulating activity induced by maintaining atachment of the dura mater and maintaining the contact with the pericranium.

7. References

(1984). *Nomina Anatomica*. Rio de Janeiro.

Abreu, A. (2002). Análise de métodos de aferição morfométrica empregados para avaliação do crescimento craniofacial: estudo em coelhos. *Faculdade de Odontologia*. Porto Alegre, PUCRS. Doutorado: 230.

Alberius, P., M. Malmberg, et al. (1990). "Variability of measurements of cranial growth in the rabbit." *J Craniofac Surg* 188: 393-400.

Alberius, P. and G. Selvik (1983a). "Roentgen sterophotogrammetric analysis of growth at cranial vault sutures in the rabbit." *Acta Anat* 117: 10.

Alberius, P., G. Selvik, et al. (1986). "Postnatal sutural growth of the calvarium: a survey with special reference to the rabbit." *Gegenbaurs Morphol Jahrb* 132(4): 519-534.

Alberius P, S. G., Fahlman MT (1986). "Postnatal Sutural Growth of the Calvarium: A Survey With Special Reference to the Rabbit." Gegenbaurs Morphol Jahrb 132(4): 519-534.

Arnaud, E., D. Marchac, et al. (2006). "Le traitement des craniosténoses: indications et techiniques." *Neurochirurgie* 52(2-3): 264-291.

Babler, W. J., P. J.A., et al. (1982). "Skull growth after coronal suturectomy, periostectomy, and dural transection." *J Neurosurg* 56(4): 529-535.

Cardim, V. L., R. F. V. Dornelles, et al. (2002). "Distracción Elástica del Tercio medio en la Disyunción Craneo-Facial." *Cir Plást Iberlatinamer* 28(4): 267-275.

Cardoso, E. S., R. P. Cançado, et al. (2007). Estudo exploratório descritivo do emprego de coelhos brancos da Nova Zelândia (*Orytolagus cuniculus*) como modelo animal na avaliação dos padrões de crescimento craniofacial. *Rev Odonto Cienc* Porto Alegre, Fac. Odonto/PUCRS. 22.

Cooper, G. M., V. K. Singhal, et al. (2006). "Intracranial volume changes in craniosynostotic rabbits: effects of age and surgical correction." *Plast Reconstr Surg* 117: 1886-1890.

David, L. R., F. Gewalli, et al. (2002). "Dynamic spring-mediated cranioplasty in a rabbit model." *J Craniofac Surg* 13(6): 794-801.

David, L. R., P. Proffer, et al. (2004). "Spring-mediated cranial reshaping for craniosynostosis." *J Craniofac Surg* 15: 810-816.

Davis, C., P. Windh, et al. (2008). "Spring-assisted cranioplasty alters the growth vectors of adjacent cranial sutures." *Plast Reconstr Surg* 123(2): 470-474.

Davis, C., P. Windh, et al. (2009). "Do expansile cranial springs erode through the cranium?" *J Craniofac Surg* 20(1): 168-170.

Dornelles, R. F. V., V. L. N. Cardim, et al. (2007). The Use of implantable springs in the dynamic expansion to cranyosinostosis treatment: 17 cases analysis. XII International Congress of the International Society of Craniofacial Surgery. . Bahia, Brasil, Abstract book: 64.

Dos Reis, F. B., R. M. Ciconelli, et al. (2002). "Pesquisa científica: a importância da metodologia." *Rev Bras Ortop* 37: 51-55.

Gault, D. T., D. Renier, et al. (1992). "Intracranial pressure and intracranial volume in children with craniosynostosis." *Plast Reconstr Surg* 90(3): 377-381.

Gewalli, F., J. P. S. Guimaraes-Ferreira, et al. (2001). "Expander elements in craniofacial surgery: an experimental study in rabbit." *Scand J Plast Reconstr Surg Hand Surg* 35(2): 149-156.

Gosain, A. K., T. C. Santoro, et al. (2002). "Osteogenesis in calvarial defects: contribution of the dura, the pericranium, and the surrounding bone in adult versus infant animals." *Plast Reconstr Surg* 112(2): 12.

Greensmith, A. L., A. D. Holmes, et al. (2008). "Complete correction of severe scaphocephaly: the melbourne method of total vault remodeling." *Plast Reconstr Surg* 121(4): 1300-1310

Guimarães-Ferreira, J., J. Miguéns, et al. (2004). "Advances in cranioysinostosis research and management." *Adv Tech Scand Neurosurg*: 29.

Haas, A. J. (1970). "Palatal expansion: just the beginning of dentofacial orthopedics." *Am J Orthod* 57: 219.

Heller, J. B., M. M. Heller, et al. (2008). "Intracranial volume and cephalic index outcomes for total calvarial reconstruction among nonsyndromic sagittal synostosis patients." *Plast Reconstr Surg* 121(1): 187-195

Hönig, J. F., U. A. Grohmann, et al. (2001). "Facial bone distraction osteogenesis for correction of malocclusin: a more than 70-year-old concept in craniofacial surgery." *Plast Reconstr Surg* 109(1): 41-44.

Ilizarov, G. A. (1989). "The tension-stress effect on the genesis and gowth of tissues: Part II. The influence of the rate and frequency of distraction." *Clin Orthop* 239: 22.

Imai, K., H. Komune, et al. (2002). "Cranial remodeling to treat craniosynostosis by gradual distraction using a new device." *J Neurosurg* 96: 654-659.

Lauritzen, C., Y. Sugawara, et al. (1998). "Spring mediated dynamic craniofacial reshaping. Case report." *Scand J Plast Reconstr Surg Hand Surg* 32(3): 331-338.

Lauritzen, C. and P. Tarnow (2003). "Craniofacial surgery over 30 years in Goteborg." *Scand J Surg* 92(4): 274-280.

Lauritzen, C. G. K., C. Davis, et al. (2008). "The evolving role of springs in craniofacial surgery: the first 100 clinical cases." *Plast Reconstr Surg* 121(2): 545-554

Losken, H. W., M. P. Mooney, et al. (1998). "Internal calvarial bone distraction in rabbits with delayed-onset coronal suture synostosis." *Plast Reconstr Surg* 102(4): 1109-1119.

Mabbutt, L. W. and V. G. Kokich (1979). "Calvarial and sutural re-development following craniectomy in the neonatal rabbit." *J Anat* 129(2): 9.

Mackenzie, K. A., C. Davis, et al. (2009). "Evolution of surgey for sagittal synostosis: the role of new technologies." *J Craniofac Surg* 20(1): 29-133.

Marchac, D. and D. Renier (1979). ""Le front flottant": traitement precoce des facio-cranioestenoses." *Ann Chir Plast* 24: 121121.

Mario, M. C. (2006). Modelo de análise de variáveis craniométricas através das Redes Neurais Artificiais Paraconsistentes. Faculdade de Medicina. São Paullo, Universidade de São Paulo. Doutor em Ciências: 176.

McCarthy, J. G., F. J. Epstein, et al. (1990). *Craniosynostosis. Plastic Surgery*. J. McCarthy. New York, Saunders Company. 4: 3013-3053.

McCarthy, J. G., J. Schreiber, et al. (1992). "Lengthening the human mandible by gradual distraction." *Plast Reconstr Surg* 89(1): 1-8.

McCarthy, J. G., E. J. Stelnicki, et al. (2001). "Distraction osteogenesis of the craniofacial skeleton." *Plast Reconstr Surg* 107(7): 1812-1827.

Molina, F. and F. Ortiz-Monasterio (1995). "Mandibular elongation and remodeling by distraction: a farewell to major osteotomies." *Plast Reconstr Surg* 96(4): 825-840.

Mooney, M., M. I. Siegel, et al. (1998). "A rabbit model of human familial, nonsyndromic unicorornalsuture synostosis I. Synostotic onset, pathology, and sutural growth patterns." *Childs Nerv Syst* 14: 236-246.

Mooney, M. P., M. I. Siegel, et al. (1998). "A rabbit model of human familial, nonsyndromic unicoronal suture synostosis II. Intracranial contents, intracranial volume, and intracranial pressure." *Childs Nerv Syst* 14: 247–255.

Mooney, M. P., M. I. Siegel, et al. (1999). "Increased intracranial pressure after coronal suturectomy in craniosynostotic rabbits." *J Craniofac Surg* 10(2): 104-106.

Ortiz-Monasterio, F., A. Fuente del Campo, et al. (1978). "Advancement of the orbits and the midface in one piece, combined with frontal repositioning for the correction of Crouzon's deformities." *Plast Reconstr Surg* 61(4): 507.

Ortiz-Monasterio, F. and F. Molina (1994). "Mandibular Distraction in Hemifacial Microsomia." *Operative Techniques in Plastic and Reconstructive Surgery* 1(2): 105-112.

Panchal, J. and V. Uttchin (2003). "Management of Craniosynostosis." *Plast Reconstr Surg* 111(6): 2032-2049.

Persing, J. A., W. J. Babler, et al. (1986). "Skull expansion in experimental craniosynostosis." *Plast Reconstr Surg* 78(5): 594-603.

Persing, J. A., J. A. Jane, et al. (1989). "Virchow and the pathogenesis of craniosynostosis: a translation of his original work." *Plast Reconstr Surg* 83(4): 738-742.

Persing, J. A., E. P. Morgan, et al. (1991). "Skull base expansion: craniofacial effects." *Plast Reconstr Surg* 87(6): 1028-1033.

Persson, K. M., W. A. Roy, et al. (1979). "Craniofacial growth following experimental craniosynostosis and craniectomy in rabbits." *J Neurosurg* 50: 10.

Putz, D. A., T. D. Smith, et al. (2001). "Cranial base changes following coronal suturectomy in craniosynostotic rabbits." *Orthod Craniofac Res* 5: 13.

Pyle, J., S. Glazier, et al. (2009). "Spring-assisted surgery - a surgeon's manual for the manufacture and utilization of springs in craniofacial surgery." *Journal of Craniofacial Surgery* 20(6): 1962-1968.

Renier, D., M. Le Merrer, et al. (2006). "Étiologie des craniosténoses." *Neurochirurgie* 52(2-3): 228-237.

Schnaider, T. B. and C. De Souza (2003). "Aspectos Èticos da experimentação animal." *Rev Bras Anestesiol* 53: 278-285.

Singhal, V. K., M. P. Mooney, et al. (1997). "Age related changes in intracranial volume in rabbits with craniosynostosis." *Plast Reconstr Surg* 100(5): 1121-1128.

Slater, B. J., K. A. Lenton, et al. (2008). "Cranial sutures: a brief review." *Plast Reconstr Surg* 121(4): 170e-178e

Smith, T. D., M. P. Mooney, et al. (1996). "Posnatal changes in the cranial base in rabbits with congenital coronal suture synostosis." *J Craniofac Genet Dev Biol* 16: 107 117.

Snyder, C. C., G. A. Levine, et al. (1973). "Mandibular Lengthening by Gradual Distraction: Preliminary Report." *Plast Reconstr Surg* 51(5): 506-508.

Tanaka, E., Y. Miyawaki, et al. (2000). "Changes in the biomechanical properties of the rat interparietal suture incident to continuous tensile force appilication." *Arch Oral Biol* 45: 1059-1064.

Tanaka, E., Y. Miyawaki, et al. (2000). "Effects of tensile forces on te expression of type III collagen in rat interparietal suture." *Arch Oral Biol* 45: 1049-1057.

Ten Cate, A., E. Freeman, et al. (1977). "Sutural development: structure and its response to rapid expansion." *Am J Orthod* 71(6): 14.

Tessier, P. (1971). "The definitive plastic surgical treatment of the severe facial deformities of craniofacial dysostosis: Crouzon's and Apert's diseases." *Plast Reconstr Surg* 48(5): 419-442.

Tubino, P. and A. Alves. (2009). "História da Medician." from http://www.unb.br/fm/hismed/arquivos/medicina_pre-colombiana.pdf.

Tung, T. H., B. R. Robertson, et al. (1999). "Successful distraction osteogenesis across a growing cranial suture without an osteotomy." *Plast Reconstr Surg* 103(2): 362-370.

Wendy, K., M. A. Fellows-Mayle, et al. (2000). "Age-related changes in intracranial pressure in rabbits with uncorrected familial coronal suture synostosis." *Cleft Palate Craniofac J* 37(4): 370-378.

Wikipedia. (2009). from http://pt.wikipedia.org/wiki/Mola.

Wong, L. and e. al. (1991). "The effect of rigid fixation on growth of the neurocranium." *Plast Reconstr Surg* 88(3): 8.

10

The Use of Cancellous Bone-Block Allograft for Reconstruction of the Atrophic Alveolar Ridge

Gavriel Chaushu and Joseph Nissan
Tel Aviv University, Tel Aviv,
Rabin Medical Center,Beilinson Hospital, Petach Tikva,
Israel

1. Introduction

The remarkable success of the traditional Brånemark implant protocol (Adell et al., 1981; Albrektsson et al., 1981; Branemark, 1983) revolutionized dentistry. The presence of uncompromised bone of adequate volume at the implant site is a major factor in the functional success of the procedure. In addition, by providing predictable support both for the implant itself and for the gingival margin and papillae, it contributes to a pleasing esthetic outcome (Belser et al., 2004; Grunder et al., 2005; Palacci & Ericsson, 2001). A good functional result is also impossible without proper biomechanics, as occlusal overload by the final prosthesis can lead to biological or mechanical complications (Rangert et al., 1995). The forces applied along the axis of the implant are distributed around the implant, resulting in high load-bearing on the peripheral supporting bone, especially in the premolar–molar region of the jaws. Furthermore, in the anterior maxillary area, the forces are exerted in significantly transverse direction which leads to a bending momentum that may adversely affect both the implant and the supporting tissues (Rangert et al., 1989). Narrow implants, implants tilted buccally, and implants with oversized clinical crowns can all be detrimental (Hsu et al., 2007).

Therefore, augmentative bone surgery is frequently a prerequisite of implant placement (Belser et al., 2004; Grunder et al., 2005; Palacci & Ericsson, 2001; Sethi & Kaus, 2001). Unlike the sinuses, the alveolar ridge does not provide a natural cavity to contain particulated grafting material (Chaushu et al., 2009); therefore, the graft must be have sufficient strength and rigidity to fixate at the recipient site and three-dimensional stability to withstand muscular forces (Moy & Palacci, 2001). Autogenous bone harvested from either extraoral or intraoral sites is regarded as the "gold standard" by some authors (Lundgren et al., 2008), and it remains the material of choice for cortical-cancellous blocks (D'Addona & Nowzari, 2001; Hunt & Jovanovic, 1999; Misch et al., 1992; Verhoeven et al., 1997). However, its use is limited by risks of donor site morbidity: immediate postoperative pain and edema, infections, hematomas, and neurosensory deficits (Nkenke et al., 2001; Nkenke et al., 2002; Raghoebar et al., 2001). A variety of alternative allogeneic, alloplastic and xenogeneic bone-grafting materials have been proposed in recent years, based on wound-healing mechanisms and bone regeneration principles, such as tissue engineering, and the osteoinductive and osteoconductive potential of different scaffolds (McAllister & Haghighat, 2007). Although cancellous bone-block allografts have been used for alveolar ridge augmentation with clinical success (Keith Jr., 2004; Keith Jr., et al., 2006; Leonetti & Koup, 2003; Lyford et al., 2003; Nissan et al., 2008; Petrungaro & Amar, 2005), there is still lack of histological evidence regarding the biological healing process.

The aim of this chapter is to describe the technique and long-term (6 years or more) outcome of dental implants placed in the atrophic alveolar process following augmentation with freeze-dried cancellous block allografts. The histological and histomorphometric findings are reported as well. The studies were performed at the Tel Aviv University in 2005 to 2011.

2. Procedure and clinical outcome by anatomical area

2.1 Anterior maxilla

2.1.1 Surgical technique

Block grafts were indicated for bony deficiencies measuring at least 3 mm horizontally and up to 3 mm vertically on computed tomography (CT) para-axial reconstruction, as recommended (Chiapasco et al., 1999). Surgery was performed under local anesthesia. The shape and size of the defect on preoperative imaging were corroborated at the recipient site (Figs. 1-3).

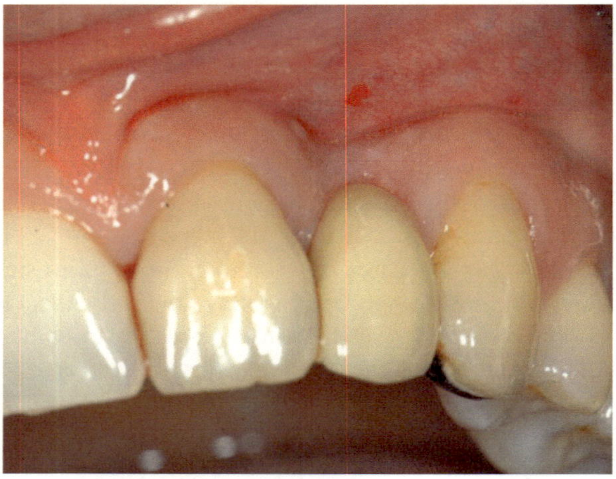

Fig. 1. Anterior maxilla. Preoperative clinical view.

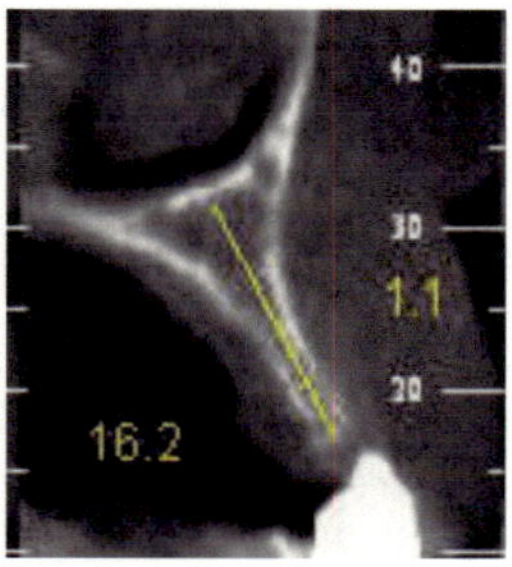

Fig. 2. Anterior maxilla. Preoperative CT.

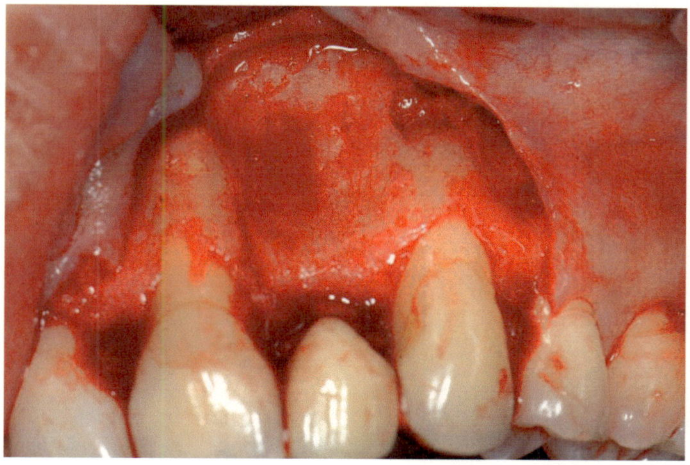

Fig. 3. Anterior maxilla. Clinical view demonstrating bone defect.

A vasoconstrictor (1:100000) was administered subperiosteally for easier separation of the bone and periosteum, and surgery was initiated about 10 minutes later. A midcrestal incision was made on the basis of the missing teeth and extended intrasulcularly around the cervical margins of the adjacent teeth up to the canines. Two releasing incisions were made on the labial aspect distal to the canines, including the papilla between the canine and the first premolar and extending away from the esthetic zone into the mobile mucosa. This exposed the buccal aspect of the alveolar ridge so the defect could be visualized directly in three dimensions. Because these maneuvers are accompanied by excessive bleeding, periosteal releasing incisions were made already at this stage, with at least 1 cm of overlapping tissue, so the site could be primarily closed under clear direct vision without the need for hemostasis.

Several techniques were used to ensure the broadest communication possible between the grafted bone and the bone marrow cavity. In the presence of noticeable cortical bone, we made multiple perforations through the cortical plate with a round bur. If the cortical bone was dense, decortication was performed. No additional preparation was used in cases presenting after trauma or surgery without evident cortex and profound bleeding.

Freeze-dried cancellous block-allografts were rehydrated in a solution of sterile saline for 45 minutes. A high-speed water-cooled fissure bur, in a handpiece was used to shape the graft to approximate the recipient bed and provide sufficient height and width. The graft was then thoroughly rinsed with sterile saline to remove residual bone particles.

The width and height of the ridge were measured with periodontal probes scaled in millimeters, and the cancellous block-graft was fitted into the defect. Once it was seated and stable, it was fixed with 1.6 mm x 10 mm bone screws (Fig. 4).

The bur was reapplied to round any sharp cortical edges and shape the block to completely conform to the defect site, and width and height measurements were repeated. Deficiencies at the edges of the graft were filled with particulate bone. The graft was covered with long-term resorbable collagen membrane (Fig. 5).

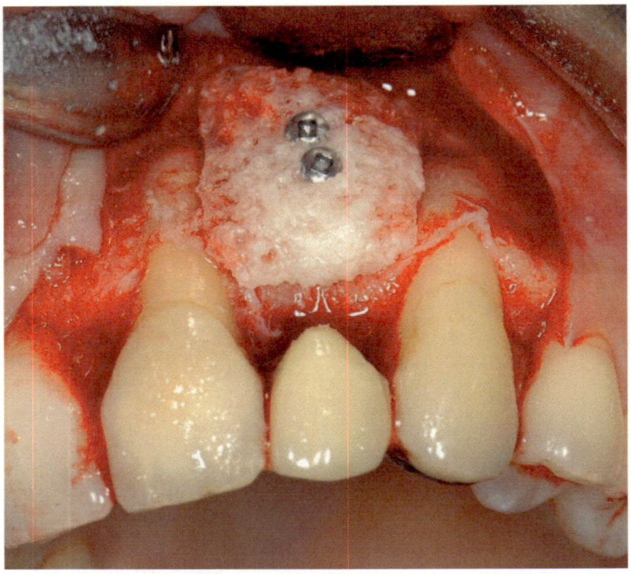

Fig. 4. Anterior maxilla. Bone block fixation.

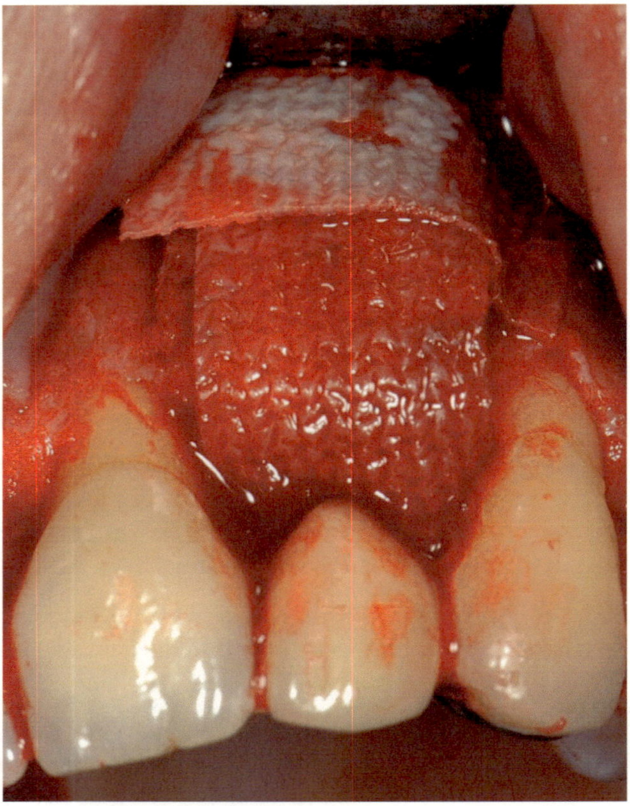

Fig. 5. Anterior maxilla. Membrane coverage.

The midcrestal incision was initially closed with interrupted or horizontal mattress sutures, as needed, and the interdental papillae and vertical incisions were secured with interrupted sutures. Patients were prescribed oral antibiotics and analgesics for 5 days postoperatively and antiseptic solution for 2 weeks. Provisional restorations were modified to prevent pressure on the healing tissues and fitted and delivered to the patient immediately after surgery.

The grafted sites were allowed to heal for 6 months. Patients were seen weekly during the first postoperative month and monthly thereafter until second-stage surgery. Each visit included a thorough clinical search for soft tissue dehiscence or color change and an overall view of the grafted ridge contour. Periapical radiographs were taken immediately after surgery and again before implant placement to corroborate graft incorporation.

For the second-stage surgery, access to the augmented ridge was obtained via incision, similar to the one used during graft placement. Surgical exposure revealed a well-integrated block graft into the surrounding bone. The fixation screws were removed, and the post-augmentation ridge was measured again to confirm bone gain. The dental implant was then placed (Fig. 6).

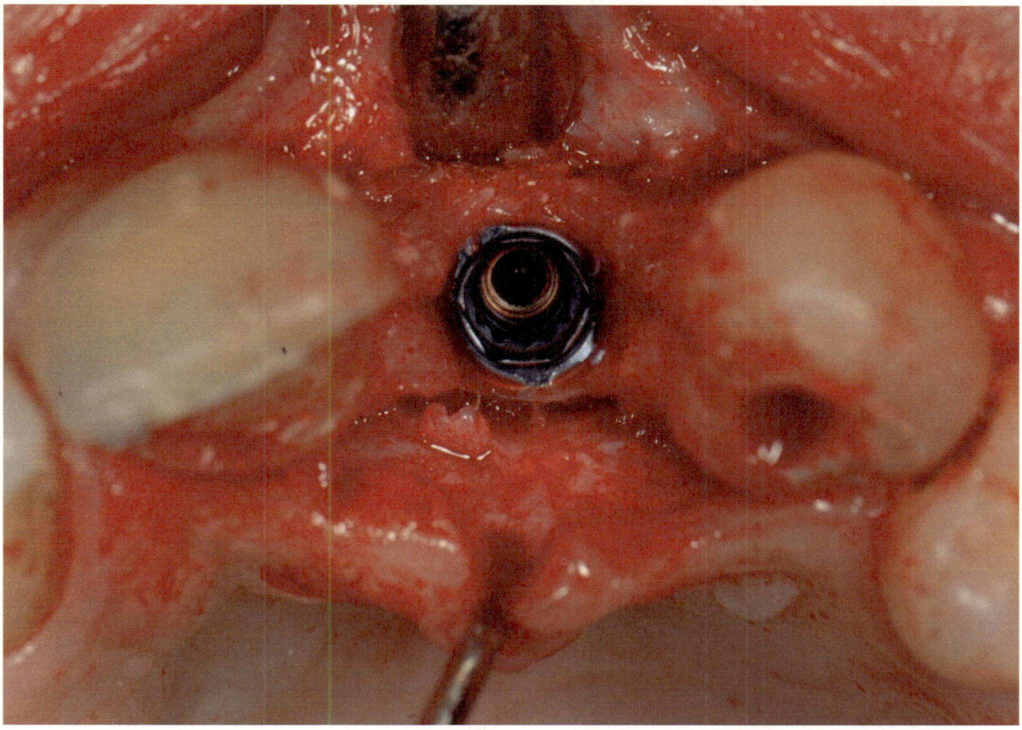

Fig. 6. Anterior maxilla. Implant placement 6 month after bone grafting.

The residual buccal thickness was measured to assess bone resorption and buccal bone dimensions. The soft tissues were allowed to mature for 3 weeks following implant exposure (Fig. 7).

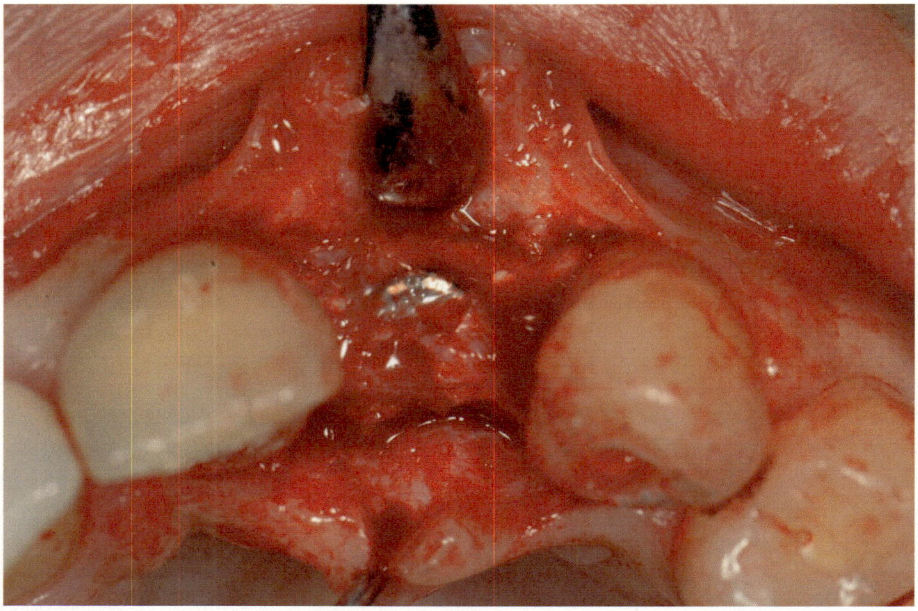

Fig. 7. Anterior maxilla. Soft tissues at second-stage surgery.

A cement-retained fixed-cement prosthesis was fabricated, and the implants were restored (Fig. 8).

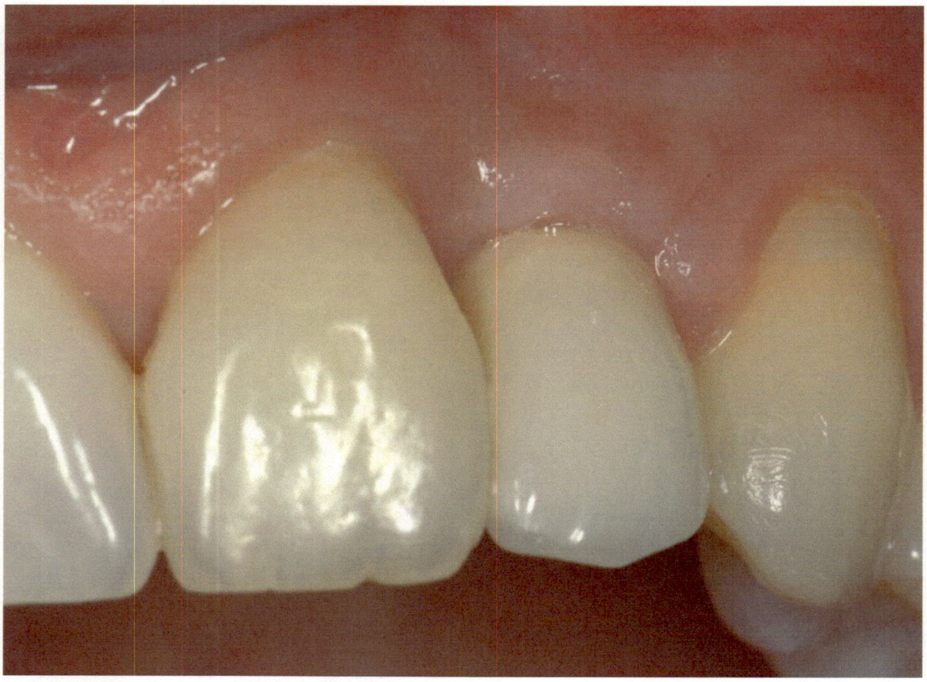

Fig. 8. Anterior maxilla. Final restoration.

Clinical and radiographic examinations were carried out at the time of restoration, every 6 months during the first year, and once a year thereafter (Fig. 9).

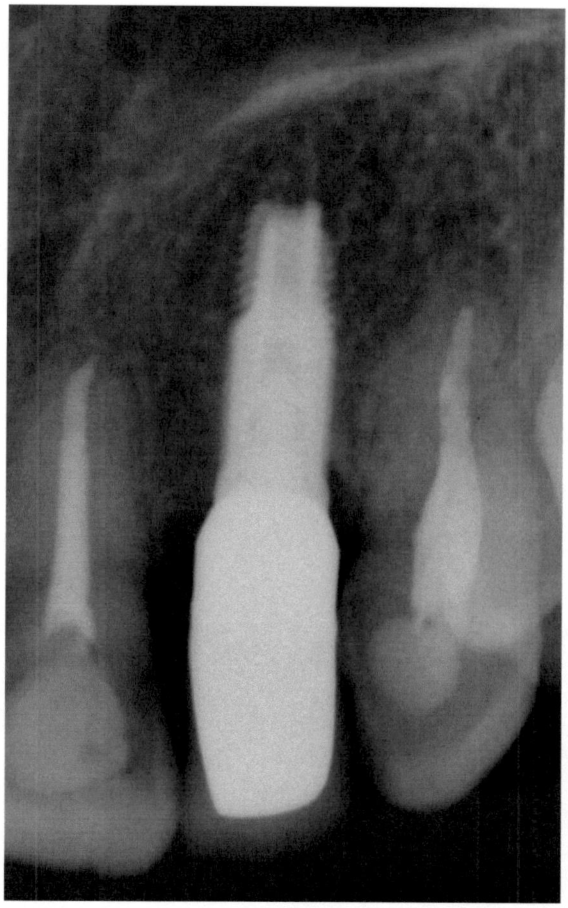

Fig. 9. Anterior maxilla. Periapical view at 4-year follow-up.

2.1.2 Clinical outcome

Use of this technique by our group yielded a greater bone gain in the horizontal dimension (5 ± 0.5 mm) than in the vertical dimension (2 ± 0.5 mm) (Nissan et al., 2011a). The buccal bone resorption rate was 0.5 ± 0.5 mm at implant placement and 0.2 ± 0.2 mm at the second-stage surgery; bone thickness buccal to the implant neck was 2.5 ± 0.5 mm at implant placement and 2.3 ± 0.2 mm at second-stage surgery. There was no evidence of vertical bone loss between the two time points. All patients received a fixed implant-supported prosthesis Two bone grafts failed, for a survival rate of 95.6%. One implant failed, caused by a vehicular accident. The implant was reinserted after 3 months and successfully osseointegrated (98% survival rate). All implants remained clinically osseointegrated at the end of the follow-up. There was no crestal bone loss around the implants beyond the first implant thread.

2.1.3 Comment

The use of dental implants in the anterior maxilla is well documented in the literature, and numerous controlled clinical trials report high overall implant survival and success rates (91.1% to 100%) (Belser et al., 2004). The bone volume in the anterior maxilla is essential from an esthetic perspective (Grunder et al., 2005): To achieve a satisfactory long-term result, the available bone thickness buccal to the implant neck should be at least 2 mm and preferably 4 mm (Spray et al., 2000). When this is not taken into consideration, the buccal bone may resorb, resulting in loss of buccal bone height followed by gingival recession. Since such bony thickness dimensions cannot be found routinely on the buccal side, augmentation procedures are indicated in almost every esthetically demanding case. Thus, even if the entire implant bony envelope is intact without thread exposure, bone grafting will be needed (Grunder et al., 2005). We found that a 2-3 mm buccal bone thickness was maintained at the second-stage surgery when cancellous block-allografts were used.

2.2 Posterior maxilla

2.2.1 Surgical technique

Cancellous block-allografts were indicated for posterior atrophic maxillas of ≤4 mm (Fig. 10).

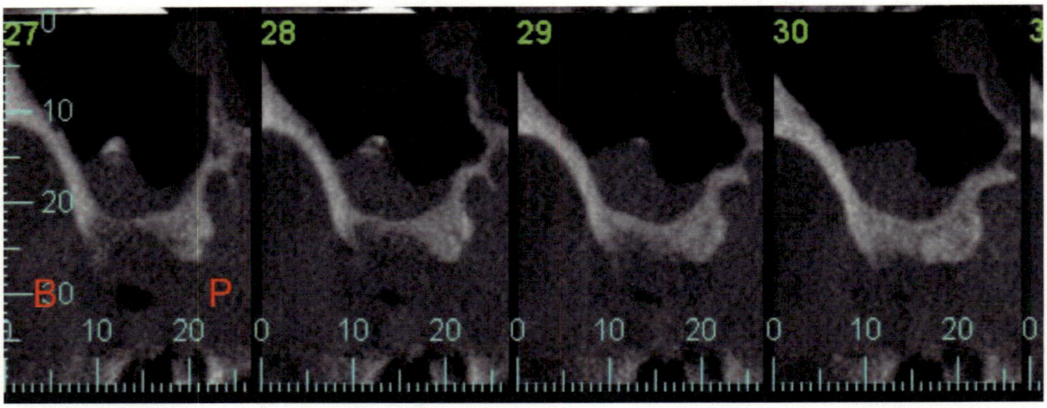

Fig. 10. Posterior maxilla. Preoperative CT demonstrating <4 mm residual alveolar ridge.

Surgery for sinus floor augmentation was performed under local anesthesia. Oral antibiotics and analgesics were administered one hour before surgery, and antiseptic mouthwash was used immediately before surgery. Two vertical releasing incisions were made to define a mucoperiosteal buccal flap extending up to the mucogingival line. The flap was raised to expose the labial bony antral wall, and a round high-speed bur with copious irrigation was used to establish a buccal window (trapezoid, rectangular or oval) for complete visibility of the entire block and the implants (Fig. 11). The membrane was then released without tension to provide an adequate compartment for the implants and the block graft. Membrane tears, if present, were left untreated.

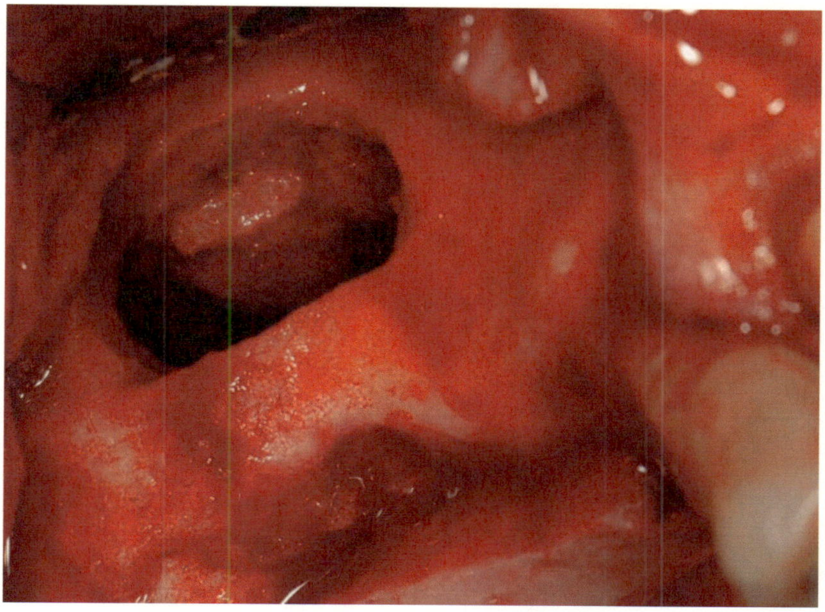

Fig. 11. Posterior maxilla. Lateral window.

Prior to graft insertion, particulate grafting material was placed against the medial and crestal aspects of the compartment created in the sinus cavity to completely adapt the block to the medial wall and floor of the sinus cavity (Fig. 12).

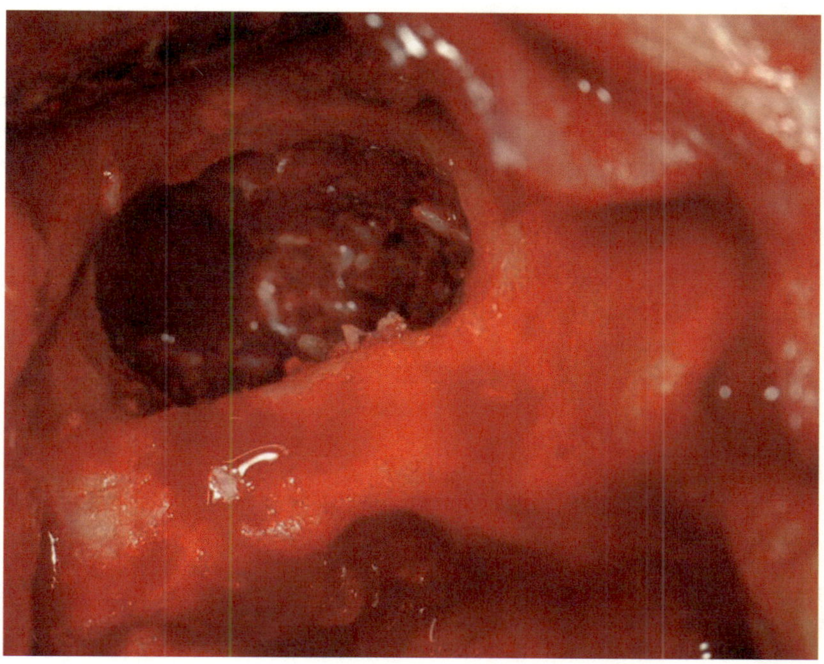

Fig. 12. Posterior maxilla. Particles prior to block insertion.

The prefabricated block (Fig. 13), soaked in sterile saline, was trimmed to adjust it to the lateral opening, inserted in a gentle-press-fit fashion into the prepared area in the sinus cavity, and pushed maximally up to the palatal wall.

Fig. 13. Posterior maxilla. Block graft following trimming.

The block was stabilized with the window frame. The implant sites were marked with a surgical stent, and osteotomies were performed according to the manufacturer's recommendation. The implants were inserted into the osteotomy sites prepared in the alveolar crest and grafted cancellous block. This locked the block to the alveolar crest on all sides, preventing its migration or a change in its position. Excessive cancellous block graft protruding lateral to the sinus bony walls was removed with a high-speed bur, and particulate grafting material was placed into any remaining voids between the block graft and the window frame. The lateral window was covered with a long-term bioresorbable collagen membrane, and the mucoperiosteal flap was closed primarily over the graft and implants.

Patients were prescribed systemic antibiotics and analgesics for 10 days postoperatively and antiseptic mouthwash twice daily for 2 weeks. They were seen weekly for the first postoperative month and monthly for the next 8 months until soft- tissue healing was complete. At 9 months, surgical re-entry was performed for implant exposure. Healing abutments were placed, and 4-6 weeks were allowed for soft tissue maturation. Impressions were made and master casts fabricated. The implants were restored with fixed partial metalloceramic.

2.2.2 Clinical outcome

All implants showed primary stability. Relatively small (5-10 mm) membrane tears were detected in 21.4% of cases, but were not treated owing to the use of the block graft. No clinical or radiological complications were recorded in any of the sinuses, including those with membrane tears, during the 9 months' healing period.

Radiographic evaluation (panoramic and periapical) prior to exposure of the implants showed what appeared to be a dense homogenous bony mass embedded in a radio-opaque area. On repeated radiographic images at the last follow-up, mean height of the augmented bone within the sinus measured 12.3 mm; no adverse reactions were noted. Implant survival rate was 94.4%. Failing implants were removed and replaced after 3 months of spontaneous healing, and all successfully osseointegrated. All patients received a fixed implant-supported prosthesis. There was no crestal bone loss around the implants beyond the first implant thread (Fig. 14) (Chaushu et al., 2009).

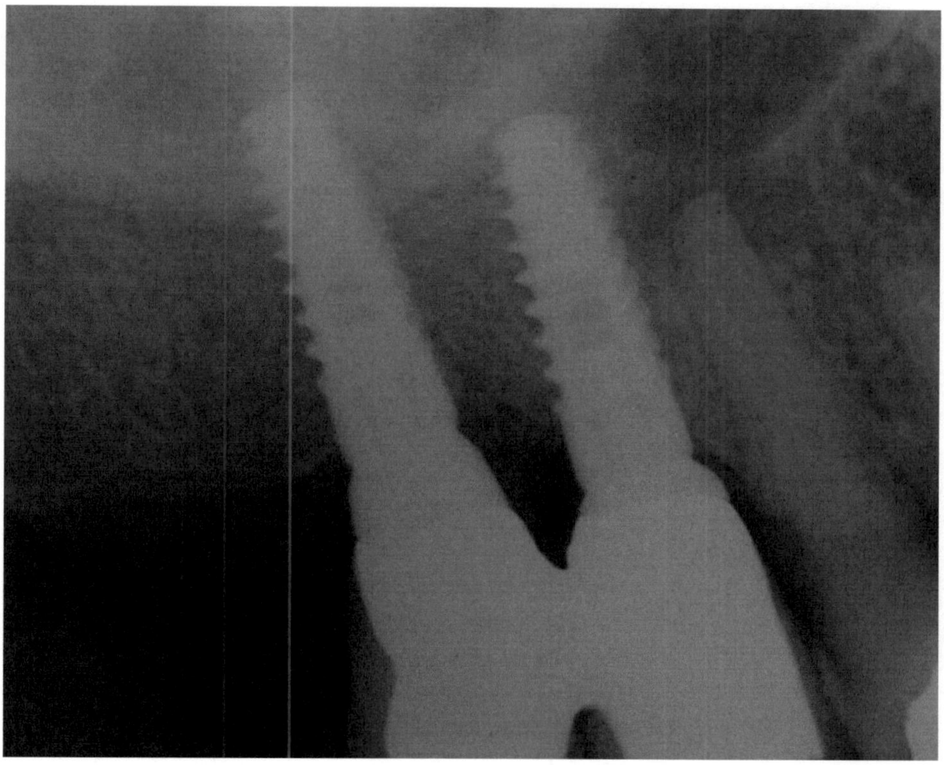

Fig. 14. Posterior maxilla. Periapical view at 3-year follow-up.

2.2.3 Comment

Simultaneous implant placement and sinus floor augmentation is advantageous to the patient because it minimizes the number of surgical interventions and shortens the time to completion of the implant-supported prosthesis (Khoury, 1999). It was once limited to cases with a minimum of 4-5 mm of alveolar bone coronally because of concerns about implant stability and accurate implant location, inclination, and parallelism (Hurzeler et al., 1996; Peleg et al., 1998; Smiler et al., 1992; Tatum et al., 1993). More recent research suggests that no specific bone height limit is necessary so long as primary implant stability is assured (Achong & Block, 2006). Nevertheless, in cases of severely atrophic posterior axillary ridges of ≤4 mm, simultaneous implant placement and sinus floor augmentation may pose a challenge (Mardinger et al., 2007). To prevent technical problems and complications, the use of cancellous block-allografts is recommended (Fig. 10). In our experience, their placement with rough surface implants was associated with a high success rate (94.4%), similar to the standards documented in the literature. Their use also abolished the need to repair membrane perforations.

2.3 Posterior mandible

2.3.1 Surgical technique

Surgery in the posterior mandible was performed under local anesthesia (Figs. 15,16). A crestal incision was centered in the remaining keratinized tissue and extended through the edentulous span, allowing for a minimum of 1-2 mm of keratinized gingiva on both sides of the flap, in most cases, slightly to the lingual side. A distal oblique releasing incision was made into the buccinator muscle posteriorly, and a vertical releasing incision mesial to the most distal tooth, on the labial aspect. A full-thickness mucoperiosteal lingual flap was initially reflected with extreme caution to prevent tears in the periosteum (Fig. 17) and then further mobilized lingually, away from the mylohyoid line.

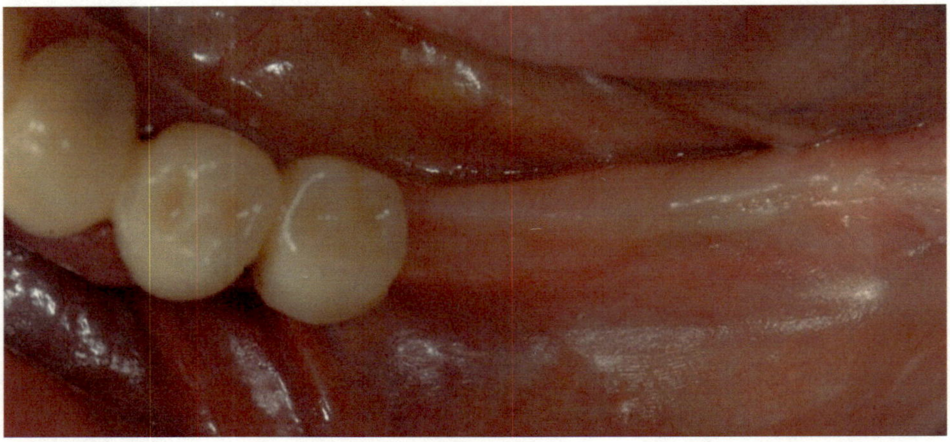

Fig. 15. Posterior mandible. Preoperative clinical view.

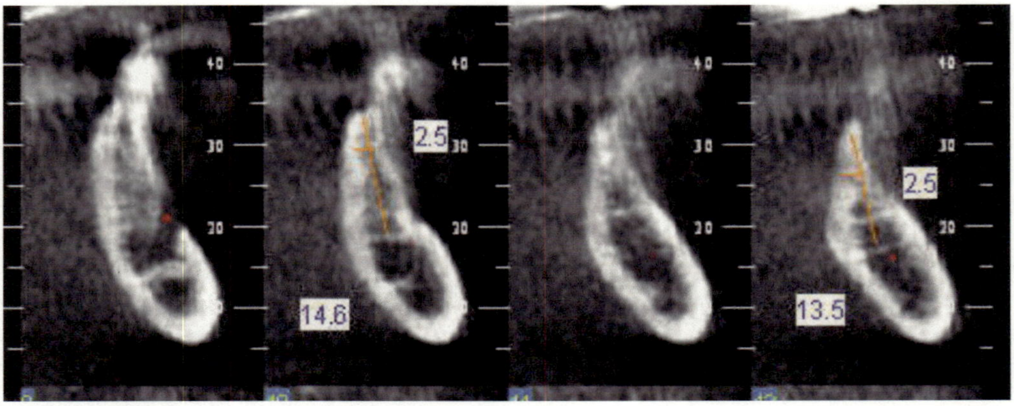

Fig. 16. Posterior mandible. Preoperative CT.

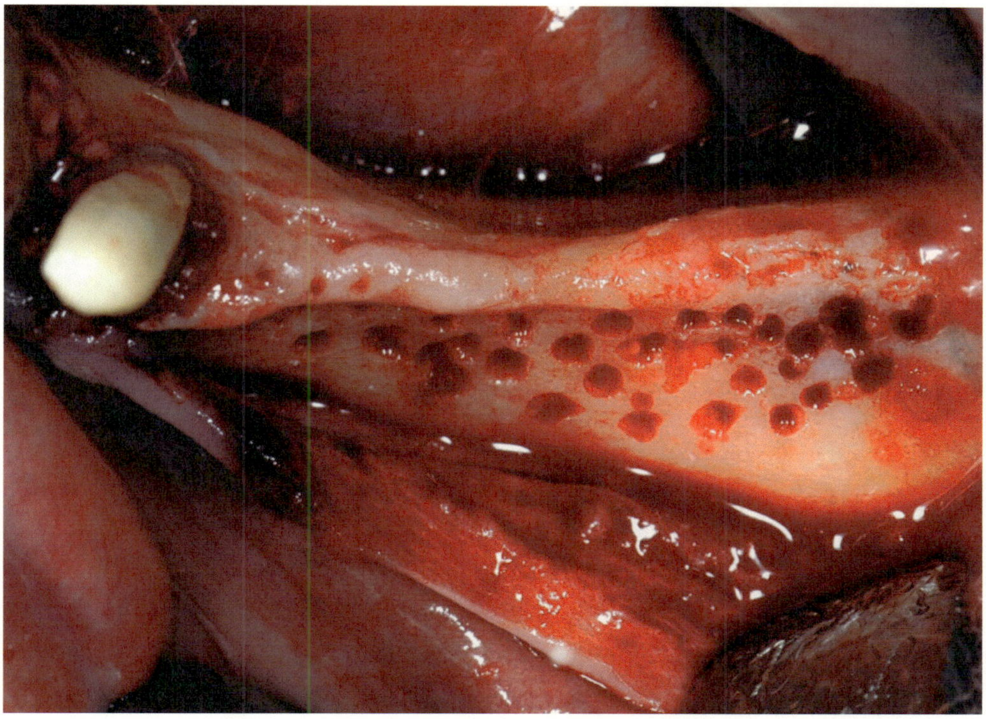

Fig. 17. Posterior mandible. Clinical view demonstrating bone defect.

The buccal aspect of the alveolar ridge was exposed via subperiosteal dissection so the defect could be visualized in three dimensions. (Visualization of the mental neurovascular bundles is mandatory.) A freeze-dried cancellous block graft was refined to fit into the defect. Once it was seated and stable, the graft was fixed with 1.6 mm x 10 mm bone screws (Fig. 18). A large round bur was used to round any sharp edges and shape the graft to completely conform to

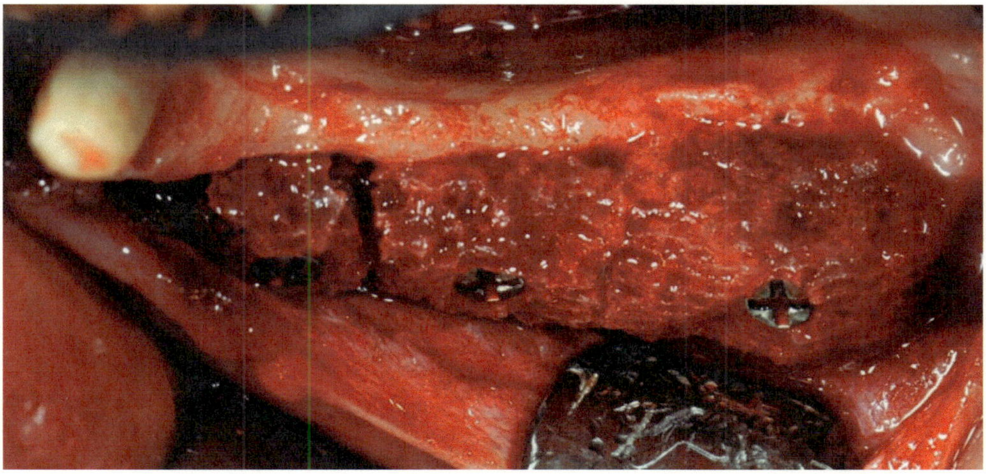

Fig. 18. Posterior mandible. Bone block fixation.

the defect site. Deficiencies at the edges of the graft were filled with particulate bone, and the graft was covered with a long-term resorbable collagen membrane (Fig. 19). The width and height of the augmented ridge were measured with a periodontal probe. The midcrestal incision was initially closed with interrupted and horizontal mattress sutures, and the vertical incision was closed with interrupted sutures.

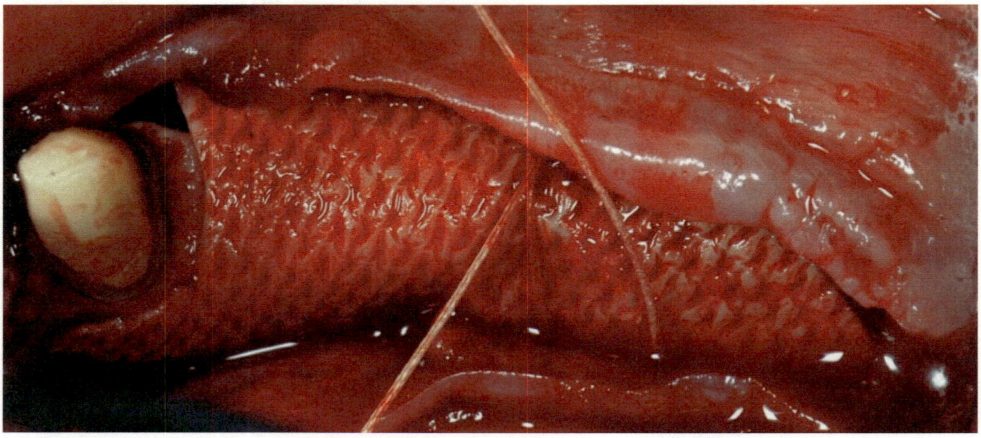

Fig. 19. Posterior mandible. Membrane coverage.

Patients were prescribed antibiotics and analgesics for 5 days postoperatively and an antiseptic solution for 2 weeks. When possible, fixed partial provisional restorations were fitted and delivered to the patient immediately after surgery. Removable provisional restorations were not used.

Patients were seen weekly during the first postoperative month and monthly thereafter for 6 months, until healing was complete. Periapical radiographs were taken immediately postoperatively and 2-3 months after surgery and evaluated particularly for signs of soft tissue dehiscence, color changes, and the overall appearance of the grafted ridge contour.

Before the second-stage surgery at 6 months, new panoramic radiographs and CT scans were obtained to determine implant width and length (Fig. 20). Access to the augmented

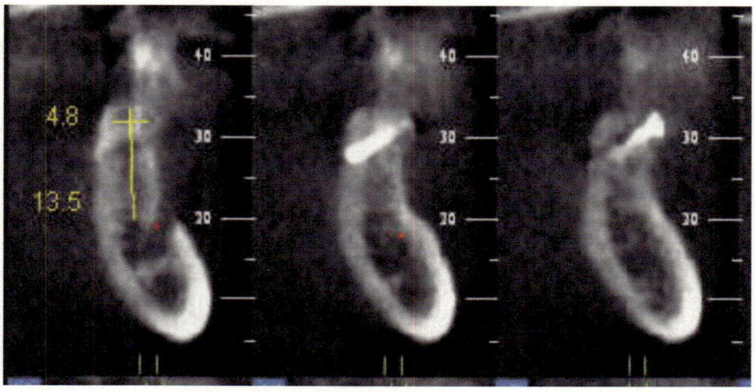

Fig. 20. Posterior mandible. CT at 6 month following bone grafting.

ridge was achieved via the initial incision. The fixation screws were removed, and the width and height of the ridge were measured with a periodontal probe. The implant sites were selected with a diagnostic template (Fig. 21). Residual buccal thickness measurements were repeated after implant placement to further evaluate bone resorption and determine the horizontal bone dimension.

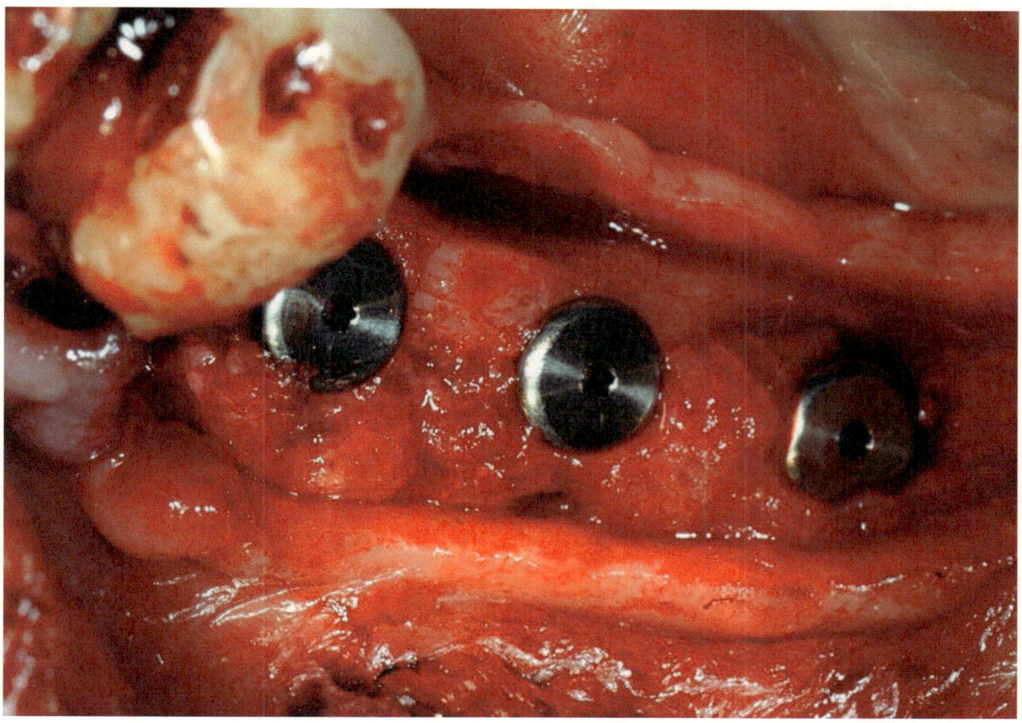

Fig. 21. Posterior mandible. Implant placement 6 months after bone grafting.

The implants were exposed 3 months later. The soft tissues are allowed to mature for 3 weeks prior to the definitive restorative phase. The implants were restored with fixed cement-retained ceramic prostheses (Fig. 22). Clinical and radiographic examinations were performed at the time of restoration, every 6 months during the first year, and once a year thereafter (Fig. 23).

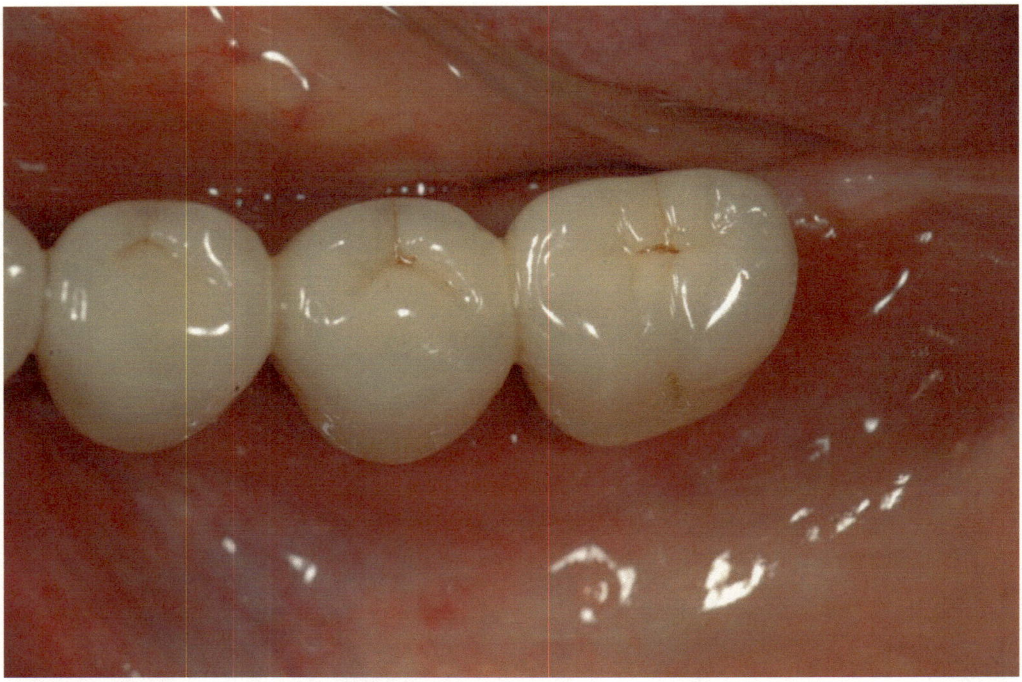

Fig. 22. Posterior mandible. Final restoration.

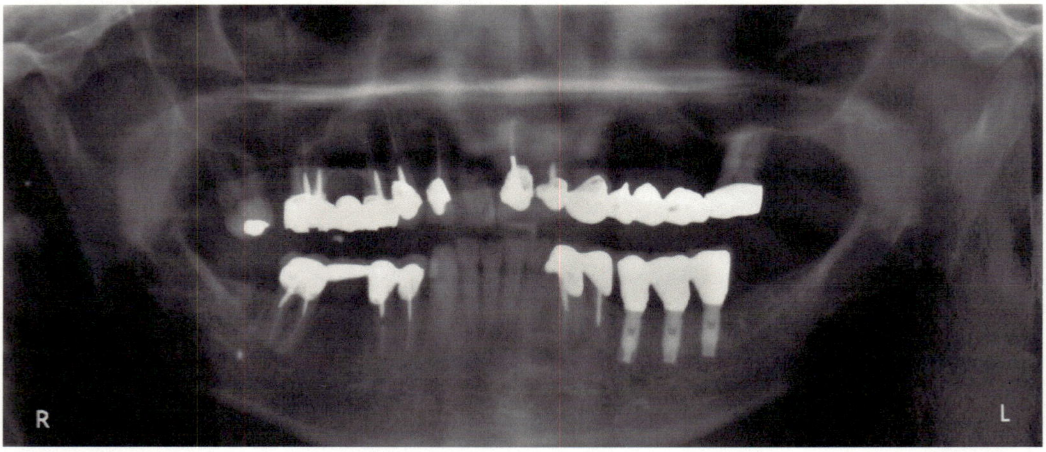

Fig. 23. Posterior mandible. Panoramic view at 5-year follow-up.

2.3.2 Clinical outcome

This technique was associated with a 79.3% block survival rate (Nissan et al., 2009). Bone width at implant placement was 7.9 ± 0.5 mm, and bone gain in the vertical dimension was 4.3 ± 1.6 mm. A nonsignificant resorption of 5% was noted from graft placement to implant placement. Mean implant diameter was 3.9 ±0. 2 mm , and mean implant length was 10.4 ± 0.7 mm (Nissan et al., 2009). Implant survival rate was 95.2%. In all cases of failure, the implants were reinserted after 2 months and successfully osseointegrated. All patients received a fixed implant-supported prosthesis. There were no late cases of loss of function. The mean crown-to-implant ratio was 0.96 ± 0.16 (Nissan et al., 2009).

2.3.3 Comment

A biomechanically stable bone implant foundation is indispensable for the long term success of fixed implant-supported prostheses in the posterior mandible (Bahat & Fontanessi, 2001b; Pikos, 2000). The increased biting forces in this region can create excessive stress (Devlin & Wastell, 1986; Gibbs et al., 1981; Mericske-Stern et al., 1995) that lead to crestal bone loss, screw loosening, occlusal material fracture, prosthesis wear and fracture, and implant failure (Goodacre et al., 2003; Misch et al., 2005, 2006; Nissan et al., 2011a, 2011b). Therefore, special attention needs to be addressed to the opposing arch and the crown-to-implant ratio (Misch et al., 2005, 2006). Treatment planning should take several factors into account: eliminating lateral interference during excursive movements; decreasing the occlusal table relative to the implant diameter or maximizing the implant diameter to minimize off-axis forces; shortening or eliminating cantilevers; increasing the number of implants (Pikos, 2000; Mische et al., 2005, 2006). Ridge augmentation enables the use of longer and wider implants that increase the surface area over which the occlusal force stresses are distributed (Bahat & Fontanessi, 2001b; Pikos, 2000). Block grafting is indicated for mandibular alveolar ridge atrophy of >3mm in the vertical and/or lateral dimension on para-axial CT reconstructions (Figs. 15,16).

3. Histological and histomorphometric observations

3.1 Histological findings

We observed newly formed vital bone, residual cancellous block-allograft bone, and connective tissue in all augmented sites. The residual cancellous block-allograft bone was identified by the presence of empty lacunae and separation lines. Newly formed bone containing viable osteocytes demonstrated intimate contact with the residual cancellous block-allograft bone. Osteoblasts were present in conjunction with newly formed bone around the residual cancellous block-allograft bone (Fig. 24).

There was no evidence of acute or chronic inflammatory infiltrate (Chaushu et al., 2010b; Nissan et al., 2011d; Nissan et al., 2011e).

3.2 Histomorphometric findings

Histomorphometric analysis of the anterior maxilla yielded the following results: mean fraction of newly formed bone, 33 ± 18%; residual cancellous block-allograft, 26 ± 17%; marrow and connective tissue, 41 ± 21%. When patients were divided by age (less or more than 40 years), statistically significant between-group differences were found for newly formed bone (younger patients: 38.6%, older patients: 19.8%, $P = 0.04$) and residual cancellous block-allograft (younger patients: 20.1%, older patients: 38.4%, $P = 0.05$). Age had no effect on the mean fraction of marrow and connective tissue (41.3% and 41.8%, respectively, $P = 0.49$) (Nissan et al., 2011e). In another study, histomorphometric analysis of the posterior maxilla yielded the following results: fraction of newly formed bone, 26.1 ±

15%; residual cancellous block-allograft, 24.7 ± 19.4%; marrow and connective tissue, 49.2 ± 20.4%. There were no statistically significant differences in the fraction of newly formed bone by patient sex or age, presence/absence of membrane perforations, or residual alveolar bone height (Chaushu et al., 2010b). In the posterior mandible, the fraction of the newly formed bone was 44 ± 28%; residual cancellous block-allograft, 29 ± 24%; and marrow and connective tissue, 27 ± 21%. Division of the patients by age (less or more than 45 years) yielded statistically significant differences in the percentage of newly formed bone (69% vs. 31%, respectively; $P = 0.05$), but not in the percentage of residual cancellous block-allograft (17% and 5%) or marrow and connective tissue (14% and 34%). There were no statistically significant histomorphometric differences in percentage of newly formed bone by patient sex or presence/absence of soft tissue dehiscence (Nissan et al., 2011d).

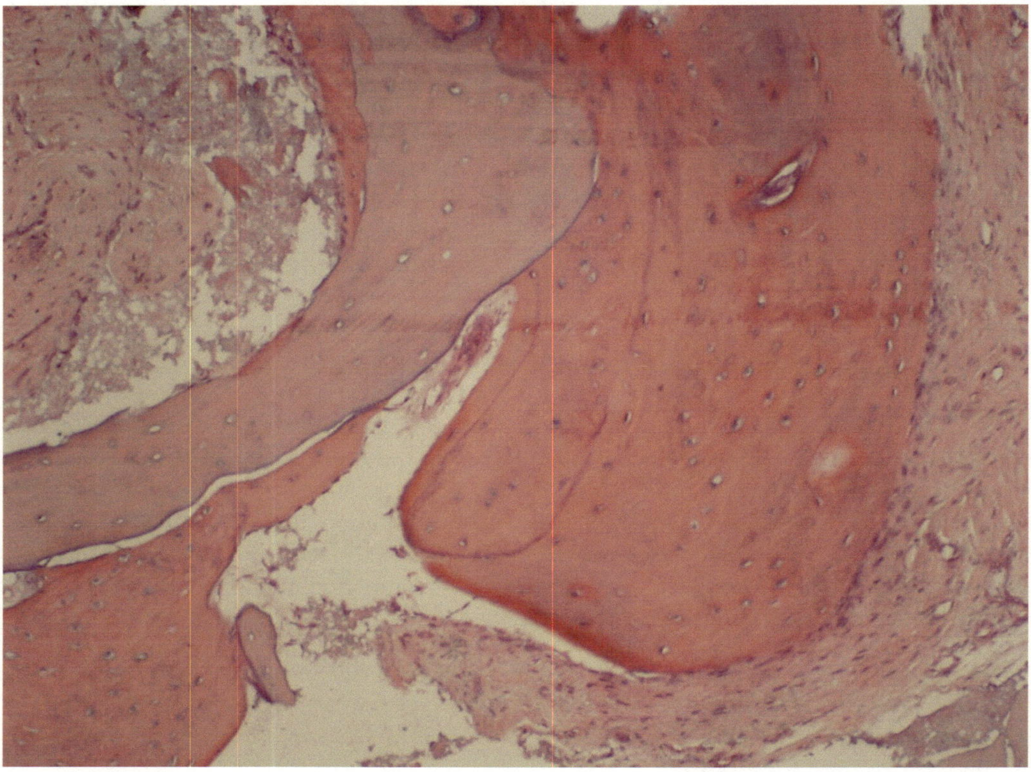

Fig. 24. Posterior mandible. Histological view of block allograft surrounded by new bone.

3.3 Comment

Our histomorphometric findings for cancellous bone allografts were close to those reported in earlier studies on the use of freeze-dried bone allograft (FDBA) materials for ridge augmentation in the atrophic anterior maxilla (28% new bone, 35% connective tissue, and 37% residual graft material; Iasella et al., 2003) and for sinus augmentation (29.1% new bone, 51.9% connective tissue, and 19% residual graft material; Kolerman et al., 2011). Accordingly, Holmquist et al. (2008) reported a mean bone content of 31.8% after placement of a morselized impacted bone allograft for reconstruction of the narrow alveolar crest in the anterior maxilla, and Froum et al. (2006), using mineralized cancellous bone allograft particles for sinus augmentation, reported rates of 28.3% new bone formation, 64.1% connective tissue and 7.6% residual graft. In the same study, using a split mouth design, values for bovine bone mineral were 12.4% new bone formation, 54.6% connective tissue, and 33% residual graft (Froum et al., 2006). Together, these results corroborate the osteoconductive property of FDBA.

The similarity between the rate of new bone formation (26.1%) in sinus augmentation with cancellous block-allograft and the reported vital bone proportion of pristine bone in the posterior maxilla (23% -28%) (Trisi & Rao, 1999; Ulm et al., 1999) may explain the high implant survival rates. Accordingly, in the posterior mandible, new bone formation following augmentation with cancellous block-allograft (44 ± 28%) was compatible to the vital bone proportion of pristine bone in the posterior mandible (56 ± 21%) (Todisco & Trisi, 2005).

Our findings of a significantly greater percentage of new bone formation in younger patients relative to older ones following remodeling with cancellous block-allograft in the anterior atrophic maxilla and posterior atrophic mandible (Nissan et al., 2011d; Nissan et al., 2011e) agree with earlier studies of the effect of age on new bone formation (Fisher et al., 2010; Matsumoto et al., 2001; Rando, 2006; Tanuka et al., 1996; Walboomers et al., 2006). Following induction of injury to the tibia marrow cavity in a rat model, MicroCT and histomorphometry revealed a reduced response with an increase in age, suggesting that the restoration of normal tissue is age-dependent (Fisher et al., 2010). Others also noted differences in the osteogenic properties of bone marrow-derived osteoblast-like cells by donor age (Walboomers et al., 2006). In a study of bone formation induced by recombinant human bone morphogenetic protein-2 in male Wistar rats, the total volume of newly formed bone declined with aging (Matsumoto, et al., 2001). The reasons for the age-related decline in tissue repair are not well understood. Suggested possibilities include impaired cellular activity (Tanaka et al., 1996) and a reduction in the total number of osteoblasts (Rando, 2006).

4. Complications

In our patients, cancellous bone block allografting was associated with soft tissue complications, namely membrane exposure (30.6%), incision line opening (30%), perforation of the mucosa over the grafted bone (14%), and infection (13%) (Chaushu et al., 2010a). Similar types of complications were reported by earlier studies as well (Bahat & Fontanesi, 2001a, 2001c; Li & Wang, 2008). Treatment was initiated as soon as possible. Necrotic soft tissue was removed, and the bone was leveled with the soft tissue using a high-speed bur. The area was immediately and thoroughly irrigated with chlorhexidine. Patients were prescribed an additional week of oral antibiotics and instructed to apply chlorhexidine gel over the affected area twice daily and to refrain from eating at the grafted site until mucosal healing was complete.

Overall, partial graft failure occurred in 7% of patients and total graft failure in 8%. The presence of complications significantly increased the risk of graft failure. In sites with membrane perforation, the total graft failure rate was 17% and the partial graft failure rate, 24%s. By contrast, total implant failure occurred in only 4% of non-membrane exposure sites and partial failure in 0%. These differences were statistically significant. Incision-line opening was associated with a 17% total graft failure rate and 20% partial graft failure rate, compared to rates of 4% and 2% failure rates, respectively, for sites in which the incision line remained intact, and mucosal perforation over the grafted bone was associated with an 11% total graft failure rate and a 26% partial graft failure compared to 8% and 4%, respectively, for sites without mucosal perforation. At sites with infection of the bone block, the total graft failure rate was 39% and the partial graft failure rate 22%, compared to rates of 3.4% and 5%, respectively, at non-infected sites.

Analysis of the possible risk factors for complications revealed no significant association with patient age or sex. Complications occurred significantly more in the mandible than the maxilla. The presence of more than one complication significantly increased the risk of infection and graft failure. Specifically, combined membrane exposure with incision line opening led to infection in 44% of cases and to total graft failure in 17%. Our findings suggest that the graft failures were not attributable to use of the cancellous block graft per se but to procedural and technical factors.

Implant failure occurred in 4.4% of cases. None of the complications had a statistically significant effect on the occurrence of implant failure.

It is noteworthy that the rate of new bone formation was similar in sinuses with (25.5%) and without (27.3%) membrane perforations (Chaushu et al., 2010b). Membrane perforations can lead to the loss of graft particles into the air chamber of the sinus and have a reported incidence of 10% to 56% (Fugazzotto & Vlassis, 2003; Jensen et al., 1994; Pikos, 2008; Shlomi et al., 2004; van den Bergh et al., 2000; Wallace et al., 2007). A split-mouth study of perforated sinus membrane repair with a resorbable collagen membrane (Proussaefs et al., 2004) found that when bone particles were use, nonperforated sites demonstrated significantly more bone formation (33.58%) than perforated sites (14.17%), with a higher implant survival rate at second-stage surgery (100% vs 69.56%). The authors suggested that the technique or materials be changed to lessen the risk posed by perforation and repair of the sinus membrane. Our histomorphometric results indicate that allogeneic block grafts may offer a good solution (Chaushu et al., 2010a,b). Using a block graft, we were able to leave membrane perforations untreated (Pikos, 1999) without compromising either survival or new bone formation and without interrupting the operative procedure.

5. Conclusions

Cancellous freeze-dried blocks can serve as a satisfactory allografting material for sinus floor augmentation and for initial stabilization of both grafts and dental implants during simultaneous procedures, even in the presence of membrane perforation. In posterior mandible reconstruction, bone grafts can increase the number, length, and diameter of implants that can be placed. The use of cancellous block allografts allows for the placement of implants of standard length and diameter, thereby improving long-term prognosis of the implant-supported reconstruction. More attention to detail and meticulous technique may prevent the progression of complications to graft failures.

6. Acknowledgment

We thank Melani Kawe and Gloria Ginzach of the Editorial Board, Rabin Medical Center, Beilinson Campus, for their editorial assistance.

7. References

Achong, RM. & Block, M.S. (2006). Sinus floor augmentation: Simultaneous versus delayed implant placement, In: *The Sinus Bone Graft,* O.T. Jensen, (Ed.), p. 60, Quintessence Publishing Co, Inc., Illinois

Adell, R.; Lekholm, U., Rockler, B. & Brånemark, P.-I. (1981). A 15-year study of osseointegrated implants in the treatment of the edentulous jaw. *International Journal of Oral Surgery,* 10, pp. 387–416

Albrektsson, T.; Branemark, P.-I., Hansson, H.A. & Lindström, J.(1981). Osseointegrated titanium implants: requirements for ensuring a long lasting, direct bone to implant anchorage in man. *Acta Orthopaedica Scandinavica,* 52, pp. 155–170

Bahat, O. & Fontanesi, F.V. (2001a). Complications of grafting in the atrophic edentulous or partially edentulous jaw. *The International Journal of Periodontics & Restorative Dentistry,* 21, pp. 487-495

Bahat, O. & Fontanessi, R.V. (2001b). Efficacy of implant placement after bone grafting for three-dimensional reconstruction of the posterior jaw. *The International Journal of Periodontics & Restorative Dentistry,* 21, pp. 221–231

Bahat, O. & Fontanesi, F.V. (2001c). Complications of grafting in the atrophic edentulous or partially edentulous jaw. *The International Journal of Periodontics & Restorative Dentistry,* 21, pp. 487-495

Belser, U.C.; Schmid, B., Higginbottom, F. & Buser, D. (2004). Outcome analysis of implant restorations located in the anterior maxilla: a review of the recent literature. International Journal of Oral & Maxillofacial Implants 19, Suppl, pp. 30-42

Branemark, P.-I. (1983). Osseointegration and its experimental background. *Journal of Prosthetic Dentistry* 50, pp. 399–410

Chaushu, G.; Mardinger, O., Calderon, S., Moses, O. & Nissan, J. (2009). The use of cancellous block-allograft for sinus floor augmentation with simultaneous implant placement in the posterior atrophic maxilla. *Journal of Periodontology* 80, pp. 422-428

Chaushu, G.; Mardinger, O., Peleg, M., Ghelfan, O. & Nissan, J. (2010a). Analysis of complications following augmentation with cancellous block allografts. *Journal of Periodontology* 81, pp. 1759-1764

Chaushu, G.; Vered, M., Mardinger, O. & Nissan, J. (2010b). Histomorphometric analysis after maxillary sinus floor augmentation using cancellous bone-block allograft. *Journal of Periodontology* 81, pp. 1147-1152

Chiapasco, M.; Abati, S., Romeo, E. & Vogel, G. (1999). Clinical outcome of autogenous bone blocks or guided bone regeneration with e-PTFE membranes for the reconstruction of narrow edentulous ridges. *Clinical Oral Implants Research* 10, pp. 278–288

D'Addona, A. & Nowzari H. (2001). Intramembranous autogenous osseous transplants in aesthetic treatment of alveolar atrophy. *Periodontology 2000,* 27, pp. 148–161

Devlin, H. & Wastell, D. (1986). The mechanical advantage of biting with the posterior teeth. *Journal of Oral Rehabilitation* 13, pp. 607-610

Fisher, M.; Hyzy, S., Guldberg, R.E., Schwartz, Z. & Boyan, B.D. (2010). Regeneration of bone marrow after tibial ablation in immunocompromised rats is age dependent. *Bone* 46, pp. 396-401

Froum, S.J.; Wallace, S.S., Elian, N., Cho, S.C. & Tarnow, D.A. (2006). Comparison of mineralized cancellous bone allograft (Puros) and anorganic bovine bone matrix

(Bio-Oss) for sinus augmentation: Histomorphometry at 26 to 32 weeks after grafting. *International Journal of Periodontics & Restorative Dentistry,* 26, pp. 543-551

Fugazzotto, P.A. & Vlassis, J.M. (2003). A simplified classification and repair system for sinus membrane perforations. *Journal of Periodontology* 74, pp. 1534-1541

Gibbs, C.H.; Mahan, P.E., Lundeen, H.C., Brehnan, K., Walsh, E.K. & Holbrook, W.B. (1981). Occlusal forces during chewing and swallowing as measured by sound transmission. *Journal of Prosthetic Dentistry,* 46, pp. 443-449

Goodacre, C.J.; Bernal, G., Rungcharassaeng, K. & Kan, J.Y. (2003). Clinical complications with implants and implant prostheses. *Journal of Prosthetic Dentistry,* 90, pp. 121-132

Grunder, U.; Gracis, S. & Capelli, M. (2005). Influence of the 3-D bone-to-implant relationship on esthetics. *International Journal of Periodontics Restorative Dentistry,* 25, pp. 113-119

Holmquist, P.; Dasmah, A., Sennerby, L. & Hallman, M. (2008). A new technique for reconstruction of the atrophied narrow alveolar crest in the maxilla using morselized impacted bone allograft and later placement of dental implants. *Clinical Implant Dentistry & Related Research,* 10, pp. 86-92

Hsu, M.L.; Chen, F.C., Kao, H.C. & Cheng, C.K. (2007). Influence of off-axis loading of an anterior maxillary implant: a 3-dimensional finite element analysis. *International Journal of Oral & Maxillofacial Implants,* 22, pp. 301-309

Hunt, D. & Jovanovic, S. (1999). Autogenous bone harvesting: A chin technique for particulate and monocortical bone blocks. *International Journal of Periodontics Restorative Dentistry,* 19, pp. 165–173

Hurzeler, M.B.; Kirsch, A., Ackerman, K.L. & Quinones, C.R. (1996). Reconstruction of the severely resorbed maxilla with dental implants in the augmented maxillary sinus: A five year clinical investigation. *International Journal of Oral & Maxillofacial Implants,* 11, pp. 466–475

Iasella, J.M.; Greenwell, H., Miller, R.L., Hill, M., Drisko, C., Bohra, A.A. & Scheetz, J.P. (2003). Ridge preservation with freeze-dried bone allograft and a collagen membrane compared to extraction alone for implant site development: a clinical and histologic study in humans. *Journal of Periodontology,* 74, pp. 990-999

Jensen, J.; Sindet-Pedersen, S. & Oliver, A.J. (1994). Varying treatment strategies for reconstruction of maxillary atrophy with implants: Results in 98 patients. *Journal of Oral & Maxillofacial Surgery,* 52, pp, 210-216

Keith, J.D., Jr. (2004). Localized ridge augmentation with a block allograft followed by secondary implant placement: A case report. *International Journal of Periodontics & Restorative Dentistry,* 24, pp. 11–17

Keith, J.D., Jr.; Petrungaro, P., Leonetti, J.A., Elwell, C.W., Zeren, K.J., Caputo, C., Nikitakis, N.G., Schöpf, C. & Warner, M.M. (2006). Clinical and histologic evaluation of a mineralized block allograft: results from the developmental period (2001-2004). *International Journal of Periodontics & Restorative Dentistry,* 26, pp. 321-327

Khoury, F. (1999). Augmentation of the sinus floor with mandibular bone block and simultaneous implantation. A six-year clinical investigation. *International Journal of Oral & Maxillofacial Implants,* 14, pp. 557-564

Kolerman, R.; Tal, H. & Moses, O. (2008). Histomorphometric analysis of newly formed bone after maxillary sinus floor augmentation using ground cortical bone allograft and internal collagen membrane. *Journal of Periodontology*, 79, pp. 2104-2111

Leonetti, J.A. & Koup, R. (2003). Localized maxillary ridge augmentation with a block allograft for dental implant placement: Case reports. *Implant Dentistry*, 12, pp. 217-226

Li, J. & Wang, H.L. (2008). Common implant-related advanced bone grafting complications: classification, etiology, and management. *Implant Dentistry*, 17, pp. 389-401

Lundgren, S.; Sjostrom, M., Nystrom, E. & Sennerby, L. (2008). Strategies in reconstruction of the atrophic maxilla with autogenous bone grafts and endosseous implants. *Periodontology 2000*, 47, pp. 143–161

Lyford, R.H.; Mills, M.P., Knapp, C.I., Scheyer, E.T. & Mellonig, J.T. (2003). Clinical evaluation of freeze-dried block allografts for alveolar ridge augmentation: A case series. *International Journal of Periodontics & Restorative Dentistry*, 23, pp. 417–425

Mardinger, O.; Nissan, J. & Chaushu, G. (2007). Sinus floor augmentation with simultaneous implant placement in the severely atrophic maxilla: technical problems and complications. *Journal of Periodontology*, 78, pp. 1872-1877

Matsumoto, A.; Yamaji, K., Kawanami, M. & Kato, H. (2001). Effect of aging on bone formation induced by recombinant human bone morphogenetic protein-2 combined with fibrous collagen membranes at subperiosteal sites. *Journal of Periodontal Research*, 36, pp. 175-182

McAllister, B.S. & Haghighat, K. (2007). Bone augmentation techniques. *Journal of Periodontology*, 78, pp. 377-396

Mericske-Stern, R.; Assal, P., Mericske, E. & Bürgin, W. (1995). Occlusal force and oral tactile sensibility measured in partially edentulous patients with ITI implants. *International Journal of Oral Maxillofacial Implants*, 10, pp. 345-353

Misch, C.E.; Goodacre, C.J., Finley, J.M., Misch, C.M., Marinbach, M., Dabrowsky, T., English, C.E., Kois, J.C. & Cronin, R.J., Jr. (2005). Consensus conference panel report: crown-height space guidelines for implant dentistry-part 1. *Implant Dentistry*, 14, pp. 312-318

Misch, C.M.; Misch, C.E., Resnik, R.R. & Ismail, Y.H. (1992). Reconstruction of maxillary alveolar defects with mandibular symphysis grafts for dental implants: A preliminary procedural report. *International Journal of Oral & Maxillofacial Implants*, 7, pp. 360–366

Misch, C.E.; Goodacre, C.J., Finley, J.M., Misch, C.M., Marinbach, M., Dabrowsky, T., English, C.E., Kois, J.C. & Cronin, R.J., Jr. (2006). Consensus conference panel report: crown-height space guidelines for implant dentistry-part 2. *Implant Dentistry*, 15, pp. 113-121

Moy, P. & Palacci, P. (2001). Minor bone augmentation procedures. In. *Esthetic implant dentistry. Soft and hard tissue management. Il*, P. Palacci & I. Ericsson (Eds.), pp. 137-158, Quintessence Publishing Co., Inc., Illinois, USA

Nissan, J.; Romanos, G.E., Mardinger, O. & Chaushu, G. (2008). Immediate nonfunctional loading of single-tooth implants in the anterior maxilla following augmentation with cancellous freeze-dried block allograft: A case series. *International Journal of Oral & Maxillofacical Implants*, 23, pp. 709-716

Nissan, J.; Ghelfan, O., Mardinger, O., Calderon, S. & Chaushu, G. (2009). Efficacy of cancellous block allograft augmentation prior to implant placement in the posterior atrophic mandible. *Clinical Implant Dentistry & Related Research*, 13, pp. 279-285

Nissan, J.; Ghelfan, O., Gross, O., Priel, I., Gross, M. & Chaushu, G. (2011a). The effect of crown/implant ratio and crown height space on stress distribution in unsplinted implant supporting restorations. *Journal of Oral & Maxillofacial Surgery*, 69, pp.1934-1939

Nissan, J.; Gross, O., Ghelfan, O., Priel, I., Gross, M. & Chaushu, G. (2011b). The Effect of Splinting Implant-Supported Restorations on Stress Distribution of Different Crown-Implant Ratios and Crown Height Spaces. *Journal of Oral &Maxillofacial Surgery*, Aug 23 [Epub ahead of print]

Nissan, J.; Mardinger, O., Calderon, S., Romanos, G.E. & Chaushu, G.(2011c). Cancellous bone block allografts for the augmentation of the anterior atrophic maxilla. *Clinical Implant Dentistry &Related Research*, 13, pp. 104-111

Nissan, J.; Marilena, V., Gross, O., Mardinger, O. & Chaushu, G. (2011d). Histomorphometric analysis following augmentation of the posterior mandible using cancellous bone-block allograft. *Journal of Biomedical Materials Research. Part A*, 97, pp. 509-513

Nissan, J.; Vered, M., Gross, O., Mardinger, O. & Chaushu, G. (2011e). Histomorphometric analysis following augmentation of the anterior atrophic maxilla using cancellous bone-block allograft. *International Journal of Oral & Maxillofacial Implants* (accepted for publication)

Nkenke, E. & Stelzle, F. (2009). Clinical outcomes of sinus floor augmentation for implant placement using autogenous bone or bone substitutes: a systematic review. *Clinical Oral Implants Research*, 20 Suppl 4, pp. 124-133

Nkenke, E.; Schultze-Mosgau, S., Radespiel-Tröger, M., Kloss, F. & Neukam, F.H. (2001). Morbidity of harvesting of chin grafts: A prospective study. *Clinical Oral Implants Research*, 12, pp. 495-502

Nkenke, E.; Radespiel-Tröger, M., Wiltfang, J., Schultze-Mosgau, S., Winkler, G. & Neukam, F.H. (2002). Morbidity of harvesting of retromolar bone grafts: A prospective study. *Clinical Oral Implants Research*, 13, pp. 514-521

Palacci, P. & Ericsson, I. (2001). Anterior maxilla classification. In. *Esthetic implant dentistry. Soft and hard tissue management. Il*, P. Palacci & I. Ericsson, (Eds.), pp. 89-100, Quintessence Publishind Co, Inc, Illinois, USA

Peleg, M.; Mazor, Z., Chaushu, G. & Garg, A. (1998). Sinus floor augmentation with simultaneous implant placement in the severely atrophic maxilla. *Journal of Periodontology*, 69, pp. 1397-1403

Petrungaro, P.S. & Amar, S. (2005). Localized ridge augmentation with allogeneic block grafts prior to implant placement: Case reports and histologic evaluations *Implant Dentistry*, 14, pp. 139-148

Pikos, M.A. (1999). Maxillary sinus membrane repair: report of a technique for large perforations. *Implant Dentistry*, 8, pp. 29-34

Pikos, M.A. (2000). Block autografts for localized ridge augmentation. Part II. The posterior mandible. *Implant Dentistry*, 9, pp. 67-75

Pikos, M.A. (2008). Maxillary sinus membrane repair: update on technique for large and complete perforations. *Implant Dentistry*, 17, pp. 24-31

Proussaefs, P.; Lozada, J., Kim, J. & Rohrer, M.D. (2004). Repair of the perforated sinus membrane with a resorbable collagen membrane: A human study. *International Journal of Oral & Maxillofacial Implants*, 19, pp. 413-420

Raghoebar, G.M.; Louwerse, C., Kalk, W.W.I. & Vissink, A. (2001). Morbidity of chin bone harvesting. *Clinical Oral Implants Research*, 12, pp. 503-507

Rando, T.A. (2006). Stem cells, aging and the quest for immortality. *Nature*, 441, pp. 1080–1086

Rangert, B.; Jemt, T. & Jörneus, L. (1989). Forces and moments on Branemark implants. *International Journal of Oral & Maxillofacial Implants*, 4, pp. 241-247

Rangert, B.; Krogh, P.H.J., Langer, B. & van Roekel, N. (1995). Bending overload and implant fracture: A retrospective clinical analysis. *International Journal of Oral & Maxillofacial Implants*, 10, pp. 326-334

Sethi, A. & Kaus, T. (2001). Ridge augmentation using mandibular block bone grafts: Preliminary results of an ongoing prospective study. *International Journal of Oral & Maxillofacial Implants*, 16, pp. 378-388

Shlomi, B.; Horowitz, I., Kahn, A., Dobriyan, A. & Chaushu, G. (2004). The effect of sinus membrane perforation and repair with Lambone on the outcome of maxillary sinus floor augmentation: a radiographic assessment. *International Journal of Oral & Maxillofacial Implants*, 19, pp. 559-562

Smiler, D.G.; Johnson, P.W., Lozada, J.L., Misch, C., Rosenlicht, J.L., Tatum, O.H. & Wagner, J.R. (1992). Sinus lift grafting and osseous implants. Treatment of the atrophic posterior maxilla. *Dental Clinics of North America*, 36, pp. 151-186

Spray, J.R.; Black, C.G., Morris, H.F. & Ochi, S. (2000). The influence of bone thickness on facial marginal bone responses: Stage 1 placement through stage 2 uncovering. *Annals of Periodontology*, 5, pp. 119-128

Tanaka, H.; Barnes, J. & Liang, C.T. (1996). Effect of age on the expression of insulin-like growth factor-I, interleukin-6, and transforming growth factor-beta mRNAs in rat femurs following marrow ablation. *Bone*, 18, pp. 473–478

Tatum, O.H.; Lebowitz, M.S., Tatum, C.A. & Borgner, R.A. (1993). Sinus augmentation, rationale, development, long-term results. *The New York State Dental Journal*, 5, pp. 43-48

Todisco, M. & Trisi, P. (2005). Bone mineral density and bone histomorphometry are statistically related. *International Journal of Oral & Maxillofacial Implants*, 20, pp. 898-904

Trisi, P. & Rao, W. (1999). Bone classification: clinical histomorphometric comparison. *Clinical Oral Implants Research*, 10, pp. 1–7

Ulm, C.; Kneissel, M., Schedle, A., Solar, P., Matejka, M., Schneider, B. & Donath, K. (1999). Characteristic features of trabecular bone in edentulous maxillae. *Clinical Oral Implants Research*, 10, pp. 459–467

van den Bergh, J.P.; ten Bruggenkate, C.M., Disch, F.J. & Tuinzing, D.B. (2000). Anatomical aspects of sinus floor elevations. *Clinical Oral Implants Research*, 11, pp. 256-265

Verhoeven, J.W.; Cune, M.S., Terlou, M., Zoon, M.A. & de Putter, C. (1997). The combined use of endosteal implants and iliac crest onlay grafts in the severely atrophic mandible: A longitudinal study. *International Journal of Oral & Maxillofacial Surgery,* 26, pp. 351–377

Walboomers, X.F.; Elder, S.E., Bumgardner, J.D. & Jansen, J.A. (2006). Hydrodynamic compression of young and adult rat osteoblast-like cells on titanium fiber mesh. *Journal of Biomedical Materials Research. Part A* 2006;76:16-24.

Wallace, S.S.; Mazor, Z., Froum, S.J., Cho, S.C. & Tarnow, D.P. (2007). Schneiderian membrane perforation rate during sinus elevation using piezosurgery: clinical results of 100 consecutive cases. *International Journal of Periodontics & Restorative Dentistry,* 27, pp. 413-419

Part 3

Applied Biotechnology and Biomaterials

11

Self-Regenerative Ability of Bone and Micro Processing of Bone-Component Material in Orthopedic Surgery Healing

T. Ohtani[1], T. Nakai[2], R. Mori[2] and Y. Uchio[2]
[1]*Tokyo Gakugei University*
[2]*Shimane University*
[1,2]*Japan*

1. Introduction

To aid healing after orthopedic surgery, bone components, such as pins or screws, are used to immobilize broken or skinned bone after a bone or joint in the human body is broken (Lima et al., 2004). Metal and plastic materials are generally used to immobilize the injured bone after surgery; however, these artificial biomaterials may have residual adverse side effects.

The use of natural rather than artificial materials seems likely to lessen adverse side effects. Natural bone material is characterized by its remodeling action and self-regenerative ability (Curry, 2002).

In this study, the development of bone components was advanced through the use of self-regenerative functions. This research was promoted by the author's collaborative group in both the surgical and the engineering fields at Shimane University (Organization for the Promotion of Project Research, Shimane University, 2008).

The project developed components for orthopedic surgery healing and created a new system to precisely manufacture bone components in surgical operating rooms. The new system is described in the following section. Fusion control during contact with bone components is important for recovery of the affected area. Furthermore, controlling the conditions that promote and delay the fusion of bone components will lead to nonconventional and more effective surgical healing techniques.

Previous research focused on titanium as a biomaterial for internal bone, and initial osteoblast adhesion and cell migration profiles on a titanium surface have been reported (Zinger et al., 2004; Itahashi et al., 1995). In particular, one study reported that self-regenerative osteoblasts could detect nanoscale holes on a titanium surface (Zinger et al., 2004). However, controlled cell migration on a natural bone surface in contact with bone material has not been investigated.

This chapter presents research (Ohtani et al., 2009) on the self-regenerative ability of bone as a result of controlled surgical healing. The next section discusses details of a new machining system for manufacturing bone components. In addition, the engineered properties of bone-component material (Ohtani et al., 2005) and the effects of anisotropic tissue on the machining of bone material (Ohtani et al., 2009) are introduced briefly in Sections 3.1–3.3.

The properties of a processed bone model surface are described based on the engineering technology used in actual surgeries. Furthermore, fundamental research on the self-regenerative ability with respect to cell migration on a processed bone model surface is investigated by focusing on random migration which allows osteoblasts to grow via regenerative action.

2. Micro processing of bone components in precision machining

The first research trial in this project (Organization for the Promotion of Project Research, Shimane University, 2008) focused on the screw-shaped bone components used for orthopedic surgery. As shown in Fig. 1, the manufacturing system for bone screws used in this trial is an improved high-precision machine (MTS4) developed jointly with Nano Corporation in Japan.

The manufacturing system includes a micro processing machine, which can be brought into the surgery room and easily operated. The machine is constructed of lightweight B4-sized materials weighing 30 kg. Doctors and nurses in the surgery room can operate the system using computer-aided drawing software that executes programs quickly. The machines can be safely operated during cutting under clean conditions, which means without using cutting fluid and coolant. The system can be placed anywhere that has a 100-VAC household power supply and can be controlled by a personal computer.

Fig. 1. The precision turning machine developed in this research project.

Fig. 2 shows examples of manufactured bone screws. The bone screws can be machined into a precise shape with a dimension of approximately 5 mm and fastened to the healing area by using a customized tool. After immobilization of the bone with the bone screw, the self-regenerative ability of the bone assimilates the bone screw with the internal bone.

In this study, it was important to clarify the effect of anisotropic bone tissue on the precise machining of the screw thread. In addition, it was important to see the effect of the formation of micro asperities in the finished surface of the machined bone screw on the assimilation period. If the assimilation period is shortened after immobilization with the bone screw, the psychological burden on the patient could be reduced.

In the following section, we introduce the material properties and effects of anisotropic bone tissue that were investigated to determine the optimum procedure to precisely machine the bone.

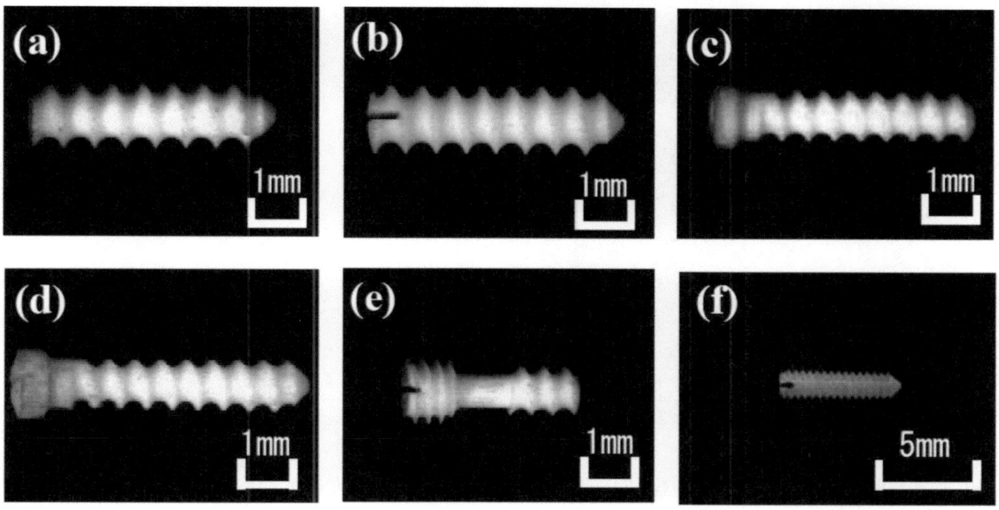

Fig. 2. Examples of manufactured bone screws.

3. Effects of anisotropic tissue on micro processing of bone components

3.1 Inhomogeneous nature of bone tissue

Biological bone has a sandwich-like structure, with clearance space on the inside and calcified compact tissue surrounding it (Gibson, 1997). Compact bone consists of an osteon lamellar layer surrounding haversian canals, through which capillaries run. Fig. 3 (a)-(c) shows optical micrographs of the outer layer of compact bovine bone, which has the characteristic microstructure of bone tissue.

The cross-sections in this figure are in the axial [Fig. 3 (a)], radial [Fig. 3 (b)], and circumferential [Fig. 3 (c)] directions, respectively. It can be seen that the inhomogeneous property of the micro tissue might affect the minute fraction of the tool tip that is in contact with the bone for precisely machined bone components.

3.2 Effect of inhomogeneous microscopic tissue on bone surface hardness

The compact bone used for the component material has an inhomogeneous structure (Fig. 3). Therefore, the effect of a small amount of this tissue on the bone surface hardness was investigated. Fig. 4 shows the results of a micro-Vickers hardness test that was conducted on the bovine bone test specimen.

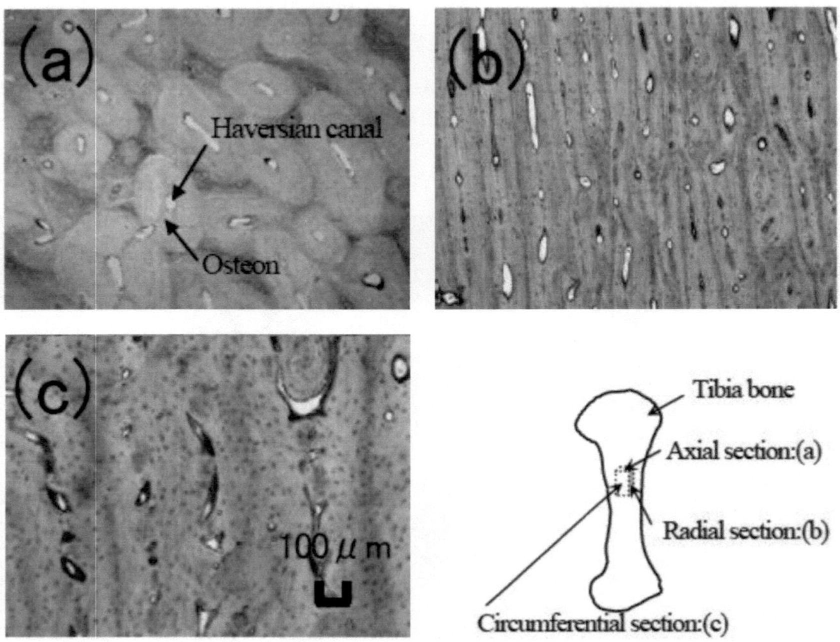

Fig. 3. Optical micrographs of the cross sections in compact bovine bone.

Vickers hardness (HV) was measured at different pressures applied via an indenter perpendicular to the axial, A (a), tangential, T (b), and radial, R (c) directions. The black circles in Fig. 4 shows the results for 30 runs. The scattered points were measured at each applied force. The white circles and error bars are the average values and standard deviations at the 95% confidence interval, respectively.

The range of HV values [Fig. 4 (a)–(c)] differed according to the applied force (F). The values were significantly lower for F = 1 N to 2 N. This result indicates that the hardness values were more heterogeneous for the applied forces between F = 1 N to 2 N, but were more homogeneous for the higher applied forces. In addition, the HV range differed for cross sections perpendicular to the A [Fig. 4 (a)], T [Fig. 4 (b)], and R [Fig. 4 (c)] directions (T > A > R for a smaller F). The average HV value for the cross sections perpendicular to the A and T directions was larger than that for the R direction. This result suggests that the anisotropic tissue differed for cross sections perpendicular to the A, R, and T directions [Fig. 3 (a)–(c)]. Thus, anisotropic tissue affects the bone surface hardness.

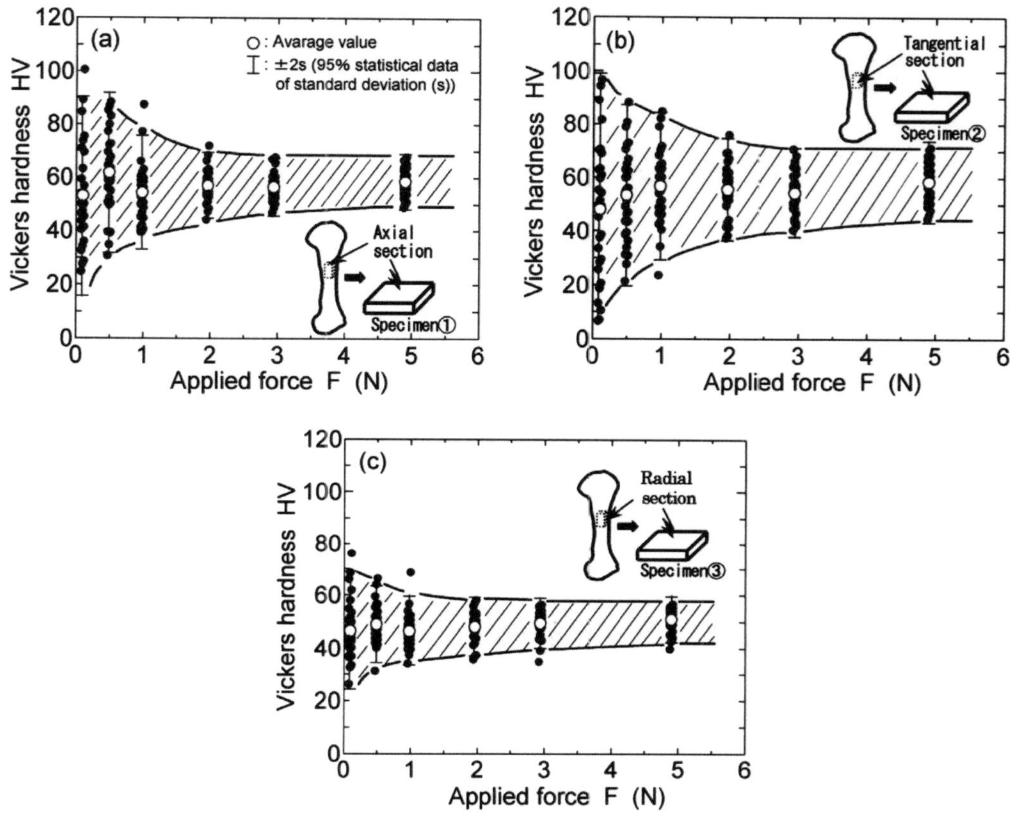

Fig. 4. Relationship between Vickers hardness (HV) and the applied force (F) in the axial (a), tangential (b) and radial (c) sections.

Fig. 5 shows the relationship between the indented depth (D) and the applied force (F) when the above data were used to investigate the heterogeneity of the hardness values displayed in Fig. 4. For this experiment, the indented depth (D) was based on the length of straight of the diagonal line indented on the bone surface, and the indentation depth of the 136° tip radius was calculated.

The indentation (D) in the cross sections perpendicular to the A, R, and T directions increased significantly with increasing F. The rate of the increase tended to decrease for F values greater than approximately 2 N. The result shown in Fig. 5 indicates that the applied force began to scatter significantly when the Vickers hardness corresponded to that at t = 10 μm.

Based on these results, it is suggested that the indentation depth affects the bone structure hardness and the heterogeneous microscopic tissue affects the hardness when its border is equal to the indentation depth.

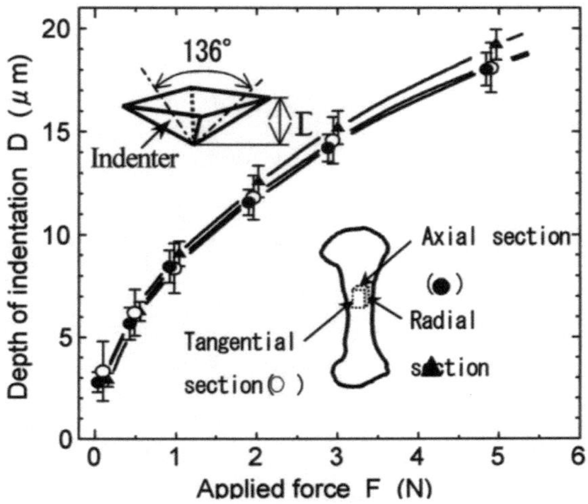

Fig. 5. Relationship between the depth of indentation (D) and the applied force (F) in the axial, tangential and radial sections.

3.3 Effect of anisotropic tissue on micro cutting

The inhomogeneity of the surface hardness was expected to affect the accuracy of the bone cutting. Fundamental bone-cutting data have been reported since the 1970s (Wiggins & Malkin, 1978). More recently, artificial joint replacement using a robotics machining system has been reported. In these more recent studies, fundamental bone-machining data were reported (Ito et al., 1983; Sugita et al., 2007); however, the effects of fractures during the cutting of calcified compact bone were not discussed in detail.

Thus, our project group has been quantitatively investigating the fracture behavior during bone cutting. With anisotropic tissue, continuous and interrupted chips are formed during the cutting process. In Fig. 6 (a)–(f), the shapes of the chips in calcified compact bone were categorized as follows: flow [Fig. 6 (a) and (b)], which generated cracks parallel to the cutting of the tool edge; shear [Fig. 6 (c) and (d)], which generated cracks along the outer edge of the osteon at the top of the tool edge; and tear [Fig. 6 (e) and (f)], which tore the osteon tissue without cutting.

The effect of the anisotropic tissue on the chip shape is observed for cutting with crack propagation. Fig. 7 shows the effects of the anisotropic tissue on bone cutting. When cutting along the osteon tissue [Fig. 7 (a)], the cut direction affected the cutting properties, and cracks propagated toward the front of the cutting tool.

It was confirmed that cracks similar to those shown in Fig. 7 (b) can occur along the outer edge of the osteon tissue, and cracks similar to those shown in Fig. 7 (c) can occur due to tearing along the osteon and not due to cutting. Thus, for precisely machined bone components, such as screws, the effect of anisotropic tissue, in particular, bone, should be considered when optimizing micro cutting conditions. In particular, the surface profiles after machining provide important information on whether the minute profile affects the self-regenerative ability of bone tissue.

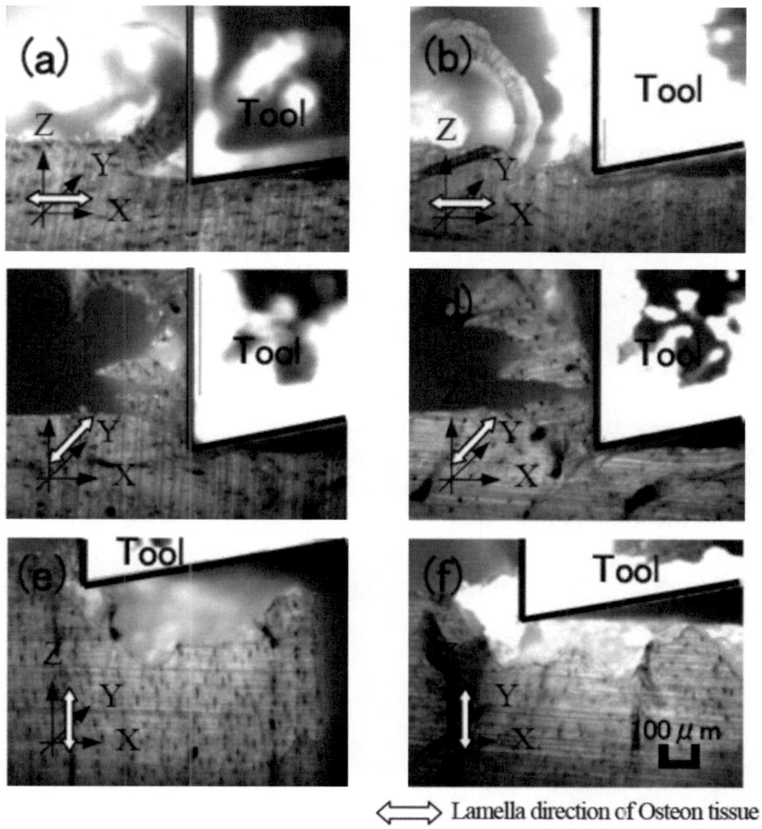

Fig. 6. Process of chip formation for the depth of cut with t=100 μm.

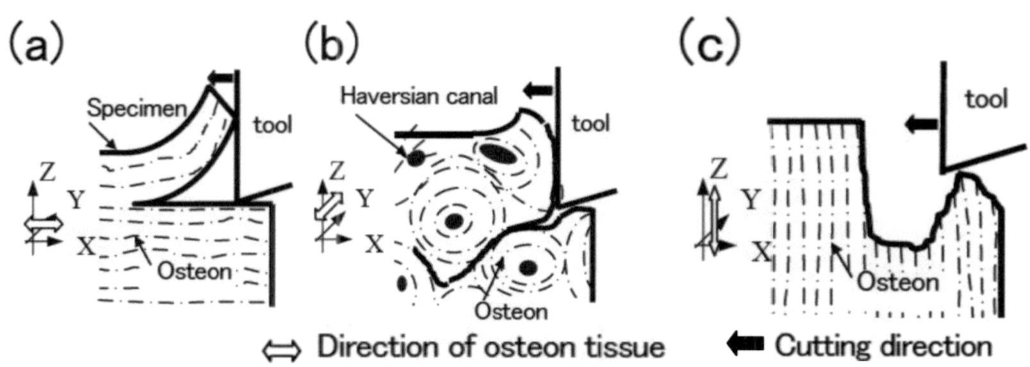

Fig. 7. Effect of anisotropic tissue in cutting of bone material with cutting.

4. Properties of surgical processing mechanisms

Different instruments are used to cut and remove bone in orthopedic surgery, and the surface profiles of bone differ according to the surgical instrument used. Fig.8 (a)-(i) show the orthopaedic surgical instruments investigated in this experiment. Destructive tools include the chisel [Fig. 8 (a)] and the clamp [Fig. 8(b)]. Saw-type tools include the manual saw [Fig. 8 (c)], the vibration saw [Fig. 8 (d)], and wire saws [Fig. 8 (e) and (f)]. Loose scratching and grinding tools include the file [Fig. 8 (g)] and revolving instruments include the steel bar and the diamond bar [Fig. 8 (h) and (i)].

Fig. 9 shows the macro-edge shape of the surgical instruments shown in Fig. 8. The chisel and clamp [Fig. 9 (a) and (b)] are used to cut and separate bone, respectively.

The tips of the saw and the wire tools [Fig. 9 (c)–(f)] can remove bone by unidirectional or reciprocating motion. The tool edges shown in Fig. 9 (g)–(i) can remove the bone surface by filing. The steel and diamond bars remove bone by mechanical revolution of the tip edge.

Fig. 10 shows scanning electron microscope (SEM) cross-sectional images of the bone surface after the tangential section was processed with the above mentioned instruments. The bone surface processed by the chisel and clamp [Fig. 10 (a) and (b)] on the lamellar layer of the osteon tissue exhibits a bumpy profile with uneven fractures. Bone tissue on the sawed surface removed by cutting shows fine debris adhered to the surface [Fig. 10 (c) and (d)]. An irregular surface was observed after wire cutting [Fig. 10 (e) and (f)] because the surface was scratched by the rugged wire edge. Fig. 10 (g)–(i) shows that the surface was gouged and scratched by the large and small edges of the tool tip.

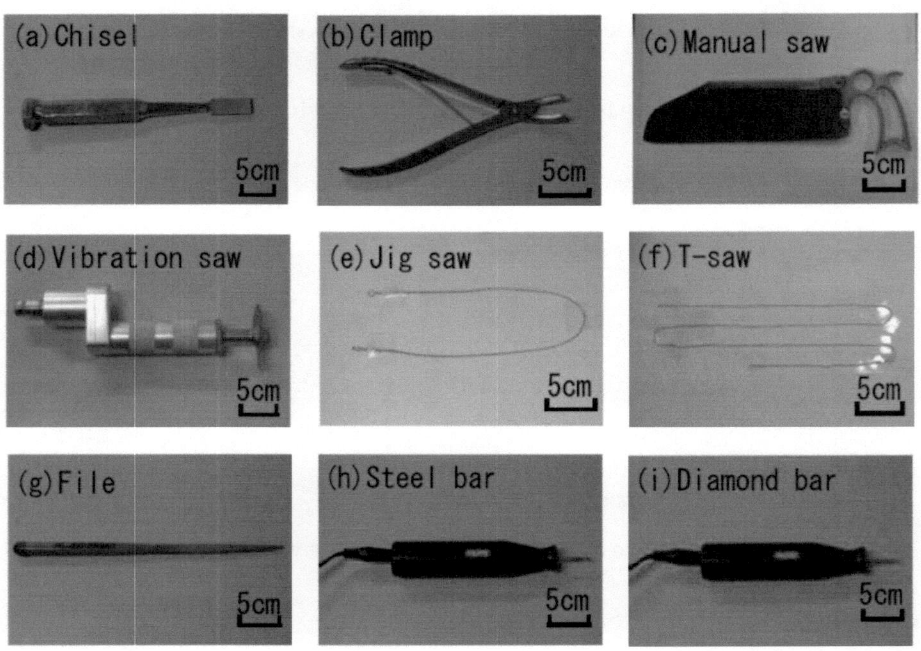

Fig. 8. Orthopedic surgical instruments used in this experiment.

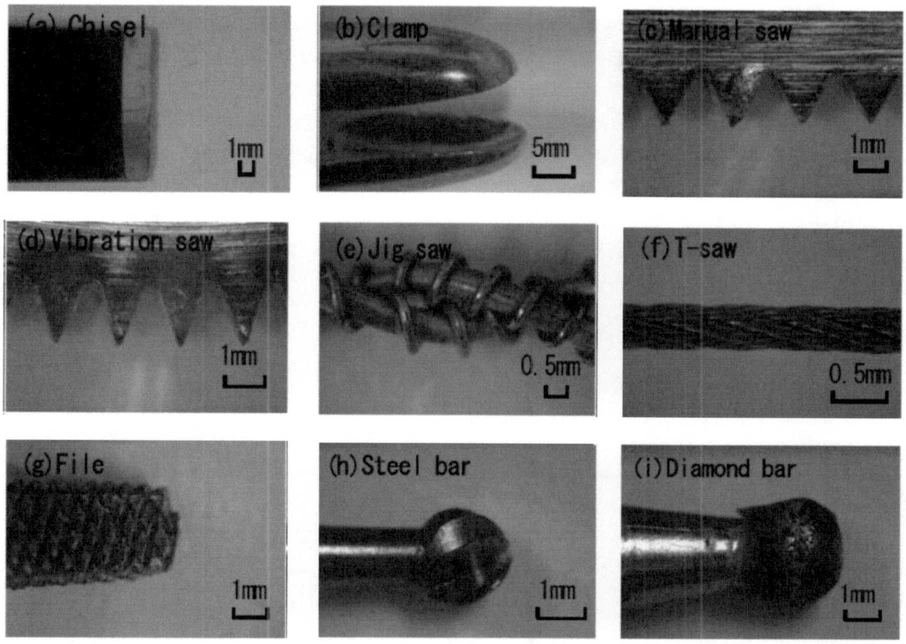

Fig. 9. Shape of edges in orthopedic surgical instruments.

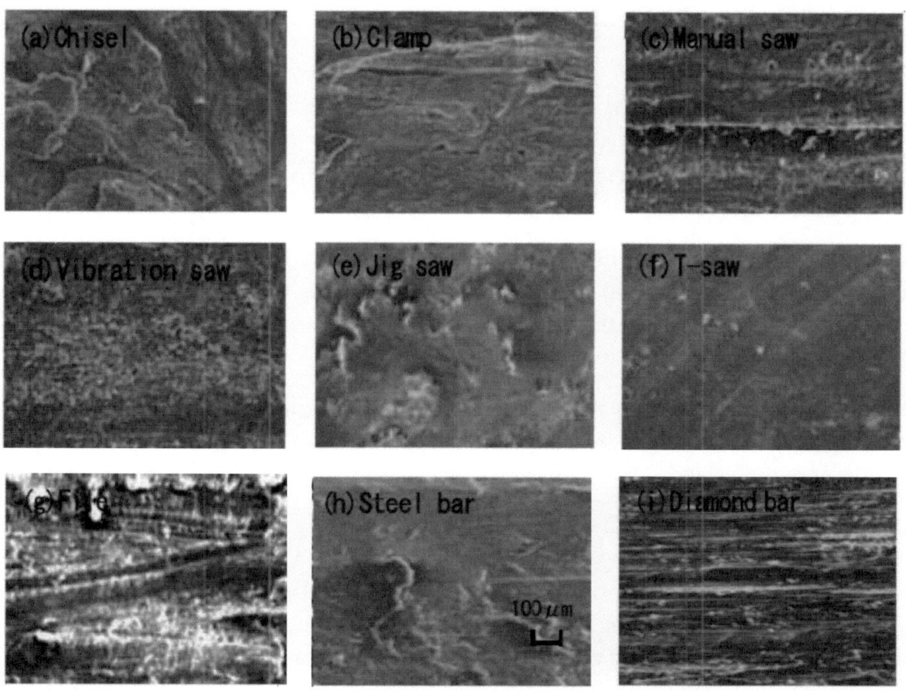

Fig. 10. Scanning electron microscope images of bone surfaces processed by orthopedic surgical instruments.

The properties of the bone surfaces processed by orthopedic surgical instruments are shown in Table 1, based on the tool tip shape. The instruments are categorized by the positive or negative rake angle of the cutting edge and the unidirectional or reciprocating motion to remove the bone material. Moreover, the table lists the properties of processed surfaces, which include directional scratches, irregular scratches, and fractures. The surface properties are related to the removal mechanism and classified as cutting, scratching due to bonded and loose abrasive grains with a negative rake angle, and cracking induced by fractures. The surgical processing mechanisms of cutting, bonded and loose abrasive grains, and cracking are related to the engineered tool tip shape.

Surgical instruments	Rake angle	Removal direction	Property of orthopedical surgical processing	Model mechanism of engineering processing
(c)Manual saw	Negative	Reciprocation	Saw cutting	Cutting
(d)Vibration saw	Negative	Reciprocation	Saw vibratory cutting	
(g)File	Negative	Reciprocation	Scratching	Bonded abrasive
(h)Steel bar	0°	Unidirectional	Scratching	
(i)Diamond bar	Negative	Unidirectional	Grinding	
(e)Jig saw	Negative	Loose direction	Loose scratching	Loose abrasive
(f)T-saw	Negative	Loose direction	Loose scratching	
(a)Chisel	Positive	Unidirectional	Cracking	Cracking
(b)Clamp	0°	Unidirectional	Cracking	

Table 1. Properties of bone surfaces processed by orthopedic surgical instruments.

5. Cell migration on the processed bone surface

A cell culture experiment was conducted to recreate the surgical processing mechanism on the processed model surface. Fig. 11 (a)–(h) shows SEM images of the model surfaces, including cut surfaces [Fig. 11 (a) and (b)], polished surfaces with bonded and loose abrasive grains [Fig. 11 (c)–(f)], a cracked surface [Fig. 11 (g)], and a control surface which was finished lubricously by turning machine [Fig. 11 (h)].

The cell migratory distance on these model surfaces was analyzed by an X-Y coordinate diagram (Fig. 12). Based on these results, cell migration on the processed model surfaces (Table 1) was characterized.

Cells on the cut surface [Fig. 12 (a) and (b)] tended to migrate in the Y direction, and the tendency increased as the cutting depth increased [Fig. 12 (a)]. Cells on the bonded abrasive surface [Fig. 12 (c) and (d)] tended to migrate in the Y direction, and the migration was similar to that on the cut surface. Cells on the loose abrasive surface [Fig. 12 (e) and (f)] tended to spread in a concentric circle with the larger abrasive grains, but spread arbitrarily with the smaller abrasive grains. Cells on the cracked surface [Fig. 12 (g)] spread arbitrarily as compared with cells spread with the loose abrasive grains, and cells on the control surface [Fig. 12 (h)] spread significantly in all directions.

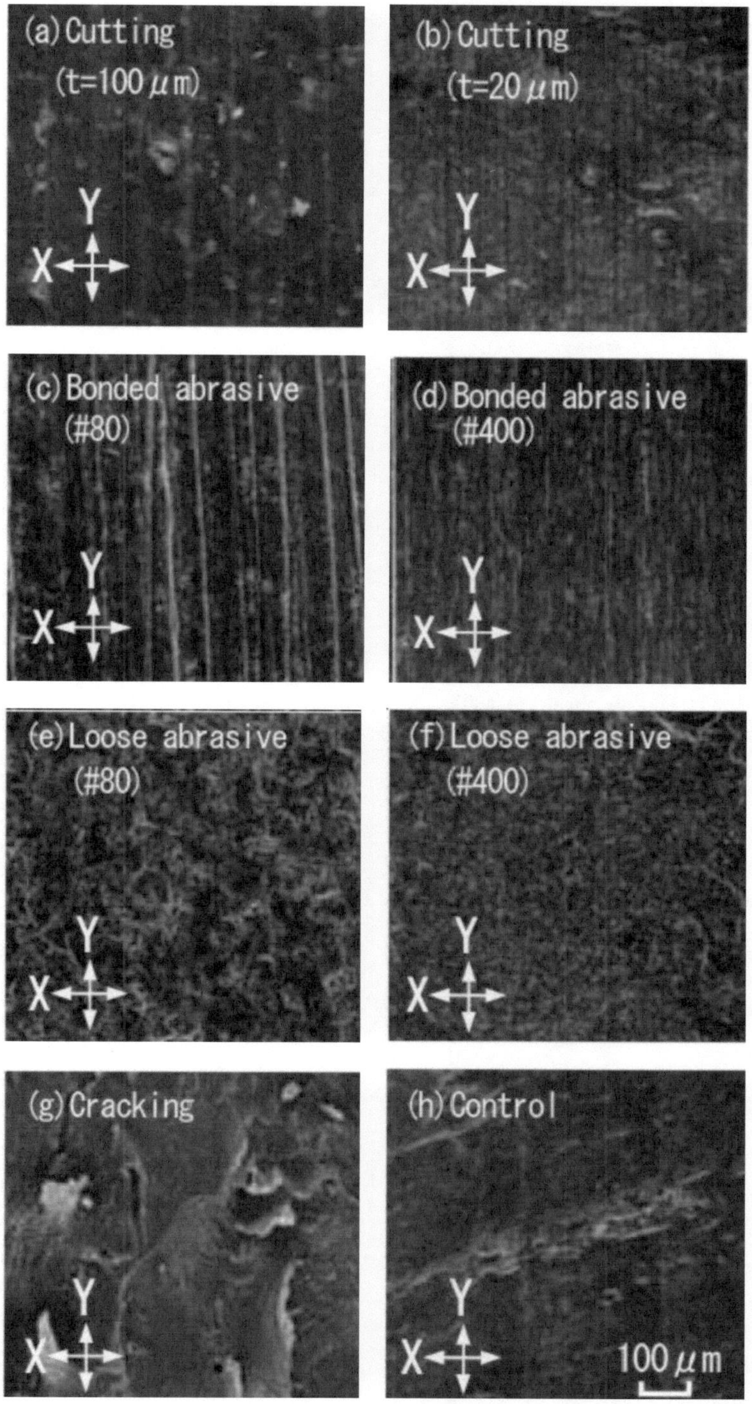

Fig. 11. Scanning electron micrographs of model bone surfaces processed by different engineering methods.

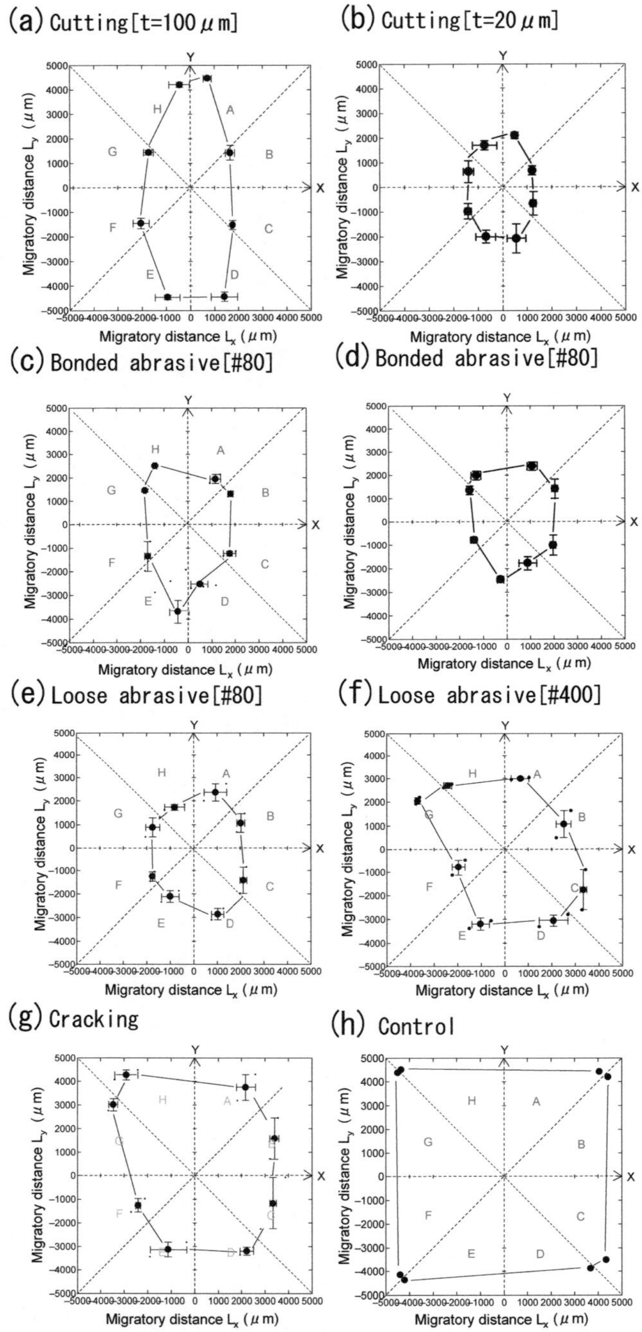

Fig. 12. X-Y diagrams of cell migratory distance on bone surfaces processed by different engineering methods: (a) and (b) by cutting [t=100 and 20 μ m], (c) and (d) by bonded abrasives[#80 and #400], (e) and (f) by loose abrasives[#80 and #400], (g) by cracking, where (f) is the control.

Based on these results, cell migration on the processed bone model surfaces was characterized by a continuous, linear trace along the tool edges on the cut and bonded surfaces. Additionally, migration on the loose abrasive surface was random because the model surface produced random, minute asperities due to scratching of the loose abrasive edge. Because splitting at the tool edge due to irregular asperities induced the cracked model surface, cell migration tended to spread in a random direction, similarly to that of the loose abrasive surface. Cells on the flat and smooth control surfaces spread randomly.

Thus, cell migration on the processed bone surfaces can be controlled by surface asperities formed by the surface processing conditions.

6. Effects of surface asperities on cell migration

To investigate the effects of processed surface asperities on cell migration, the cell migratory distances L_X and L_Y, with respect to the X-Y coordinates, are shown in Fig. 13. The results show the maximum cell migratory distance on each surface for the area from A to H (shown in Fig. 12). Each cell culture experiment was conducted five times. If the plotted data lie on a diagonal dashed line, as shown in Fig. 13, the osteoblast cells migrated equally in both the X and Y directions.

In the cut model-surface data in Fig. 13 (a), cell migration in the X-Y coordinates can be classified into two primary groups. One group tended to migrate in the Y direction, and the tendency was notable when the cut depth was $t = 100$ µm. With the bonded abrasive surface [Fig. 13 (b)], one group migrated in the Y direction with a magnitude similar to that of the cut surface, and the scattered data tended to be larger than that of the cut surface. The data for the loose abrasive and cracked surfaces [Fig. 13 (c)] were randomly scattered around the diagonal line. This result indicates that cell migration was affected by surface asperities, and osteoblast cells were distributed randomly as the surface asperities changed from a regular to a mixed arrangement.

Fig. 14 shows the relationship between the average maximum height $(R_Z)_X$ and the average cell migratory distance L_X measured with respect to the roughness profile (Fig. 13). The values of L_X for cell migration tended to decrease in the X direction for all mechanical removal actions, except for the cracked surface, as the values of $(R_Z)_X$ increased.

Research on cell proliferation on a titanium surface shows that the amount of proliferation was remarkable on a surface with multiple holes of approximately 30 µm (Zinger et al., 2004) and in surface asperities from R_Z = 20–30 µm (Itahashi et al., 1995) as compared with sub-micron holes and smooth surfaces with $R_Z < 1$ µm. Thus, these previous reports suggest that cell migration on the processed bone surface is affected by surface asperities, similarly to that on titanium, and that cell migration control is improved on mechanically removed surfaces with larger asperities.

In this project, cell migration seemed to be more affected by the surface properties of bone, such as roughness, unlike migration on titanium and cracked surfaces. This result should be studied in further detail in the future.

Thus, manipulating the processing method and surface profile can control the migration of osteoblast cells. These results provide fundamental knowledge for promoting surgical healing and controlling the local contact between bone materials.

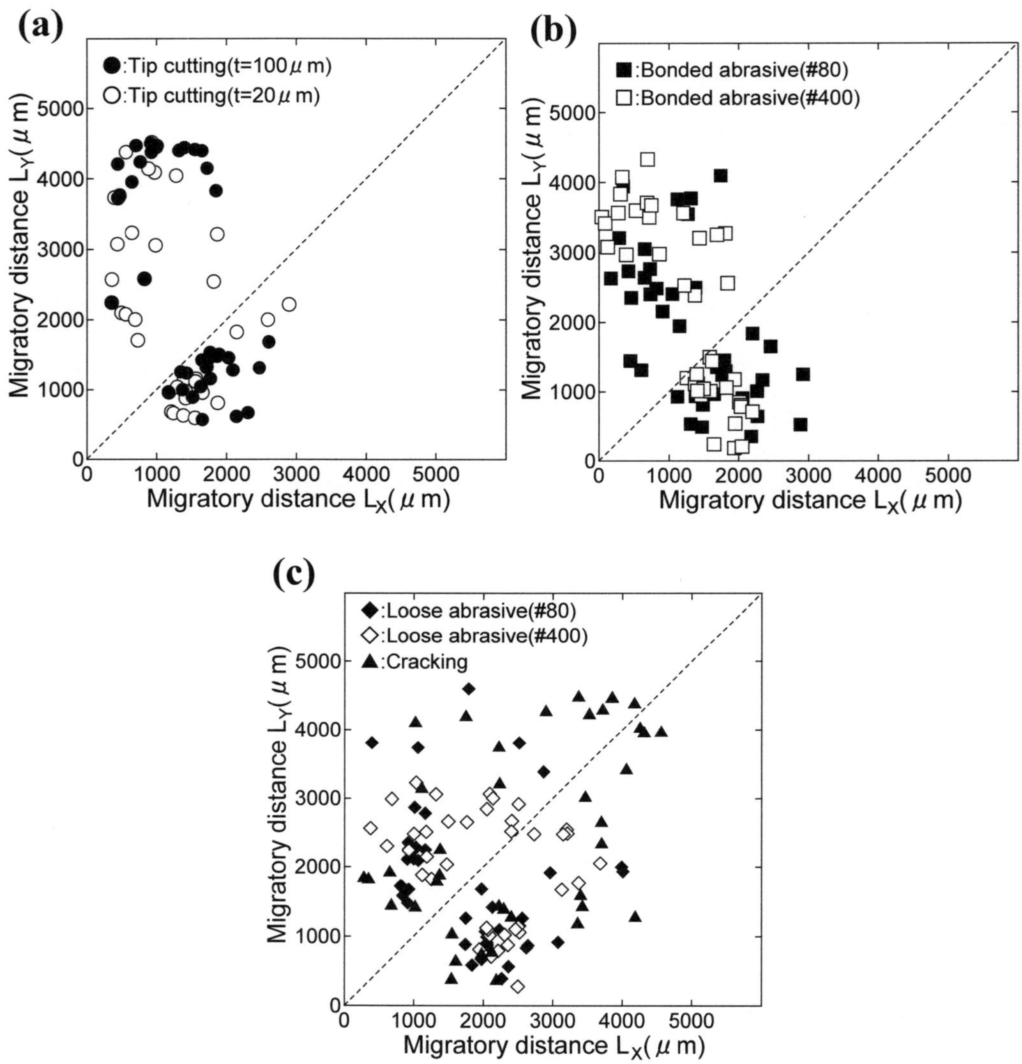

Fig. 13. Relationship between the migratory distances L_X and L_Y on the cutting surface. (a) Cutting surface, (b) bonded abrasive surface, (c) loose abrasive and cracking surfaces

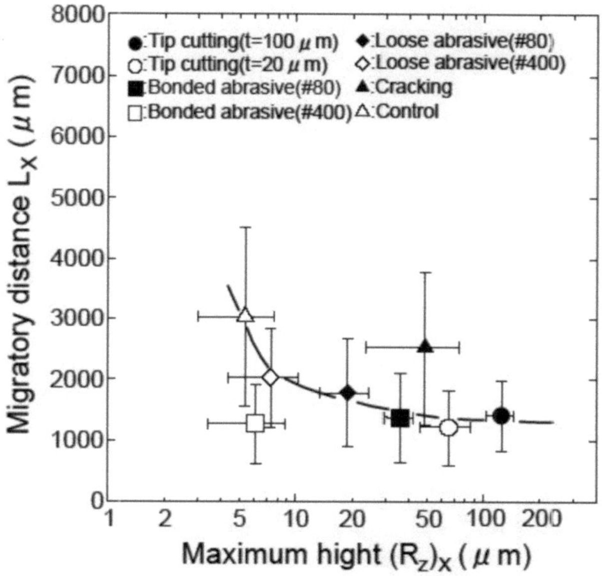

Fig. 14. Relationship between the migratory distance L_X and the maximum height $(R_Z)_X$.

7. Conclusions

This research project developed bone components using the self-regenerative ability of bone and created a system to precisely manufacture bone screws in a surgical operating room. Bone material processing was investigated, including the cutting mechanism and bone properties such as hardness. The bone processing method used in the actual surgical front was also classified in terms of the engineering technology. A cell culture experiment was conducted on a model surface produced experimentally by various engineering methods. Orthopedic surgery processing was investigated and classified into four processing mechanisms: cutting, grinding, polishing, and cracking. The cell culture results showed that cell migration on the bone surface was influenced by surface asperities, which were formed by surface processing conditions. The experimental results showed that cell migration along the groove tended to spread randomly according to the changes in the groove track. Cell migration was affected by the groove track on the surface. This result was consistent with previous studies on a titanium surface. Cell migration was also controlled by the roughness of the processed surface. The results of this study can be used to promote healing and locally immobilize bone by using the appropriate bone processing method in orthopedic surgical operating rooms.

8. Acknowledgment

The authors thank Mr. A. Hayashi, president of Nano Corporation, for supporting part of this research and Mr. T. Hasegawa, Mr. T. Kawasaki, and Mr. S. Takai for their extensive assistance with the experimental apparatus.

9. References

Curry, D.(2002), *Bones Structure and Mechanics*, p.15, Princeton University Press.

Gibson, J, Ashby F, Cellular solids, (1997),pp.429-432, Cambridge University Press.

Itahashi, H., Miyazaki, T., Fujimori, S, Lee, W., Early cellular responses of osteoblast-like cells in vitro on titanium surface with different micromorphlogies, *Dental Materials Journal*, Vol.14, No.1, (1995), pp. 136-141.

Itoh S., Ito Y., Shikita T., Basic study on Bone Cutting Forces for developing Surgical Instrument, *Bulletin of JSME*, Vol.26, No.222, (1983), pp.2295-2301.

Ohtani T., Mori R., Uchio Y., Hayashi A., Effect of microscope tissue on bone machinability in cutting process, *Proceedings of the third internationa conference on leading edge manufacturing in 21st century*, Vol.2, (2005), pp.529-534.

Ohtani T., Nakai T., Kawasaki Y., Mori R., Uchio Y., Cell Migratory Characterization in Micro-Processing Surface of Bone material for Surgery Healing Control, *Transactions of the Japan Society of Mechanical Engineers*, Vol.75 , No.749, Series C,(2009), pp. 191-196.

Ohtani, T., Nakai, T., Kawasaki, Y., Mori, R., Uchio, Y., Effect of anisotropic tissue on chip formation in micro cutting of bone, *Journal of the Japan Society of Mechanical Engineers*, Vol.75, No.749, Series C, (2009), pp.197-202.

Project promote organization in Shimane University, *Annual report of 2007 year for cooperative project in field of medical, engineering and agriculture(in Japanese)*, (2008), pp.79-90, Shimane University.

Sugita N., Mitsuichi M., Difference of Cutting Phenomena due to Material and Hydrious Property of Bone, *Journal of the Japan Society of Precision Engineering*, Vol.73, No.3, (2007), pp.377-382.

Wheeler, L., Mcloughlin, W., Biomechanical assessment of compression screws, *Clinical Orthopedics and Related Reserch*, No.350, (1998), pp.237-245

Wiggins K., Malkin S., Orthogonal Marching of bone, *Transaction of the ASME*, Vol.100, (1978), pp.122-130.

Zinger, O., Anselme, K., Denzer, A., Haberasetzer, P., Wieland, M., Jeanfils, J., Hardouin, P., Landolt, D., Time-dependent morphology and adhesion osteoblastic cells on titanium model surface featuring scale-resolved topography, *Biomaterials*, Vol.25, (2004), pp.2695-2711

12

Microstructure and Biocompatibility of Hydroxyapatite Porous Ceramics Designed by a Partial Dissolution-Precipitation Technique with Supersonic Treatment for Bone Regeneration

Toshiyuki Akazawa[1], Masaru Murata[2], Yasuhiko Tabata[3] and Manabu Ito[4]

[1]*Industrial Technology Research Dept., Hokkaido Research Organization,*
[2]*School of Dentistry, Health Sciences University of Hokkaido,*
[3]*Dept. of Biomaterials, Institute for Frontier Medical Sciences, Kyoto University,*
[4]*Graduate School of Medicine, Hokkaido University,*
Japan

1. Introduction

Hydroxyapatite (HAp: $Ca_{10}(PO_4)_6(OH)_2$) granules and ceramics have been clinically applied for bioactive and substituted materials of hard tissues in dental and medical fields because of excellent biocompatibility and osteoconduction (Oyane, 2010; Yoshikawa, 2009; 2010). In recent years, orthopedists and oral surgeons require the high capability of commercial HAp products for biomaterials, such as improvement of bio-absorption at implanted regions and early incorporation into bio-metabolic system (Artzi et al., 2004; Okuda et al., 2007; 2008). Coping with the required capability of HAp products, it will be necessary to design the biomimetic ceramics that have smooth body-fluid-permeation, appropriate mechanical strength, high surface area, and high affinity for cells.

Satisfying these points, in the previous studies (Akazawa et al., 2005, 2006a; 2006b), we developed functionally graded HAp (designated as fg-HAp) derived from calcined bovine bone; in which grain size and crystallinity of HAp gradually changed from surface regions to bulk parts by utilizing bovine bone calcined at 1073K (b-HAp). HAp porous ceramics made of animal bone can inherit physical and chemical properties of natural resources, such as pore structure and small amounts of metal ions. The fg-HAp ceramics exhibited macro-pore sizes of 100-800 μm, porosities of 60-80%, and specific surface areas of 30-40 $m^2 \cdot g^{-1}$.

Bone morphogenetic protein (BMP) accelerates osteoinduction, and it has been added to various scaffold materials due to the growth factor ability to induce bone and cartilage in intramuscular and subcutaneous tissues (Murata et al., 2009a; 2009b; 2010). Interconnected porous HAp can be a superior carrier of recombinant human BMP-2 (rhBMP-2) in ectopic and orthotopic sites. Gradations in the grain size and crystallinity of the HAp plus rhBMP-2-addition were used to achieve the desired bio-absorption and osteoinduction characteristics.

In first vivo-experiment, the fg-HAp and rhBMP-2-loaded fg-HAp (designated as rhBMP-2/fg-HAp) ceramics were implanted into the subcutaneous tissue of the back region in rats (Akazawa et al., 2005; Murata et al., 2007). At 4 weeks after the implantation, for the fg-HAp, body-fluid permeated into bulk regions of HAp through micro-pores, while for the rhBMP-2/fg-HAp, fast bio-absorption and osteoinduction were recognized.

In second vivo-experiment, the rhBMP-2/fg-HAp ceramics were implanted into the subcutaneous tissue of the back region in nude mice to look into release characteristics of rhBMP-2 from the ceramics (Tazaki et al., 2006; Kawakami et al., 2007). At 7 days after the implantation, retention percentage of ^{125}I-labeled rhBMP-2 was about 60%, suggesting that fg-HAp ceramics are ideal osteoinductive scaffolds.

In third vivo-experiment, the rhBMP-2/HAp ceramics were implanted on the periosteum of biparietal bone in rats (Hino et al., 2008). At 4 weeks after the implantation, degradation and bio-absorption of HAp, bone formation involving HAp fragmented, giant cells around HAp surface, and osteoblasts around new bone were observed. The periosteum played a role of tissue boundary film possible for separation and excision.

If a dissolution-precipitation process for b-HAp porous structure change microstructure of HAp in a very short treatment-time to make up biomimetic fg-HAp ceramics, the process will be expected a design and control-technology of living tissues for bone-regenerative therapy (Akazawa et al., 2007).

Hence, focusing on development and application of the supersonics, ultrasonic echo, ultrasonic knife, ultrasonic microscope, ultrasonic cleaning, ultrasonic disintegration, ultrasonic sterilization and ultrasonic therapy for fracture have been widespread in the medical and dental field (Akazawa et al., 2010a). Supersonic wave more than the frequency of 20 KHz can bring bubble-cavitation and make hot spot. In the hot spot, many chemical reactions will be activated by the formation of radical groups and the locally rising temperature (Rae et al., 2007; Yasuda et al., 2009).

Surface structure design of b-HAp ceramics by a supersonic treatment might easily produce new scaffolds which control the bio-absorption rate and the adsorption ability for proteins or cells. Concerning the commercial HAp products synthesized from reagents, modification of microstructure may be able to improve bio-absorption and tissue-affinity of HAp ceramics by selecting partial dissolution-precipitation conditions (Akazawa et al., 2009c). Provided that a vigorous dissolution-precipitation process for medical HAp products vary microstructure of HAp in a short treatment-time and construct biomimetic HAp ceramics, most of surgeons or medical users will adopt the process as a design and control-technology before the implant-operation for the purpose of advancement of therapeutic grade (Akazawa et al., 2007; 2009b).

In first section, advantage and effectiveness of the fg-HAp ceramics for bone-regeneration together with the supersonic techniques will be introduced and discussed in detail. The spongy b-HAp ceramics using bovine bone were partially dissolved by the two ways, such as stirring for hours and supersonic treatment for minutes. The fg-HAp ceramics originated from bovine or porcine bone were designed by the partial dissolution-precipitation techniques along with the stirring-supersonic treatment. Effects of supersonic treatment-condition on the biomimetic microstructure and biocompatibility of fg-HAp ceramics were investigated.

In the second section, the partial dissolution-precipitation techniques along with the supersonic treatment will be applied for surface modification of commercial HAp ceramics. Porous and dense HAp ceramics which have been sold as commercially medical products or research products were easily modified by the partial dissolution-precipitation techniques. Effects of supersonic dissolution-condition on the microstructure of the HAp ceramics were clarified. From in vivo-experiments, histology and histomorphometry of the partially-dissolved and precipitated HAp (designated as PDP-HAp) ceramics were evaluated in contrasting with those of non-treated HAp ceramics.

2. Surface function design of animal bone-originated HAp ceramics by a partial dissolution-precipitation technique with supersonic treatment

2.1 Dissolution characteristics of spongy bone-originated HAp ceramics

Bovine and porcine bones (Hokkaido Meat Packer Product Industry Ltd. Co.) were used as starting materials and cut to some pieces of the cubic samples. They were boiled to eliminate marrow and most of collagen components from animal bone. They were sequentially calcined at 773-1473K to obtain crystalline HAp (b-HAp) ceramics.

The pulverized cortical b-HAp was completely dissolved into a HNO_3 aqueous solution, while the spongy b-HAp was partially dissolved in the HNO_3 aqueous solution saturated with respect to apatite by the two different ways, such as stirring at 293K and 300rpm or/and supersonic treatment at 120W and 38kHz (Akazawa et al., 2009a; 2009c).

For dissolution characteristics of the ceramics, the dissolution efficiencies were calculated from difference in weight of the ceramics before and after the stirring or supersonic treatment in the HNO_3 aqueous solutions.

Figure 1 shows dissolution efficiencies of spongy b-HAp ceramics obtained by the calcination of bovine bone and the supersonic treatment (Akazawa et al., 2009a; 2009c). The dissolution efficiency decreased with increasing calcination temperature of animal bone. This phenomenon would be caused by the decrease in surface area contacting with the HNO_3 aqueous solution because b-HAp grains grew and surface area of grain boundary decreased with increasing calcination temperature, although mechanical strength of b-HAp increased by sintering between the particles.

At 1073K, the dissolution efficiency of b-HAp by the supersonic treatment was significantly higher than that by the stirring. For example, the dissolution efficiency for 5 min-supersonic dissolution was almost same as that for 15h-stirring dissolution.

Scanning electron microscopic (SEM) and Digital microscopic (DM) photographs of fg-HAp ceramics were taken to observe the morphology and microstructure. Figure 2 shows SEM photographs of the b-HAp ceramics derived from bovine bone and treated by the supersonic dissolution (Akazawa et al., 2009a; 2009c). For the b-HAp calcined at 1073K, grains with less than 1 μm and micro-pores were seen, while for the b-HAp calcined at 1473K, large plate-like grains with 3-8 μm were observed. In both the cases after the supersonic treatment, angle-free grains, a lot of micro-pores, and developing micro-cracks were recognized.

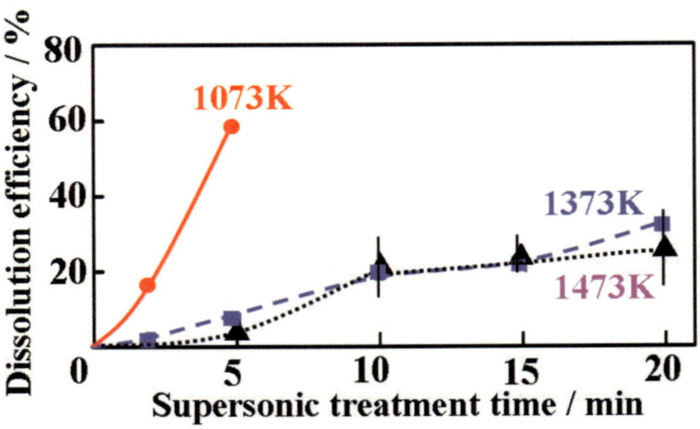

Fig. 1. Dissolution efficiencies of spongy b-HAp ceramics calcined at 1073-1473K for 24h in air and partially-dissolved by supersonic treatment at 120W and 38 kHz in the HNO_3 aqueous solutions saturated with respect to apatite.
●: b-HAp ceramics calcined at 1073K.
■: b-HAp ceramics calcined at 1373K.
▲: b-HAp ceramics calcined at 1473K.

2.2 Surface characteristics and microstructure of fg-HAp ceramics derived from animal bone

Comparing with only the stirring dissolution, the stirring-supersonic dissolution shortened partial dissolution time to one tenth and it made up micro-pores and micro-cracks that are efficient for body fluid permeation. The b-HAp ceramics calcined at 1073K derived from bovine or porcine bone were partially dissolved in the HNO_3 aqueous solution saturated with respect to apatite by the stirring for 15 min, the supersonic treatment for 2 min, and the stirring for 13min. The dissolution efficiencies of b-HAp gave 4% for the first process, 34% for the second, and 41% for the third, respectively.

By gradually adding a NH_3 aqueous solution at pH 9-11 and 298K into the solution, HAp microcrystals were precipitated on the surfaces and pore-walls of spongy b-HAp. The modified spongy b-HAp was aged for 24h, washed, and dried at 323K to form fg-HAp ceramics (Akazawa et al., 2009c).

The crystalline phase of the samples was identified by Micro-X-ray diffraction (Micro-XRD) using Cu Kα1 radiation. Micro-XRD patterns of the fg-HAp ceramics showed that in both cases, the degree of crystallinity of HAp single phase gradually distributed better from the pore surface layer to bulk region of the b-HAp body structure. The lattice parameters of HAp phase in the bulk region were a = 0.942 nm and c = 0.688 nm, whose values were in good agreement with those of the JCPDS card (9-432).

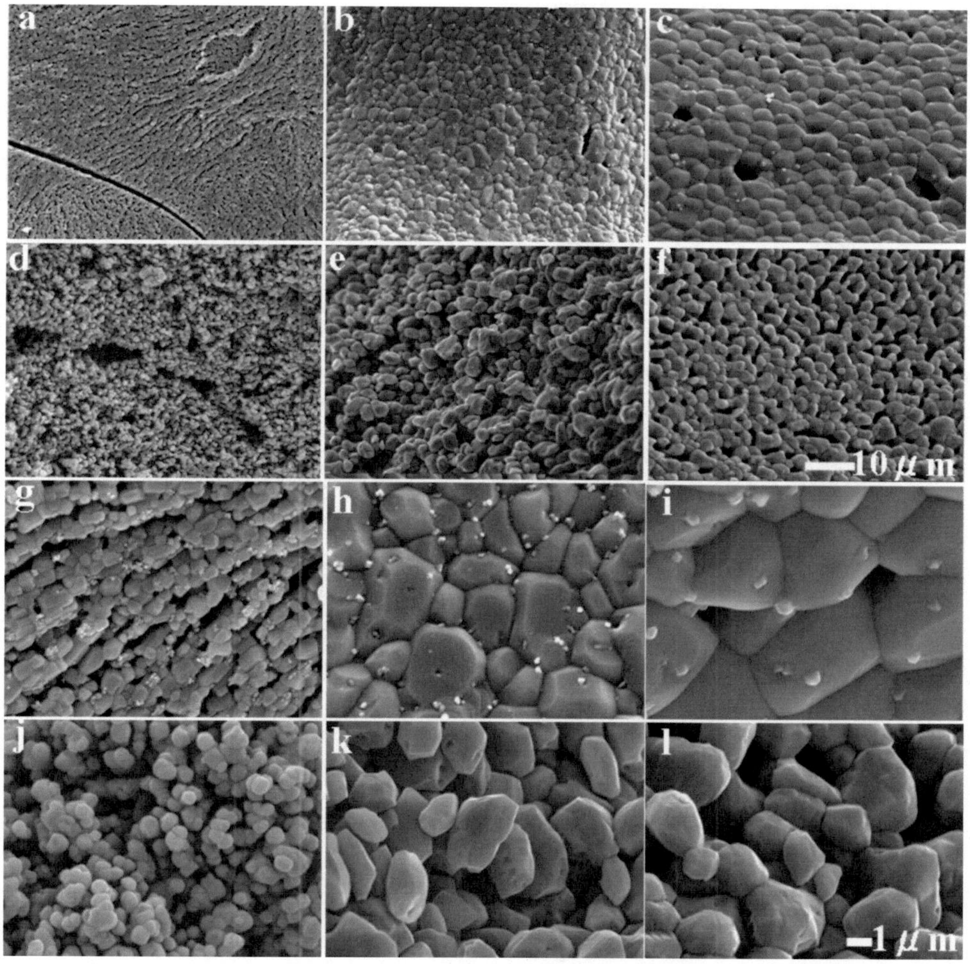

Fig. 2. SEM photographs of spongy b-HAp ceramics using bovine bone calcined at 1073-1473K for 24h in air and partially-dissolved by supersonic treatment at 120W and 38 kHz in the HNO_3 aqueous solution saturated with respect to apatite.
a: b-HAp ceramics calcined at 1073K (x2000 magnification).
b: b-HAp ceramics calcined at 1273K (x2000 magnification).
c: b-HAp ceramics calcined at 1473K (x2000 magnification).
d: b-HAp ceramics calcined at 1073K and partially-dissolved for 5 min (x2000 magnification).
e: b-HAp ceramics calcined at 1273K and partially-dissolved for 20 min (x2000 magnification).
f: b-HAp ceramics calcined at 1473K and partially-dissolved for 20 min (x2000 magnification).
g: b-HAp ceramics calcined at 1073K (x10000 magnification).
h: b-HAp ceramics calcined at 1273K (x10000 magnification).
i: b-HAp ceramics calcined at 1473K (x10000 magnification).
j: b-HAp ceramics calcined at 1073K and partially-dissolved for 5 min (x10000 magnification).
k: b-HAp ceramics calcined at 1273K and partially-dissolved for 20 min (x10000 magnification).
l: b-HAp ceramics calcined at 1473K and partially-dissolved for 20 min (x10000 magnification).

The composition ratios of Ca^{2+} and PO_4^{3-} ion (Ca/P) were determined by electron probe microanalysis (EPMA). The quantitative analyses of inorganic components were measured by using inductively coupled plasma (ICP). The fg-HAp was Ca^{2+}-deficient HAp with the (Ca/P) ratios of 1.64-1.66 containing small amounts of Na^{+} and Mg^{2+} ions.

Figure 3 shows DM and SEM photographs of the fg-HAp ceramics derived from bovine bone (Akazawa et al., 2009a; 2009c). Spherical moss-like grains with 1μm in size, which consisted of about 100 nm needle-like microcrystals and many micro-pores, were recognized.

Evaluating surface structure, the specific surface areas and pore size distribution were measured from N_2-adsorption at 77K. The porosities of the ceramics were estimated by the water displacement method. The fg-HAp ceramics exhibited porosities of 60-80%, macro-pore sizes of 100-800 μm, micro-pores of 10-160 nm, and specific surface areas of 30-50 $m^2 \cdot g^{-1}$.

Fig. 3. DM and SEM photographs of fg-HAp ceramics obtained by the partial dissolution-precipitation of spongy b-HAp calcined at 1073K for 24h using bovine bone. The dissolution-precipitation involves a stirring for 15 min in the HNO_3 aqueous solution saturated with respect to apatite, a supersonic treatment at 120W and 38 kHz for 2 min, a stirring for 13 min, and a precipitation at 298K and pH 10.5 for 24h.
a: DM image of fg-HAp ceramics.
b: SEM image of fg-HAp ceramics (x300 magnification).
c: SEM image of fg-HAp ceramics (x2000 magnification).
d: SEM image of fg-HAp ceramics (x10000 magnification).

Based on these results above, it is found that the supersonic dissolution-precipitation is an effective process to shorten the partial dissolution time of b-HAp ceramics and design the biomimetic microstructure of fg-HAp ceramics that accelerate degradation-bio-absorption and body fluid permeation of the ceramics in a living body.

2.3 Bone-bonding ability of fg-HAp ceramics in a biomimetic environment

To preliminary evaluate bone-bonding ability, fg-HAp ceramics originated from bovine or porcine bone were soaked at 309.5K and pH 7.40 for 4 weeks in a simulated body fluid (SBF) which means the solution where ion concentrations and pH nearly equal to those in human plasma. Microstructure of the fg-HAp ceramics was shown in Figure 4 (Akazawa et al., 2006c; 2009a).

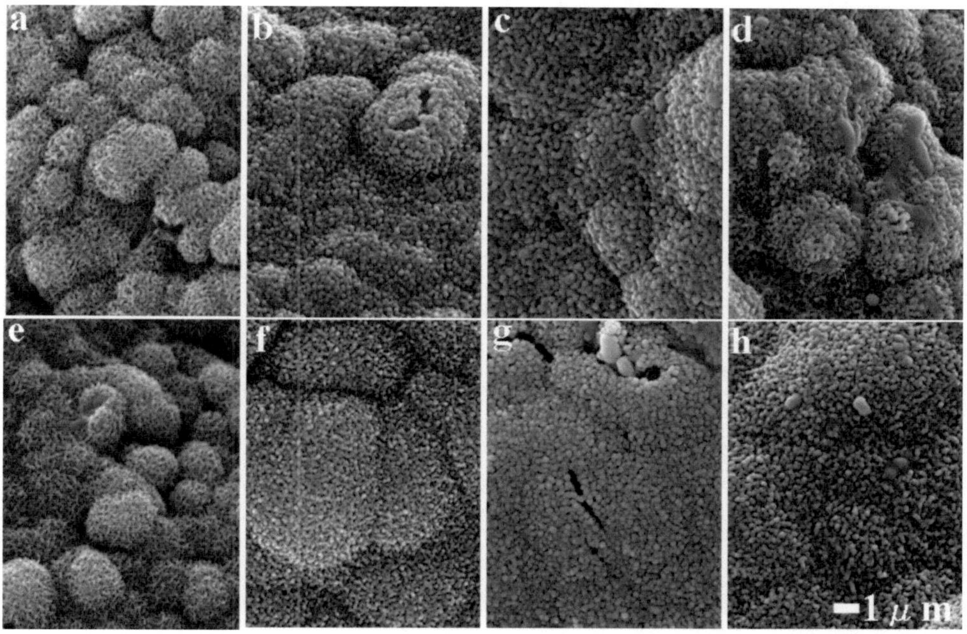

Fig. 4. SEM Photographs of the fg-HAp ceramics soaked at 309.5K and pH 7.40 in a SBF solution after the partial dissolution-precipitation with supersonic treatment.
a: bovine bone-originated fg-HAp ceramics at 1 day after the soaking.
b: bovine bone-originated fg-HAp ceramics at 3 day after the soaking.
c: bovine bone-originated fg-HAp ceramics at 5 day after the soaking
d: bovine bone-originated fg-HAp ceramics at 7 day after the soaking
e: porcine bone-originated fg-HAp ceramics at 1 day after the soaking.
f: porcine bone-originated fg-HAp ceramics at 3 day after the soaking.
g: porcine bone-originated fg-HAp ceramics at 5 day after the soaking.
h: porcine bone-originated fg-HAp ceramics at 7 day after the soaking.

Even at 1 day after the soaking, microstructure of fg-HAp changed from small grains to dense cocoon-like ones by rapid precipitation of HAp microcrystals, suggesting that the fg-HAp keeps high bone-bonding ability because ideal biomaterials bond to living bone through an apatite layer formed on the surfaces in a living body. The fg-HAp surfaces will give larger amounts of proteins adsorbed and higher adsorption heats for proteins than the b-HAp surfaces.

2.4 Biocompatibility and bio-absorption of fg-HAp ceramics

The fg-HAp ceramics having sufficient strength for operation were implanted into the subcutaneous tissues of back region in 4-week-old male Wistar rats (Akazawa et al., 2006c; 2009a). At 3 and 8 weeks after the implantation, the sample blocks were explanted, fixed in neutral buffered formalin, decalcified with formic acid, embedded in paraffin, sectioned and stained with hematoxylin and eosin. They were histologically evaluated using an optical microscope.

Figure 5 shows photomicrographs of HE sections at 3 week after the implantation of fg-HAp ceramics (Akazawa et al., 2006c; 2009a). At 3 weeks after the implantation, body fluid well permeated and diffused into bulk regions of HAp through micro-pores of the ceramics. Microcracks in bulk regions of the ceramics formed by the stirring-supersonic treatment would result in auto-degradation and body fluid permeation.

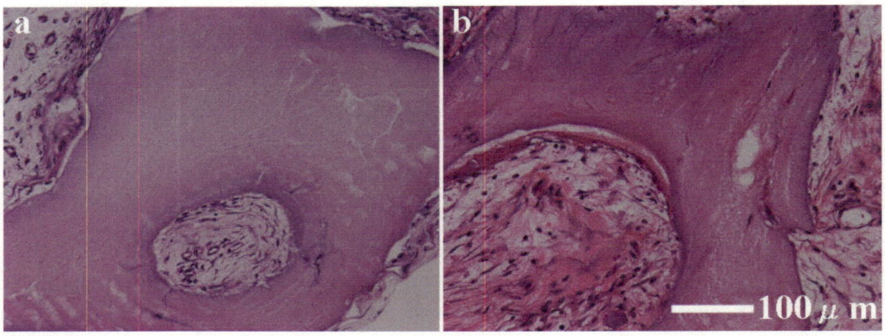

Fig. 5. Photomicrographs of HE sections at 3 week after the implantation of fg-HAp ceramics obtained by the partial dissolution-precipitation of spongy b-HAp calcined at 1073K using animal bone.
a: bovine bone-originated fg-HAp ceramics.
b: porcine bone-originated fg-HAp ceramics.

At 8 weeks, the fg-HAp ceramics containing body fluid were degraded from both the surface layer of the ceramics and bulk region, and many giant cells appeared on the surface and inside of the bulk structure of the ceramics. Surface- and bulk-degradations of the fg-HAp proceeded, so that a total size of the ceramics dramatically decreased.

2.5 Conclusions

Biomimetic microstructure of fg-HAp ceramics was effectively designed using bovine or porcine bone calcined at 1073K (b-HAp) by the partial stirring-supersonic dissolution and precipitation of spongy b-HAp ceramics. The partial dissolution processes of spongy b-HAp ceramics by the stirring at 293K for 28 min in a HNO_3 aqueous solution and the supersonic treatment at 120W and 38 kHz for 2 min produced convenient micro-pores and micro-cracks for body fluid permeation. The fg-HAp ceramics exhibited macro-pore sizes of 100-800 μm, porosities of 60-80% and specific surface areas of 30-50 $m^2 \cdot g^{-1}$.

After the soaking at 309.5K and pH 7.40 in a SBF solution, microstructure of fg-HAp changed from small grains to dense cocoon-like ones, suggesting bone-bonding ability and biocompatibility of the ceramics. The fg-HAp porous ceramics were implanted into the subcutaneous tissues of back region in rats. At 4 weeks after the implantation, body fluid well permeated from surface into bulk regions of the ceramics and auto-degradation of fg-HAp was recognized.

Based on the results of the soaking tests in a SBF solution and the animal experiments, it was found that the fg-HAp ceramics prepared by the supersonic treatment would be biomimetic materials that indicate fast bio-absorption and excellent biocompatibility for bone-regenerative therapy.

3. Surface function design of commercial HAp ceramics by a partial dissolution-precipitation technique with supersonic treatment

3.1 Partial dissolution characteristics of porous HAp products

Commercial HAp of medical products (HOYA Corporation: APACERAM (G-S-10 granules, AX A-1 ceramics (85% in porosity), B-4-1010 ceramics (55% in porosity) (Sakamoto et al., 2007)) were used as starting materials. The G-S-10 granules were completely dissolved in 0.39N-HNO_3 aqueous solutions to be 1.7 or 3.4×10^{-2}N-HNO_3 aqueous solutions containing 4.98 or 9.96×10^{-2}M-Ca^{2+} and 2.98 or 5.96×10^{-2}M-PO_4^{3-} ions. Then, the porous AX A-1 or B-4-1010 ceramics were cut to 5×5×5 mm-cubic samples and each of the samples was impregnated into the HNO_3 aqueous solutions containing Ca^{2+} and PO_4^{3-} ions saturated with respect to apatite and partially-dissolved by the supersonic treatment at 120W, 38 kHz, 293-315K and pH 1.0 for 5-75 min (Akazawa et al., 2009b; 2010a). Consequently, degradation of porous ceramics was controlled and frame network of porous ceramics was substantially preserved through the dissolution process.

Figure 6 shows partial dissolution characteristics for the porous HAp products treated by the supersonic treatment (Akazawa et al., 2009b; 2009c; 2010a).

Along with supersonic treatment-time, liquid temperature of the HNO_3 aqueous solutions raised from 293K to 315K because of accumulation of supersonic energy, which implies that the dissolution condition became harder environment.

Dissolution efficiencies of all the HAp products increased with supersonic treatment-time and mechanical strength decreased due to the strong acid etching at pH 1.0. Higher concentration of the HNO_3 aqueous solutions gave higher dissolution efficiency. Dissolution efficiencies for the AX A-1 were much higher than those for the B-4-1010 because the AX A-1 had higher porosity and surface area than the B-4-1010.

In comparison with the stirring dissolution, for the AX A-1 treated by the supersonic dissolution in 1.7×10^{-2}N-HNO_3 aqueous solutions, the dissolution efficiency at the supersonic treatment-time of 15 min gave 46%, whose value was same as that at the stirring time of 6h. Microstructure of the AX A-1 at the supersonic treatment-time of 30 min was similar as that at the stirring time at 20h, suggesting an advantage of the supersonic dissolution method. The supersonic dissolution shortened partial dissolution time to about one thirtieth and it made up larger micro-pores and micro-cracks (Akazawa et al., 2009b; 2009c; 2010a).

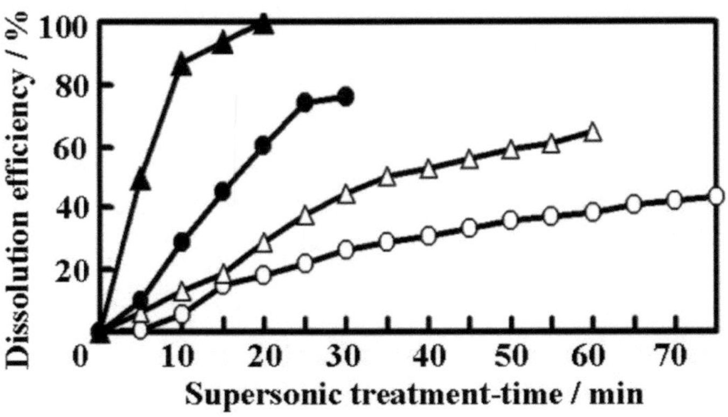

Fig. 6. Partial dissolution characteristics for porous HAp products treated by the supersonic technique at 120W, 38kHz, 293-315 K and pH 1.0 in the HNO_3 aqueous solutions containing Ca^{2+} and PO_4^{3-} ions saturated with respect to apatite.
▲:AX A-1 ceramics treated in 3.4 $\times 10^{-2}$N-HNO_3 aqueous solutions.
●:AX A-1 ceramics treated in 1.7$\times 10^{-2}$N-HNO_3 aqueous solutions.
△:B-4-1010 ceramics treated in 3.4 $\times 10^{-2}$N-HNO_3 aqueous solutions.
○:B-4-1010 ceramics treated in 1.7$\times 10^{-2}$N-HNO_3 aqueous solutions.

For the B-4-1010, strong connection between the sintered grains was kept by the supersonic treatment. However, in 3.4$\times 10^{-2}$N-HNO_3 aqueous solutions, some craters 100-300 μm in diameter, which composed of independent grains, were produced by bubble-cavitation during the supersonic treatment.

Based on these results above, it was suggested that the bioceramics with suitable dissolution efficiency and mechanical strength would be easily designed using commercial HAp products by the supersonic treatment techniques in various acid solutions.

3.2 Surface characteristics of partially dissolved and precipitated HAp ceramics

The AX A-1 ceramics with high porosity were selected and partially dissolved in 1.7$\times 10^{-2}$N-HNO_3 aqueous solution by the stirring of 300 rpm at 298K and pH 1.0 for 30 min and the subsequently supersonic treatment at 120W, 38 kHz and 293-308K for 25 min. Nano-crystals consisting of calcium phosphate were reprecipitated on the pore-wall surface in the macro-pores and micro-pores of the ceramics by adding an NH_3 aqueous solution. They were aged at pH 9-10 and 298K for 24h under a bubbling of N_2 gas in the solution supersaturated with respect to apatite. The modified HAp was carefully washed with distilled water, and dried at 323-393 K to fabricate the PDP-HAp ceramics (Akazawa et al., 2009b; 2009c; 2010).

The quantitative analyses of Ca^{2+} and PO_4^{3-} ions in the solutions were conducted using ICP. The concentrations of Ca^{2+} and PO_4^{3-} ions in the solution after the partial dissolution were 7.93$\times 10^{-2}$M and 4.75$\times 10^{-2}$M, respectively. From Micro-XRD patterns of the PDP-HAp ceramics, single phase of HAp was identified in both the surface layer and bulk region. The (Ca/P) ratios were 1.64-1.66. The qualitative analysis of carbonate ions was carried out by Fourier transform-infrared spectroscopy and carbonate ions were not detected.

Fig. 7. SEM photographs of HAp ceramics before and after the PDP-treatment involving supersonic dissolution. PDP-HAp ceramics were obtained by the stirring of 300rpm at 298K and pH 1.0 for 30min in 1.7×10^{-2}N-HNO_3 aqueous solutions saturated with respect to apatite and the supersonic treatment at 120W, 38 kHz and 293-308K for 25 min and the precipitation at 298K and pH 9-10 for 24 h.
a: commercial HAp ceramics (APACERAM AX A-1, HOYA Co.) (x300 magnification).
b: commercial HAp ceramics (APACERAM AX A-1, HOYA Co.) (x5000 magnification).
c: commercial HAp ceramics (APACERAM AX A-1, HOYA Co.) (x10000 magnification).
d: PDP-HAp ceramics (x300 magnification).
e: PDP-HAp ceramics (x5000 magnification).
f: PDP-HAp ceramics (x10000 magnification).

Figure 7 shows SEM photographs of the PDP-HAp ceramics involving the supersonic treatment (Akazawa et al., 2009b; 2009c; 2010a). Microstructures of the PDP-HAp changed to spherical moss-like grains forming needle-like nano-crystals, micro-pores, and micro-cracks on the grains, which would be efficient for body fluid-permeation and adsorption for plasma-proteins or cells. The PDP-HAp ceramics exhibited macro-pore sizes of 50-200 μm, porosities of 80-90%, specific surface areas of 1-2 $m^2 \cdot g^{-1}$.

3.3 Partial dissolution characteristics of dense HAp products

To further investigate effects of supersonic dissolution on the microstructure of HAp ceramics, commercial and dense HAp pellets of research products (HOYA Corporation: CELLYARD HAp pellets (13 mm in diameter, no open pore) (Ogawa et al., 2003)) were used as starting materials and treated by the supersonic technique at 120W, 38 kHz, and pH 1.0.

For the HAp pellets, many micro-cracks were induced on the almost flat surface by the supersonic treatment for a short time, although the pellets gave low dissolution efficiencies of 1.8-3.8% in 1.7×10^{-2}N-HNO_3 aqueous solution saturated with respect to apatite.

The microstructure of HAp ceramics before and after the supersonic dissolution-treatment was shown in Figure 8 (Akazawa et al., 2010a). The HAp pellets that were dense HAp ceramics with 95% in relative density had submicron-grains and closed pores.

At the supersonic treatment-time of 10-20 min, dense HAp ceramics were significantly etched to form uneven surface and make up many craters 5-10 μm in diameter. Especially, at 20 min, deep and long crinkle-like cracks were formed by accumulation of supersonic energy. Propagation of micro-cracks near the grain boundary in the ceramics would result in inter-and-trans-granular destruction and facilitate bio-degradation in vivo-experiments. The micro-cracks will be very necessary for normal bone-metabolism or bone-remodeling because there are some micro-cracks in various parts and regions of a living bone.

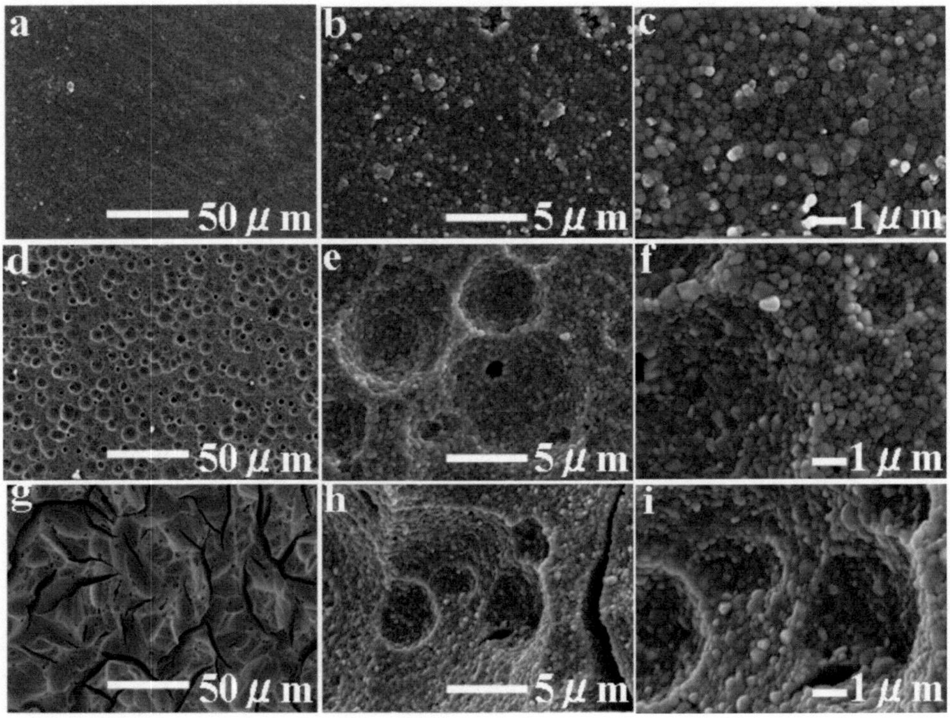

Fig. 8. SEM photographs of dense HAp products treated by the supersonic technique at 120W, 38 kHz, 293-300K and pH 1 in 1.7×10^{-2}N-HNO_3 aqueous solutions saturated with respect to apatite.
a: commercial HAp pellets (CELLYARD, HOYA Co.) (x500 magnification).
b: commercial HAp pellets (CELLYARD, HOYA Co.) (x5000 magnification).
c: commercial HAp pellets (CELLYARD, HOYA Co.) (x10000 magnification).
d: commercial HAp pellets treated by the supersonic dissolution for 10min (x500 magnification).
e: commercial HAp pellets treated by the supersonic dissolution for 10min (x5000 magnification).
f: commercial HAp pellets treated by the supersonic dissolution for 10min (x10000 magnification).
g: commercial HAp pellets treated by the supersonic dissolution for 20min (x500 magnification).
h: commercial HAp pellets treated by the supersonic dissolution for 20min (x5000 magnification).
i: commercial HAp pellets treated by the supersonic dissolution for 20min (x10000 magnification).

3.4 Bio-absorption and osteoconduction of PDP-HAp ceramics

Twenty-four adult female of Japanese white rabbits were used and bone defects were bilaterally made at the medial condyle of femur. The HAp products and PDP-HAp ceramics that have sufficient strength for operation were implanted into the bone defects. Eight rabbits were sacrificed at 4, 8, and 16 weeks after the implantation, respectively.

Investigating bio-affinity and biocompatibility of HAp ceramics, both the PDP-HAp and AX A-1 ceramics were implanted into the bone defects in rabbits. Actually, the PDP-HAp ceramics had mechanical strength enough for handling of operation or implantation, in comparison with that of the AX A-1 ceramics.

Micro-computed tomography (Micro-CT) scanning in the axial plane was taken to evaluate the bio-absorption of the implants. Figure 9 shows Micro-CT photographs of different HAp ceramics implanted in the medial condyle of femur in rabbits (Akazawa et al., 2010a; Ding et al., 2010; 2011). At 4-16 weeks after the implantation, the total volume of the PDP-HAp gradually tended to decrease and the shape changed from cubic to angle-free materials, however, those of the AX A-1 were almost similar. Bio-absorbability of the PDP-HAp ceramics will be important for maintaining bone inside implanted ceramics.

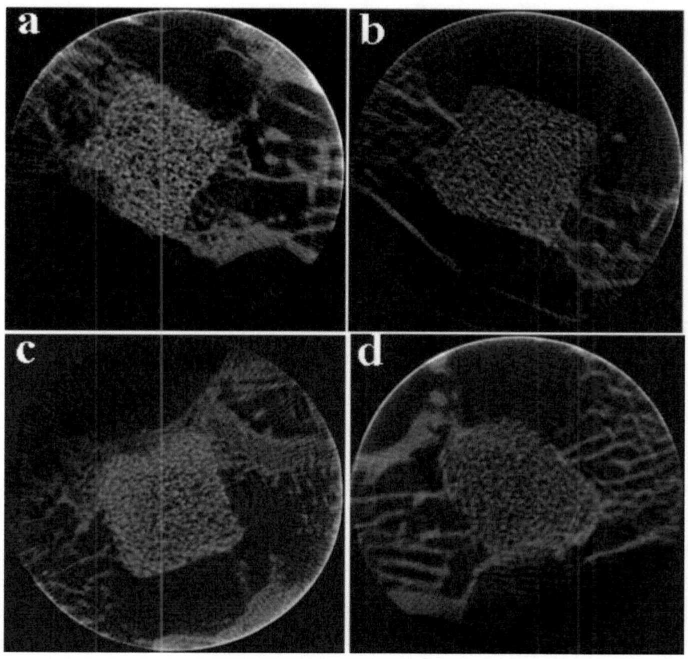

Fig. 9. Micro-CT photographs of different HAp ceramics implanted into the medial condyle of femur in rabbits.
a: AX A-1 ceramics at 4 weeks after the implantation.
b: AX A-1 at 16 weeks after the implantation.
c: PDP-HAp at 4 weeks after the implantation.
d: PDP-HAp at 16 weeks after the implantation.

The extracted specimens that were not decalcified but stained with toluidine blue O or tartrate-resistant acid phosphatase (TRAP) were observed by light microscopy.

Figure 10 shows photomicrographs of different HAp ceramics implanted into the medial condyle of femur in rabbits and stained with toluidine blue O (Akazawa et al., 2010a; Ding et al., ; 2011). Although superior osteoconduction for the PDP-HAp and AX A-1 was observed, at 16 weeks, the amounts of bone tissue present for the PDP-HAp were larger than those for the AX A-1, suggesting that the PDP-HAp ceramics induced biological bone-remodeling balanced.

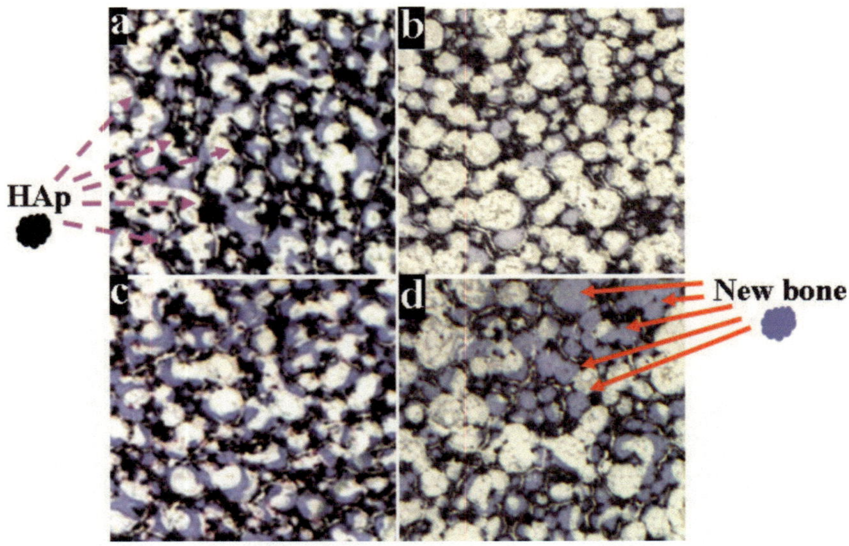

Fig. 10. Photomicrographs of different HAp ceramics implanted into the medial condyle of femur in rabbits and stained with toluidine blue O.
a: AX A-1 ceramics at 4 weeks after the implantation.
b: AX A-1 ceramics at 16 weeks after the implantation.
c: PDP-HAp at 4 weeks after the implantation.
d: PDP-HAp at 16 weeks after the implantation.

At 8 and 16 weeks, other extracted specimens stained with TRAP for both the ceramics indicated multinuclear osteoclasts. For the PDP-HAp, a lot of osteoclasts were directly adsorbed on the surface of HAp and tissue fusion around the ceramics was smoothly advanced. These results can be explained by the body fluid-permeation of PDP-HAp from the surface layers into the bulk regions of the ceramics through the micro-pores or micro-cracks. The partial dissolution-precipitation involving the supersonic treatment resulted in surface modification of the AX A-1 ceramics to a sort of fg-HAp structure which can stimulate the activity or differentiation for osteoclasts and accelerate the bio-absorption of HAp ceramics. Moreover, the surface layers of the PDP-HAp, such as nano-crystals may disintegrate on solid-liquid interface in vivo.

Accordingly, it was found that the partially supersonic dissolution-precipitation technique enhanced the bio-absorption ability of HAp and would activate the bone-metabolic system.

3.5 Conclusions

Microstructure of commercial porous HAp ceramics drastically modified to the fg-HAp structure by the specific dissolution-precipitation techniques, which involved a stirring of 300 rpm at 298K in the HNO_3 aqueous solutions containing Ca^{2+} and PO_4^{3-} ions or /and a supersonic treatment at 120W and 38 kHz in the same solutions.

The dissolution efficiency of porous HAp products by the supersonic treatment drastically increased with time, depending on the porosity of ceramics and the concentration of HNO_3 aqueous solution. For even dense HAp products, enhancement of micro-pores and propagation of micro-cracks were recognized by the supersonic technique at 120W, 38 kHz, and pH 1.0 for 10-20 min. Commercial dense HAp pellets were significantly etched to form uneven surface and make up many craters and thunder-like cracks. The partially supersonic dissolution-treatment will be used as a convenient and effective preparation technology for production of bio-absorbable and bioactive ceramics.

After the stirring of 30 min and the subsequently supersonic treatment of 25 min, HAp nano-crystals with the (Ca/P) of 1.64-1.66 were successfully precipitated on the pore wall surface in the macro-pores and micro-pores of the ceramics at 298K and pH 9-10.

The PDP-HAp ceramics that gave macro-pore sizes of 50-200 μm and porosities of 85-90% were implanted into the bone defects at the medial condyle of femur in Japanese white rabbits. At 8 and 16 weeks after the implantation, the PDP-HAp ceramics exhibited more excellent bio-absorption and tissue-affinity than commercial HAp products because of smooth body-fluid-permeation and effective surface nature for cell-adsorption. Superior osteointegration of the PDP-HAp induced biological bone-remodeling balanced.

The PDP-HAp ceramics may be clinically applied for bone tissue-substituted materials with excellent osteointegration when they are completely bio-absorbed in vivo by optimizing the partial dissolution-precipitation conditions..

4. Biomaterials controlling biological balance and sonochemistry

Microstructure and chemical nature-control of biomaterials is an inevitable and great challenge for bone-regenerative medicine therapy to increase therapeutic grade and improve quality of lives of patients. In these days, we are focusing on development and application of the bioabsorbable and biomimetic materials synchronized with living tissues and reformed by the partial dissolution-precipitation techniques with supersonic treatment (Akazawa et al. 2010a).

Concerning an ideal implantation cure of biomaterials, it will important for the cell or growth factor-loaded bioceramics and the gene-transferred biomaterials that were coated with a thin HAp layer having protein and DNA to keep donating suitable amounts of physical and chemical stresses for a certain period of time to biological tissues near the implanted region (Akazawa et al. 2010b). This phenomenon would bring high cell-activity and excellent bone-remodelling of the biological tissues. Moreover, balance control of autonomic nervous system and immune system must be necessary for various cells to accept biomaterials as life environment without allograft rejection and continue receiving moderate stress.

From the viewpoints of creating biomaterials controlling biological balance, the partial dissolution-precipitation method using calcium phosphate aqueous solutions may be efficient for design and preparation of microstructure promoting degradation-bio-absorption and body fluid permeation as a wet synthetic technology which can add biomimetic function into the b-HAp ceramics. Also, commercial bioceramics would be easily modified to reconstruct microstructure and improve bio-absorption and osteoconduction characteristics by optimizing the partial dissolution-precipitation conditions.

Particularly, sonochemistry related to the supersonic dissolution is expected to be applied for bone-regenerative medicine as a simplified design and preparation technology of absorbable and biomimetic bioceramics in the future.

5. References

Akazawa, T., Murata, M., Sasaki, T., Tazaki, J., Kobayashi, M., Kanno, T., Matsushima, K.,Itabashi, K., & Arisue, M. (2005). Bio-absorption and osteoinduction innovation of bone morphogenetic protein-supported functionally graded apatites originated from cattle bone. *J. Am. Ceram. Soc.,* 88.,12., 3545-3548.

Akazawa, T., Murata, M., Arisue, M., Kanno, T., & Kobayashi, M., (2006a). *Japanese Patent,* 3718723.

Akazawa, T., Murata, M., Sasaki, T., Tazaki, J., Kobayashi, M., Kanno, T., Matsushima, K., & Arisue, M. (2006b). Biodegradation and bioabsorption innovation of the functionally graded cattle-bone-originated apatite with blood compatibility. *J. Biomed. Mater. Res.,* 76A., 1., 44-51.

Akazawa, T., Murata, M., Tazaki, J., Nakamura, K., Itabashi, K., Kanno, T., & Kobayashi, M. (2006c). Surface structure design and characterization of bioabsorbable and functionally graded apatites originated from bovine bone. *Key Engineering Materials,* 309-311, 1051-1054.

Akazawa, T., Murata, M., Hino, J., Nakamura, K., Tazaki, J., Kikuchi, M., & Arisue, M. (2007). Materials design and application of demineralized dentin/apatite composite granules derived from human teeth. *Archives of BioCeram. Res.,* 7., 25-28.

Akazawa, T., Murata, M., Tazaki, J., Nakamura, K., Hino, J., Ito, K., Yamamoto, M., Tabata, Y., Takahata, M., & Ito, M., (2009a). Biomimetic microstructure and biocompatibility of functionally graded hydroxyapatite derived from animal bone by a supersonic dissolution-precipitation method.*Bioceramics,* 22., 155-158.

Akazawa, T., Murata, M., Hino, J., Tazaki, J.,Ito, K., Nakamura, K., Takahata, M, Abe, Y., Xianjun, D., Itoh, M., Nakajima, T., & Sakamoto, M. (2009b). Microstructure and bio-absorption characteristics of hydroxyapatite modified by a partially supersonic dissolution-precipitation technique. *Archives of BioCeram. Res.,* 9., 123-126.

Akazawa, T., Nakamura, K., Itabashi,K., Inano, H., Horikawa, H., Takahashi,H., Yoshinari, S., Murata, M., Tazaki, J., Hino, J., Tabata, Y., Yamamoto, M., Hanawa, T., Yamachika, H., Ohmori, T., &Kikuchi, M. (2009c). Development of bio-absorbable and biomimetic apatite and its application for bone-regenerative therapy. *Reports of the Hokkaido Indus. Res. Inst.,* 308., 47-55.

Akazawa, T., Murata, M., Takahata, M., Xianjun, D., Abe, Y., Nakamura, K., Hino, J., Tazaki, J., Ito, K., Ito, M., Iwasaki, N., Minami, A., Nakajima, T., & Sakamoto, M. (2010a). Characterization of microstructure and bio-absorption of the hydroxyapatite ceramics modified by a partial dissolution-precipitation technique using supersonic treatment. *J. Ceram. Soc. Jpn.*, 118., 6., 535-540.

Akazawa, T. (2010b). Biomaterials controlling biological balance . *Journal of Japanese Society for Biomaterials*, 28., 85-86.

Artzi, Z., Weinreb, M., Givol, N., Rohrer, MD., Nemcovsky, CE., Prasad, HS., & Tal, H. (2004). Biomaterial resorption rate and healing site morphology of inorganic bovine bone and beta-tricalcium phosphate in the canine: a 24-month longitudinal histologic study and morphometric analysis. *Int. J. Oral Maxillofac. Implants.*, 19., 3., 357-368.

Ding, X., Takahata, M., Akazawa, T., Abe, Y., Murata, M., Komatsu, M., Iwasaki, N., Ito, M., & Minami, A. (New Orleans, March 6-9, 2010). Supersonic modification of crystal surface using partial dissolution and precipitation treatment improves bioabsorbability of synthetic hydroxyapatite. Proceeding of 56th Annual Meeting for Orthopaedic Res. Soc.

Ding, X., D., Takahata, M., Akazawa, T., Komatsu, M., Abe, Y., Iwasaki, N., Murata, M., Ito, M., & Minami, A. (2011). Improved bioabsorbability of synthetic hydroxyapatite through partial dissolution-precipitation of its Surface. *J. Mater. Sci.: Mater. Med.*, DOI 10.1007 s10856- 011-4291-x.

Hino, J., Murata, M., Akazawa, T., Tazaki, J., & Arisue, M. (2008). Bone induction by biomimetic functionally graded hydroxyapatite with rhBMP-2 on rat skull periosteum. *Int. J. Oral Maxillofac. Implants.*, 21., 327-337.

Kawakami, T., Kuboki, Y., Tanaka, J., Hijikata, S., Akazawa, T., Murata, M., Fujisawa, R., Takita, H., & Arisue, M. (2007). Regenerative Medicine of Bone and Teeth—with special references to biological principles, problems and their indicators—. *J. Hard Tissue Biology*, 16., 95-113.

Murata, M., Akazawa, T., Tazaki, J., Ito, K., Sasaki, T., Yamamoto, M., Tabata, Y., & Arisue, M. (2007). Blood permeability of a novel ceramic scaffold for bone morphogenetic protein-2. *J Biomed Mater Res*, 81B., 2., 469-476.

Murata, M., Akazawa, T., Tazaki, J., Ito, K., Hino, J., Kamiura, Y., Kumazawa, R., & Arisue, M. (2009). Human Dentin autograft for bone regeneration - Automatic pulverizing machine and biopsy -. *Bioceramics 22*, 22., 745-748.

Murata, M., Kawai, T., Kawakami, T., Akazawa, T., Tazaki, J., Ito, K., Kusano, K., & Arisue, M. (2010a). Human acid-insoluble dentin with BMP-2 accelerates bone induction in subcutaneous and intramuscular tissues. *J. Ceram. Soc. Jpn.*, 118., 438-441.

Murata, M., Akazawa, T., Takahata, M., Ito, M., Tazaki, J., Hino, J., Nakamura, K., Iwasaki, N., Shibata,T., & Arisue, M. (2010b). Bone induction of human tooth and bone crushed by newly developed automatic mill. *J.Ceram. Soc. Jpn.*, 118., 434-437.

Ogawa, T., Sugo, K., & Yamamoto, A. (2003). Ceramic hydroxyapatite "CELLYARD" for cell culture. *Cell*, 35., 474-476.

Okuda, T., Ioku, K., Yonezawa, I., Minagi, H., Kawachi, G., Gonda, Y., Murayama, H., Shibata, Y., Minami, S., Kamihira, S., Kurosawa, H., & Ikeda, T. (2007). The effect of the microstructure of b-tricalcium phosphate on the metabolism of subsequently formed bone tissue. *Biomaterials*, 28., 2612-2621.

Okuda, T., Ioku, K., Yonezawa, I., Minagi, H., Gonda, Y., Kawachi, G., Kamitakahara, M., Shibata, Y., Murayama, H., Kurosawa, H., & Ikeda, T. (2008), The slow resorption with replacement by bone of a hydrothermally synthesized pure calcium-deficient hydroxyapatite. *Biomaterials*, 29., 2719-2728.

Oyane, A., (2010). Development of apatite-based composites by a biomimetic process for biomedical appications. *J. Ceram. Soc. Jpn.*, 118., 77-81.

Rae, J., Ashokkumar, M.,Eulaerts, O., Von Sonntag, C., Reisse, J., & Grieser, F. (2005). Estimation of ultrasound induced cavitation bubble temperatures in aqueous solutions. *Ultrason. Sonochem.*, 12., 325-329.

Sakamoto, M., Nakasu, M., Matsumoto, T., & Okihara, H. (2007). Development of superporous hydroxyapatite and their examination with a culture of primary rat osteoblasts. *J. Biomed. Mater. Res.*, 82A, 238-242.

Tazaki, J., Akazawa, T., Murata, M., Yamamoto, M., Tabata, Y., Yoshimoto, R., & Arisue, M. (2006). BMP-2 dose-response and release studies in functionally graded HAp. *Key Engineering Mater.*, 309-311., 965-968.

Yasuda, K. (2009). Decomposition of chemical compounds by ultrasound and development of sono chemical reactor. *THE CHEMICAL TIMES*, 212., 2-7.

Yoshikawa, H. (2009). Development and clinical application of interconnected porous hydroxyapatite ceramics for bone tissue engineering. *J. Jpn. Soc. Biomaterials*, 27., 17-23.

Yoshikawa, H. (2010). Development and regeneration of bone. *J. Jpn. Soc. Biomaterials*, 28., 241-247.

13

Development and Evaluation of Superporous Ceramics Bone Tissue Scaffold Materials with Triple Pore Structure A) Hydroxyapatite, B) Beta-Tricalcium Phosphate

Michiko Sakamoto and Toshio Matsumoto
R&D Department PENTAX New Ceramics Division HOYA Corporation
Japan

1. Introduction

Hydroxyapatite ceramics (HAp) have been used as bone graft materials, due to their excellent biocompatibility and osteoconductivity. Moreover, they had a wide range of porosity for using various bone defect parts. For example, HAp from 0 to 15 % porosity with high strength was useful as an ilium spacer, intervertebral spacer which requires high strength. HAp from 30 to 40% porosity was useful as spinous process spacer for laminoplasty which requires bone formation and middle strength. Furthermore, HAp from 40 to 60% porosity was useful for the calvarias plate which requires good bone formation (Fig.1). However, due to the mechanical strength, the porosity was limited to 60% or less. Recently, regenerative therapy using cellular activity to cure bone defects has been developed, requiring HAp with porous spaces which cells can populate. We have developed superporous HAp (HAp-S) and designed the porosity, pore size and pore structure of HAp-S for managing directly-opposing factors of both high porosity and mechanical strength. In this study, the properties of HAp-S were examined in vitro using primary rat calvarial osteoblasts culture and an implant model using the canine femoral defect model, ilium defect model and rat calvarial defect model.

Moreover, we report on Beta-tricalcium phosphate (Beta-TCP) ceramics, which have been used as bone tissue scaffold like HAp. Since 1999 in Japan, Beta-TCP ceramics have also been widely used as bone tissue scaffold materials in many areas of surgery due to their biocompatibility and biodegradation. Beta-TCP is gradually degraded in the bone tissue and replaced with natural bone. However, if the degradation rate is faster than the bone formation rate, the mechanical strength of Beta-TCP may become poor in the bone tissue. This study evaluated the degree of osteointegration of superporous Beta-TCP ceramics (TCP-S) with the same high porosity and triple pore structure as HAp-S (Apaceram –AX). In our previous study, HAp-S, which has smaller pores in HAp-Ss artificial bone made by two types of surfactants, was most suitable as an artificial bone. However, biodegradation of TCP is different in characteristics from non-biodegradation. We investigated whether the most suitable parameter of HAp applies to TCP using a canine femoral defect model (short term). The most suitable TCP was then selected as an artificial bone and the long-term osseous implantation test, biomechanical examination and the micro-CT measurement were evaluated.

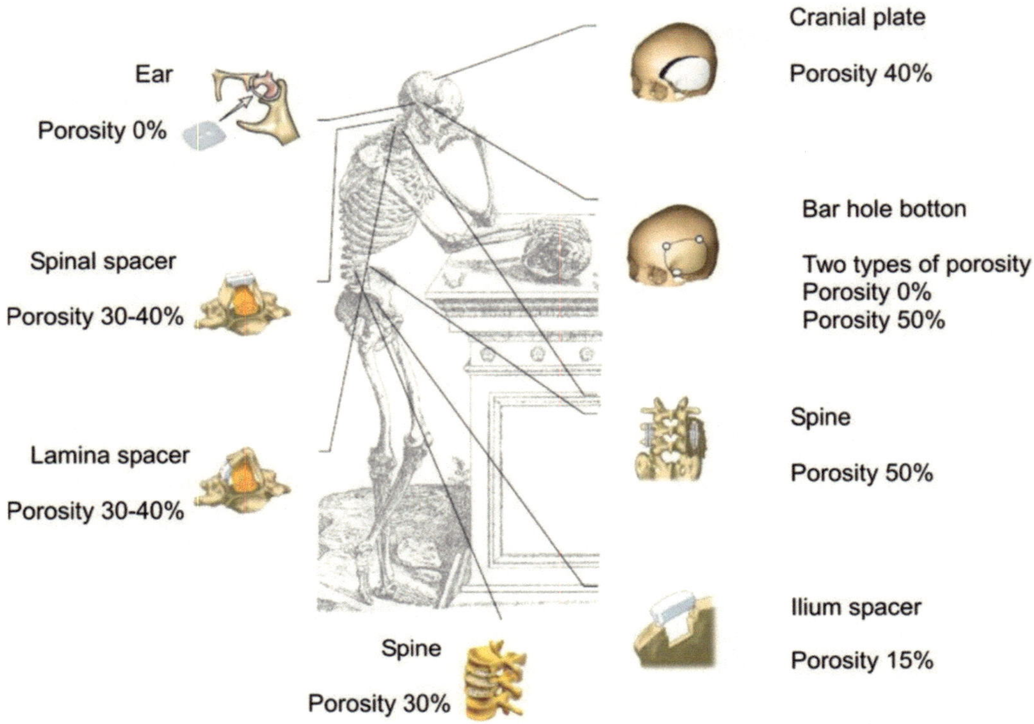

Fig. 1. Schematic drawing presenting the potential usage of HAp in various degrees of porosity HAp from 0 to 15 % porosity with high strength was useful as an ilium spacer, intervertebral spacer which requires high strength. HAp from 30 to 40% porosity was useful as spinous process spacer for laminoplasty which requires bone formation and middle strength. Furthermore, HAp from 40 to 60% porosity was useful for the calvarial plates which require good bone formation.

A) Superporous hydroxyapatite ceramics

2. Material and methods

2.1 Formation of superporous hydroxyapatite

An aqueous solution of phosphoric acid was poured into a suspension of purified calcium hydroxide, which resulted in hydroxyapatite slurry. The slurry was sprayed and dried using the spray-dry method to produce a fine spheroid powder. HAp powder was homogenized in water. A water soluble polymer was added to the slurry as a binder. A surfactant was used for bubble formation and micro air bubbles were introduced into the HAp slurry. The slurry was gelated and dried. The green block was then shaped into a disk and rod and sintered at 1200°C in air.

2.2 Measurement of compression strength

The compression strength of the HAp-S and HAp50 were measured by the recommended method of JIS at the speed of 0.5 mm/min by Autograph DSS (Autograph DSS/5000, Shimadzu Co., Kyoto, Japan).

2.3 Scanning Electron Microscopy (SEM) observation

The surface and inside structures of HAp-S and HAp50 were observed by SEM (S-4200, Hitachi Co., Tokyo, Japan).

2.4 In vivo: Animal test

2.4.1 Canine femoral defect model

HAp-S 4 mm in diameter and 12 mm in length was implanted at 2 sites in the left and right femurs of three male beagle dogs following ISO10993-6. Four weeks after implantation, animals were euthanized and the left and right femurs of three beagles were dissected out, and the effects and changes in the surrounding tissue and implanted material were histologically examined using Toluidine Blue (T.B) staining. The comparative material was HAp with 50% Porosity (HAp-50).

2.4.2 Canine ilium defect model (biomechanical testing)

HAp-S 4 mm in diameter and 6 mm in length was implanted at 1 site in the ilium of male beagles for 4 following ISO10993-6. Four weeks and 13 weeks after implantation, the ilium of 3 beagles was excised. The operative segments were gently trimmed off all the soft tissue. Before biomechanical testing, each specimen was fixed using resin. They were evaluated by nondestructive compression strength testing (until compressed 0.5mm by adding load) using a biomechanical testing machine (858 Mini Bionix II, MTS System Co., Minneapolis, MN, USA). Moreover, the effect and change of HAp-S were investigated histologically by T.B staining.

2.4.3 Rat calvarial defect model

The periosteum of the calvaria of Wistar rats was removed, and in the calvaria, defects measuring 4 mm in diameter and 1 mm in depth were made. HAp-S matching the defect in size was implanted and the periosteum was replaced. 12 weeks after implantation, the calvarias of 3 rats were excised, and the effects and changes in the surrounding tissue and implanted material were histologically investigated..HAp-50 was used as control material porosity 50%.

2.5 In vitro

Newborn Wistar rats were sacrificed by chloroform and their calvarias were removed aseptically. They were minced, washed with PBS, and digested by a 0.1% collagenase solution in a digestion chamber at 37°C for 90 min. They were filleted with a cell strainer and the filtrates were centrifuged at 1000 rpm for 5 min. The osteoblasts were suspended in 10 ml of MEM containing 100 U/mL penicillin, 100 µg/mL streptomycin, 0.25µg/mL fungison, and

15% FBS. The cell number was counted and adjusted to 7 x 10^5 counts of osteoblasts / pellet. The osteoblasts were laid on HAp-S and HAp-50% pellets. They were incubated in the culture medium described earlier at 37°C in a CO_2 incubator. The pellets of 3, 7 and 14 days of culture were estimated by alkaline phosphatase activity and ALP staining.

2.5.1 Alkaline phosphatase (ALP) activity

The pellets with the cells were washed with PBS, crushed by a cryo-press, and transferred into tubes. Five hundred μL of 10% Triton-X 100 aqueous solution, and 250 μL of 2-1amino-2-methyl-1-propanol buffer were added, followed by the addition of 250 μL of substrate. The mixture was kept at 37°C for 15 min. The reaction was stopped with 250μL of 1N NaOH, and the absorbance was measured at 450 mm.

2.5.2 ALP staining

The pellets with the cells were washed with PBS, fixed for 30 s with a 60% acetone-citrate buffer, washed with distilled water for 45 s, and reacted with a dye mixture for 30 min in the dark. The dye mixture was composed of 48 mL of a fast violet B salt solution and 2 mL of a 0.25% naphthol AS-MX phosphate alkaline solution. After the reaction, the pellets were washed with distilled water and photographed using a digital surface microscope.

3. Results

3.1 Characteristics of superporous hydroxyapatites

The porosity of HAp-S measured by its volume density was 85%,. The porosity was the highest among the inorganic ceramics as bone substitute material. The compression strength of HAp-S was approximately 2 MPa which is manageable during surgery (Table 1).

Sample name	HAp-S	HAp-50
General name	Hydroxyapatite	Hydroxyapatite
Molecular formula	$Ca_{10}(PO_4)_6(OH)_2$	$Ca_{10}(PO_4)_6(OH)_2$
Porosity	85%	50%
Pore structure	Consist of Spherical macro pore and micro pore	
Macro pore	50～300μm	50～500μm
Interconnected pore	50～100μm	—
Micro pore	0.5～10μm	0.5～10μm
Compression strength	2.0MPa	30.0MPa

Table 1. Characteristic of HAp-S and HAp-50.

3.2 SEM observation

In HAp-S, the size of the macro pores was from approximately 50 to 300 μm (Table 1). Under high magnification, there were many interconnecting pores among the macro pores and micro pore gaps between secondary particles of HAp ceramics on the pore walls. The size of the interconnecting pores and micro pores was from 50 to 100 μm, and from 0.5 to 10 μm, respectively (Table 1). In brief, HAp-S had the triple pore structure (Fig.2 a,b).

In HAp-50, the size of the macro pores averaged between 50 and 500 μm (Table 1). There were micro pore gaps between secondary particles of HAp ceramics on the pore walls. The size of the micro pores varried from 0.5 to 10 μm .HAp-50 had the interconnecting pore structure with the macro pores and micro pores (Fig.2 c) and d)).

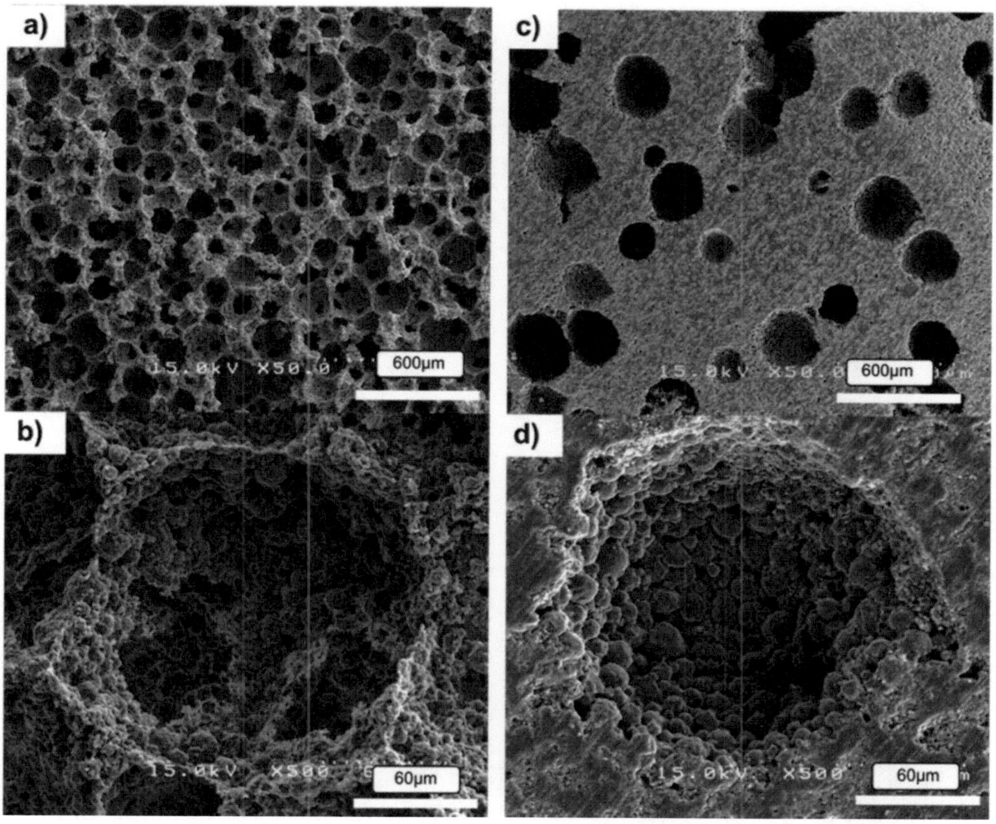

Fig. 2. SEM images of HAp-S a), b) and HAp-50 c),d). a), c): ×50 b), d): ×500
a): The size of the macro pores was from approximately 50 to 300 μm.
b): There are many interconnecting pores among the macro pores and micro pore gaps between secondary particles of HAp ceramics on the pore walls. The size of the interconnecting pores and micro pores is from 50 to 100 μm, and from 0.5 to 10 μm, respectively. HAp-S had the triple pore structure.
c): The size of the macro pores was from approximately 50 to 500 μm .
d): There are micro pore gaps between secondary particles of HAp ceramics on the pore walls. The size of the micro pores is from 0.5 to 10 μm.
HAp-50 had the interconnecting pore structure with the macro pores and micro pores.

3.3 In vivo animal test

3.3.1 Canine femoral defect model

T.B staining in HAp-S 4 weeks after implantation showed new bone formation in the entire implanted material area (Fig.3). Moreover, tissue of the surrounding implanted area did not

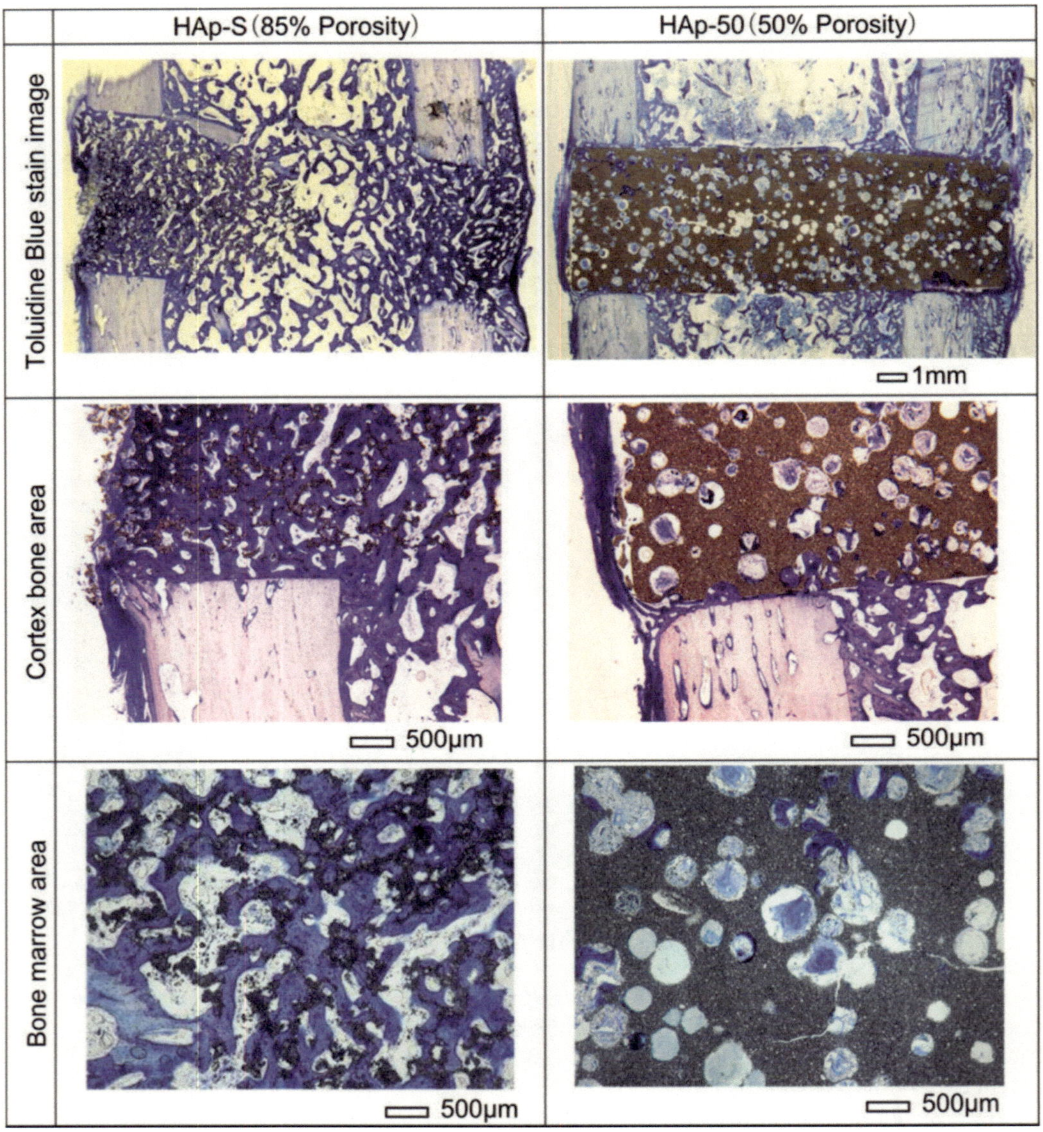

Fig. 3. T.B staining of femur of beagle at after 4 weeks implanted.
T.B staining in HAp-S 4 weeks after implantation showed new bone formation in the entire implanted material area. Moreover, tissue of the surrounding implanted area did not show any abnormalities. The magnified image of the cortex in HAp-S showed new bone formation not only in the boundary area between the original cortex bones, but also inside HAp-S. On the other hand, although T.B staining of HAp-50 showed new bone formation in the surrounding material and pores on the surface, new bone formation was not observed in the pores inside the material.

show any abnormalities. The magnified image of the cortex in HAp-S showed new bone formation not only in the boundary area between the original cortex bones, but also inside HAp-S. On the other hand, although T.B staining of the comparative material HAp-50 showed new bone formation in the surrounding material and pores on the surface, new bone formation was not observed in the pores inside the material. T.B staining in HAp-S 4 weeks after implantation showed new bone formation in the entire implanted material area. Moreover, tissue of the surrounding implanted area did not show any abnormalities.

The magnified image of the cortex in HAp-S showed new bone formation not only in the boundary area between the original cortex bones, but also inside HAp-S.On the other hand, although T.B staining of HAp-50 showed new bone formation in the surrounding material and pores on the surface, new bone formation was not observed in the pores inside the material.

3.3.2 Canine ilium defect model (biomechanical testing)

The compression strength of the HAp-S 4 weeks after implantation was 6-fold higher than that before implantation. Thirteen weeks after implantation, the compression strength of the HAp-S was over 8-fold higher than that before implantation. The strength of the bone defect was recovered at the early stage with the period of implantation (Fig.4-1). T.B staining of HAp-S showed the HAp-S material surrounding new bone formation tissue (Fig.4-2). Moreover, because the formation of bone tissue among the inside materials was tightly bonded by the micro pores of HAp-S.

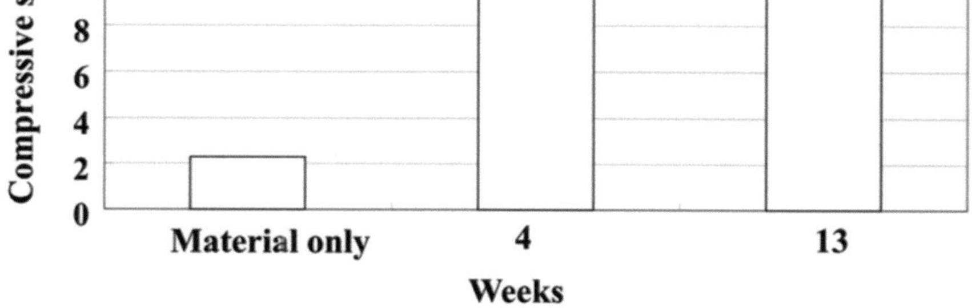

Fig. 4-1. Time course of Compressive strength of HAp-S after implanted.
The compression strength of the HAp-S 4 weeks and 13 weeks after implantation was 6-fold and 8-fold higher than that before implantation. The strength of the defect recovered at the early stage with the period of implantation.

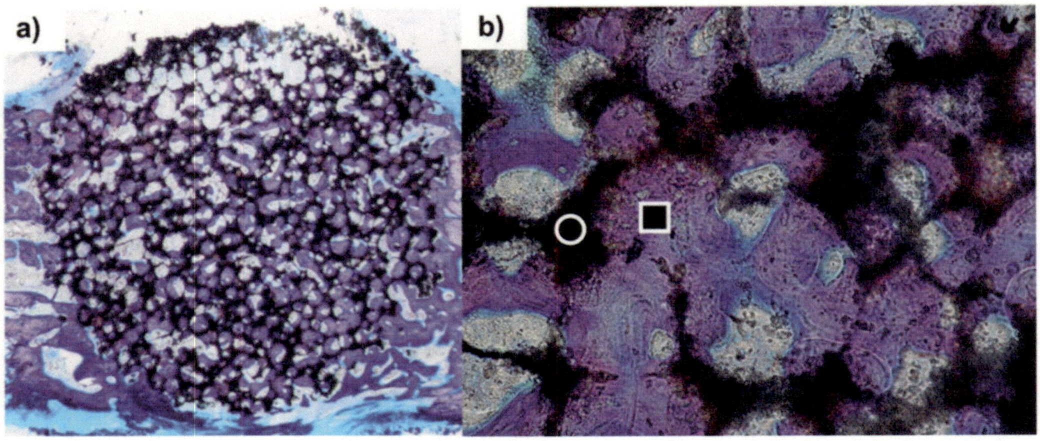

Fig. 4-2. TB staining of illium of beagle at after 4 weeks implanted.
a) Whole image of HAp-S b) Enlarged image of a) ○ : HAp-S □: New bone
Some part of HAp-S particles was surrounded by newly formed bone closely.

3.3.3 Rat calvarial defect model

H.E staining of HAp-S showed that the HAp-S material had excellent new bone formation tissue not only in the surrounding area, but also the entire HAp-S area. Moreover, tissue surrounding the implanted area did not show any abnormalities. On the other hand, H.E staining of the comparative material HAp-50 showed new bone formation in the surrounding material (Fig.5).

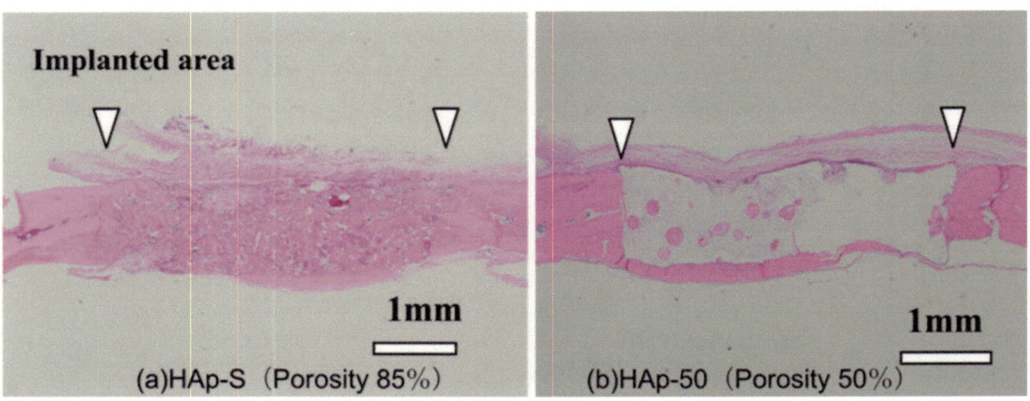

Fig. 5. HE staining of calvaria of rat at after 12 weeks implanted.
a) HE staining of HAp-S showed that the HAp-S material had excellent new bone formation tissue not only in the surrounding area, but also the entire HAp-S area.
b) HE staining of the comparative material HAp-50 showed new bone formation in the surrounding material.

3.4 In vitro test

ALP activity of HAp-S was higher than that of HAp-50 during the entire culture period (Fig.6-1). ALP staining of HAp-S was showed staining on the surface and inside of HAp-S. In contrast, ALP staining of HAp-50 was showed on the upper surface.(Fig.6-2)

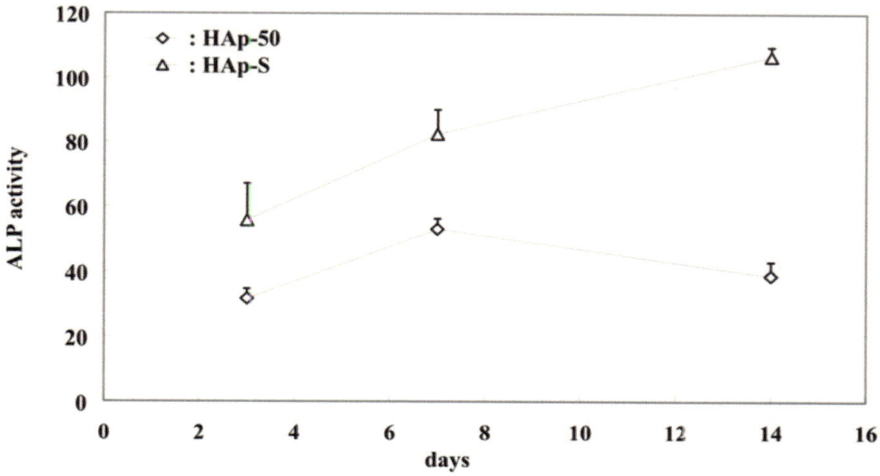

Fig. 6-1. Time course of ALP activities. △ : HAp-S ◇ : HAp-50
ALP acticity of HAp-S is higher than that of HAp-50 during the entire culture period.

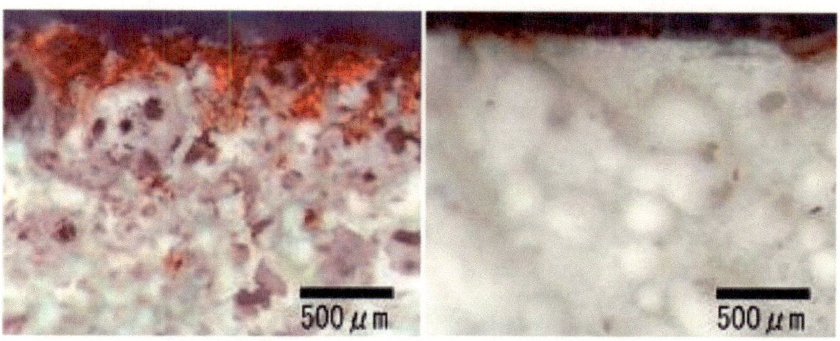

(a)HAp-S (85% Pososity) (b)HAp-50 (50% Porosity)

Fig. 6-2. ALP staining of sectioned sample at 14days after seeding.
ALP staining of HAp-S showed staining on the surface and inside of HAp-S. In contrast, ALP staining of HAp-50 showed that only the surface.

4. Discussion

Recently, regenerative therapy using cellular activity to cure bone defects has been developed.Cellular activity required HAp with porous spaces that allows cellular penetration in order to obtain the most suitable porosity, optimal pore size and favorable pore structure, we considered the following aspects:

1. Porosity:We reviewed materials and facilities requiring the addition of a bubble method. In the conventional method, bubble formation is created by mixing and naturally adding air to the slurry. This time, we added surfactant and formed unprecedented micro bubbles in the HAp slurry.
2. Pore size:Adding surfactant allowed uniformity of the macro pores and interconnecting pores. If these pore sizes are too large, the inside of pores will be covered with fibrous tissue. As a result, bone formation was prevented by fibrous tissue. If these pore sizes are too small, cell cannot permeate to the inside of the materials and bone formation will be limited to the surface layer. Therefore, determining the ideal pore size was required for good bone formation. We chose the surfactant which created more uniform micro bubbles. Moreover, it was important to keep adding bubbles to the slurry without crushing until the final shape was obtained.
3. Pore structure: To present a scaffold that is necessary for more excellent bone formation, we investigated the condition of HAp powder and achieved porosity of the pore structure. Macro pores were necessary to provide a comfortable space for cells to penetrate. Interconnecting pores were required to allow penetration to the inside of HAp. Micro pores were required as the super bonding space for cells(Fig.7).

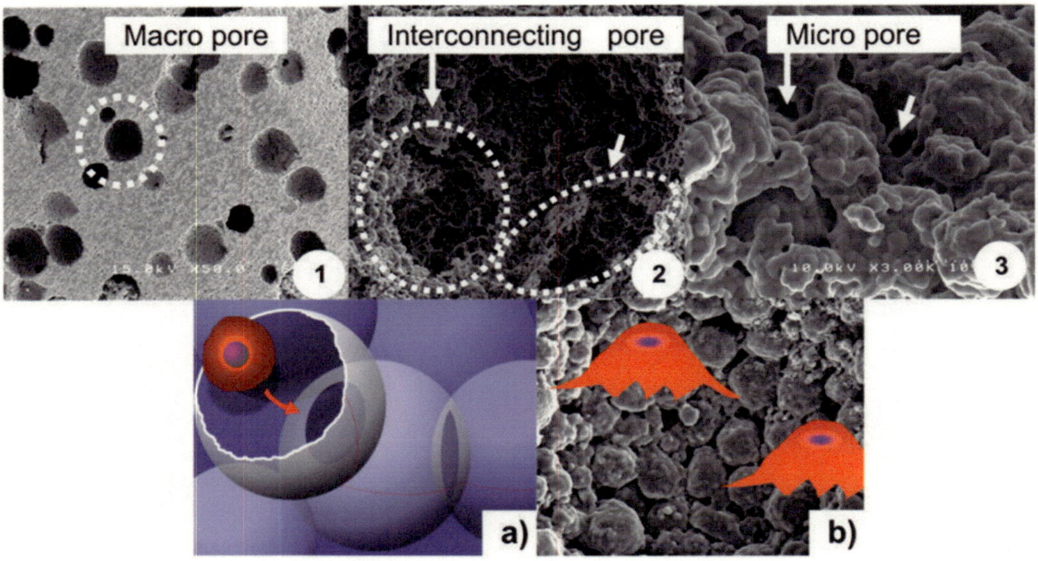

Fig. 7. Images 1,2,3) & illustrations a,b) of triple pore structure
1: Macro pores were necessary to provide a comfortable space for cells to penetrate.
2: Interconnecting pores were required to allow penetration to the inside of HAp(a)).
3: Micro pores were required as the super bonding space for cells (b)).

In the above improvements, the porosity of HAp-S was 85%, and its compression strength was approximately 2.0MPa. Although the porosity of HAP-S was the highest among the commercial inorganic ceramics artificial bones, mechanical property was enough during surgery. In the in vivo test, HAp-S had better new bone formation tissue in not only the surrounding area, but also the entire area of HAp-S than that of HAp-50. Moreover, HAp-S showed regenerative bone tissue, integration with the surrounding tissue and recovery of the compression strength of the defect area in the early stage. These results strongly suggest that the controlled triple pore structure, namely, uniform macro pores, many interconnecting pores and micro pores on the pore wall, advances osteogenesis.

5. Conclusion

HAp-S was confirmed to have the most suitable pore structure for bone regeneration as the bone substitute material, and as being a useful material for application to bone disease.

In 2006, HAp-S (APACERAM-AX, HOYA Co., Tokyo, Japan) was approved by the MHLW as a bone substitute in orthopedics for spine fusion and tumors, as well as the dental field. The clinical data on the follow-up more than 5 year after the treatment with HAp-S showed favorable results in all cases.

B) Superporous Beta-tricalcium phosphate ceramics

6. Materials and method

6.1 Formation of superporous Beta-tricalcium phosphate (TCP-S)

Beta-TCP powder was made by wet mixing and the spray-dry method. Beta-TCP powder was homogenized in water. A water soluble polymer was added to the slurry as binder. Two types of surfactant were used to form inner pores. One was anionic (AN), and the other was nonionic (NI). The surfactant was used as bubbled formation in the Beta-TCP slurry, which was gelated and dried. The dried block was shaped into blocks and rods and, then sintered at 1,100°C in air.

Sample name	TCP-S(AN)	TCP-S(NI)
General name	Beta-tricalcium phosphate	Beta-tricalcium phosphate
Molecular formula	$Ca_3(PO_4)_2$	$Ca_3(PO_4)_2$
Porosity	75%	75%
Pore structure	Consist of Spherical macro pore and micro pore	
Macro pore	50～300μm	50～500μm
Interconnected pore	50～100μm	50～100μm
Micro pore	0.5～10μm	0.5～10μm
Compression strength	6.0MPa	5.0MPa

Table 2. Characteristics of TCP-S(AN) and TCP-S(NI)

6.2 Measurement of compression strength

The compressive strength of two types of TCP-Ss was measured at 1 mm/min of test speed by Autograph AGS-H (SHIMADZU Co., Kyoto, Japan).The test sample was 10×10×20 mm in size.

6.3 SEM observation

The surface and inner structures of TCP-Ss were observed by SEM (S-4200, Hitachi Co., Tokyo, Japan).

6.4 Animal experiments

6.4.1 Canine femoral defect model (short term)

Adult beagles weighing 10~13 kg were used. After anesthetization, the integument and fascia of the femur were incised using an electrical surgical knife. Holes were drilled in both femora of beagles 4 mm in diameter and 12 mm in depth according to the implant shape and two types of TCP-Ss were implanted in each hole. The hole positions were approximately 40 mm (distal position) and 60 or 70 mm (proximal position) from the apophysis of the femur. At 1, 2 and 3 weeks postoperatively, the femora were harvested. A T.B. staining specimen was prepared and the histopathology examined.

6.4.2 Canine femoral defect model (long term)

Adult beagles weighing 10~13 kg were used. After anesthetization, the integument and fascia of the femur were incised using an electrical surgical knife. Holes were drilled in both femora of beagles 4 mm in diameter and 12 mm in depth according to the implant shape and TCP-Ss were implanted in each hole. The hole positions were approximately 40 mm (distal position) and 60 or 70 mm (proximal position) from the apophysis of the femur. For the 26-week implantation test, we used very small markers made of dense Hydroxyapatite 0.8 mm in diameter and 3.5 mm in length. The marker was implanted 55 mm from the apophysis of the femur. At 2, 4, 13 and 26 weeks postoperatively, the femora were harvested. T.B. staining specimens were prepared and the histopathology examined.

6.4.3 Canine femoral defect model (biomechanical testing)

We tested the change in compressive strength of TCP-S after implantation 2, 4 and 8 weeks postoperatively in the same beagle model. The sample for the compressive strength test was trimmed to 12 mm in length. Then, the compressive strength of the materials was analyzed using 858Mini Bionix (MTS, cross head speed 0.5 mm/min, destructive test, displacement about -1.5~2.0 mm). The animal experiment was in conformity with ISO10993-6:2007.

6.4.4 Rat calvarial defect model (micro-CT measurement)

After the periosteum was removed, a defect 4 mm in diameter and 1 mm in length was made in the calvarias of Wistar rats. Then, TCP-S the same size as the defect was implanted and the periosteum was returned. Four weeks after implantation, the calvarias of 3 rats were excised and scanned by micro-CT (Skyscan1172, SKYSCAN Co., Kontich, Belgium). The ratio of bone formation and absorption of material in TCP-S was calculated.

7. Result

7.1 Characteristics of superporous TCP

The compressive strength of AN of TCP-S 10x10x20 mm in size was approximately 6 MPa at 75% porosity and that of NI of TCP-S 10x10x20 mm in size was approximately 5 MPa at 75% porosity, respectively. The uniform pore structures gave a compressive strength higher than other commercial products with the same composition and porosity.This compressive strength had been strong enough to handle.

7.2 SEM observation

The porosity of these materials is about 75%. The pore size of AN of TCP-S is approximately 200 μm and that of NI of TCP-S is approximately 300 μm. The macro pores of both were very uniform. Moreover, the two types of TCP-Ss had many interconnecting pores and micro pores on the pore wall (Fig.8). Both TCP-Ss had a uniform triple pore structure with 75% porosity.

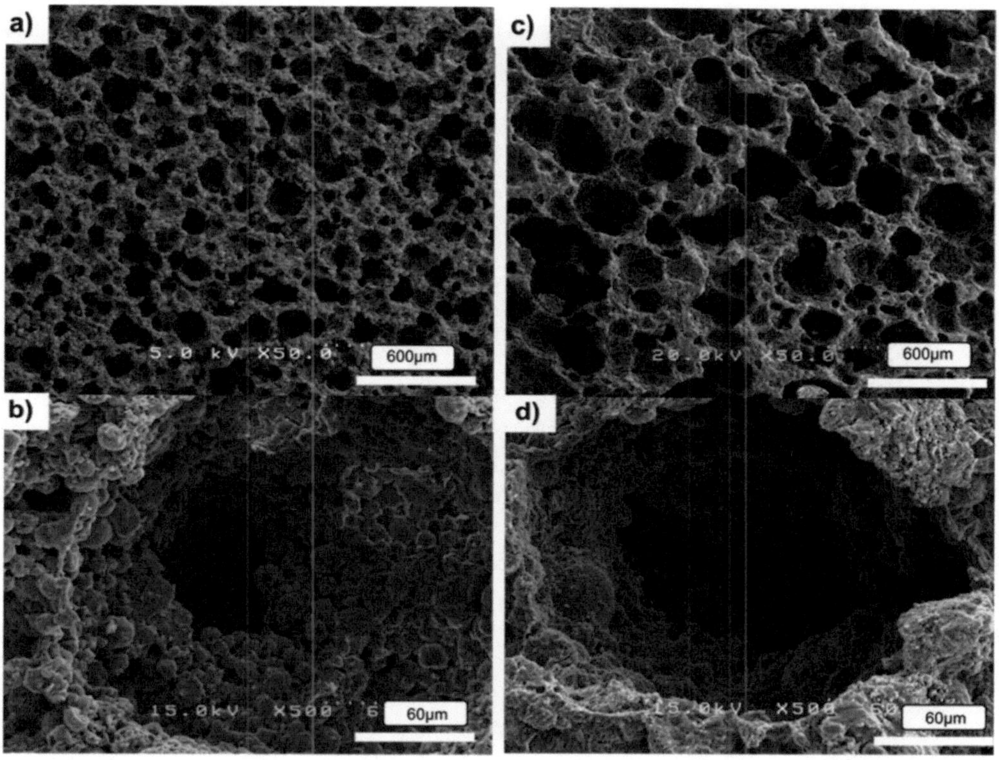

Fig. 8. SEM images of TCP(AN) and TCP(NI).
a), b): TCP(AN) c),d): TCP(NI) a), c): ×50 b), d): ×500
a): The size of the macro pores was from approximately 50 to 500 μm.
c): The size of the macro pores was from approximately 50 to 300 μm.
b)d): There are many interconnecting pores among the macro pores and micro pore gaps between secondary particles of TCP ceramics on the pore walls. The size of the interconnecting pores and micro pores is from 50 to 100 μm, and from 0.5 to 10 μm, respectively. TCP-S(AN) and TCP-S(NI) had the triple pore structure.

7.3 Animal experiments

7.3.1 Canine femoral defect model (short term)

At 1 week postoperatively, moderate proliferation of the periosteum was observed at the outer femur surface of AN and NI of TCP-Ss. New bone was not observed in either material. At 2 weeks postoperatively, the AN specimen showed new bone not only at the area adjacent to the cortex bone but also the inner pores of the specimen. In contrast, although the NI specimen showed new bone at the area adjacent to the cortex bone, new bone was observed only slightly inside the specimen. At 3 weeks postoperatively, new bone formation on both specimens was better than that at 2 weeks. The AN specimen showed better early bone formation, and in an extended image, some parts of the TCP-S particles were closely surrounded by newly formed bone. On the other hand, the NI specimen showed that there was a gap between the material and new bone tissue(Fig. 9 and 10). The TCP-S of AN showed just TCP-S.

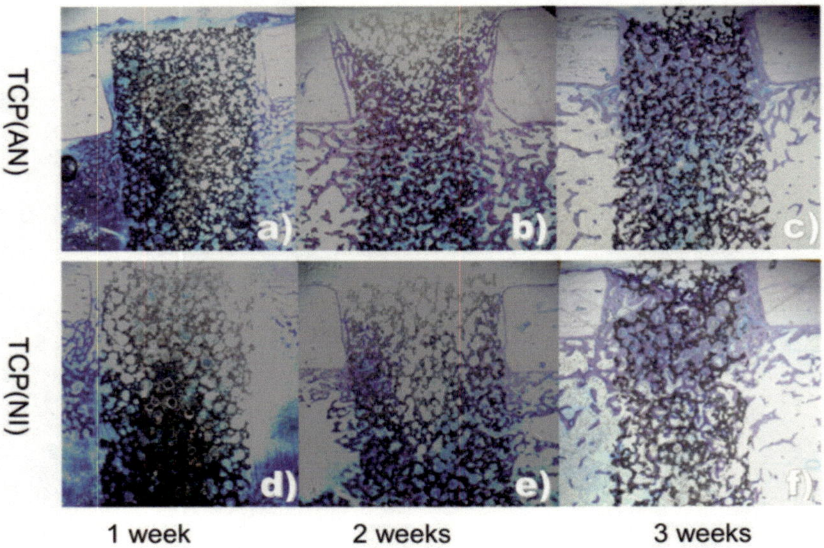

Fig. 9. T.B staining of femur of beagle at after 1,2 and 3 weeks implanted.
a), b),c): TCP(AN) d),e),f): TCP(NI)
a), d):1week b),e):2weeks c),f):3weeks
At 1 week postoperatively, moderate proliferation of the periosteum was observed at the outer femur surface of AN and NI of TCP-Ss. At 2 weeks postoperatively, the AN specimen showed new bone not only at the area adjacent to the cortex bone but also the inner pores of the specimen. In contrast, although the NI specimen showed new bone at the area adjacent to the cortex bone, new bone was observed only slightly inside the specimen. At 3 weeks postoperatively, new bone formation on both specimens was better than that at 2 weeks.

7.3.2 Canine femoral defect model (long term)

At 2 weeks postoperatively, osteoblasts and osteoclasts increased in number. The TCP-S specimen showed newly formed bone not only at the area adjacent to the cortex bone but also the inner pores of TCP-S. The triple pore structure of TCP-S was suitable for penetrating into the inner pores for fibrous tissue, bone tissue and so forth at the very early stage. At 4 weeks postoperatively, newly formed bone was expressed better than that at 2 weeks, especially in the cortex bone. Remnant TCP-S in the cortex bone and medullar was slight. Some parts of the TCP-S particles were closely surrounded by the newly formed bone. This result depended on the controlled triple pore structure, namely macro pores, interconnecting pores and micro pores, and each pore size of TCP-S. At 13 weeks postoperatively, osteoblast and osteoclast further increased in number greater than that of 4 weeks. Newly formed bone in the cortex bone was more mature than in the medullar. Moreover, the mature bone tissue in the cortex bone showed the same lamellar pattern as natural bone. Remnant TCP-S in the cortex bone and medullar was much less than that of 4 weeks (Fig. 11). At 26 weeks, TCP-S implanted in

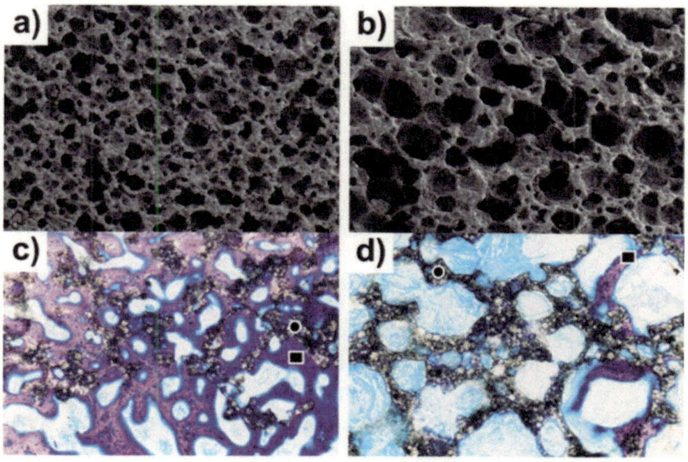

Fig. 10. SEM images of TCP-Ss and T.B staining of femur of beagle at after 3 weeks implanted.
a), c): TCP(AN) b),d): TCP(NI)
a), b):SEM image
c), d): T.B. staining(Enlarged image of Fig.9-c and 9-f) ○ : TCP-S □: New bone
The AN specimen showed better early bone formation, and in an extended image, some parts of the TCP-S particles were closely surrounded by newly formed bone. On the other hand, the NI specimen showed that there was a gap between the material and new bone tissue.

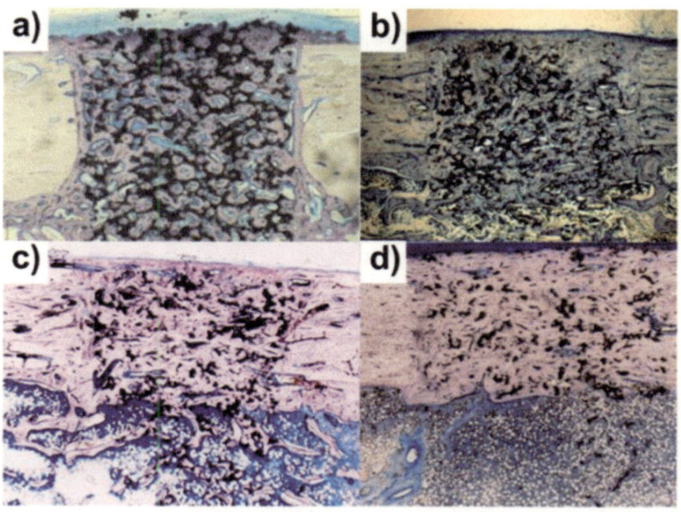

Fig. 11. T.B staining of femur of beagle at after 4,8,13 and 26 weeks implanted.
a): 4 weeks b): 8 weeks c): 13 weeks d): 26 weeks
At 4 weeks postoperatively, newly formed bone was expressed better than that at 2 weeks,

especially in the cortex bone. At 13 weeks postoperatively, newly formed bone in the cortex bone was more mature than in the medullar. At 26 weeks, TCP-S implanted in the cortex changed to cortex bone completely and that in the medullar changed to medullar completely, respectively.

the cortex changed to cortex bone completely and that in the medullar changed to medullar completely, respectively. Remnant TCP-S in the cortex bone was scarcely observed and remnant TCP-S in the medullar was not observed. TCP-S was gradually degraded in the bone tissue and replaced with natural bone completely at 26 weeks. These results of the animal study suggest that TCP-S with a triple pore structure had good osteogenesis and bioresorption in the early stage after implantation.

7.3.3 Femoral defect model (biomechanical testing)

The compressive strength of TCP-S in the distal position at 2, 4 and 8 weeks postoperatively was 40.2 MPa, 71.2 MPa and 81.7 MPa, respectively. On the other hand, in the proximal position, the compressive strength of TCP-S at 2, 4 and 8 weeks postoperatively was 36.2 MPa, 82.7 MPa and 89.7 MPa, respectively (Fig. 3). The compressive strength at 2 weeks increased approximately 8 times more than that of the material only. Moreover, during 4 and 8 weeks postoperatively, the compressive strength increased approximately 16 times more than that of the material only. The compressive strength of TCP-S implanted in defect bone was much higher than the material strength in the early stage. This result suggests that TCP-S showed early recovery of strength at the bone defect.

7.3.4 Rat calvarial defect model (micro-CT measurement)

In the Micro-CT image of 4 weeks after implantation, excellent bone formation in the TCP-S pellet was exhibited from the surroundings of the original bone. Moreover, Micro-CT analysis revealed that the ratio of bone formation was more than 50% in implantation at 4 weeks. In addition, the absorption of TCP materials was mean 3.9% with the ratio of absorbent material TCP-S.In the results of rat, at 4 weeks after implantation, the ratio of new bone formation in the TCP-S was higher than the materials absorption.

8. Discussion

Artificial bone consisting of HAp that was manufactured in Japan from about 1985, and artificial bone consisting of β-TCP featuring substitution of absorption bone that was not present in HAp manufactured from about 1999 have attracted attention. We investigated whether the most suitable parameter of HAp applies to TCP from the following aspects.

1. Pore size: In the absorption material of TCP-S, the bone formation in materials having a macropore greater than 300 μm produced a gap between the materials and new bone formation tissue. This is considered due to fibrous tissue entering and inhibiting bone formation. Materials and the new bone contacted closely, and the materials that had macropores less than 300 μm showed good bone formation. Regardless of the material properties, pore structure and pore sizes suitable for bone formation has been found to be common.

2. Strength recovery of the filling department: If the degradation rate is faster than the bone formation rate, the mechanical strength of Beta-TCP may become poor in the bone tissue. TCP-S with a triple pore structure showed that the compressive strength of TCP-S implanted in defect bone was much higher than the material strength in the early stage, because TCP-S was uniform and had the most suitable size and pore structure for excellent bone formation.
3. Bone formation and absorption of material: TCP-S bone formation generated more quickly than the absorption of materials in the earlier stage in the test of rat. TCP-S has a good balance with bone formation and material absorption.

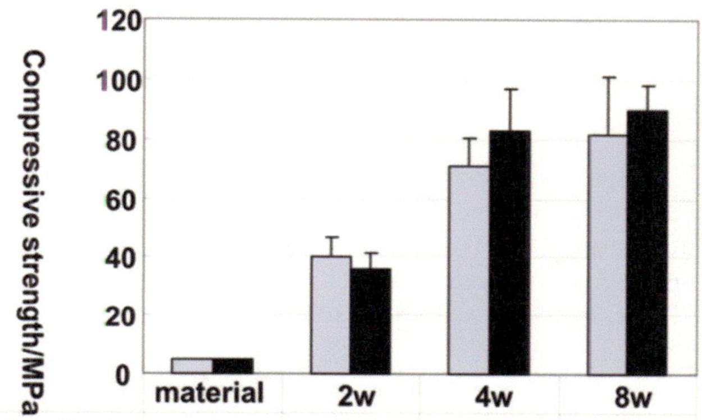

	material	2w	4w	8w
□ TCP-S in distal	5	40.2	71.2	81.7
■ TCP-S in proximal	5	36.2	82.7	89.7

Fig. 12. Results of Compressive strength of TCP-S after implanted.
During 4 and 8 weeks postoperatively, the compressive strength increased approximately 16 times more than that of the material only. The compressive strength of TCP-S implanted in defect bone was much higher than the material strength in the early stage.

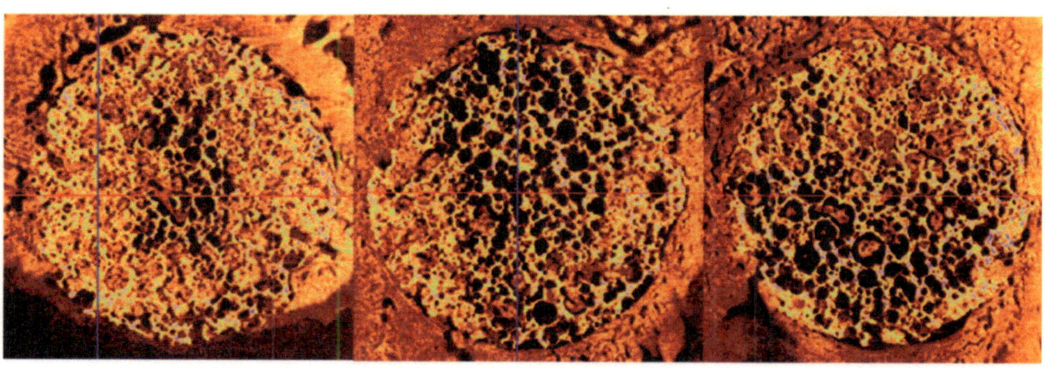

Fig. 13. Micro-CT image of calvaria of rat at after 4weeks implanted.
Excellent bone formation in the TCP-S pellet was exhibited from the surroundings of the original bone.

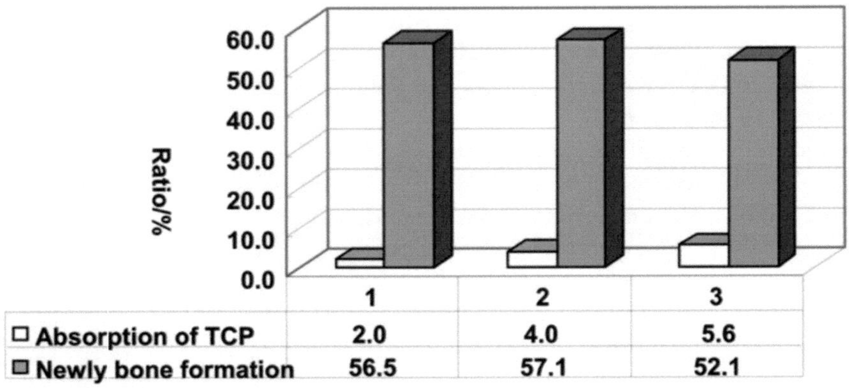

Fig. 14. Micro-CT analysis of TCP-S pellet at after 4 weeks implanted.
The ratio of bone formation was 56.5%, 57.1% and 52.1% in implantation at 4 weeks. In addition, the absorption of TCP materials was mean 3.9% with the ratio of absorbent material TCP-S. In the results of rat, at 4 weeks after implantation, the ratio of new bone formation in the TCP-S was higher than the materials absorption.

9. Summary and conclusion

We developed porous beta-TCP ceramics (TCP-S) with a controlled triple pore structure, namely macro pores, interconnecting pores and micro pores. TCP-S showed good osteogenesis, bioresorption and recovery of strength in the early stage after implantation. TCP-S is useful for bone graft materials. In 2010, TCP-S (SUPERPORE, HOYA Co., Tokyo, Japan) was approved by the MHLW as a bone substitute in orthopedics. The clinical data on the follow-up more than 1 year after the treatment with TCP-S showed favorable results in all cases.

The triple pore structure with a macropore of 50~300 µm, an interconnecting pore of 50~100 µm, and a micro-pore of 0.5~10 µm showed strength recovery in osteoplasty at an early stage of the filling region without being affected by the composition of the artificial bone and it was the most suitable bone filling material.

10. References

A.Myoi, H.Yoshikawa,.(2009),Clinical issues of bone substitutes,*J Jpn Orthop Assoc*,83, 463-468

I.Ono et.al.,(2004),Studies on the dynamics of bone formation in porous hydroxyapatite ceramic.*The 13th Research Council Meeting of Japan Society of Plastic and Reconstructive Surgery*,Vol.13,p61

M.Kato et.al.,(2006),Lumbar posterolateral fusion with beta-tricalcium phosphate,*The Central Japan Association of Orthopaedic Surgery & Traumatology*,49(6),1063-1064

M.Motomiya, K.Kadoya, M.Ito, et.al.,(2006),The Potentiality of novel high porous hydroxyapatite for better spinal arthrodesis. *The Orthopaedic Research Society*,#p.875

M.Motomiya, M.Ito, M.Takahata,et.al.,(2007),Effect of hydroxyapatite porous characteristics on healing outcomes in rabbit posterolateral spinal fusion model. *Eur Spine J.* 16,2215-2224.

M.Motomiya,K.Kadoya,M.Ito et.al.,(2005),The effect of the presence or absence of bone union interconnecting pore and porosity of hydroxyapatite., *J.Jpn.Orthop*,Vol.79, No.8,S918

M.Nakasu,H.Okihana,M.Sakamoto et.al.,(2005),Bone Formation of Highly Porous Hydroxyapatite with Different Pore Size Ranges,*CERAMICS JAPAN*,Vol.40, No.10, 828-830

M.Ogiso,(2003),Titanium and apatite as implant materials,*Japanese Society for Biomaterial*,Vol.21,No.6,450-456

M.Ogiso,S.Mineno,T.Matsumoto,(2004),Osteoanagenesis method that uses super-minute bone dust and porous apatite. Bone regeneration method using microscopic bone powder and porous apatite. *J.Jpn.Orthop.*,Vol.78,No.8,S799

M.Ozawa et.al.,(2003/2004),Appropriate use of β-TCP.*Orthopaedic Ceramic Implants*,Vol 23/24,47-50

M.Sakamoto et.al.,(2006),Biomechanical evaluation of ultra porous hydroxyapatite ceramics on in vivo study.*20thEuropean Conference on Biomaterials*,T54

M.Sakamoto et.al.,(2008),Evaluation of porous beta-TCP ceramics with triple pore structure as bone tissue scaffold, *Orthopaedic Ceramic Implants*,Vol.28.5-8

M.Sakamoto et.al.,(2009),Development and evaluation of porous Beta- TCP ceramics with triple pore structure. *The Orthopaedic Research Society*, #p.1376

M.Sakamoto, M.Nakasu, T.Matsumoto et al.,(2007),Development of superporous hydroxyapatite and their examination with a culture of primary rat osteoblasts.*J Biomed Mater Res*, 82A, 238-242

M.Sakamoto, T.Matsumoto, M.Nakasu et al.,(2005),Development and evaluation of porous hydroxyapatite ceramics with high porosity.*European Conference on Biomaterials*,No.19,p43

M.Sakamoto,T.Matsumoto,M.Nakasu et.al.,(2004),Evaluation of porous Hydroxyapatite ceramics with high porosity as bone tissue scaffold,*The 24th Annual Meeting of The Japanese Society of Orthopaedic Ceramics and Implants*,Vol.24,p34

M.Sakamoto,T.Matsumoto,T.Nakajima et.al.,(2006),Functional evaluation of porous hydroxyapatite bone regeneration using ultra-high porosity ceramics,*The Japanese Society for Regenerative Medicine*, Vol.5, No.19, p214.

M.Takahata, M.Ito, M.Motomiya et al.,(2006),A novel technique to generate autogenous graft bone using ultra-high porous hydroxyapatite scaffold on in vivo ilium in preparation for spinal interbody fusion. *The Orthopaedic Research Society*,#p.1731

N.Okii, S.Nishimura, K.Kurisu et al.,(2001),In vivo Histological Changes Occurring in Hydroxyapatite Cranial Reconstruction -Case Report-,*Neurologia Med Chir (Tokyo)*,41,100-104

N.Yamazaki, M.Hirano, K.Nanno et al., (2009),A Comparative Assessment of Synthetic Ceramic Bone Substitutes With Different Composition and Microstructure in Rabbit Femoral Condyle Model,*J Biomed Mater Res Part B*, 91B, 788-798,

S.Hirabayashi,K.Kumano,(1999),Contact of hydroxyapatite spacers with split spinous processes in double-door laminoplasty for cervical myelopathy,*J Orthop Sci* 4, 264-268

T.Matsumoto,(2006),"Hoc nunc os ex ossibus meis"., *Japanese Society for Biomaterial*, Vol.24,No.1,39-40,2006

T.Matsumoto,M.Sakamoto et.al.,(2004),Pore structure of high-porous hydroxyapatite ceramics. *The Society of Inorganic materials,JAPAN*,Vol.108th,p46

T.Nakajima, et al.,(2010),In vivo histological study of posterolateral spinal fusion using superporous hydroxyapatites as a bone graft extender,*The 39th Annual Meeting of the Japanese Society for Spine Surgery and Related Research*,Vol.1,No.4,949

T.Ueno, A.Ota, N.Shirasu et al.,(2009),A case of usage ultra-high porous hydroxyapatite for alveolar bone augmentation,*J Okayama Dental Soc*,28,65-69

14

Histocompatibility of Acellular Matrix Bone with Osteoblast and Vascular Endothelial Cells

Gang Rui, Xuhong Jin and Shan Lin
The First Affiliated Hospital of Xiamen University
China

1. Introduction

The culture of stem cells is an essential element in tissue engineering.The base for the repairing of bone defect in tissue engineering is still its vascularization. Co-culture of the osteoblast and the vascular endothelial cell (VEC) can not only promote the proliferative activity of the osteoblast but can also accelerate the revascularization of the regenerating bone. Thus it can help to increase the ability of regenerating bone in the repairing of bone defect. The bone source and the site for supplying bone were limited in autografting, meanwhile it is difficult in the modeling of self-tissue, therefore its clinical application is limited. There is certain immunologic rejection in the isotype heterogenous bone grafting and the tissue engineering bone provides a new way for solving this problem. In this experiment, the osteoblast and the vascular endothelial cell were compounded by using the acellular rib of pig. Then the tissue engineering bone was constructed and implanted into the animal. Observe and study the biological compatibility, try to find out the pathological process of in vivo ossification, bone defect repairing and revascularizarion for the heterogenous bone grafting. Thus it will provide the theological base for the clinical application of tissue engineering bone. This chapter is divided into 3 parts: Part I: The culture of the seed cell in tissue engineering and the study of cellular compatibility, Part II: Study the effect of heterogenous bone acellular matrix (HBACM) on the biological compatibility between the osteoblasts and the vascular endothelial cells, Part III The experimental research of implanting the compound cell of heterogenous bone acellular matrix into the animal body.

2. General methods (Parts I - III)

2.1 Part I: The culture of the stem cells in tissue engineering and the study of cellular compatibility

The osteoblast (A group) and vascular endothelial cell (B group) of the rabbit was cultured, and then both of the cells were co-cultured (C group). The morphology and the growth of the cells in the 3 groups were observed under the inverted phase contrast microscope. Immunocytochemical stain of type I collagen was used to identify the osteoblast. Immunocytochemical stain of the vascular factor VIII was used to identify the VEC. The sections were made by the co-cultured cells. Three days later, observe the mixed growth of the two kinds of cells after HE stain and Masson trichrome stain. Test the activity of ALP by using the method of paranitrobenzene phosphate. Observe whether the vascular endothelial would affect the ALP activity produced by the osteoblast. Analyse the growth and proliferation of the cells in the 3 groups.

2.2 Part II: Study the effect of heterogenous bone acellular matrix (HBACM) on the biological compatibility between the osteoblasts and the vascular endothelial cells

The fresh rib of the pig was defatted, deantigened and decellular by hydrogen peroxide solution, chloroform/methanol and Triton X-100, then it was made into the framework material of the acellular matrix and was observed under the scanning electron microscope. And then, the three groups of rabbit cells were compounded with the acellular rib of the pig individually. The adhesion of the cells with the material, the distribution of the cells in the pores of the material, and the growth, differentiation and proliferation of the cell were observed under the inverted phase contrast microscope and the scanning electron microscope. The biological compatibility was observed after HE stain, toluidine blue and Masson trichrome stain, after 5 days co- cultured. The cell cycle and the ploidy of the co-cultured three groups were tested by flow cytometer to understand the toxicity of the material to the cell on the 1d, 3d, 5d and 7d.

2.3 Part III: The experimental research of implanting the compound cell of the heterogenous acellular matrix into the animal body

The three groups of rabbit cells were marked with BrdU, and then they were compound with the decellular rib of the pig to repair the bone defect in the body. 27 New Zealand big-ear-rabbit weighted 2.5kilogram were selected and they were grouped into 3 groups with 9 in each group. Resect the 1.5cm long radius of the double forelimbs of the rabbit and prepare animal model of bone defect. The tissue engineering bone which had been compounded and cultured for 1 week by the heterogenous bone acellular matrix and the cells of each group was implanted to the left side. The heterogenous bone acellular matrix was implanted to the right side as a control. Select the material at 3 weeks, 6 weeks and 12 weeks to observe bone fracture. The development and the prognosis of the tissue engineering bone were observed by naked eye. X-ray examination was performed to understand the formation,change and moulding of the callus at the site with bone defect. The Brdu marked cells were traced and tested to understand the survival of seed cell in the experimental animal. The pathological process of in vivo ossification, bone defect repairing and revascularization were observed after HE, toluidine blue and Masson stain. And the immunohistochemistry of type I collagen was used to understand the secretion of type I collagen by each group of cells at various phase.The border between the tissue engineering bone and the normal bone was selected and the area of the vessel in each unit was measured by image analyzer.

3. Conclusion

1. The ossification activity of the osteoblasts was significantly increased ($P< 0.01$) by the vascular endothelial cells. The co-cultured cells had strong potency of proliferation.

2. The heterogenous bone acellular matrix had good biological compatibility, with low antigenicity, no cytotoxity or neoplastia ; The co-cultured cells had strong potency of proliferation in the extracellular matrix of the heterogenous bone acellular matrix.
3. The major ossification of the tissue engineering bone in the body was the cells cultured in vitro. There was synergistic action between the osteoblast and the vascular endothelial cell whether in vivo or in vitro. Fracture healing could be accelerated and the repairing of bone defect could be promoted by using the co-cultured cell as seed cell.

4. Part I: The culture of the seed cell in tissue engineering and the study of cellular compatibility

4.1 Introduction

The culture of seed cell is an essential element in tissue engineering. Osteoblast is the seed cell in tissue engineering, but the base for the repairing of bone defect in tissue engineering is still its vascularization. Co-culture of the osteoblast and VEC can not only promote the proliferative activity of the osteoblast but it they can also accelerate the revascularization of the regenerating bone. Thus, it can help to increase the ability of regenerating bone in the repairing of bone defect. To this end, we conducted related studies and reported as follows.

4.2 Materials and methods

4.2.1 Cell culture

4.2.1.1 The culture of osteoblast (group A) The calvaria cultures were taken from two 2-week-old newborn rabbits and the soft tissue was removed. After several times of rinsing in PBS, the soft tissue was cut into blocks of one mm^3 in size, which were then digested in 1:1 of 0.25% trypsin and 0.1% collagenase at 37 °C for 10 min. After the digestion terminated, the solution was centrifuged at 1200 r/min for 5 min and the supernatant was discarded. Then, the cells was suspended in 20% of fetal calf serum DMEM and inoculated in culture flasks, which were kept in the cell incubator at 37 °C with 5% of CO_2 and saturated humidity. The tissue blocks were digested using 1:1 of 0.25% trypsin and 0.1% collagenase at 37 °C for 20 min. The cells were collected via centrifugation and were inoculated in culture flasks. The above steps were repeated twice. Then, the culture medium was replaced every two to three days. The cells that became monolayer were used for subculture.

4.2.1.2 The culture of VEC (group B) The bilateral kidney cortex of the above two 2-week-old newborn rabbits were cut into fragments, which were then digested with 0.25% of trypsin at 37 °C for 20 min. After the digestion terminated, the solution was centrifugal at 1200 r/min for 5 min and the supernatant was discard. Then, the cells was suspended in 20% of fetal calf serum DMEM and inoculated in culture flasks, which were kept in the cell incubator at 37 °C with 5% of CO_2 and saturated humidity. Then, the culture medium was replaced every two to three days. The cells that became monolayer were used for subculture.

4.2.1.3 Co-culture of osteoblast and VEC (group C) Primary cultured osteoblast, which was in logarithmic growth phase, was inoculated by the ratio of 1:1 in the culture flasks. 20% of fetal calf serum DME were added in the flasks and kept in cell incubator at 37 °C with 5% of

CO_2 and saturated humidity. Then, the culture medium was replaced every two to three days. The cells that became monolayer were used for subculture.

4.2.2 The index of observation

4.2.2.1 Observation under the inverted phase contrast microscope Observe morphology and the growth of the cells in the three groups under the inverted phase contrast microscope.

4.2.2.2 Immunocytochemical stain The cells of group A and group B were inoculated with the density of 2×10^5 cells per well into the six-pore board with a built-in glass slide for making cell climbing slices. Three days later, immunocytochemical stain of type I collagen to identify the osteoblast for group A. Immunocytochemical stain of the vascular factor Ⅷ to identify the VEC for group B.

4.2.2.3 Histological observation The cells of group C were inoculated with the density of 2×10^5 cells per well into the six-pore board with a built-in glass slide for making cell climbing slices. Three days later, observe the mixed growth of the two kinds of cells after HE stain and Masson trichrome stain.

4.2.2.4 Test of the activity of alkaline phosphatase (ALP) The cells of groups A, B and C were respectively inoculated with the density of 3×10^3 cells per well into 96-pore board (12 pores each group). Test the activity of ALP by using the method of paranitrobenzene phosphate via a machine of UV / fluorescence / visible high efficiency analyzer (Perkin Elmer, USA). Observe whether the VEC would affect the ALP activity produced by the osteoblast.

4.2.2.5 Test of the cellular activity by the method of methythiazolyl tetrazolium bromide (MTT) The cells of groups A, B and C were digested and inoculated with the density of 4×10^3 cells per well into 96-pore board. 12 pores were taken on the 1st, 3rd, 5th and the 7th day, respectively. Added 5 mg ml^{-1} of MTT and reacted for four hours. After the liquid was discarded, dimethyl sulfoxide was added and oscillated for 10 min. An enzyme mark instrument (Perkin Elmer, USA) was used to obtain the adsorption at the wavelength of 570 nm and to analyze the growth and proliferation of the cells in the three groups.

4.2.3 Statistical methods

The result of the experiment was expressed by mean±standard deviation. The analysis of variance and q test of the measurement data among groups were done by using SPSS 10.0 software packet. This difference had statistic meanings ($P<0.01$).

4.3 Result

4.3.1 Observation under the inverted phase contrast microscope

Primary cultured osteoblast and VEC began to adhere to the wall four hours later and was adherent completely in 48 hours. Osteoblast displayed various shapes, some were spindle, and some were long triangles, with massive sharp horn like (Fig. 1a). VEC showed ovoidal and changed to typical 'flagstone' appearance when it fussed to monolayer (Fig. 1b). When VEC and osteoblast were co-cultured, they grew closely. There was no contact inhibition, rejection and phagocytosis. The volume of osteoblast was large and the volume of VEC was relatively small (Fig. 1c).

4.3.2 Immunocytochemical stain

The immunocytochemical stain of osteoblast type I collagen and VEC factor VIII were both positive (Figs. 2 and 3). The cytoplasm revealed yellow staining and the cell nuclear blue staining. Results show that the primary cultured cells were osteoblast and VEC.

4.3.3 Histochemistry stain

The HE staining and Masson staining of the cells of group C showed the mixed growth of two types of cells with good cellular compatibility. The volume of osteoblast was large and the volume of VEC was relatively small (Fig. 4). VEC appeared red by Masson trichrome stain (Fig. 5).

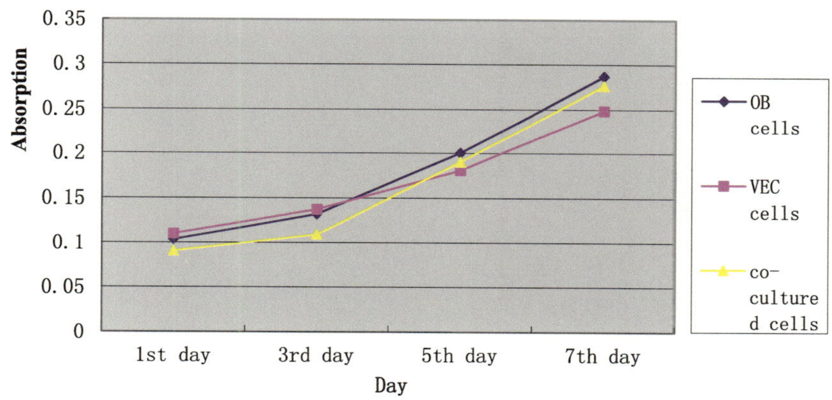

Fig. 1. Curves of cells viability in different groups

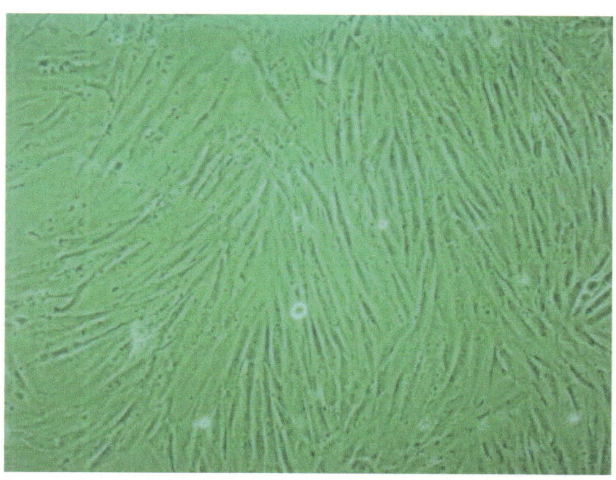

Fig. 1a. Observation under the inverted phase contrast microscope (100x). OB displayed various shapes, some were spindle, typical and some were long triangle, with massive sharp to monolayer.

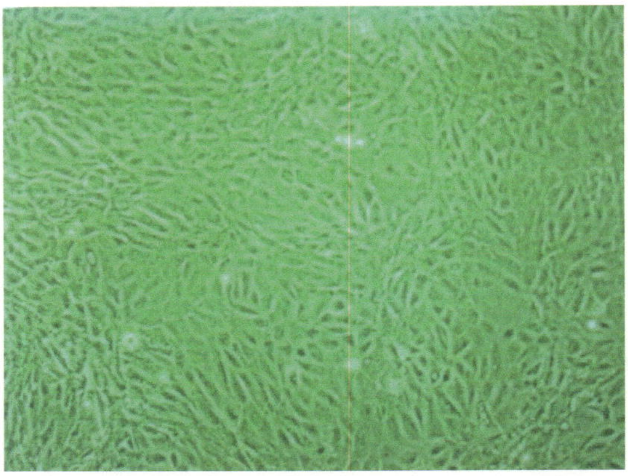

Fig. 1b. Observation under the inverted phase contrast microscope (100×). VEC showed ovoidal and 'flagstone' appearance horn like.

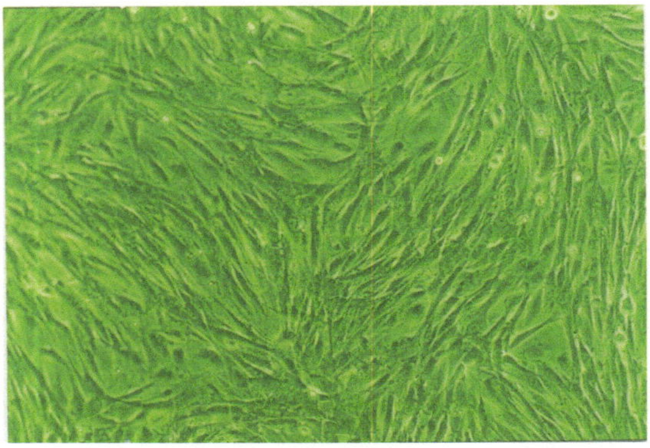

Fig. 1c. Observation under the inverted phase contrast microscope (100×). When VEC and osteoblast were co-cultured, cytoplasm revealed they grew closely. Nuclear blue.

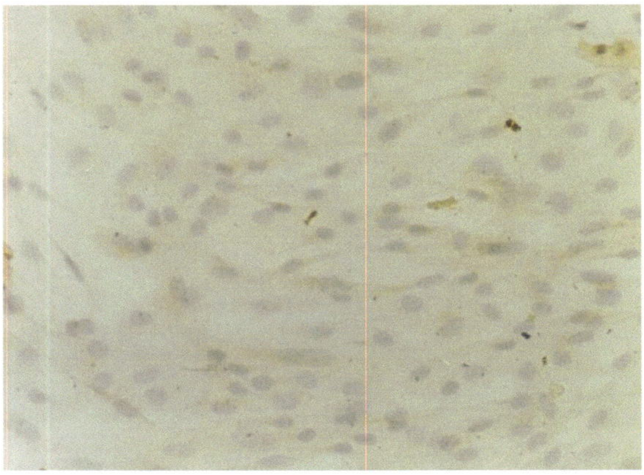

Fig. 2. The immunocytochemical stain of osteoblast type I collagen (100×). The result was positive. The cytoplasm yellow staining and the cell nuclear blue staining.

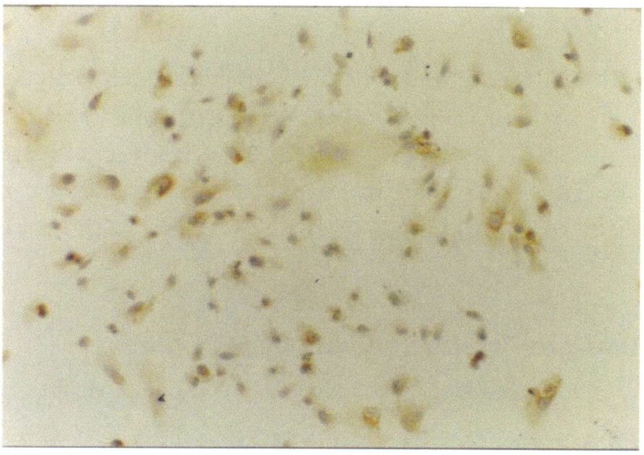

Fig. 3. The immunocytochemical stain of VEC factor VIII(100×). The result was positive. The cytoplasm revealed yellow staining and the cell nuclear blue staining.

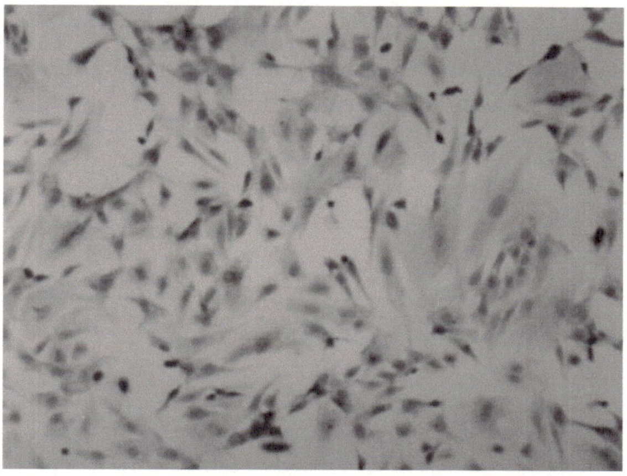

Fig. 4. HE staining of the cells of group C (100×). The HE staining showed the mixed growth of two types of cells with good cellular compatibility.

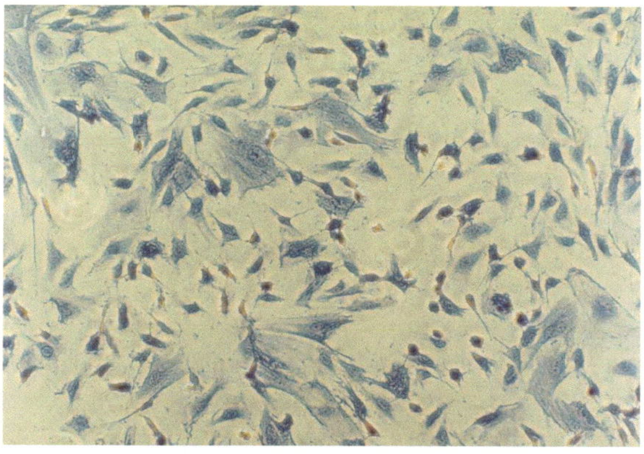

Fig. 5. Masson staining of the cells of group C (100×). The Masson staining showed the mixed growth of two types of cells with good cellular compatibility.

4.3.4 Test of the activity of ALP

After three days of growth, the activity of ALP in each group was as follows. Group A (0.1568 ± 0.0132) U/L, group B (0.1148 ± 0.0145) U/L, group C (0.5018 ± 0.0873) U/L and the F value was 202.811 (P <0.01). The ALP activities from high to low in order were group C, group A and group B. The ALP activity of group A was higher than that of group B (P<0.05). The ALP activity of group C was obviously higher than that of group A and group B (P<0.01).

4.3.5 Test of the cellular proliferation activity by MTT method

The activity of the cells is listed below (Table 1, Fig. 1). On the first day, group B> group A > group C, the difference was statistically significant (P <0.01, P <0.05). On the third day, the activity of group A and B was higher than that of group C, the difference was statistically significant (P <0.01) and there was no significant difference between group A and group B (P> 0.05). On the fifth day, results show that there was no significant difference among group A, B and C. However, on the seventh day, the activity of group A was higher than that of group B (P <0.01), and the activity of group C was higher than that of group B (P <0.01), the difference was statistically significant and there was no significant difference between group A and group C (P> 0.05).

Groups	Culture time			
	1st day	3rd day	5th day	7th day
A	0.1037±0.0085	0.1323±0.0093	0.2003±0.0191	0.2869±0.0132
B	0.1203±0.0194**	0.1375±0.0055	0.1808±0.0099	0.2485±0.0116**
C	0.0803±0.0047*ΔΔ	0.1093±0.0129**ΔΔ	0.1902±0.0203	0.2765±0.0178ΔΔ

Table 1. Comparison of different cells viability (Compared with group A: * P<0.05, **P<0.01; Compared with group C: ΔP <0.05, ΔΔ P<0.01.)
Results show that the cells of group C proliferated slowly in the early stage and faster in the later stage.

Table 1. Comparison of different cells viability

4.4 Discussion

Osteoblast is the progenitor cell in tissue engineering; while VEC is rarely used as seed cell in tissue engineering. The three basic processes of bone graft are the vascularization of the graft, bone regeneration and bone-terminal fusion. The ultimate goal is to repair bone defects. The vascularization of the graft is the key step, which affects the total process of transplantation and repair, and it play a decisive role on the mode and effect of bone regeneration and fusion. Zhu, et al indirectly cultured osteoblast and VEC using nest dish culture method, in which the mediums of the two kinds of cells contact with each other, while the two kinds of cells do not contact. Results show that the medium of VEC can enhance the proliferation of osteoblast. However, in this paper, VEC was directly co-cultured with osteoblast. We observed whether the cellular compatibility of the two kinds of cells was good or not; what's the impact on the osteogenic activity of the osteoblast.

From the experimental results, immunocytochemical stain of type I collagen and the vascular factor VIII identified the osteoblast for group and the VEC for group B. Observation under the inverted phase contrast microscope, HE staining and Masson trichrome staining showed that the two kinds of cells mixed grew well, and the cells were well distributed. There was no contact inhibition, rejection and phagocytosis, explaining that osteoblast and VEC are with good cellular compatibility.

Brighton et al and Katagiri et al find in the experiments that the epidermal cells, endothelial cells and osteoblast may be the bone progenitor cells, which can behavior the characteristics of osteoblast after culture. For example, they can synthesize ALP, osteocalcin and type I collagen. The internal environment of body is maintained by the synergy of a variety of cells. And in vitro environment, osteoblast and VEC might have effects of mutual synergies. Lu et al find that the TGF-β release system can significantly promote bone marrow stromal cell proliferation and differentiation to osteoblast. Moreover, the number of surface cells, the activity of ALP and the osteocalcin production was significantly higher than that of the control group that is without the system. The bone morphogenetic protein, vascular endothelial growth factor (VEGF) and tetracycline have similar regulation with TGF-β to osteoblast. VEC can synthesize and secrete VEGF, thus significantly promote the activity of ALP in osteoblast.

Seen from the test of ALP activity by paranitrobenzene phosphate method, the ALP activity of group A was obviously higher than that of group B ($P < 0.05$); while the ALP activity of group C was remarkable higher than that of groups A and B ($P < 0.01$). Results show that VEC can significantly enhance the ALP activity in osteoblast, it further explained that VEC could increase the function of osteoblast, including the proliferation of osteoblast, and could promote osteoblast to transform to bone cell. The results were consistent with that of Ref.. On the other hand, osteoblast can secrete VEGF, FGF and other pro-angiogenic factors, which affect VEC to promote angiogenesis. In the bone marrow hemopoietic microenvironment, osteoblast plays a central role by producing colony-stimulating factor to promote the regeneration of hematopoietic stem cells and VEC.

Results of the MTT test show that activity of VEC on the 1st day and the 3rd day was strong. It began to decline on the 5th day. However, osteoblast began to proliferate fast on the 5th days. The proliferation activity of co-culture cells was lower than osteoblast and VEC on the 1st and the 3rd day. From the 3rd day, the two types of cells have good cellular compatibility and started into the logarithmic growth phase, thus demonstrating strong proliferation activity. On the 5th days, the cell proliferation activity obviously accelerated, and has no significant difference with osteoblast and VEC. On the 7th day, the proliferation activity of osteoblast and co-culture cell was remarkably higher than that of VEC. Whether these were due to the reason that the two kinds of cell competed for living sites on the board from the 1st to the 3rd day after co-culture, which affected the cell proliferation activity. It needs to further study. The experiment in this paper showed that the co-culture cells proliferated slowly in the early stage and faster in the later stage. Osteoblast and the co-culture cells are with strong proliferation, the proliferation of VEC is comparatively weak, indicating strong potency of proliferation of co-culture cells.

The experimental results show that, there was good cellular compatibility between the osteoblast and VEC. VEC could increase the function of osteoblast, and promote the proliferation of osteoblast. The VEC cultured in vitro may help the revascularization of the tissue engineering bone. Thus, it can provide normal nutritional, and the regulation of nerves and body fluid. Therefore, the co-cultured cells between osteoblast and VEC are expected to become the seed cells of the tissue engineering bone.

4.5 Conclusion

The conclusion section should be an independent paragraph. A reader who choose to read onlt the conclusion must be able to reasonably understand the topic. The conclusion should start with a sentence or two that describe the experiment before concluding results.

1. There was good cellular compatibility between the osteoblast and the VEC.
2. The ossification activity of the osteoblast was significantly increased by the VEC.
3. The co-cultured cells had strong potency of proliferation.

5. Part II: Study the effect of heterogenous bone acellular matrix (HBACM) on the biological compatibility between the osteoblasts and the vascular endothelial cells

5.1 Introduction

Tissue engineering is a science that uses cell biology and engineering principles to research and develop biological alternatives for the repair and improvement of wounded tissue and function. The histocyte that is cultured in vitro and with high concentration is amplified and is absorbed in a biocompatible extracellular matrix (ECM) that is degradable and can be absorbed by human body. The seeded cells continue to grow and reproduce during the degradation and the absorption process in the scaffold, and develop to corresponding new tissue and organ with original special function and morphology, then the expected purpose of wounds reparation and reconstruction function was attained. In this experiment, the osteoblast and the vascular endothelial cell (VEC) were compounded by using the acellular rib of pig. Observe and study the biological compatibility. Thus it will provide the theological base for the clinical application of tissue engineering bone.

5.2 Materials and methods

5.2.1 Preparation for the framework material for HBACM

Fresh pork ribs were selected and the soft tissue around was removed. Then the soft tissue was made into skeletons in size of 2.0cm × 0.5cm × 0.5cm. They were treated as the following steps: (1) the skeletons were immersed in 15% of H_2O_2 in at 37 °C for four hours and were washed repeatedly Double-distilled water; (2) the skeletons in the upper step were immersed in 1% of Triton X-100-Tris-HCl solution (V/V) with protease inhibitors added in it, and was oscillated at 4 °C for 48 hours; (3) circulating flushed for 12 hours in PBS; (4) the skeletons in the upper step were immersed into chloroform/methanol solution (3:1) at 4 °C overnight; (5) immersed in 0.02% Tris / EDTA (pH 8.0) solution, and was oscillated at 4 °C for 48 hours; (6) washed with distilled water continuously for 48 hours, then the acellular disposal completed (7) the skeletons were freeze-dried and sterilized with ethylene oxide, for use, and then observe the degree of acellular under scanning electron microscope.

5.2.2 Cell culture

The co-culture osteoblast and VEC, the identification of osteoblast, VEC and the co-cultured cells were the same as the former.

5.2.3 Compound of the cell with HBACM

Experimental groups were as follows:

Group A: osteoblast compound with HBACM;
Group B: VEC compound with HBACM;
Group C: osteoblast + VEC compound with HBACM.

5.2.4 The indexes of observation

5.2.4.1 Observation under the inverted phase contrast microscope Observe the adhesion of the cells with the material, the distribution of the cells in the pores of the material, and the growth, differentiation and proliferation of the cell.

5.2.4.2 Histological observation The three groups of cells were compounded and cultured with the materials for 5 days. After being fixed with alcohol and stained directly by Masson trichrome, they were observed under optical microscope. Three of the materials in each group were taken and embedded in paraffin after decalcification to determine on histological sections. Then observe the biological compatibility after HE stain, toluidine blue and Masson trichrome stain.

5.2.4.3 Observation under the scanning electron microscope The three groups of cells were compounded and cultured with the materials for 5 days. One of the compound materials in each group were taken and rinsed in PBS for thrice, fixed with 2.5% glutaraldehyde, dehydrated with alcohol gradiently, and then fixed in iso-amyl acetate. After dried and coated with gold, the samples were observed by using a scanning electron microscope (JSM-5900) to test the adhesion, growth, proliferation and the matrix secretion of the cells on the HBACM.

5.2.4.4 Test by flow cytometer The three groups of cells were compounded and cultured on the culture plate with HBACM. Three compounds of cellular material were selected individually from each group on the 1d,3d,5d and 7d. Then they were digested. Collect and count the cells. The cells were staining in PI dye and were detected using a flow cytometer (Coulter Corporation, U.S.) to analyze the effect of HBACM on on cell cycle and DNA content. Then the DNA index (DI values), $DI = 1.0 \pm 0.15$ was calculated as normal diploid cells. Analyze the cell cycle and the ploidy by flow cytometer to test the toxicity of the material to the cell.

5.2.5 Statistical methods

The result of the experiment was expressed by mean±standard deviation. The analysis of variance and q test of the measurement data among groups were done by using SPSS 10.0 software packet.

5.3 Result

5.3.1 Observation of HBACM under scanning electron microscope

The structure of the bone trabecular of HBACM was integrated, the bone lacunas were empty, and no residual cellular components were found (Fig. 1).

5.3.2 Observation under the inverted phase contrast microscope

When the three groups of cells were co-cultured with HBACM for 12 hours, the cells could adhere and grow in the pores and on the surface of HBACM. With the culture time gone, grew, differentiate and proliferate along the edge of HBACM pore gradually. The cells gradually extended, stretched across the pores, and the pseudopodium tended to contact with each other from the 5th day co-cultured with HBACM (Fig. 2).

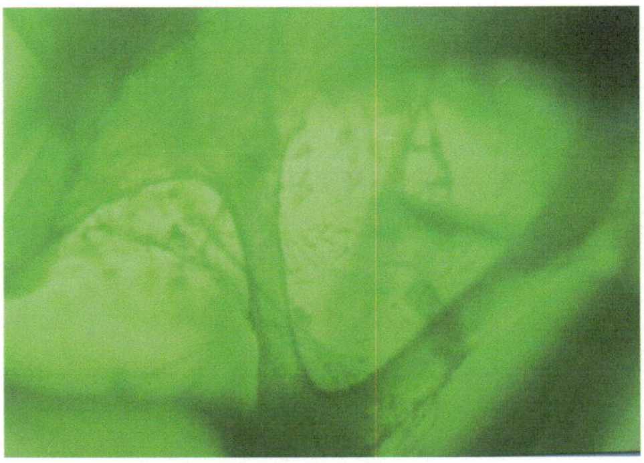

Fig. 1. Observation under the inverted phase contrast microscope (100×) when the osteoblast was co-cultured with materials for 5 days. The cells gradually extended, stretched across the pores.

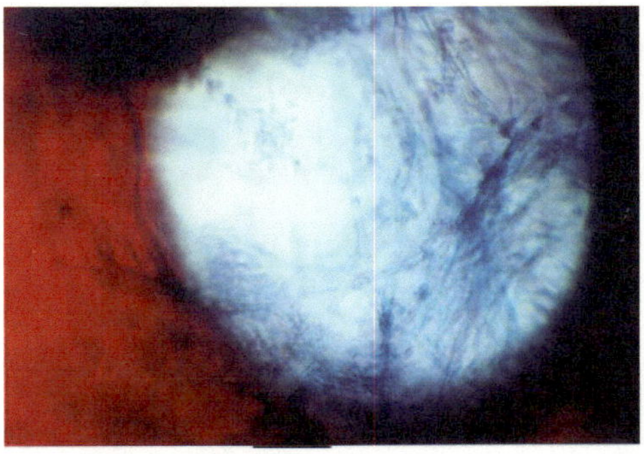

Fig. 2. Observation under the inverted phase contrastmicroscope (100×) after Masson staining for compound of the co-cultured cells and the material. The cells and the materials attached closely and connected into a network.

5.3.3 Histological observation

The HBACM of compound cell was directly Masson stained. Results show that the cells in each group tightly adhered with materials. They gathered into a group, connected into a network, and there were a large number of proliferations of the cell on the surface and in the pores of HBACM. The number of cells adhesion on HBACM in osteoblast + VEC group was the maximum (Fig. 3). Stained histological sections showed that the cells were closely attached along the edge of materials, and the histocompatibility of cells and HBACM was good.

5.3.4 Scanning electron microscope results

After cultured with HBACM for five days, the cells were tightly adhered on the surface of HBACM. Osteoblast appeared spindle-shaped or polygonal, the adjacent cells connected to each other with processes (Fig. 4). VEC was oval-shaped, with angular protrusions, and was closely attached to the materials (Fig. 5). The co-cultured cells were showing their own form, and the two kinds of cells tightly combined with each other. The cell compatibility and compatibility between the cells and HBACM were good, with network collagen attached around (Fig. 6). A plenty of cells adhered on the surface of the materials of osteoblast + VEC group, and much collagen formed. However, less collagen formed on the surface of VEC group.

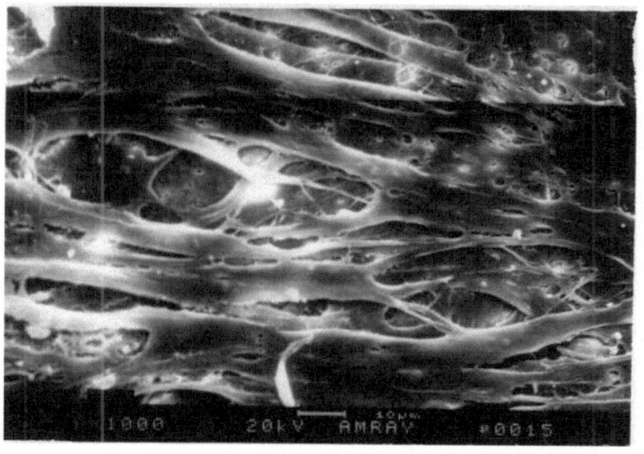

Fig. 3. The compound of osteoblast cells and the material (scanning electron microscope, 1000×)

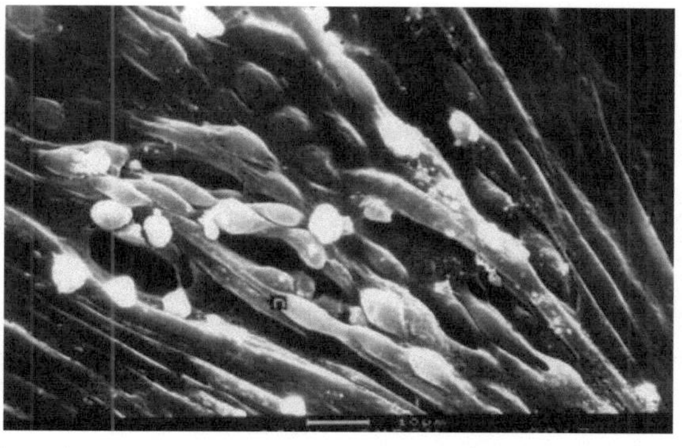

Fig. 4. VEC compound with the material (scanning electron microscope, 1200×)

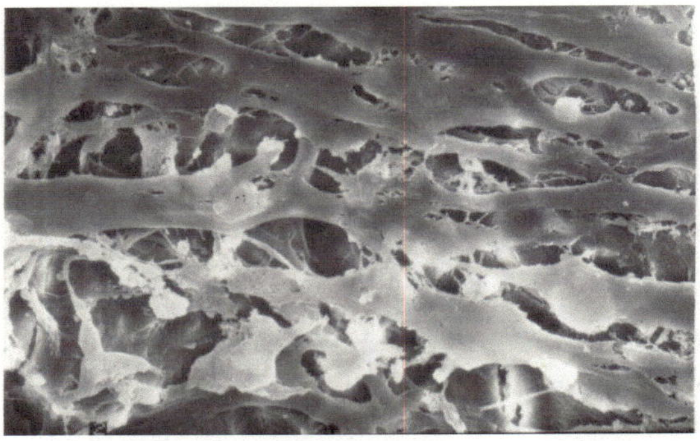

Fig. 5. The co-cultured cell compound with the material (scanning electron microscope, 1000×)

5.3.5 Test by flow cytometer

The three groups of cells were compounded and cultured with the material for one, three, five, seven days, respectively. The testing results of cell cycle and ploidy level are listed Table 1. The S phase cells of the co-cultured group on the 1st and the 3rd day was higher than that of the single cell group, no heteroploid cells were found. The effect of HBACM on the cell cycle and the ploidy of the three groups of cells are listed on Table 1.

Cell cycle \ Group		Group A	Group B	Group C	Control group
Pre-synthesis stage of DNA (G_1, %)	1 day	67.4	71.5	57.9	68.8
	3 day	74.5	80.5	70.2	73.9
	5 day	67.1	66.3	68.2	66.7
	7 day	53.3	69.9	67.8	63.6
Synthesis stage of DNA (S, %)	1 day	24.6	18.7	29.5	20.6
	3 day	12.3	12.1	20.6	13.5
	5 day	19.1	15.7	19.7	17.8
	7 day	23.2	16.8	21.0	19.9
Pre-mitosis and mitosis (G_2/M, %)	1 day	8.0	9.8	12.7	10.6
	3 day	13.2	7.4	9.2	12.6
	5 day	13.8	18.0	12.1	15.5
	7 day	23.5	13.3	11.3	16.5
DNA content of G_1 stage	1 day	93.0	100.4	101.9	98.2
	3 day	97.5	91.5	105.6	96.8
	5 day	119.7	108.8	112.3	113.7
	7 day	118.2	117.6	122.4	119.8
DNA index	1 day	0.95	1.02	1.04	
	3 day	1.01	0.95	1.09	
	5 day	1.05	0.96	0.99	
	7 day	0.99	0.98	1.02	

Table 1. The influence of the HBACM to the cell and DNA index of three group cells

5.4 Discussion

As a kind of artificial extracellular matrix (ECM), scaffolds provide three-dimensional space for the growth, reproduction and metabolism of the cells. The biocompatible of scaffold directly affects the adhesion, growth and proliferation of the cell. Different kinds of ECM are widely used nowadays, they are: biodegradable polymer; ceramic material; composite materials and natural extracellular matrix material (NECM).

Heterogeneous bones, which are obtained from animals such as pigs and cattle, have a rich resource and lower cost, and are easy to access and process. Moreover, the bone tissue retains the dense pore structure naturally and is with good effect of scaffold, which provides three-dimensional space for the proliferation, differentiation and the osteogenesis of the cells. At the same time, the bone tissue is with high bone induction activity and degradation rate matched well with the growth rate of new bone. Thus, it has great potential for developing as bone substitute material. However, if the untreated heterogeneous bones were implanted into human body, there will be a strong immune rejection. Hence, a variety of physical, chemical treatments are used to eliminate the antigen as far as possible, and simultaneously retain the useful biologically active substances and its bone induction activity, as well as its mechanical strength. Those are also the research focus of the heterogeneous bones tissue engineering transplantation. This study focused on the transplantation of the heterogeneous bones and the elimination of the cells antigen, and then used the heterogeneous bones as a superior scaffold for bone tissue engineering.

5.4.1 Biological compatibility of HBACM

Bone extracellular matrix is the intercellular substance of calcified bone tissue, and is consist by organic and inorganic ingredients. The organic ingredients are mainly: collagen fibers and some amorphous material, which are both the secretion of bone cells. Inorganic components are mainly calcium salt, which exist mostly as hydroxyapatite crystals. Low-permeability method was selected to produce HBACM in this experiment. The cells were burst and the membrane structure was damaged. Other methods were also used to extract the cell membrane, cytoplasm, organelles and other ingredients. Only the cell was extracted, and the inorganic and organic ingredients were retained. Results show that the cells of the three groups could be attached to the HBACM and grew. The reasons maybe as follows: the natural pore system was not significantly damaged; the original trabecular bone, the trabecular bone gap and the lumen system was retained; the three-dimensional structure of the bone salt still existed. This natural structure was conducive to cell adhesion, growth and provide generous interior space and surface for the secretion of ECM. Therefore, the acellular bone matrix showed excellent biocompatibility.

5.4.2 Antigenicity of HBACM

It is generally believed that transplantation antigen of heterogeneous bones present in bone cells, hematopoietic cells, white blood cells, red blood cells, plasma, blood vessels and bone matrix. To transplant heterogeneous bones, we should reduce or eliminate the antigenicity of transplanted bones. After a series of physical and chemical treatment, the rib of pig was made as HBACM, which did not contain cells under the scanning electronic microscopy. Thus, it was of low antigenicity. Observation from the inverted phase contrast microscope, the three groups of cells grew along the edge of the pores of the material. The cells gradually extended, stretched across the pores, and the pseudopodium tended to contact with each other, they gradually grew, differentiated and proliferated. Stained histological sections and the result of scanning electron microscopy showed that the cells were closely attached, gathered into a group, and connected into a network. These indicated that the compatibility between cells and the compatibility between cells and HBACM were good, without any mutual exclusiveness. These further explained that there were none or extremely low antigenicity when using this method to prepare acellular bone matrix.

5.4.3 Cytotoxicity of HBACM

Flow cytometry could be used to determine DNA content and proliferation activity of the cells, and could analyse the cell cycle and the ploidy level. When the body suffers from cancer or precancerous lesions with malignant tendency, the DNA content of the cells changes abnormally, leading DNA aneuploid. Flow cytometry can detect aneuploid cells by measuring DNA content, which has been an effective tool making cancer for the diagnosis of tumor cytology. In this study, flow cytometry was used to determine the ploidy level and the DNA content of the three groups of cells that co-cultrued with acellular matrix scaffold in vitro. No aneuploid cells were found, indicating the scaffolds were without cytotoxicity and tumorigenesis. We could see from the result of cell cycle analysis that the cells of each group had normal growth cycle in the acellular matrix scaffolds. The S phase of the co-cultured group on the 1st and the 3rd day was higher than that of the single cell group, indicating high proliferation of the cells on scaffold materials when osteoblast was co-cultured with VEC.

With the research and development of biodegradable scaffolds with three-dimensional and multi-space structure, the bioactive bone transplantation materials formed by the three kinds of cells and acellular matrix bone shows a broad application prospects.

5.5 Conclusion

It is preferred not to use abreviations in the conclusionsor at least use the fuul description once, at the first conclusion.

1. Heterogenous bone acellular matrix has good biological compatibility.
2. Heterogenous bone acellular matrix was with low antigenicity.
3. Heterogenous bone acellular matrix had no cytotoxity or neoplastia.
4. The co-cutured cells had strong potency of proliferation. heterogenous bone acellular matrix

6. Part III: Implanting the compound cell of heterogenous bone acellular matrix in vivo: An animal study

6.1 Introduction

Autografting and isotype heterogenous bone grafting have been widely used in clinic for many years. Autografting is the best method of bone graft because of the bioactive molecules, living cells, and blood supply in the grafted tissue. However, the bone source and the site for supplying bone were limited in autografting, meanwhile it is difficult in the modeling of self-tissue, and therefore its clinical application is limited. There is certain immunologic rejection in the isotype heterogenous bone grafting, which worried many people. At the same time, the bone source is limited. Since the 1980s, with the rapid development of cell culture technology, tissue engineering bone provides a new way for solving this problem. In this study, heterogenic acellular extracellular matrix was compound with cell to construct tissue engineering bone, which was implanted into the animal. Then the bone defect repairing was observed.

6.2 Materials and methods

6.2.1 Mark the cell with BrdU (5 - bromodeoxyuridine)-labeled cells

400 µL of BrdU was added into 200 mL of serum-free F_{12} culture medium. 35 bottles of 3 generations and 5 generations of rabbit osteoblast, VEC and the co-cultured cells were taken with the culture medium being discarded, and then, the serum-free F_{12} culture medium marked by BrdU were added (6mL per bottle). The bottles were kept in the cell incubator at 37 °C with 5% of CO_2 and saturated humidity for one hour. After that, the supernatant was discarded and the cells were digested and collected, which was washed for thrice with culture medium to remove BrdU that had not incorporated cells.

6.2.2 The compound cell of heterogenous bone acellular matrix

30 blocks compound cell of heterogenous bone acellular matrix (pork ribs) with the size of 2.0cm × 0.5cm × 0.5cm were divided into three groups (10 blocks each group). Group A was compound with osteoblast, Group B was compound with VEC, and the Group C was compound with co-cultured cells. The three groups of cells marked with BrdU were inoculated with the density of 6×10^5 cells per block into HBACM, respectively. One block of each group was taken for in vitro testing, and the remaining 27 blocks were used for implanting into the animal body.

6.2.3 Animal experiments

6.2.3.1 Animal grouping 27 New Zealand big-ear rabbits weighted 2.5 kilogram were selected and they were grouped into three groups with nine in each group. Groups of A, B, C were HBACM compound with osteoblast, VEC and co-cultured cells marked by BrdU. Animal grouping is listed as follows.

Time / Group	3rd week	6th week	12th week	Total
osteoblast group	3	3	3	9
VEC group	3	3	3	9
Co-cultured cells	3	3	3	9
Total	9	9	9	27

Table 1. Animal grouping of the New Zealand rabbits

6.2.3.2 Preparation for animal model of bone defect Resect the 1.5cm long radius of the double forelimbs of the rabbit and prepare animal model of bone defect. The tissue engineering bone which had been compounded and cultured for one week by the heterogenous bone acellular matrix and the cells of each group was implanted to the left side. The heterogenous bone acellular matrix was implanted to the right side as a control. Select the material at 3 weeks, 6 weeks and 12 weeks to observe bone fracture.

6.2.4 The index of observation

6.2.4.1 Observation of gross morphology Three, six and 12 weeks after the operation, three animals in each group were sacrificed and the radius were dissected. Observe the development and the prognosis of the tissue engineering bone at various phases by naked eye.

6.2.4.2 X-ray examination of bone defect After three, six, and 12 weeks of the operation, samples were taken, respectively. The X-ray (CR) radiography was used to observe the healing of bone defects, and in order to understand the conjugation of the implanted tissue engineering bone and the peripheral normal radius, also to understand the formation, change and moulding of the callus at the site with bone defect.

6.2.4.3 Trace and test the BrdU marked cells The paraffin sections was prepared in three and six weeks, respectively, for BrdU immunohistochemistry stain, to detect BrdU marked cells, then to understand the survival of seed cell in the experimental animal. After HBACM and BrdU marked cells was co-cultured in vitro for one week, one tissue engineering bone in each group was taken for the production of paraffin section, in order to detect the compound cell of heterogenous bone acellular matrix in vitro.

6.2.4.4 Routine histological examination About 1.5 cm of the samples was selected from three experimental animals at various phases. The samples were taken with the implanted bone tissue engineering as the center and some normal bone on both ends. After fixed, decalcified, dehydrated and embedded, the samples were made into paraffin sections for HE, toluidine blue, Masson staining.

Pathological changes (such as growth and differentiation of the cells in vivo, inflammatory cell infiltration, degradation of the scaffold, rebuilding and remodeling of callus, recanalization of canal, and the revascularization) of HBACM compound with the cells in each group were observed under optical microscope.

6.2.4.5 Immunohistochemistry of type I collagen The paraffin sections of the sixth weeks was taken for type I collagen immunohistochemistry stain, to understand the secretion of type I collagen by each group of cells at various phase.

6.2.4.6 Image analysis Masson stains at the 3rd and the 6th week were selected and the border between the tissue engineering bone and the normal bone was detected by an image analyzer to measure the area of the vessel in each unit. The revascularization of tissue engineering bone in each group was observed.

6.2.5 Statistics method

The result of the experiment was expressed by mean±standard deviation. The analysis of variance and q test of the measurement data among groups were done by using SPSS 10.0 software packet.

6.3 Result

6.3.1 Observation of gross morphology

The samples were taken and observed at the 3rd, 6th, and 12th week, respectively. Results show that the speed and degree of repairing the bone defect by tissue engineering bone from high to low in order were: that of the co-cultured cell group, that of the osteoblast group, that of the VEC group and that of the simple framework material group (Figs. 1a, 1b, 1c, and 1d).

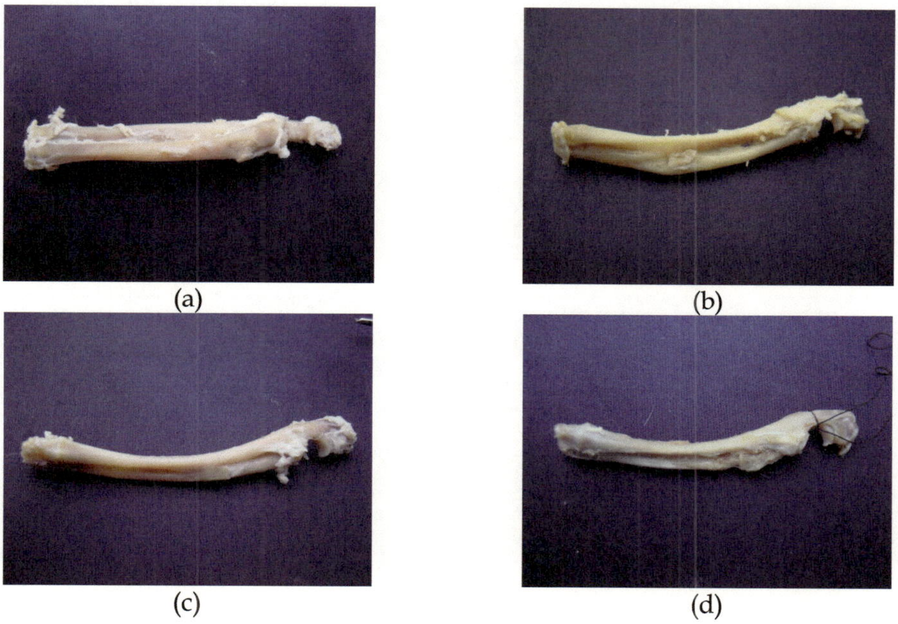

Fig. 1. Observation of gross morphology in the 6th week: (a) osteoblast group, (b) VEC group, (c) co-cultured cell group, (d) simple framework material group

6.3.2 X-ray examination of bone defects

We could see from the X-ray examination that the speed and degree of repairing the bone defect by tissue engineering bone from high to low in order were: that of the co-cultured cell group, that of the osteoblast group, that of the VEC group and that of the simple framework material group (Figs. 2a, 2b, 2c, and 2d).

6.3.3 Trace and test the BrdU marked cells

One week after compound in vitro, we could see from the immunohistochemical staining result that cells compound and grew well with HBACM, the cells attached to the pores of HBACM, closely adhere to the edge of HBACM and grew (Fig. 3). When the bone was implanted for three weeks and six weeks, the marked cells could still be detected in HBACM in vitro, and the new cartilage could also be observed (Figure 4).

6.3.4 Routine histological examination

In the 3rd week, a large number of cartilages formed in tissue engineering bone. On the juncture with the normal bone, the formation of cartilage zone could be seen. osteoblast and osteoclasts coexisted in the pores of HBACM, in which the capillary grew (Fig. 5). In the 6th weeks, the pores in HBACM gradually grew confluent, the blood vessel increased in the pores, and the cartilage cells gradually transformed to bone cells (Fig. 6). In the 12th weeks, the marrow between tissue engineering bone and normal bone recanalized, and the fracture healed (Fig. 7).

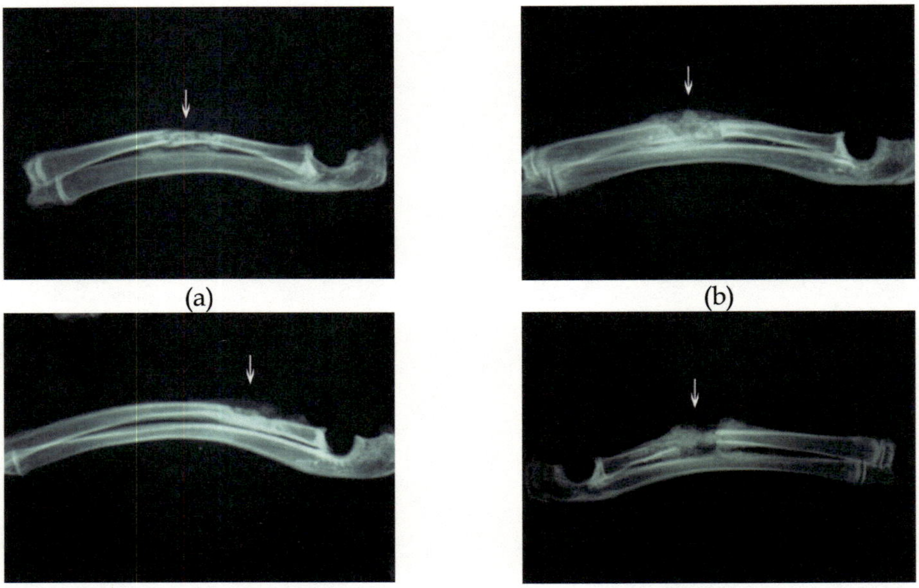

(a) (b) (c) (d)

Fig. 2. X-ray examination in the 3rd week: (a) osteoblast group, (b) VEC group, (c) co-cultured cell group, (d) simple framework material group

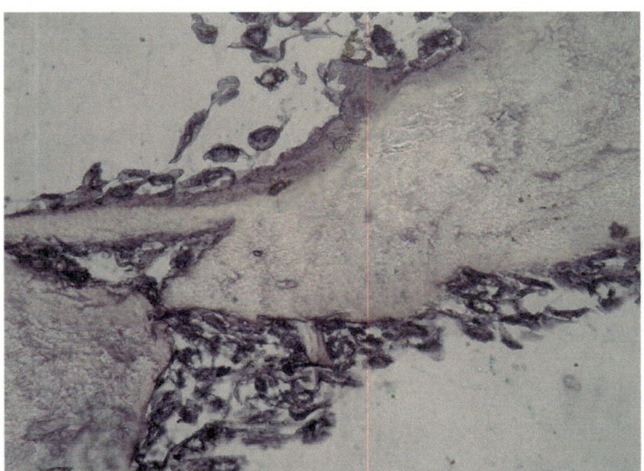

Fig. 3. BrdU immunohistochemical staining (400×) osteoblast closely adhere to the edge of HBACM and grew.

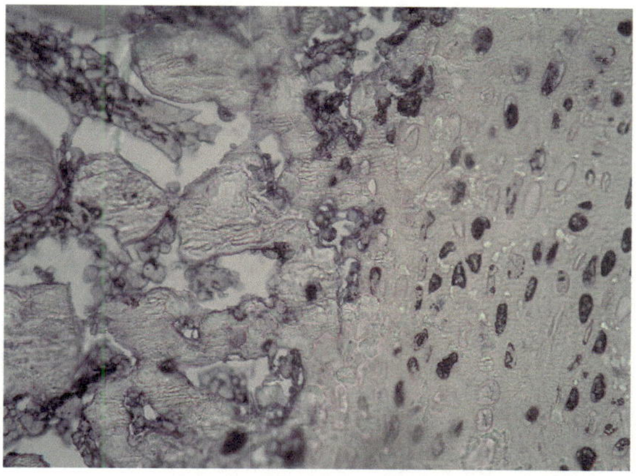

Fig. 4. BrdU immunohistochemical staining (400×). When the bone was implanted for three weeks, marked cells could still be detected in HBACM, and the new cartilage could also be observed.

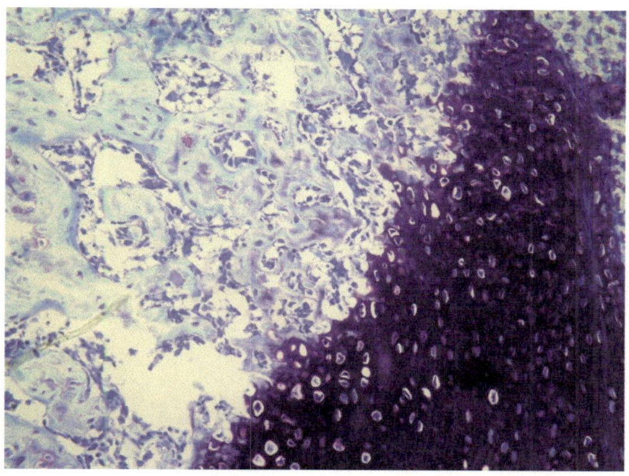

Fig. 5. Toluidine blue in the 3rd week (200×) On the juncture with the normal bone, the grewformation of cartilage zone could be seen in the pores. osteoblast and osteoclasts coexisted in the pores.

6.3.5 Immunohistochemistry of type I collagen

After the three groups of cells compound with HBACM were implanted in vivo, type I collagen was expressed (Fig. 8), while VEC group was lower expressed.

6.3.6 Image analysis

The results of image analysis in the 3rd week were as follows. osteoblast group (41.20 ± 7.37), VEC group (47.54 ± 5.71), co-cultured cell group (50.54 ± 3.86), co-cultured cells group> osteoblast group, the difference was statistically significant ($P < 0.05$). In the 6th weeks, osteoblast group (28.52 ± 6.52), VEC group (50.24 ± 10.11), co-cultured cell group (66.98 ± 7.72), co-cultured cell group> VEC group> osteoblast group, the difference was statistically significant ($P < 0.01$).

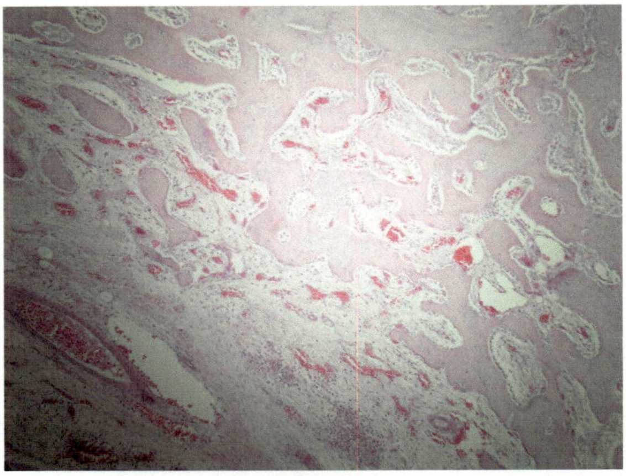

Fig. 6. HE stain in the 6th week (100×). In the 6th weeks, the pores in HBACM gradually confluent, the blood vessel increased of HBACM.

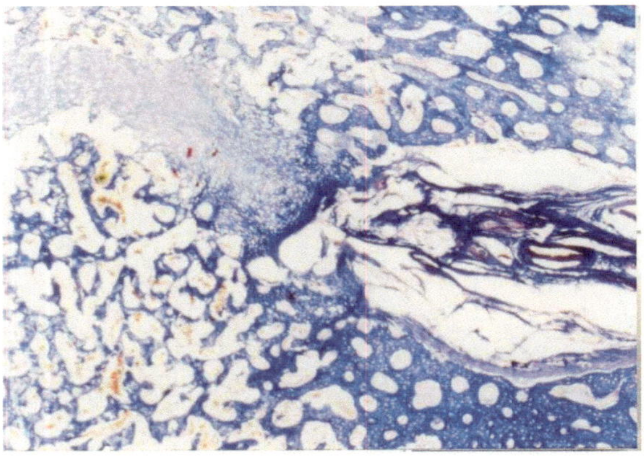

Fig. 7. In the 12th weeks, medullary cavity formed in the tissue engineering bone.

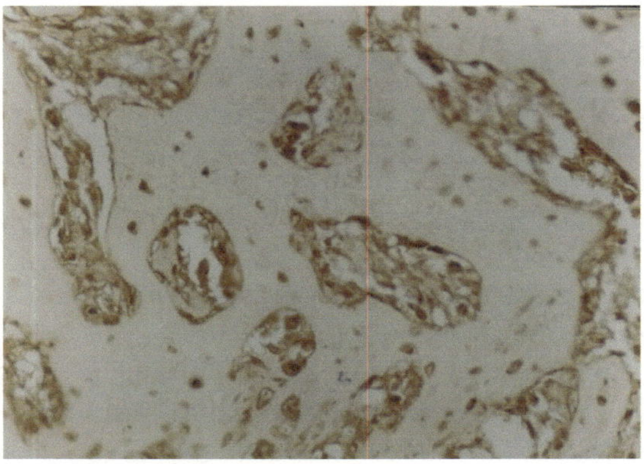

Fig. 8. Immunohistochemical of type I collagen in the 6th week

6.4 Discussion

As the research development of cell biology, molecular biology, and biological engineering and materials science, tissue engineering was born as a new discipline. "Tissue engineering" was put forward in 1987 by the National Science Foundation (USA), which refers to the science that uses cell biology and engineering principles to research and develop biological alternatives for the repair and improvement of wounded tissue and function. The basic premise of modern tissue engineering is to guide the tissue repair and regeneration by the application and the control of cell tissue and the micro-environment. It is a thoroughly new therapeutic mode to repair bone defects by using the methods and tools of tissue engineering, which has broad application prospects. With the rapid development of the current research of bone tissue engineering, tissue engineering bone has become one of the most promising tissue engineering achievements that will be come into use in clinical applications, which will lead a revolutionary change in the treatment of the fracture healing and the bone defects caused by trauma, tumor resection and malformation correction.

6.4.1 Trace and test the Brdu marked cells

The structure of BrdU is similar with thymidine, it is characteristic is that the methyl, which connects the fifth carbon atom and pyrimidine ring of phase Ⅲ in thymine, is substituted by bromide atom. During the synthesis process, BrdU can specifically incorporate into DNA at S phase, which is the same as thymine. Simultaneously, there is no cross-reaction between BrdU and thymine, and it is without radioactive material contamination. Therefore, BrdU has been used to mark cells in both in vivo and in vitro, and has been widely applied into the research of cell kinetics, as its high accuracy, rapidity and safety for marking cells. Generally speaking, it can be detected in vivo within eight weeks, and will not be longer than 12 weeks. In this study, BrdU was used to mark cells that cultured in vitro. After compound the marked cells with HBACM, in vitro test and in vivo test for implantation at the 3rd week and at the 6th week were done. Results show that cells compound and grew well with HBACM, the cells attached to the pores of HBACM, closely adhere to the edge of HBACM and grew. When the bone was implanted for three weeks and six weeks, the marked cells cultured in vitro could still be detected in HBACM, and the new cartilage could also be observed. These indicated that the histocompatibility between HBACM and cells was good; and it further indicated that the major ossification of the tissue engineering bone in the body was indeed due to the cells cultured in vitro.

6.4.2 Pathological process of bone tissue engineering repairing bone defects

Fracture healing is a complex process, which is continuous and is the result of the mutual synergistic effect between osteoblast and osteoclasts. The research results in this experiment showed the pathological process after the tissue engineering bone was implanted into animal body. The endochondral bone formation is the main effect during the fracture healing process. Under the effect of osteoblast, HBACM gradually osteoblasted outward; gradually transformed the fibrous tissue among the fracture ends and the pores of HBACM into cartilage. Then the pores grew confluent. At the same time, the surface of bone trabecula in HBACM began to change into small cavities, which gradually interconnected to form larger cavities. Vascular formed, the chondrocyte gradually proliferated, calcified and ossified in the larger cavities. The fracture healing process completed when the larger cavities grew confluent and recanalized with autogenous marrow cavity.

6.4.3 Comparison of the osteogenic ability of bone tissue engineering in vivo

Type I collagen is the major extracellular matrix secreted by osteoblast. It is expressed from the beginning of cell proliferation, and achieved to the maximum at synthesis stage of matrix. The results showed that after the three groups of cells compound with HBACM were implanted in vivo, type I collagen was expressed (Fig. 8), while VEC group was lower expressed. From the gross specimen and X-ray results, we could see that the speed and degree of repairing the bone defect by tissue engineering bone from high to low in order were: that of the co-cultured cell group, that of the osteoblast group, that of the VEC group and that of the simple framework material group. These indicated that fracture healing could be accelerated and the repairing of bone defect could be promoted by using the co-cultured cell as seed cell. Simultaneously, the simple framework material group could also achieve fracture healing, which indicated that the HBACM might contain bone morphogenetic protein, with a strong cross-species induction of osteogenic activity; namely, induced undifferentiated mesenchymal cells differentiate into cartilage or bone, its effect was non-species-specific and could repair bone defects.

6.4.4 Comparison of revascularization ability of bone tissue engineering in vivo

There has been a significant progress in the research of bone tissue engineering. However, with the experiment object becoming large-scale, the research of bone tissue engineering bone is facing the key problem -- rapid vascularization. Like other types of transplanted bone, adequate blood supply is the decisive factors to ensure tissue engineering bone to survive in vivo. Current research on the reconstruction of the blood supply in the new tissue engineering bone is still at the initial stage. The main methods are as follow: (1) combined transplantation of VEC and osteoblast (2) the use of VEGF to promote the growth of blood vessel (3) the application of microsurgical techniques in the revascularization of tissue engineering bone, including ①osteoblast + biomaterials + tissue flap with vascular pedicle embedded; ②osteoblast + biomaterials + vascular bundle implantation. Zhengfu Fan, et al prepared a deep facial flap animal model with a nameless vascular pedicle in forelimb. When the facial flap was applied to cover the biomaterials of compound cells to repair the 1.5 cm of radial bone defect, the speed of revascularization was obviously higher than that of the control group without facial flap after the repairing of bone defect was finished. Casabona, et al further designed a bio-engineered prefabricated flap. Human bone marrow stromal cells which were cultured for two weeks were delivered into the latissimus dorsi of athymic mice by a porous hydroxyapatite ceramic model. Eight weeks after the implantation, histologic examination revealed the presence of spongious bone tissue with rich blood supply. A simple myocutaneous flap was thus transformed into a composite osteomyocutaneous flap. Although the flap can cover any area or any known blood vessels, and had the possibility of preshaping the graft to the exact characteristics of the defect. However, two operations should be taken in vivo. The image analysis showed that the number of revascularization of tissue engineering in co-cultured cells group in the body in the 3rd week was higher than that of the osteoblast group, the difference was statistically significant ($P < 0.05$). In the 6th week, the number of revascularization from high to low in order was: that of the co-cultured cells group, that of the

VEC group, that of the osteoblast group, and the difference was statistically significant (P <0.01). The combined transplantation of VEC and osteoblast were used in this experiment and the same result was obtained, which made the new bone tissue engineered bone to entirely replace the autogenous bone graft entirely possible, and provide a excellent experience for tissue engineered bone to be implanted in vivo.

It can be seen, the activity of osteoblast could be enhanced, the fracture healing could be accelerated, the repairing of bone defect could be promoted and the revascularization of tissue engineering bone could be improved by using the co-cultured cell osteoblast and VEC as seed cell. There was mutual synergistic action in vitro and mutual promoted effects in vivo between the osteoblast and the VEC

6.5 Conclusion

This in vivo animal study suggests that:

1. The osteogenesis process of the regenerating bone was the formation of endochondral bone regeneration.
2. The major ossification of the regenerationg bone occurred to the cells cultured in vitro.
3. There was synergistic action between the osteoblast and the VEC whether in vivo or in vitro.
4. Fracture healing could be accelerated and the repairing of bone defect could be promoted by using the co-cultured cell as seed cell.

7. References

[1] Enncking WF, Campanacci DA. Retrieved human allografts: a clinical- pathological study [J]. J Bone Joint Surg (Am), 2001, 83: 971-986.

[2] Boone DW. Complications of iliac crest graft and bone grafting alternatives in foot and ankle surgery [J]. Foot Ankle Clin, 2003, 8: 1-14.

[3] Xiao Y, Mareddy S, Crawford R. Clonal characterization of bone marrow derived stem cells and their application for bone regeneration [J]. Int J Oral Sci, 2010, 2(3): 127-135.

[4] Prioleau MN,Gendron MC,Hyrien O.Replication of the chicken beta-globin locus:early-firing origins at the 5'HS4 insulator and the rho-and beta A-globin genes show opposite epigenetic modifications [J]. Mol Cell Biol, 2003, 23: 3536-3549.

[5] Hakeda Y.Action of glucocortieoid on bone-forming and bone-resorbing cells [J]. Clin Calcium, 2006, 16: 1817-1822.

[6] Grishchuk EL, Ataullakhanov FI. In vitro assays to study the tracking of shortening microtubule ends and to measure associated forces [J]. Methods Cell Biol, 2010, 95: 657-676.

[7] Okada E, Yamanaka M, Ishikawa O. New insights into the mechanism of abnormal calcification in nephrogenic systemic fibrosis-gadolinium promotes calcium deposition of mesenchymal stem cells and dermal fibroblasts [J]. J Dermatol Sci, 2011, 62(1): 58-63.

[8] Isaac J, Nohra J, Lao J, Jallot E, et al. Effects of strontium-doped bioactive glass on the differentiation of cultured osteogenic cells [J]. Eur Cell Mater, 2011, 21: 130-143.

[9] Zhiming Yang, Zhengfu Fan, Huiqi Xie, et al. Vascularization and its relationship to osteogenesis in transplantation of tissue engineered bone to repair seg ment defect of long bone [J]. Chinese Journal of Microsurgery. 2002; 25 (2): 119

[10] Takahashi N.Bone and bone related bioehemieal examinations.Bone and collagen related metabolites.Regulatory mechanisms of osteoclast differentiation and function [J]. Clin Calcium, 2006, 16: 940-947.

[11] Tao Sheng-xiang, Chen Zhen-guang, Xie Yun, et al. An experimental study on revascularization after vascularized bone allograft [J]. Chinese Journal of Microsurgery, 2008, (2):119-121.

[12] Bishop AT, Pelzer M. Vascularized bone allotransplantation: current state and implications for future reconstructive surgery [J]. Orthop Clin North Am, 2007, 38:109 - 122.

[13] Weinan Zhu, Zhiming Yang, Xiuqun Li, et al. Experimental study on rabbit periosteal osteoblasts and renal vascular endothelial cells indirect co-culture in vitro [J]. Chinese Journal of Reparative and Reconstructive Surgery. 2002; 16(5): 307

[14] Rui gang, Wang Zhesheng, Lv Gang, et al. Study on biological behavior of osteoblast and vascular endothelial cell cultuer [J]. Chinese Journal of Reparative and Reconstructive Surgery. 2004, 18 (4): 327-330

[15] Brighton CT, Lorich DG, Kupcha R, et al. The pericyte as a possible osteoblast progenitor cell [J]. Clin Orthop, 1992, (275): 287

[16] Katagiri T, Yamaguchi A, Komaki M, et al. Bone morphogenetic progenetic protein-2 converts the differentiation pathway of C2C12 myoblasts into the osteoblast lineage [J]. J Cell Biol, 1994, 127(6 Pt 1): 1755-1766

[17] Lu L, yaszemski MJ, Mikos AG. TGF-betal release from biodegradable polymer microparticles: its effects on marrow stromal osteoblast function [J]. J Bone Joint Surg Am, 2001; 83-A Suppl 1(Pt 2): S82

[18] Oldham JB, Lu L, Zhu X, et al. Biological ctivity of rhBMP-2 released from PLGA microspheres [J]. J Biomech Eng, 2000; 122(3): 289

[19] Murphy WL, Peters MC, Khon DH, et al. Sustained release of vascular endothelial growth factor from mineralized poly(lactide-co-glycolide) scaffolds for tissue engineering. Biomaterials [J], 2000; 21(24): 2521

[20] Schwartz Z, Lohmann CH, Wieland M, et al. Osteoblast proliferation and differentiation on dentin slices are modulated by pretreatment of the surface with tetracycline or osteoblasts [J]. J Periodontol, 2000; 71(4): 586

[21] Zhiming Yang. Basis of tissue engineering and clinic [M]. Chengdu: Sichuan science and technology press, 2000: 134-134

[22] Winn SR, Hollinger JO. An osteogenic cell culture system to evaluate the cytocompatibility of Osteo-et, a calcium sulfatebone void filler [J]. Biomaterials. 2000; 21: 2413-2425.

[23] Burg KJL, Porter S, Kellam JF. Biomaterial developments for bone tissue engineering [J]. Biomaterials, 2000; 21: 2347-2359.

[24] Grane GM,Lshaug SL,Mikos AG,et al. Bone tissue engineering [J]. Nature Medicine. 1995, 1(12): 1322-1324

[25] Chen VJ, Ma PX. The effect of surface area on the degradationrate of nano-fibrous poly (L-lactic acid) foams [J]. Biomaterials, 2006, 27(20):3708-3715.

[26] Cheung AY, Lau KT, Lu TP, et al. A critical review on polymer-based bio-engineerd materials for scaffold development [J]. Composites: Part B 2007;38:291-300.

[27] Wiria FE, Leong KF, Chua CK,et al. Poly-epsilon-caprolactone/hydroxyapatite for tissue engineering scaffold fabrication via selective laser sintering [J]. Acta Biomater. 2007; 3(1): 1-12.

[28] Zhiming Yang. Tissue engineering and Experiment Surgery [J]. Chinese Journal of Experiment Surgery, 2002, 19(2): 103-104

[29] Linbo Wu, Jiandong Ding. Advances in Fabrication Methodology and Technology of Three-Dimension Porous Scaffold for Tissue Engineering [J]. Journal of Functional Polymers, 2003, 16(1): 91-96.

[30] Hillsley MV, Frangos JA. Reviews:Bone tissue engineering: The role of interstitial fluid flow [J]. Biotechnology and Bioengineering, 1994, 43: 573-581.

[31] Suzuki M, Guimaraes MV, Marin C, et al. Histomorphometric evaluation of alumina-blasted/acid-etched and thin ion beam-deposited bioceramic surfaces: an experimental study in dogs [J]. J Oral Maxillofac Surg. 2009; 67(3): 602-607.

[32] Mikos AG, Sarakinos G, Leite SM, et al. Lamimated three-dimensional biodegradable foams for use in tissue engineering [J]. Biomaterials, 1993, 14: 323-330

[33] Cheung HY, Lau KT, Lu TP, et al. A critical review on polymer-based bio-engineerd materials for scaffold development [J]. Composites: Part B 2007; 38: 291-300.

[34] Daculsi G, Goyenvalle E, Cognet R, et al. Osteoconductive properties of poly (96L/4D-lactide)/beta-tricalcium phosphate in long term animal model [J]. Biomaterials. 2011, 32(12): 3166-77.

[35] Sirin Y, Olgac V, Dogru-Abbasoglu S, et al. The influence of hyperbaric oxygen treatment on the healing of experimental defects filled with different bone graft substitutes [J]. Int J Med Sci. 2011, 8(2):114-25.

[36] Ginebra MP, Espanol M, Montufar EB,et al. New processing approaches in calcium phosphate cements and their applications in regenerative medicine [J]. Acta Biomater. 2010, 6(8): 2863-2873.

[37] Di Sha, Yingling Jiang, Jingheng Gao, et al. Preparation of porcine long bone acellular bone extracellular matrix (ABECM) and preliminary measurement of its biomechanical and histological quality [J]. Chinese Journal of Practical Aesthetic and Plastic Surgery. 2000, 12, 11 (6): 331-334

[38] Brown KLB, Montreal M. Bone and cartilage transplantation in orthopedic surgery [J]. J Bone Joint Surg (Am). 1982; 64: 270-279

[39] Isaac J, Nohra J, Lao J, Jallot E, et al. Effects of strontium-doped bioactive glass on the differentiation of cultured osteogenic cells [J]. Eur Cell Mater, 2011, 21: 130-143.

[40] Merkel De, Mccuire WL, Ploidy, Proliferation activity and prognosis DNA flow cytometry of solid tumors [J]. J Cancer, 1990, 65(5); 1194-1205

[41] Shan Lin, Qingshui Yin, Yu Zhang, et al. Prepartion and cytotoxicity evaluation of the digital coral hydroxyapatite artifical bone in vitro [J]. Chinese Journal of Bone and Joint Injury, 2011, 26 (2): 23-26.

[42] Shan Lin, Qingshui Yin, Yu Zhang, et al. Prepartion and Characteristics of digital coral hydroxyapatite artifical bone [J]. The Orthopedic Journal of China, 2010, (24): 2039-2041.

[43] Ewers R, Goriwoda W, Schopper C, et al. Histologic findings at augmented bone areas supplied with two different bone substitutematerials combined with sinus floor lifting [J]. Report of one case: Clin Oral Implants Res, 2004, 15(1): 96-100.

[44] Khan SN, Cammisa FP Jr, Sandhu HS, et al. The biology of bone grafting [J]. J Am Acad Orthop Surg, 2005, 13: 77-86.

[45] Zijderveld SA, ten Bruggenkate CM, van Den Bergh JP, et al. Fractures of the iliac crest after split-thickness bone grafting for preprosthetic surgery: report of 3 cases and review of the literature [J]. J Oral Maxillofac Surg, 2004, 62: 781-786.
[46] Shan Lin, Qingshui Yin, Yu Zhang, et al. Prepartion and cytotoxicity evaluation of the digital coral hydroxyapatite artifical bone in vitro [J]. Chinese Journal of Bone and Joint Injury, 2011, 26 (2): 23-26.
[47] Shan Lin, Qingshui Yin, Yu Zhang, et al. Prepartion and Characteristics of digital coral hydroxyapatite artifical bone [J]. The Orthopedic Journal of China, 2010, (24): 2039-2041.
[48] Taboas J M, Maddox P H,et al. Indirect Solid Free Form Fabrication of Local and Global Porous, Biomimetic and Composite 3D Polymer-ceramic Scaffolds [J]. Biomaterials, 2003, 24: 181-194.
[49] Ginebra MP, Espanol M, Montufar EB,et al. New processing approaches in calcium phosphate cements and their applications in regenerative medicine [J]. Acta Biomater. 2010, 6(8): 2863-2873.
[50] Mistry S, Kundu D, Datta S, et al. Comparison of bioactive glass coated and hydroxyapatite coated titanium dental implants in the human jaw bone [J]. Aust Dent J. 2011; 56(1): 68-75.
[51] Yunyu Hu, Prospection of repair of osteoarticular injury [J]. J Trauma Surg. 2000, 3 (2): 129-130
[52] Kuo YC, Yeh CF. Effect of surface-modified collagen on the adhesion, biocompatibility and differentiation of bone marrow stromal cells in poly(lactide-co-glycolide)/chitosan scaffolds [J]. Colloids Surf B Biointerfaces, 2011, 82(2): 624-31.
[53] Meng B, Chen J, Guo D, et al. The effect of titanium particles on rat bone marrow stem cells in vitro [J]. Toxicol Mech Methods, 2009, 19 (9): 552-558.
[54] Grishchuk EL, Ataullakhanov FI. In vitro assays to study the tracking of shortening microtubule ends and to measure associated forces [J]. Methods Cell Biol, 2010, 95: 657-676.
[55] Okada E, Yamanaka M, Ishikawa O. New insights into the mechanism of abnormal calcification in nephrogenic systemic fibrosis-gadolinium promotes calcium deposition of mesenchymal stem cells and dermal fibroblasts [J]. J Dermatol Sci, 2011, 62(1): 58-63.
[56] Isaac J, Nohra J, Lao J, Jallot E, et al. Effects of strontium-doped bioactive glass on the differentiation of cultured osteogenic cells [J]. Eur Cell Mater, 2011, 21: 130-143.
[57] Zhiming Yang, Tissue Engineering. Chemical Industry Press: Beijing, 2002, 210
[58] Dan Jin, Guoxian Pei, Bin Chen, Yu Qin. Angiogenesis and neurotization study of bone tissue engineering [J]. Journal of Clinical Rehabilitative Tissue Engineering Research. 2001, 8 (5): 18-19
[59] Zhengfu Fan, Zhiming Yang. Study on the vascularization of the transplants and the relationship of vascularization and osteogenesis in transplantation of tissue engineered bone to repair segmental defect of long bone. Proceedings of the Second National Academic Conference of tissue engineering. 378-386.
[60] Casabona F, Martin I, Muraglia A, et al. Prefabricated engineered bone flaps: an experimental model of tissue reconstruction in plastic surgery [J]. J Plast Reconstr Surg. 1998, 101(3): 577